X-RAY DIAGNOSTICS OF ASTROPHYSICAL PLASMAS

X-RAY DIAGNOSTICS OF ASTROPHYSICAL PLASMAS

Theory, Experiment, and Observation

Cambridge, Massachusetts 15 – 17 November 2004

EDITOR
Randall K. Smith
NASA Goddard Space Flight Center
Greenbelt, Maryland

SPONSORING ORGANIZATIONS
The National Science Foundation through a grant for the Institute for Theoretical Atomic, Molecular and Optical Physics at Harvard University and Smithsonian Astrophysical Observatory
The Chandra X-Ray Center
The Constellation-X Mission
The Astro-E2 Guest Observer Facility at NASA Goddard Space Flight Center

AMERICAN INSTITUTE OF PHYSICS

Melville, New York, 2005
AIP CONFERENCE PROCEEDINGS ■ VOLUME 774

Editor:

Randall K. Smith
NASA Goddard Space Flight Center
Code 662.0
Greenbelt, MD 20771
USA
E-mail: randall.k.smith.1@gsfc.nasa.gov

Cover Image: Created by Karen Smale, combining the Chandra image of M87 (NASA/CXC/W. Forman et al.) with the XMM/Newton spectrum from Peterson et al. 2003, APJ, **590**, 207.

L.C. Catalog Card No. 2005926943
ISBN 0-7354-0259-0
ISSN 0094-243X
Printed in the United States of America

TABLE 3. Fe L-shell lines with peak emissivity $\Lambda > 6 \times 10^{-17}$ erg cm^3/s

Ion	λ(Å)	E (keV)	Label*	UL†	LL**	log(Λ)	Label
Fe XXIV	10.619	1.1676		6	1	-16.00	$3p\ ^2\mathrm{P}_{3/2} \rightarrow gnd$
Fe XXIII	10.981	1.1291	Be9	15	1	-15.96	$2s3p\ ^1\mathrm{P}_1 \rightarrow gnd$
Fe XXIII	11.019	1.1252	Be8	13	1	-16.14	$2s3p\ ^3\mathrm{P}_1 \rightarrow gnd$
Fe XXIV	11.029	1.1242	Li4	7	2	-16.17	$3d\ ^2\mathrm{D}_{3/2} \rightarrow 2p\ ^2\mathrm{P}_{1/2}$
Fe XXIV	11.176	1.1094	Li3	8	3	-15.91	$3d\ ^2\mathrm{D}_{5/2} \rightarrow 2p\ ^2\mathrm{P}_{3/2}$
Fe XVII	11.254	1.1017	5D	93	1	-16.11	$2p^5 5d\ ^3\mathrm{D}_1 \rightarrow gnd$
Fe XXIV	11.432	1.0845	Li1	4	3	-16.19	$3s\ ^2\mathrm{S}_{1/2} \rightarrow 2p\ ^2\mathrm{P}_{3/2}$
Fe XXIII	11.736	1.0564	Be2	20	5	-15.62	$2s3d\ ^1\mathrm{D}_2 \rightarrow 2s2p\ ^1\mathrm{P}_1$
Fe XXII	11.770	1.0534	B13	21	1	-15.55	$2s^2 3d\ ^2\mathrm{D}_{3/2} \rightarrow gnd$
Fe XVII	12.124	1.0226	4C	71	1	-15.67	$2p^5 4d\ ^1\mathrm{P}_1 \rightarrow gnd$
Fe XXIII	12.161	1.0195	Be1	12	5	-15.87	$2s3s\ ^1\mathrm{S}_0 \rightarrow 2s2p\ ^1\mathrm{P}_1$
Fe XVII	12.266	1.0108	4D	59	1	-15.71	$2p^5 4d\ ^3\mathrm{D}_1 \rightarrow gnd$
Fe XXI	12.284	1.0093	C10	40	1	-15.30	$2s^2 2p3d\ ^3\mathrm{D}_1 \rightarrow gnd$
Fe XXI	12.393	1.0004	C8	40	2	-16.05	$2s^2 2p3d\ ^3\mathrm{D}_1 \rightarrow 2s^2 2p^2\ ^3\mathrm{P}_1$
Fe XXII	12.754	0.9721	B4	23	6	-15.99	$2s2p_{1/2}3s \rightarrow 2s2p^2\ ^2\mathrm{D}_{3/2}$
Fe XX	12.846	0.9652	N31	58	1	-15.55	$2s^2 2p_{1/2} 2p_{3/2} 3d_{3/2} \rightarrow gnd$
Fe XX	12.864	0.9638	N31	56	1	-15.63	$2s^2 2p_{1/2} 2p_{3/2} 3d_{5/2} \rightarrow gnd$
Fe XXI	12.822	0.9670	C4	83	7	-16.18	$2s2p_{1/2} 2p_{3/2} 3d_{5/2} \rightarrow 2s2p^3\ ^3\mathrm{D}_1$
Fe XX	12.824	0.9668	N31	60	1	-15.94	$2s^2 2p_{1/2} 2p_{3/2} 3d_{3/2} \rightarrow gnd$
Fe XX	12.912	0.9602	N30	51	1	-16.13	$2s^2 2p_{1/2} 2p_{3/2} 3d_{5/2} \rightarrow gnd$
Fe XX	12.965	0.9563	N29	48	1	-16.06	$2s^2 2p_{1/2} 2p_{3/2} 3d_{3/2} \rightarrow gnd$
Fe XX	13.061	0.9493	N26	42	1	-16.16	$2s^2 2p^2_{1/2} 3d_{5/2} \rightarrow gnd$
Fe XX	13.385	0.9263	N18	111	6	-16.18	$2p^3(^5S)3d\ ^4\mathrm{D}_{7/2} \rightarrow 2s2p^4\ ^4\mathrm{P}_{5/2}$
Fe XIX	13.462	0.9210	O26	74	1	-16.12	$2s^2 2p^3(^2\mathrm{D})3d\ ^3\mathrm{S}_1 \rightarrow gnd$
Fe XIX	13.497	0.9186	O25	71	1	-15.88	$2s^2 2p^3(^2\mathrm{D})3d\ ^3\mathrm{P}_2 \rightarrow gnd$
Fe XXI	13.507	0.9179	C3	42	7	-15.94	$2s2p^2_{1/2} 3s \rightarrow 2s2p^3\ ^3\mathrm{D}_1$
Fe XIX	13.518	0.9172	O24	68	1	-15.54	$2s^2 2p^3(^2\mathrm{D})3d\ ^3\mathrm{D}_3 \rightarrow gnd$
Fe XX	13.767	0.9006	N14	19	1	-16.26	$2p^2(^3\mathrm{P})3s\ ^4\mathrm{P}_{5/2} \rightarrow gnd$
Fe XIX	13.795	0.8988	O19	53	1	-15.93	$2s^2 2p^3(^4\mathrm{S})3d\ ^3\mathrm{D}_3 \rightarrow gnd$
Fe XVII	13.825	0.8968	3A	33	1	-15.82	$2s2p^6 3p\ ^1\mathrm{P}_1 \rightarrow gnd$
Fe XXI	14.008	0.8851	C1	28	7	-16.03	$2s^2 2p_{1/2} 3p_{1/2} \rightarrow 2s2p^3\ ^3\mathrm{D}_1$
Fe XVIII	14.208	0.8726	F20	56	1	-15.39	$2s^2 2p^4(^1\mathrm{D})3d\ ^2\mathrm{D}_{5/2} \rightarrow gnd$
Fe XVIII	14.208	0.8726	F20	55	1	-15.65	$2s^2 2p^4(^1\mathrm{D})3d\ ^2\mathrm{D}_{3/2} \rightarrow gnd$
Fe XVIII	14.256	0.8697	F19	52	1	-16.09	$2s^2 2p^4(^1\mathrm{D})3d\ ^2\mathrm{S}_{1/2} \rightarrow gnd$
Fe XX	14.267	0.8690	N9	54	6	-16.05	$2s2p^2_{1/2} 2p_{3/2} 3s \rightarrow 2s2p^4\ ^4\mathrm{P}_{5/2}$
Fe XVIII	14.373	0.8626	F17	49	1	-15.80	$2s^2 2p^4(^3\mathrm{P})3d\ ^2\mathrm{D}_{5/2} \rightarrow gnd$
Fe XVIII	14.534	0.8531	F15	41	1	-15.92	$2s^2 2p^4(^3\mathrm{P})3d\ ^2\mathrm{F}_{5/2} \rightarrow gnd$
Fe XIX	14.664	0.8455	O13	15	1	-16.06	$2s^2 2p^3(^2\mathrm{D})3s\ ^3\mathrm{D}_3 \rightarrow gnd$
Fe XVII	15.014	0.8258	3C	27	1	-14.70	$2p^5 3d\ ^1\mathrm{P}_1 \rightarrow gnd$
Fe XIX	15.079	0.8222	O8	11	1	-16.00	$2s^2 2p^3(^4\mathrm{S})3s\ ^5\mathrm{S}_2 \rightarrow gnd$
Fe XIX	15.198	0.8158	O5	82	6	-16.09	$2s2p^2_{1/2} 2p^2_{3/2} 3s \rightarrow 2s2p^5\ ^3\mathrm{P}_2$
Fe XVII	15.261	0.8124	3D	23	1	-15.25	$2p^5 3d\ ^3\mathrm{D}_1 \rightarrow gnd$
Fe XVII	15.453	0.8023	3E	17	1	-16.12	$2p^5 3d\ ^3\mathrm{P}_1 \rightarrow gnd$
Fe XVIII	15.625	0.7935	F11	9	1	-15.78	$2s^2 2p^4(^1\mathrm{D})3s\ ^2\mathrm{D}_{5/2} \rightarrow gnd$
Fe XVIII	15.824	0.7835	F9	7	1	-15.99	$2s^2 2p^4(^3\mathrm{P})3s\ ^4\mathrm{P}_{3/2} \rightarrow gnd$
Fe XVIII	16.004	0.7747	F6	5	1	-15.71	$2s^2 2p^4(^3\mathrm{P})3s\ ^2\mathrm{P}_{3/2} \rightarrow gnd$
Fe XVIII	16.071	0.7715	F4	4	1	-15.62	$2s^2 2p^4(^3\mathrm{P})3s\ ^4\mathrm{P}_{5/2} \rightarrow gnd$
Fe XIX	16.110	0.7696	O2	37	6	-15.89	$2s^2 2p_{1/2} 2p^2_{3/2} 3p_{1/2} \rightarrow 2s2p^5\ ^3\mathrm{P}_2$
Fe XVIII	16.159	0.7673	F3	64	3	-16.04	$2s2p^5 3s\ ^2\mathrm{P}_{3/2} \rightarrow 2s2p^6\ ^2\mathrm{S}_{1/2}$
Fe XVII	16.780	0.7389	3F	5	1	-15.03	$2p^5 3s\ ^1\mathrm{P}_1 \rightarrow gnd$
Fe XVII	17.051	0.7271	3G	3	1	-14.96	$2p^5 3s\ ^3\mathrm{P}_1 \rightarrow gnd$
Fe XVII	17.096	0.7252	M2	2	1	-14.98	$2p^5 3s\ ^3\mathrm{P}_2 \rightarrow gnd$
Fe XVIII	17.623	0.7035	F1	29	3	-15.77	$2s^2 2p^4 3p\ ^2\mathrm{P}_{3/2} \rightarrow 2s2p^6\ ^2\mathrm{S}_{1/2}$

* From Brown et al. 1998 ApJ, 502, 1015

† ATOMDB v1.3.1 upper level; see `http://cxc.harvard.edu/atomdb/WebGUIDE`

** ATOMDB v1.3.1 lower level

Contents

I: GENERAL THEORY

II: IRON L- AND K- SHELL DIAGNOSTICS

V: XAFS AND CHARGE EXCHANGE DIAGNOSTICS

VI: X-RAY DIAGNOSTICS FOR NEUTRON STARS

VII: DIELECTRONIC RECOMBINATION AND OTHER PLASMA PROCESSES

VIII: X-RAY SPECTRAL ANALYSIS OF INDIVIDUAL SOURCES

IX: FUTURE MISSIONS, PLANS, AND PROSPECTS

Preface

On November 15 – 17, 2004, an international group of astronomers and physicists convened in Cambridge, Massachusetts to discuss the recent progress in analyzing high-resolution X-ray spectra from a range of astrophysical sources. The presentations varied from observational astrophysics to theoretical atomic physics, but each connected to the theme of the workshop. The wide range of the workshop allowed these scientists to exchange ideas, and many new collaborations were formed.

The workshop was particularly timely as the existing high-resolution X-ray spectroscopy missions (*Chandra, XMM/Newton*) have expanded the amount of data available, and more is anticipated from Astro-E2, Constellation-X, and XEUS. As a result, this proceeding should help both experienced scientists and newcomers to the field explore the wide range of research available.

I gratefully acknowledge the crucial assistence of Kate Kirby and Nancy Brickhouse in developing the original concept for the workshop, and later serving on the Scientific Organizing Committee along with Jay Bookbinder, Claude Canizares, Jelle Kaastra, Tim Kallman, Duane Liedahl, and John Raymond, who were instrumental in selecting the topics and speakers for the workshop.

Financial support for the conference was provided by the National Science Foundation through a grant for the Institute for Theoretical Atomic, Molecular and Optical Physics at Harvard University and Smithsonian Astrophysical Observatory. Additional financial support was provided by grants from the Chandra X-ray Center, the Constellation-X mission, and the Astro-E2 mission.

Special thanks are due to Verity Parris for creating the XDAP website and for so smoothly handling the registrations as they came in from all over the world. An artistic thank you goes to Karen Smale who designed the cover of this proceedings. I also thank Lisa Bastille who created the workshop booklet and handled the myriad last-minute details with such aplomb. Lauren Boice's assistance was also invaluable throughout the conference and in preparing the proceedings. Finally, thanks to everyone who participated and attended, making this workshop such a sucesss.

Randall Smith
Greenbelt, MD
March, 2005

Rolf Mewe: a career devoted to X-ray spectroscopy

Jelle S. Kaastra* and Rolf Mewe†

*SRON, Sorbonnelaan 2, 3584 CA Utrecht, The Netherlands
†deceased May 4, 2004

Abstract. An overview of the life and work of Rolf Mewe (1935-2004) as an X-ray spectroscopist is given. He was one of the pioneers in the field of X-ray spectroscopy. His work illustrates nicely how this field developed from the early days up to the present high-resolution era. His plasma emission codes, developed by him and collaborators over several decades, is one of the most widely used. His thorough knowledge of the field, as well as his ability and enthousiasm to cooperate with many colleagues, made his career a succes. He will be missed by all of us for his work and personality.

INTRODUCTION

Rolf Mewe (born Amsterdam, July 20, 1935; died Houten, May 4, 2004) was one of the pioneers in the field of X-ray spectroscopy. This contribution (starting from the next section) consists mainly of a write-up by himself of the popular talk that he gave on the occasion of his official retirement on July 7, 2000. After his unexpected and sudden death, we found this document on his computer. Maybe in some parts dated, it gives a nice personal overview of how Rolf looked back to his career.

Although his scientific interest was focused on stellar coronae and the Sun, he had a much broader interest; however, his work on plasma emission codes that spanned more than three decades is perhaps the most lasting part of his work. He was not the only person to develop this field, more or less parallel to him others did similar work (we only mention here John Raymond et al., M. Landini et al., K. Masai et al., but there are more groups). I believe Rolf's models and codes were in particular so succesful because of several reasons. He had a good intuition for what was the most important physics, and was handy in producing useful interpolation formulae and approximations. He was not afraid of metaculously and carefully extracting the quantities needed from large, inhomogeneous sets of publications, and he had a good eye for critically evaluating and comparing the numbers from various sources. Maybe more importantly, and what was in particularly appreciated, was his ability to organize cooperations between people with various specialisations and to work in a team; many young scientists liked him because of his deep interest not only in their scientific work but also in their personal happiness.

After his official retirement in 2000, he kept on working three days a week at SRON, happy that he finally got the good data to compare with his models (thanks to Chandra and XMM-Newton), and stimulating the young people at SRON with his advise and interest. His sudden death, shortly after he had changed home, hit us like a bomb; we expected him to arrive at work when we heard the message that his heart had stopped

CP774, *X-Ray Diagnostics of Astrophysical Plasmas*, edited by Randall K. Smith

beating. We miss him and remember him for his broad knowledge, his warm personality, and great interest in people. However we will continue our X-ray spectroscopy building further on the foundations laid by him.

PRE-HISTORY

Already in October 1952, shortly after I started my academic studies, my father (born 8 Dec. 1898) took me to a series of lectures on the field of space travel and provoked my interest in this field. Unfortunately, he died already on 14 Jan. 1957, even before the launch of the first Spoetnik on 14 Oct. 1957. The last sentences of his notes of these lectures were: "The expectation is that mankind will be able in about 25 à 50 years to land on the Moon. The incentive is the passion for adventures. The importance is probably only scientific (astronomy)".

In 1956 I started my practical exercise in physics on the spectroscopical Zeeman laboratory in Amsterdam though most of the people gave me the advice to choose the "modern" nuclear physics rather than such an "oldfashioned" branch of science as spectroscopy. However, I found there an excellent teacher of spectroscopy in the person of "mister" Klinkenberg (he attended my lecture as a smart 85 years outstanding old man!). On an uttermostly charming way he teached me the principles of spectroscopy. When as a young student I entered his room with some new idea he did not say immediately "I do not agree with you" but rather "I do not completely understand you, explain me, why do you assume ..., could it perhaps be that ..., etc", and imparted in this way a lot of his fabulous knowledge to me.

During my service in the army (1959–1961) I studied as co-worker of the scientific advisory office of the General Staff of the land-forces among others the effects of radio-active radiation from nuclear explosions and wrote an in French written confidential report of the General Staff about this topic. It was therefore my intention to continue my scientific work in the field of radio-active dosimetry and applied for a job at the radiobiological laboratory of TNO in Rijswijk. However, I was not accepted but instead my dear friend Johan Broerse made a brilliant career at that institute.

I came at the institute of plasma physics "Rijnhuizen" (located in a beautiful park with a castle) where I did from 1961 until 1970 research on gas discharges and hot fusion plasmas and finished my thesis work in 1969 (promotores Kees Braams and Henk van Bueren; to my great joy my ushers Fred de Vries and Anton Postma were also present at my talk).

SPACE RESEARCH

The early beginning

On 1 Juli 1970 I started in the interpretation group "INTEGRO" of Kees de Jager who gave me the simple order to develop the X-ray spectroscopy on the laboratory for space research ("LRO") in Utrecht (now called SRON = Space Research Organization Netherlands). I was so naïve to promise him to try this, still ignorant of the fact that my predecessor had answered this question by saying this is really impossible! The next day at the lab my division head Leen de Feiter introduced me to Bert Brinkman with the request to help this man with the theoretical interpretation of solar X-ray observations to be described in his coming thesis. In fact, this gave me the motivation to make a quick start with the set-up of the X-ray spectroscopy and this was also the true beginning of a fruitful 30-years co-operation with Bert in many X-ray projects (he as PI, I as project scientist). My impact in Bert's work had as result that Bert could defend his thesis in 1972 to get his Ph.D. as the first scientist on LRO and I had the honour to accompany him together with our colleague Hans Imhof as his usher.

What is a spectrum and why space research?

In my life I have studied many spectra but what is a spectrum? It is for example a unique collection of lines. Think of the bar code which tells you in a unique way the sort and price of an article in the supermarket. In daily life we can sometimes observe so pretty in the sky the sparkling colors of the rainbow (or of two rainbows) as the spectrum of the scattered light of the Sun. With a suitable instrument ("spectrograph") we can distinguish absorption lines as Fraunhofer did for the first time more than a century ago.

We could place a "thermometer" along a spectrum. The wavelength of radiation produced by an object is usually related to its temperature. The visible light in the spectrum (wavelength 5000 Å) tells us something about emitting gas as hot as e.g. the photosphere of the Sun, the visible surface layer with a temperature of nearly 6 000 Kelvin (K). Radiation of much higher photon energy (shorter wavelength, e.g., 50–5 Å) (*X-ray radiation*[1]) originates from much hotter gases (1 to 10 million K, such as for example the tenuous corona around the Sun and other stars, while infrared radiation (50 micron) and radio radiation (1 cm) are often (but not always) associated with very cool gas (say 100 K).

Of all radiation from the Universe only the visible light penetrates the Earth atmosphere and some part of the radio radiation. X-rays are absorbed in the atmosphere of the Earth and can therefore be measured only with instruments that are brought above the atmosphere by rockets or satellites at a height of about 100 or more km. I was mainly involved in the study of X-ray spectra of hot cosmic sources. Measuring such radiation is a matter of space research.

Intermezzo: the Sun

Shortly after the 2nd World War German V-2 rockets were available to do space research and the first experiments (with a photographic plate) detected in 1948 X-rays from the Sun which should come from the tenuous and hot (few million K) "corona" around the Sun. This corona is in visible light a million times weaker than the solar

---

[1] Röntgenstrahlung (X-ray radiation) was discovered on 8 November 1895 by the German physicist Wilhelm Conrad Röntgen.

FIGURE 1. ANS satellite. Photograph: SRON

photosphere and therefore only visible during a total solar eclipse as the faint but magnificent shine around the "black" Sun. An image of the Sun in X-rays shows a complicated picture of many kinds of loop structures consisting of hot gases contained in magnetic fields.

These first experiments showed that the Sun emits only a tiny fraction (10^{-7}) of his total emission as X-rays. The Sun placed at a distance of the closest star would be so faint that around the 1950s one expected nothing to see with the instruments available at that time. In the early period all X-ray studies were therefore entirely dedicated to the Sun and one cherished scarcely any hope to detect X-ray sources outside our solar system. Therefore, it was really a surprise that on 18 June 1962 with a rocket flight intended to detect fluorescent X-rays from the Moon one discovered in Scorpius the tremendously strong X-ray source Scorpius X-1, 10 000 times stronger than the whole radiative output of the Sun. This discovery opened an entirely new window on the X-ray Universe.

All the striking it was that already on 1 Oct. 1961 our space research laboratory in Utrecht was established by Kees de Jager and other people when only the Sun was known as a cosmic X-ray source. Clearly he expected to detect in future other cosmic X-ray sources as well!

From ANS to AXAF

When I entered on 1 July 1970 the space research lab in Utrecht about 20 cosmic X-ray sources were known but still no corona around another star than the Sun. On the basis of an otherwise wrong but optimistic model Kees and I expected to detect in the near future many stellar coronae.

In the past 30 years I was engaged in more than 20 different projects but I shall make only a limited choice out of all these (mainly X-ray) projects and I begin with the Astronomical Netherlands Satellite (ANS, the name of a pretty girl friend one of my daughters thought). We all were very proud of this tiny (1 m) satellite (Fig. 1). This pride was justified in view of the fact that a special series of post stamps issued in 1999 exhibiting the ten most important events occurred in the Netherlands during the last century also showed the ANS representing the Dutch space science!

ANS

After the launch of ANS on 30 August 1974 the ANS science team headed by Bert Brinkman and further consisting of John Heise, Ton den Boggende, Hans Schrijver, Ed Gronenschild and I got a tiring but very fascinating time. We discovered various types of new X-ray sources such as flare stars (UV Ceti and YZ CMi), a star corona (Capella), a dwarf nova (SS Cygni), gigantic X-ray explosions from neutron stars ("bursters"), and X-rays from a white dwarf (Sirius B). We made a complete map of the extended supernova remnant Cygnus Loop. About 70 papers and 3 theses (Ed Gronenschild, Ton den Boggende and John Heise; Kees de Jager promotor and I co-promotor of the first two) appeared.

Because I was deeply involved in the *discovery of X rays from Sirius with ANS* I will dwell a little more upon this topic.

With a visual magnitude of -1.46 Sirius is the brightest object on the sky seen in the winter in the stellar constellation Great Dog and 8.7 light years away. The main star Sirius A has a very faint companion Sirius B ten thousand times weaker than the main star. Sirius B is as heavy as the Sun but as small as the Earth. It is the prototype of a socalled white dwarf representing the end phase of a relatively light star like the Sun when all thermonuclear processes have ended. Its density is enormous: a matchbox full of matter is as heavy as a big truck. The binary has a period of about 50 years and the maximum angular distance projected on the sky is about 11 arcseconds (at the time of the ANS observation). Note that one arcsecond = 5 mm seen at a distance of 1 km.

Did I expect X-rays from Sirius? Honestly speaking not, because a star like Sirius A with a hot photosphere of 10 000 K generates only little convection and therefore little acoustic flux. No corona was expected around Sirius A ... but we forgot the white dwarf! But I was simply curious to know what the brightest star on heavens would do. However there was a problem to observe such a bright star because the star sensor that should recognize two bright guide stars in order that the satellite could determine and fix its pointing direction would be overwhelmed by the strong radiation from Sirius. Though the people of Philips Company thought it would still be possible to measure on Sirius I was not quite sure. On my request the people at the operations centre in Darmstadt made a temporary change in the software in order that the satellite got first an offset and then made a slew towards the position of Sirius. Another problem was that the X-ray detector was also sensitive to ultraviolet (UV) radiation which is copiously emitted by Sirius A but this problem could be solved by an additional measurement with an UV filter that transmitted only UV light (above about 1100 Å). By three reaction wheels the satellite could be stabilized along three axes (one axis continously directed perpendicularly to the Sun around which the the satellite could rotate). In this way the telescope could be pointed alternatively on the source ("On") and off the source ("Off") (Fig. 2).

What was the result of the observation on 4 April 1975? The normal mode failed every time but my "experimental" test mode was successful in pointing to the source. To my surprise I found a net signal (at the 5σ level) after correcting for the background and

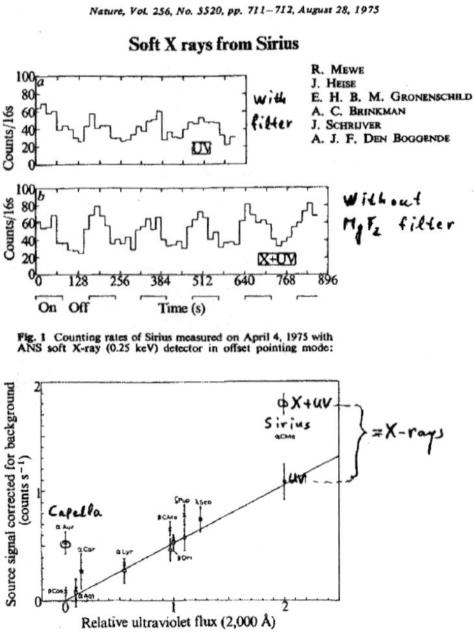

FIGURE 2. Sirius X-rays discovered with ANS. Annotation by R. Mewe, figure from Mewe et al. (1975) [1], reprinted with permission from Nature (http://www.nature.com/).

subtracting from the total (X+UV) signal the UV signal. Conclusion: ANS discovered X-rays from Sirius! A few months later I reported this discovery on an international congress (COSPAR) and published the result in the journal Nature ([1]). (Fig. 2).

But what was the source of the X rays? On the basis of theoretical arguments I excluded Sirius A as the source and thought that the white dwarf could perhaps be the X-ray source. I put forward the possibility of a very dense and compact (1 m high) hot corona around the white dwarf Sirius B. Joe Cassinelli (Wisconsin) who was just at that time guest on our lab, had calculated a few years earlier together with Böhm that sufficient acoustic flux could be generated in the hot He II/III ionization zone in the photosphere of Sirius B to form such a hot and dense corona. With Tony Hearn of the Utrecht Observatory I predicted on the basis of his minimum flux coronal model that the observations would be consistent with a hot (few MK), dense (million times denser than the solar corona), and small (1 m high) corona around the white dwarf Sirius B. Shortly thereafter Mike Lampton (Berkeley) and I wrote a review paper on the heating of coronae around white dwarfs. Unfortunately, ANS measured only in one restricted energy band (0.25 keV, wavelength 50 Å), hence not a real spectrum so that we could not test conclusively the white dwarf coronal model.

On 4 October 1977 Mike Lampton confirmed my discovery with an instrument on board the first High Energy Astrophysical Observatory (HEAO-1) and detected the same X-ray flux as was measured by the ANS in the soft X-ray band. Moreover, he measured a spectrum in five energy channels between 0.16 and 0.4 keV. This spectrum declined so steeply towards high energies that this could not be reconciled with my coronal model but rather with a photospheric model developed by Shipman in 1976. This model assumes that the outer atmosphere of Sirius B consists practically of hydrogen that just in the measured wavelength region is nearly transparent for X-ray radiation. Therefore the measured X rays emerge from deeper layers below the visible outer photosphere (temp. only $\sim$20 000 K) that are much hotter (e.g., $\sim$100 000 K) and therefore copiously emit soft X-ray radiation. Such an explanation is based on the assumption that the X-rays indeed come from Sirius B rather than from Sirius A which at that time was not proved. We had to wait for HEAO-2 that had an X-ray telescope that could resolve the two components in Sirius (angular separation between A and B about 10 arcseconds during the period 1976–1982).

EINSTEIN

In the year I became head of the division INTEGRO (until 1992) the second High Energy Astrophysical Observatory (HEAO-2), later baptized "EINSTEIN" was launched on 12 November 1978. This X-ray mission caused a tremendous break-through in the field of X-ray astronomy because this observatory had for the first time an imaging telescope which suppressed the background with a factor 1000 or more compared to previous missions. One of its instruments, the High Resolution Imager with an angular resolution of about 2 arcseconds could indeed establish that Sirius B is the major X-ray source and that Sirius A should be at least 100 times weaker (the UV sensitivity of the detector prevented to measure the actual X-ray flux from Sirius A).

This mission carried out for the first time an extended survey during which more than 4000 stellar X-ray coronae were discovered.

In this context I want to mention that about two years before the launch of EINSTEIN a close and fruitful co-operation began with Professor Kees Zwaan of the Utrecht Observatory who unfortunately died on 16 June 1999. Together we submitted with success various proposals to observe with EINSTEIN coronae of cool stars. We participated in the first important paper of the EINSTEIN stellar survey by Vaiana et al. (ApJ 244 (1981), 163). One year later (in 1979) the Ph.D. student Karel Schrijver joined us in our research on stellar coronae which finally resulted in 1986 in his thesis (cum laude). He is now a prominent scientist in this field.

Our laboratory had in EINSTEIN a spectrometer with transmission gratings which had neasured nice X-ray spectra of stellar coronae (Capella; analysis published in ApJ together with the Ph.D. student Ed Gronenschild) and of other sources, e.g., Sirius B, the compact X-ray binary Sco X-1, and the dwarf nova EX Hydrae. On EX Hydrae I have measured with the Imaging Proportional Counter (IPC) on EINSTEIN more than 15 hours and discovered the 67 min. period - earlier detected in visible light - also in X-rays.

EXOSAT

With the advent of the European X-ray Observatory Satellite (EXOSAT; launch 26 Mei 1984) our laboratory was able to continue the spectroscopy with our transmission gratings. We have analyzed the spectra of three stars: Capella, σ^2 CrB, and Procyon, and of three white dwarfs: Sirius B, HZ 43, and Feige 24, which resulted in theses of Karel Schrijver and Frits Paerels, respectively.

Development of X-ray spectroscopy at SRON

On behalf of the interpretation of the measured X-ray spectra we have developed in the past 30 years a series of spectral models. Many people has helped me with this job, to name a few:

- Hans Schrijver 1972–80
- Ed Gronenschild 1975–85
- Janusz Sylwester (Wrocław) 1978–84
- Karel Schrijver (Palo Alto) 1979–98
- Jim Lemen (Palo Alto) 1983–89
- Bert van den Oord 1985–98
- Frans Alkemade 1989–93
- Piotr Majer (Wrocław) 1989–90
- Jelle Kaastra 1988–
- Hans Nieuwenhuijzen 1992–
- Duane Liedahl (Berkeley) 1994–
- Manuel Güdel (Zürich) 1995–
- Ken Phillips (Rutherford) 1996–
- Ton Raassen 1999–

$\Downarrow$

SPEX 2000

Finally this will result in a new code SPEX 2000[2] to which especially Jelle Kaastra has contributed for a big part.

In the beginning the spectral code carried only my name but later also the names of various of my co-workers. However, in the course of time we became more anonymous and one could think of the following scenario:

1972: Mewe	
1975: Mewe-Gronenschild	$\Longleftarrow$ Ed Gronenschild
1985: Mewe-Gronenschild-vd Oord	$\Longleftarrow$ Bert vd Oord
1990: MEKA	$\Longleftarrow$ Jelle Kaastra
1994: MEKAL	$\Longleftarrow$ Duane Liedahl
2000: KAL??	$\Longrightarrow$ Rolf Mewe
2000: RAASKAL??	$\Longleftarrow$ Ton Raassen

N.B. "raaskallen" = to rave

[2] Rolf refers here to SPEX Version 2.0, see http://www.sron.nl/divisions/hea/spex, J.S.K.

Intermezzo SCIAMACHY

In the years 1989–1995 I became involved in several orientating studies in the field of Earth atmospheric research. A peculiar detail: One of my scratch diagrams to indicate in a bird's eye view which trace gases could be measured in the atmosphere with various instruments was later used as a design of the poster for SCIAMACHY (= SCanning Imaging Absorption Spectrometer for Atmospheric Cartography). This satellite contains two high-resolution (UV, visible, and IR) spectrometers measuring perpendicularly to the Earth in nadir and along the limb (launch 2002).

YOHKOH, EUVE, ASCA, SAX, Spectrum X-γ

In the period after EXOSAT in which we had no own flying X-ray instrument (except in BeppoSax) I was involved as guest observer in a series of X-ray missions such as the Japanese solar mission YOHKOH (1991), the American Extreme Ultraviolet Explorer (EUVE; 1992), the Japanese Advanced Satellite for Cosmology and Astrophysics (ASCA; 1993), the Italian-Dutch Satellite per Astronomia in Raggi X (BeppoSAX; 1996) and the Russian-Danish Spectrum X-γ (launch ??). In all cases we have studied many spectra of cosmic X-ray sources or made spectral simulations with our spectral code SPEX.

Our evolving view on the X-ray sky

At the time I entered our institute in July 1970 the X-ray sky known to us consisted of only a twenty sources (e.g., Sco X-1 and the Crab nebula), but this picture drastically changed with the advent of the first X-ray satellite (UHURU) in December 1970 which rapidly extended the number of known X-ray sources to about 500. The EINSTEIN mission has enlarged this number since 1980 with a factor of ten or so. A full survey of the German ROSAT satellite (1 june 1990) has brought this number to nearly 200 000! Indeed a rapid evolution during my career at SRON.

Towards AXAF and XMM: the fantastic year 1999

Already more than ten years I have made many simulations that should exhibit how beautiful the spectra to be measured in future with our high-resolution X-ray spectrometers on board the NASA Advanced X-ray Astrophysical Facility (AXAF) and the ESA X-ray Multi Mirror mission (XMM) would be (Fig. 3).

In 1999 it was so far. First we had the launch on 23 July (planning was 20 July, my birthday) of AXAF, now baptized CHANDRA (after Chandrasekhar an Indian-American physicist who got the Nobel price in 1983) in the space shuttle Columbia on Cape Kennedy in Florida. Then on 10 December (planned was 8 Dec., my father's

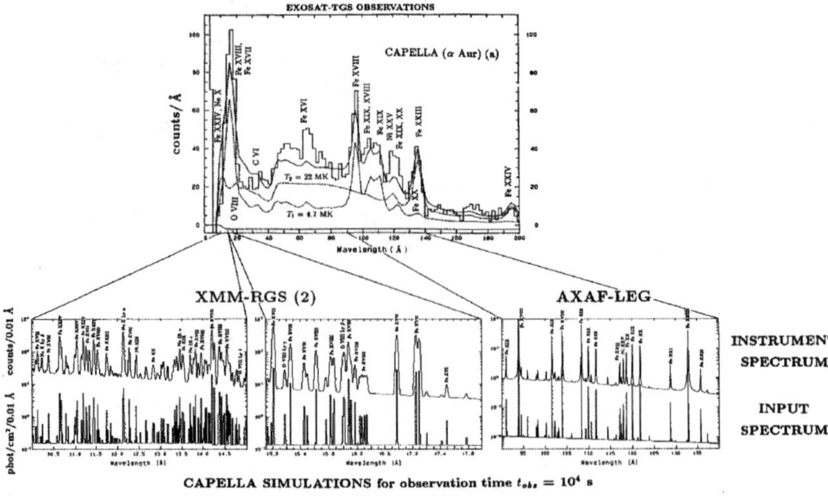

FIGURE 3. Spectral simulations for AXAF and XMM. Figure produced by R. Mewe and H. Braun.

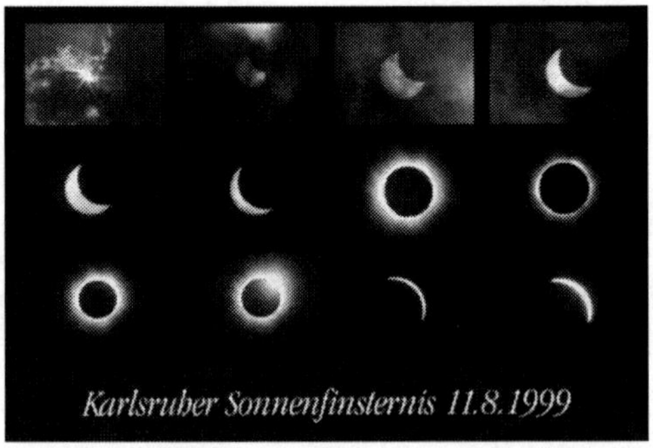

FIGURE 4. Solar eclipse. Poster, copyright F. Fiedler, Institut für Meteorologie und Klimaforschung – Forschungszentrum Karlsruhe, Universität Karlsruhe. Printed by Wilhelm Stober GmbH, Eggenstein, Germany.

birthday) XMM-Newton was lifted off with an Ariane-504 rocket from Kourou in French Guyana. Both launches were successful.

In the mean time we could admire on 11 August a Solar eclipse. Together with Hans Nieuwenhuijzen and our wives we enjoyed in the little village Schaidt (in the German Pfalz near Karlsruhe) this spectacle. The sky was cloudy and the Sun was only partly seen from time to time but just during the 2 min. of total eclipse we were lucky that a hole in the clouds appeared around the Sun that stayed during a few minutes! I could make a complete series of pictures of the whole phenomenon (comparable to those shown in Fig. 4).

In the mean time we are now in the year 2000 in the midst of the analysis of many

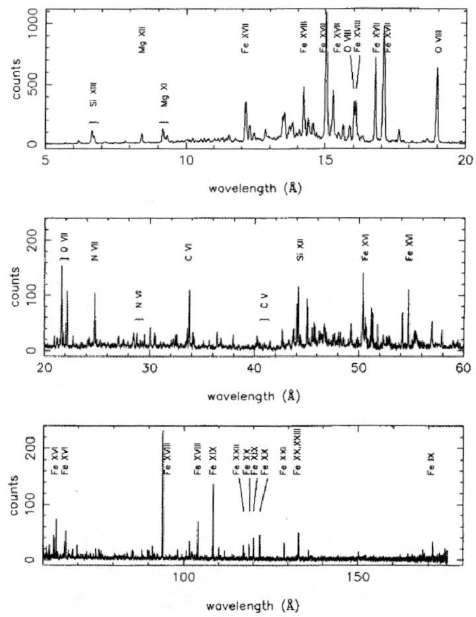

FIGURE 5. CHANDRA-LETGS "first-light" spectrum of Capella. From A.C. Brinkman et al. (2000) [2]; Copyright The American Astronomical Society.

beautiful spectra taken with our transmission gratings on CHANDRA and our reflection gratings on XMM-Newton of various cosmic sources such as stellar coronae (Fig. 5), compact X-ray binaries, white dwarfs, and the active galactic nucleus (AGN) NGC 5548. The observation of NGC 5548 with our instrument on CHANDRA by Jelle Kaastra and me yielded a nice spectrum with a series of absorption lines indicating an absorbing gas of a few million K and it was put on the headlines of NASA under the title "Chandra reads the Cosmic Bar Code of Gas Around a Black Hole". Our instrument also measured the two components of Sirius (now at an angular distance of 4 arcseconds and very well separated). The spectra exhibit that Sirius B is the bright X-ray source and that Sirius A probably emits only visible and UV light that "leaks" through the filters of the detector.

CONCLUSION AND FUTURE

The analysis of all these spectra can now very well be performed with SPEX and gives me the opportunity to continue my scientific work on a part-time base. Thanks to the way I always could do my work at SRON I am very motivated to stay a few years more here. I wish to thank my directors Kees de Jager and Johan Bleeker for the opportunity they gave me to develop my scientific work in all freedom. The presents and financial contributions of SRON colleagues and friends from abroad are highly appreciated. Finally, also on behalf of my wife Joke, children, grandchildren, and friends, I want

FIGURE 6. Rolf Mewe in 1998

to thank the organization which had offered us a splendid "farewell" party.

Rolf Mewe De Meern, 31 December 2000

ACKNOWLEDGMENTS

Jelle Kaastra expresses his thanks to Hans Nieuwenhuijzen for providing the many photographs for his talk, and Hans Braun for scanning these photos. SRON is supported financially by NWO, the Netherlands Organization for Scientific Research.

REFERENCES

1. Mewe, R., Heise, J., Gronenschild, E. H. B. M., Brinkman, A. C., Schrijver, J., and den Boggende, A. J. F., *Nature*, **256**, 711–712 (1975).
2. Brinkman, A. C., Gunsing, C. J. T., Kaastra, J. S., van der Meer, R. L. J., Mewe, R., Paerels, F., Raassen, A. J. J., van Rooijen, J. J., Bräuninger, H., Burkert, W., Burwitz, V., Hartner, G., Predehl, P., Ness, J.-U., Schmitt, J. H. M. M., Drake, J. J., Johnson, O., Juda, M., Kashyap, V., Murray, S. S., Pease, D., Ratzlaff, P., and Wargelin, B. J., *ApJ*, **530**, L111–L114 (2000).

I: GENERAL THEORY

An Historical Overview of Thermal X-ray Spectral Models

John C. Raymond

Center for Astrophysics, 60 Garden St., Cambridge, MA 02138

Abstract. Theoretical models of the X-ray spectra emitted by hot astrophysical plasmas have been calculated by many authors over the past 3 decades. In that time the computer power available and the quality of the atomic data that go into the calculations have improved enormously. This overview summarizes the development of model codes, the philosophies behind them, the directions they have taken, and some of the likely future developments.

PACS: 95.30.Dr, 95.75.Fg, 95.85.Nv

OVERVIEW

Stellar astrophysicists begin with the assumption of Local Thermodynamic Equilibrium (LTE) and move toward the more complex NLTE situation when forced to by the lower densities high in the atmosphere. X-ray spectroscopists start with the assumption of coronal equilibrium, in which all ions are in their ground states, and they move reluctantly toward the more complex situation where excited levels are significantly populated only when forced to by high densities in solar or stellar flares or in accreting systems.

Astronomers began computing theoretical models of the X-ray emission spectra of hot, low density plasmas back in the late 60s and early 70s. The computational efforts arose from two different cultures. On the one hand, solar physicists had high resolution UV and X-ray spectra to interpret. They were used to using the intensity ratios of carefully chosen pairs of spectral lines to infer densities, temperature distributions and elemental abundances, and they were able to select strong lines with reliable atomic rates for analysis. Jordan (1969) computed the ionization state of coronal plasma, and Landini & Monsigniori Fossi (1970) computed the 1-100 Å spectrum, which was gradually updated and extended to longer wavelengths (e.g. Landini et al. 1985), eventually leading to the Arcetri Spectral Code (Landi & Landini 2002), which gave way to the CHIANTI data base and suite of codes (Young et al. 2003).

On the other hand, astronomers had X-ray spectra of very low resolution. Proportional counters and various metallic or plastic filters gave $\Delta E/E \sim 1$. Worse yet, it was not possible to invert a proportional counter pulse height spectrum to get a photon spectrum. It was necessary to fold a model spectrum through the proportional counter response and fit it to the observations. It was therefore necessary to include all the emission that contributed significantly, whether accurate atomic data were available or not. Cox & Tucker (1969) computed the ionization state and total cooling rate of hot astrophysical plasma as a function of temperature. Tucker & Koren (1971) and Mewe (1972) presented

CP774, *X-Ray Diagnostics of Astrophysical Plasmas,* edited by Randall K. Smith
2005 American Institute of Physics 0-7354-0259-0/05/$22.50

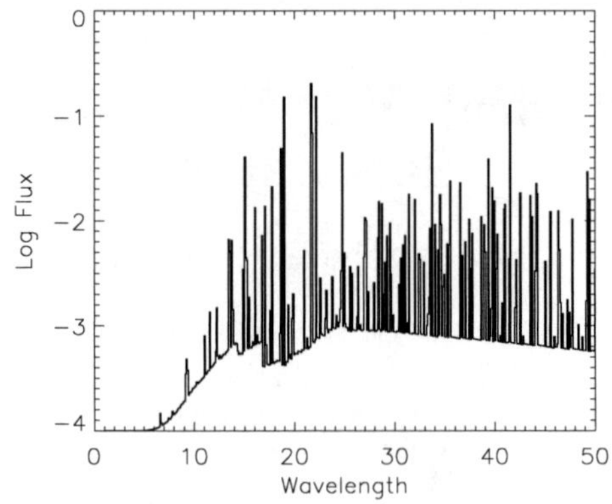

FIGURE 1. Sample model spectrum included to enliven the text.

tables of the intensities of strong emission lines. Both papers led to general purpose codes that included more spectral lines, updated atomic data and automatic binning of line and continuum emission (e.g., Raymond & Smith 1976; Mewe, Gronenschild & van den Oord 1985), eventually giving way to the MEKAL code of Mewe, Kaastra and Liedahl (Phillips et al. 1999) and to the APEC/APED database and codes (Smith et al. 2001). A number of other codes were written in the 1970s and 80s (Kato 1976; Gaetz & Salpeter 1983; Hamilton, Chevalier & Sarazin 1983) and expanded over the years (e.g. Borkowski et al. 2001).

For a time, the authors of various codes each made different choices among different sets of ionization balance and excitation rates available in the literature, so that comparison of different codes gave the users an idea of the accuracy to be expected. Such comparisons could be deceiving, in that the the Boltzmann factor $e^{h\nu/kT}$ determines the general shape of the spectrum, so that fits based on different codes tended to give very nearly the same temperatures, even though lines dominate the spectrum and different authors chose different atomic data for those lines. Significantly different emission measures were derived with the different codes, however. Figure 1 shows one model spectrum. It is left to the reader to guess the temperature.

The current large database efforts tend to use the same atomic data for the strong lines. For instance, APED uses rates from CHIANTI for many transitions, and both allow the user to choose among similar sets of ionization balance tabulations. Therefore, one should not assume that close agreement of different codes implies that highly accurate rates are known.

IONIZATION BALANCE

Wolfgang Lotz (1964) published a formula for the ionization rate coefficient of any ion based on fits to measured rates of a number of neutral and singly ionized species, and it performs remarkably well when extrapolated to more highly ionized elements. Distorted Wave calculations (e.g. Younger 1982) eventually supplanted the Lotz formula. For some isoelectronic sequences it is necessary to add the contribution of innershell excitation followed by autoionization (e.g. Pindzola et al. 1986). Current ionization balance calculations, such as that by Mazzotta et al. (1998), make use of laboratory measurements that extend to ions as high as Fe XVI (Gregory et al. 1986).

The radiative recombination rates are computed from photoionization cross sections and the detailed balance relation for recombination to the ground state. Recombination to excited levels is usually computed in the hydrogenic approximation, but the use of detailed balance plus photoionization cross sections from the upper levels seems to have only a small effect on the total rate (Arnaud & Raymond 1992). Burgess (1964) pointed out that dielectronic recombination dominates in many cases, and for many years the DR rates dominated the uncertainties in the overall ionization state. A unified approach to radiative and dielectronic recombination has been used to compute total recombination rates for many ions (e.g. Nahar & Pradhan 1997), but it does not lend itself as easily to computation of the photon spectrum produced in the recombination process. Charge transfer can alter the ionization state, but in ionization equilibrium it is most important at temperatures too cold to produce strong X-ray emission.

EMISSION LINE EXCITATION

The early models were often limited to semiclassical or Born approximation excitation cross sections. Unfortunately, these tend to be best far above threshold, while X-ray emission lines are generally produced by electrons close to threshold. The basic reason in that ion concentrations tend to peak at temperatures where the ionization potential $\chi \sim 5-10kT$, and transitions involving a change in principal quantum number typically have thresholds E above $\chi/2$. Therefore, E is typically $2.5-5kT$. More accurate Distorted Wave calculations have become widely available, especially with the HULLAC (Klapisch et al. 1977) and M.-F. Gu's FAC (e.g. Savin et al. 2003) codes that can handle large numbers of transitions. The Close Coupling method embodied in the RMATRIX code used, for instance, in the Iron Project (Hummer et al. 1993) handles the resonances in excitation cross sections that can dominate the total excitation rates in forbidden or intercombination transitions. It produces very accurate results provided that enough states can be included in the calculation.

The excitation rates for strong lines of simple ions now seem to be in good shape, but the more complicated L-shell ions of Si, S and Fe may not be as well understood. As more accurate direct excitation rates become available, other processes must be considered. Cascades from higher-lying levels and recombination to excited levels were found to be important for some ions long ago (e.g., Loulergue & Nussbaumer 1975; Mewe & Schrijver 1978). Dielectronic recombination satellites and inner shell excitations were also recognized early on (Gabriel 1972). In general, these processes have been considered in detail for the strongest lines of some of the dominant isoelectronic sequences, but they have received less attention in other cases. The current codes include detailed treatments of recombination to excited levels and of satellite lines for the H-like and He-like ions, but not for most other ions.

CONTINUUM PROCESSES

The radiative recombination continuum is computed in the same calculation as the radiative recombination rate, and it should be as accurate as the photoionization cross section that goes into it. Recombination to exited levels has not generally been included because of the lack of good photoionization cross sections from excited levels. It should be noted that in some cases the recombination rate to a low-lying excited level can greatly exceed that to the ground state. For instance, radiative recombination to the 2s2p ^3P level of a Be-like ion can dominate over that to the $2s^2$ ^1S ground state by virtue of the 9:1 ratio of statistical weights. While codes such as that of Raymond & Smith (1977) took such effects into account, they were limited to approximate treatments by the lack of available data.

Two-photon continua are produced by decays of the metastable 2s ^2S levels of H-like ions and the 1s2s ^1S levels of He-like ions. The continuum shapes are accurately known, and the total emissivities are computed along with those of the strong emission lines. Because they upper levels are metastable, the intensities of these continua are density dependent. While this dependence was included in the Raymond & Smith and Mewe codes long ago, tabulated spectra in the low density limit are often used to fit spectra of stellar coronae dense enough to shift power from the two-photon continua to the resonance lines.

Bremsstrahlung is the simplest of the continuum processes, and reasonable accurate emission spectra have been available for many years (Karzas & Latter 1961). Until recently it was somewhat questionable to use the available model codes at temperatures above 10^8 K, however, because relativistic effects in the bremsstrahlung cross section were not included. New calculations by Itoh et al. (2000) account for relativistic effects, and those cross sections are used in APEC.

CONSISTENCY

To some extent, consistency may be more important than absolute accuracy. It is obviously important that the recombination radiation agree with the recombination rates used in the ionization balance calculation. It is also important to use total cooling rates consistent with the spectral calculation, both in terms of atomic rates and elemental abundances, in simulations of cooling flows or coronal loops.

A more subtle case arose when Arnaud & Raymond (1992) computed the ionization state of Fe using newly available atomic data. However, they did not compute ionization states for the other elements, because much less new data were available. For some years, the Arnaud & Raymond Fe ionization balance was combined with the Arnaud & Rothenflug (1985) ionization balance for the other elements, causing a mismatch when comparing Fe with other elements. Fortunately, the ionization balance of Mazzotta et al. (1998) is a consistent set.

SPECIAL APPLICATIONS

Density: The early codes included density sensitivity by computing level populations by a matrix inversion for up to about 6 low-lying metastable levels. Modern ones invert a matrix for all the levels of a given ion, sometimes over 1000, which automatically takes care of the cascade contributions to line intensities. High resolution X-ray spectra make these effects important as density diagnostics. The density sensitivity of the ionization balance is not generally considered, but it can be significant. Density sensitivity of the dielectronic recombination rate arises because the highly excited intermediate state is vulnerable to collision processes (Burgess & Summers 1969), and this sensitivity was included in the calculations such as Jordan (1969). Modern rates are available in a series of papers including that of Zatsarinny et al. (2004). The electric fields involved can modify the density effects somewhat (Jacobs et al. 1976; Reisenfeld et al. 1992; Böhm et al. 2003). At high densities, the populations of metastable levels become significant, and the ionization and recombination rates from these levels can be much different from those of the ground states. Vernazza & Raymond (1979) considered the effects of metastables at solar coronal densities.

Photoionization: Photoionized plasmas represent another extension of the original codes. In principle, one merely computes a photoionization rate using the same cross sections used in the recombination calculation by Raymond (1993), the XSTAR code (e.g. Kallman et al. 2004), and the CLOUDY code (see Ferland 2003). A difficulty is that fits to excitation rates and dielectronic recombination rates can fail at low temperatures, and that large disagreements among rate calculations can arise (Savin & Laming 2002). Because the Radiative Recombination Continua (RRCs) are detectable, consistency between the photoionization rate and the recombination continuum is especially important.

Time-Dependent Ionization: The ionization states in some astrophysical settings are far from equilibrium, and time-dependent ionization was included in early on in models of shocked gas in QSOs (Daltabuit & Cox 1972), supernova remnants (Itoh 1977), and solar flares (Mewe & Schrijver 1980). There is no great difficulty in computing such models, but they depend on several parameters and are therefore more difficult to tabulate. Until recently, time-dependent models were mainly used by their authors, but now that NEI and shock models are available in XSPEC (Borkowski et al. 2001) they are more widely used.

Non-Maxwellian Tails: All the existing codes assume a Maxwellian velocity distribution for the electrons, but in situ measurements of electron velocities in the solar wind always show departures from Maxwellian distributions. If the departure only occurs at very high energies it does not affect the spectrum. If it occurs at $\sim$ 10 kT, or 1-2 times the typical ionization potential of ions found at temperature T, it affects the ionization state. It must occur at even lower energies to significantly affect the excitation rates. A commonly used parameterization is the κ distribution, and it has been used for some UV spectral models (Anderson, Raymond & van Ballegooijen 1996), but other parameterizations have been used for some X-ray computations (Dzifčáková & Kulinová 2001. There is no great difficulty (but considerable inconvenience) in incorporating non-Maxwellian distributions in a standard spectral code, the main difficulty arising because excitation cross sections that include resonances are often integrated over Maxwellians for convenient tabulation. So far, there are only a few cases where a non-Maxwellian distribution is thought to be severe enough to justify the effort and where X-ray diagnostics are good enough to constrain it in interesting ways (Seely, Feldman & Doschek 1987). In other cases, it tends to add a poorly constrained free parameter.

FUTURE

The current generation of thermal X-ray models seems to be converging in several ways. The advantages of a database structure are apparent for the large numbers of transitions now included. Both CHIANTI and APEC strive to cover very broad ranges in wavelength, density and temperature, making it possible to consistently model simultaneous observations in different wavelength bands, to cover situations from the ISM to stellar coronae, and to deal with severe departures from ionization equilibrium. Both codes also strive to maintain an ease of use that in the old days was referred to as "blackbox."

This last consideration leads to some possible divergence as special applications call for specialized codes. Indeed, the old codes were used by their developers for many special applications such as time-dependent models, but these were not made part of the standard distribution because of the effort needed to document these capabilities and make them robust. One major need will be a "faster, cheaper better" code that can be integrated into hydro codes and still contain enough detail to reliably match current X-ray and UV observations. Others may include a more general purpose code that includes photoionization and radiative transfer in spectral lines, a general purpose time-dependent code, and a code that includes non-Maxwellian distributions in a way amenable to standard fitting techniques.

That said, the major thrust is still likely to be improvement of ionization, recombination and excitation rates beyond the typical 20-30% now common. Continued benchmarking of theoretical rates against laboratory measurements is especially crucial in establishing the level accuracy of the codes.

REFERENCES

1. Anderson, S.W., Raymond, J.C. & van Ballegooijen 1996, ApJ, 457, 939
2. Arnaud, M. & Raymond, J. 1992, ApJ, 398, 394
3. Arnaud, M., & Rothenflug, R. 1985, A&AS, 60, 425
4. Böhm, Böhm, S., Müller, A., Schippers, S., Shi, W., Eklöw, N., Schuch, R., Danared, H. & Badnell, N. R. 2003, A&A, 405, 1157
5. Borkowski, K.J., Arnaud, K.A., Dorman, B., Hughes, J.P., Sarazin, C.L., & Smith, R.A. 2001, BAAS, 199, 126.19
6. Burgess, A. 1964, ApJ, 139, 776
7. Burgess, A., & Summers, H.P. 1969, ApJ, 157, 1007
8. Cox, D.P., & Tucker, W.H. 1969, ApJ, 157, 1157
9. Daltabuit, E., & Cox, D.P. 1972, ApJL, 137, L13
10. Dzifčáková, E. & Kulinová, A. 2001, Sol. Phys. 203, 53
11. Ferland, G.J. 2003, ARAA, 41, 517
12. Gabriel, A. 1972, MNRAS, 160, 99
13. Gaetz, T.J., & Salpeter, E.E. 1983, ApJS, 52, 155
14. Gregory, D.C., Meyer, F.W., Müller, A., & Defrance, P. 1986, PRA, 34, 3657

15. Hamilton, A.J.S., Chevalier, R.A., & Sarazin, C.L. 1983, ApJS. 51, 115
16. Hummer, D.G., Berrington, K.A., Eissner, W., Pradhan. A.K., Saraph, H.A. & Tully, J.A. 1993, A&A, 279, 298
17. Itoh, H. 1977, PASJ, 29, 813
18. Itoh, N., Sakamoto, T., Kusano, S., Nozawa,S., & Kohyama, Y. 2000, APJS, 128, 125
19. Jacobs, V.L., Davis, J., & Kepple, P.C. 1976, PhRvL, 207, 605
20. Jordan, C. 1969, MNRAS, 142, 501
21. Kallman, T.R., Palmeri, P., Bautista, M.A., Mendoza, C., & Krolik, J.H. 2004, ApJS, 155, 675
22. Karzas, W.J., & Latter, R. 1961, ApJS, 6, 167
23. Kato, T. 1976, ApJS, 30, 397
24. Klapisch, M., Schwab, J.L., Fraenkel, B.S., & Oreg, J. 1977, J.Opt.Soc. Am., 67, 148
25. Landi, E., & Landini, M. 2002, A&A, 384, 1124
26. Landini, M., & Monsigniori Fossi, B. 1970, A&A, 6, 468
27. Landini, M., Monsigniori Fossi, B., Paresce, F., & Stern, R.A. 1985, ApJ, 289, 709
28. Lotz, W. 1964, ApJS, 14, 207
29. Loulergue, M., & Nussbaumer, H. 1975, A&A, 45, 125
30. Mazzotta, P., Mazzitelli, G., Colafrancesco, S. & Vittorio, N., A&AS. 133, 403
31. Mewe, R. 1972, Sol. Phys. 22, 459
32. Mewe, R., Gronenschild, E.H.B., ṽan den Oord, G.H.J. 1985, A&AS, 62, 197
33. Mewe, R. & Schrijver, J. 1978, A&A, 65, 99
34. Mewe, R. & Schrijver, J. 1980, A&A, 87, 261
35. Nahar, S.N., & Pradhan, A.K. 1997, ApJS, 111, 339
36. Nakayama, M., & Masai, K. 2001, A&A, 375, 328
37. Phillips, K.H.J., Mewe, R., Harra-Murnion, L.K., Kaastra, J.S., Beiersdorfer, P., Brown, G.V., & Liedahl, D.A. 1999, A&AS. 138, 381
38. Pindzola, M.S., Griffin, D.C., Bottcher, C., Younger, S.M., & Hunter, H.T.
39. Raymond, J.C. 1993, ApJ, 412, 267
40. Raymond, J.C., & Smith, B.W. 1977, ApJS, 35, 419
41. Reisenfeld, D.B., Raymond, J.C., Young, A.R., & Kohl, J.L. 1992, ApJ, 339, 37
42. Savin, D.W. & Laming, J.M. 2002, ApJ, 566, 1166
43. Savin, D.W., et al. 2003, ApJS, 127, 421
44. Seely, J.F., Feldman, U., & Doschek, G.A. 1987, ApJ, 319, 541
45. Smith, R.K., Brickhouse, N.S., Liedahl, D.A., & Raymond, J.C. 2001, ApJL. 556, L91
46. Tucker, W.H., & Koren, M. 1971, ApJ, 168, 283
47. Vernazza. J.E., & Raymond. J.C. 1979, ApJL, 228, L89
48. Young, P.R., Del Zanna, G., Landi, E., Dere, K.P., Mason, H.E. & Landini, M. 2003, ApJS, 144, 135
49. Younger, S.M. 1982, Phys. Rev. A, 26, 3177
50. Zatsarinny, O., Gorczyca, T.W., Korista, K., Badnell, N.R. & Savin, D.W. 2004, A&A, 426, 699

Dr. Charles Alcock opening the workshop.

A primer on atomic theory calculations (for X-ray astrophysicists)

F. Robicheaux

Department of Physics, Auburn University, AL 36849-5311

Abstract. This paper gives a general overview of different types of atomic calculations with an emphasis on how the methods work. The focus will be on the methods that are mostly used to generate data for plasma physics. The topics will be limited to atomic structure and electron scattering processes. Several important topics (in particular photo-ionization and ion-atom and ion molecule scattering) will not be reviewed. The main methods include mean field approximation, perturbation theory, and configuration interaction/close coupling. This paper gives a general picture of how the different methods work and describes the strengths and limitations of the different methods.

Keywords: atomic theory
PACS: 31.10.+z,34.10.+x

INTRODUCTION

Calculation of atomic properties has a 90 year history, going back to the calculation of the spectra of hydrogen-like ions by Bohr. After the advent of relativistic quantum mechanics, the equations used to describe the interaction between charged particles were sufficiently accurate for almost all purposes. These equations can be expressed in the form of a wave equation. In principle, all of the atomic properties needed for any application can be calculated.

Atomic calculations have achieved stunning accuracy when the number of particles is small. However, the effort required to achieve a level of accuracy can increase dramatically with the number of particles. Thus, the properties of atoms and ions with one electron outside of a closed shell can be computed very accurately but it may be impossible with current theoretical techniques and computation resources to calculate the properties of atoms and ions with two or more open shells. The limitations are not due to the underlying equations but are due to limitations of computational resources and techniques.

Due to the approximate nature of calculations, it might be argued that it would be better to appeal directly to experiment. There are two problems with this suggestion. First, it is too difficult to obtain all of the data needed for most applications. Second, experiment is simply another *approximate* way of obtaining atomic properties. There are many examples where the experiments were in error (correct answer was well outside error bars) or where the experiments have unacceptably high uncertainty. Currently, experiment and theory are jointly exploring many atomic properties. Progress in understanding the uncertainties and errors in the two methods can be achieved by comparing the results in a few key systems. This paper gives a general overview of the strengths

CP774, *X-Ray Diagnostics of Astrophysical Plasmas,* edited by Randall K. Smith

and weaknesses of the more popular computational techniques.

PHYSICAL EFFECTS

This section gives a brief description of the physical effects that play a role in atomic properties.

Mean Field

The main feature of atoms and ions is the heavy nucleus with charge Ze surrounded by light electrons with charge $-e$ where e is roughly 1.6×10^{-19} C. When a particle is in a spherically symmetric potential, its angular momentum and the projection of the angular momentum on a direction is conserved. It is standard to use letters for the angular momentum quantum number: s, p, d, f, g, ... refer to particles with angular momentum $\ell = 0, 1, 2, 3, 4, \ldots$. Each angular momentum has $2\ell + 1$ projections on an axis. Thus each state of a given total angular momentum has $2 \times (2\ell + 1)$ possibilities where the $2\times$ arises from the spin of the electron.

In atomic physics, the principal quantum number is the total number of nodal surfaces (plus 1) in the object's wave function. Thus, the 1s state has no nodes and has angular momentum 0 while the 4d state has 3 nodal surfaces (one in radius and two in angles) and has an angular momentun 2. For hydrogen-like ions without relativity, the energy only depends on the principal quantum number: $\mathrm{E} = -13.6\,\mathrm{eV}\,\mathrm{Z}^2/\mathrm{n}^2$. The only reason that calculations for many electron atoms can be done at all is due to the structure of the energy with principal quantum number. The electrons in the $n = 1$ shell are very tightly bound and are not strongly perturbed by electrons in higher n shells. Conversely, the $n = 1$ electrons move so rapidly their main effect on the electrons in the higher energy shells is to provide a spherical charge distribution which gives an additional spherical potential. This argument can be repeated for the $n = 2$ electrons and so on.

At an approximate level, each electron experiences a spherically symmetric potential due to the nucleus and an average over the other electrons. This leads to a standard property of atoms. Electrons with high angular momentum do not approach the nucleus as closely as atoms with low angular momentum due to the centrifugal barrier. The region near the nucleus has the most negative potential energy because it is least screened by the other electrons. Thus, the low angular momentum states of a given principal quantum number are more deeply bound than high angular momentum states. For example, a 3s electron is more deeply bound than a 3p electron is more deeply bound than a 3d electron. This effect can be so large that states with higher principal quantum number can be more deeply bound than states of lower principal quantum number; for example, the 4s state in K is more deeply bound than the 3d by a couple eV. This effect is less pronounced in highly charged ions because the potential from the ion dominates.

Correlation

The next main effect usually goes by the name of correlation and only refers to the correlation between electrons. Generically correlation is simply the quality that one parameter depends on the value of a second parameter. In atoms, correlation is somewhat more complicated. This is because a correlation exists even at the mean field level due to the fact that the electrons' wave function must be anti-symmetric under exchange of two electrons. For example, suppose two electrons both have spin up and one electron is in the 1s state and the other is in the 2s state. The spatial part of the wave function must be

$$\psi(\vec{r}_1,\vec{r}_2) = \phi_{1s}(\vec{r}_1)\phi_{2s}(\vec{r}_2) - \phi_{2s}(\vec{r}_1)\phi_{1s}(\vec{r}_2) \tag{1}$$

where the ϕ's are the one electron orbitals. Note that which state electron 1 is in depends on which state electron 2 is in and vice versa. Since this correlation is present even when there is no interaction between the electrons, we typically do not include it under the heading: correlation.

Nontrivial correlation arises due to the Coulombic electron-electron interaction:

$$V(\vec{r}_1,\vec{r}_2) = \frac{e^2}{4\pi\varepsilon_0|\vec{r}_1 - \vec{r}_2|}. \tag{2}$$

This interaction can cause a pair of electrons to each change angular momentum and energy. When the mean field approximation is moderately accurate, the form of this potential controls what sort of correlation is present. Typically, one and two electron changes from the main composition dominate. For example, the $1s^22s^22p^6$ state has some admixture of $1s^22s^22p^43d^2$ but much less of $1s^22s^22p^23d^4$; as another example, the $1s^22s^22p^5$ state has some admixture of $1s^22s^12p^53d^1$ but much less of $1s^22s^12p^43d^3$.

The form of the interaction also controls the decay of atomic resonance states. For example, the $1s^22s^12p^6$ state is at the same energy as $1s^22s^22p^4Ed$ where the Ed is meant to indicate a free electron. Transitions where three electrons change character from the dominant part of the wave function are typically weaker than those where two electrons change character.

Often, atomic physicists become sloppy in their description of a process due to the (generally) weak correlation. For example, often photons are described as having caused a two electron transition. A classic case is the absorption of one photon by helium causing a transition from the $1s^2$ ground state to the $2s2p$ resonance state. The one photon can only be absorbed by one electron; it is the correlation in the initial and the final state that gives the appearance of a two electron transition.

Relativity

The average velocity of an electron in a state with principal quantum number n is roughly $v \sim c(\alpha Z/n)$ where $\alpha \simeq 1/137$ is the fine structure constant. For light atoms, relativity is a minor effect but it becomes increasingly important with increasing nuclear charge. Depending on their size and the desired accuracy of the calculation, relativistic effects can be treated by using the relativistic wave equation which includes many relativistic effects to all orders or can be treated by using a nonrelativistic wave equation but account for relativity using perturbation theory.

At the level of $(v/c)^2$ there are three relativistic effects. (1) The spin-orbit interaction which is the interaction of the magnetic dipole moment of the spin with the electric field created by the nucleus and other electrons. (2) The mass-velocity term arises from the power series expansion of the relativistic kinetic energy: $KE = p^2/2m - p^4/8m^3c^4 ....$ (3) The Darwin term arises from an effective spread of the electron over a distance $\sim h/mc$.

At higher order in v/c there are contributions similar to those described above. Also, "QED" effects arise. The largest effects are due to the self energy and vacuum polarization as well as the Breit interaction.

STRUCTURE METHODS

In this section, I describe some of the methods used to compute atomic structure. Accurate wave functions and energy levels for bound states underly almost all atomic calculations. Clearly, the energy levels are needed to determine wavelengths of light emitted by atoms. They are also used in scattering calculations since the position of resonances can strongly change scattering rates. Finally, the accurate wave functions are needed for computing transition rates between states.

Hartree/Dirac-Fock

The Hartree/Dirac-Fock method is a mean field method for computing structure with a proper treatment of the anti-symmetry between electrons. The Hartree-Fock method is based on the nonrelativistic Schrödinger equation (with perhaps approximate relativistic interactions) while the Dirac-Fock method is based on the relativistic Dirac equation. In both methods, the wave function is approximated as a sum of a simple product of orbitals that gives the correct symmetry. For example, the $1s2s\ ^3S$ state of a helium like ion has the spatial part of the wave function with the form

$$\psi(\vec{r}_1,\vec{r}_2) = R_{1s}(r_1)R_{2s}(r_2) - R_{2s}(r_1)R_{1s}(r_2) \tag{3}$$

where the two R functions only depend on the distance of the electron from the nucleus. The equations for the two R functions are determined through the use of a variational principle for the energy; this means the error in the energy might be smaller than you expect since it is quadratic in the error of the wave function (roughly speaking 10% error in the wave function translates into a 1% error in the energy).

Treatment of the atomic physics at this level has several advantages: there are well developed (and available) computer programs, the energies and wave functions can be found very fast, and the approximate energies and wave functions can be used as a starting point which will be fixed through more sophisticated calculations. The difficulty with this method is the equations are nonlinear which can sometimes cause convergence problems and only the energy is variational. Because only the energy is variational, transition rates (or any other parameters that depend on the wave function) are much less accurate. However, all of the methods described here have this drawback so I will not repeat it further down.

Perturbation Theory

Perturbation theory can be used whenever corrections to an approximate wave function are small. The Hamiltonian, H, determines the wave equation that must be solved. If one can find a simplified Hamiltonian, H_0, for which the wave functions can be accurately determined and *if* this simplified Hamiltonian is sufficiently close to the exact Hamiltonian, then expressions for the wave function and energies can be found as a "power series" in $H - H_0$. For example the energy for the a-th state is

$$E_a = E_{a,0} + \langle \psi_{a,0} | H - H_0 | \psi_{a,0} \rangle + \sum_{b \neq a} \frac{|\langle \psi_{a,0} | H - H_0 | \psi_{b,0} \rangle|^2}{E_{a,0} - E_{b,0}} + \ldots \qquad (4)$$

where $\psi_{a,0}$ and $E_{a,0}$ are the eigen-state and eigen-energy for the simplified Hamiltonian H_0. This procedure is often used when including relativistic effects: H_0 is the full nonrelativistic Hamiltonian and $H - H_0$ are the relativistic corrections. This procedure is also used when including correlation: H_0 is the Hartree/Dirac-Fock Hamiltonian and $H - H_0$ gives non-mean field corrections.

The advantages of this methods are that there are well developed prescriptions and programs that use this method, the results can be very accurate when $H - H_0$ is small, and there are procedures to include higher order correlations than typically included in the two methods described below. Also, this method takes the reasonable approach of treating large effects differently from small effects. The difficulty with this method is that you must be able to find an H_0 such that the $H - H_0$ is small and the existing programs have not been applied to generic atomic states with more than one open shell.

MCHF/MCDF

The multi-channel Hartree-Fock (MCHF) and multi-channel Dirac-Fock (MCDF) methods approximate the wave function by a superposition of antisymmetrized product wave functions; the Hartree/Dirac-Fock methods use a single antisymmetrized product wave function. The MCHF/MCDF methods allow for the approximate inclusion of correlation in the wave function. As an example, the ground state of a Helium-like ion can be approximately represented by a wave function which is the superposition of $1s^2$, $2s^2$ and $2p^2$ states. The equations for the orbitals are determined by a variational principle with the equations for the orbitals coupled together, i.e. the equation for the $1s$ orbital depends on the $2s$ and $2p$ orbitals, etc. The coefficients for each term in the sum are also determined from a variational principle. Calculations have included thousands of product wave functions in the sum although smaller numbers are more typical.

The advantages of this method are that there exist well tested and easy-to-use programs that perform the calculations, you obtain the most accurate energies and wave functions for the fewest terms in the approximate wave function, and it can be very accurate since the orbitals are determined for the particular state of interest. The difficulties of this method are that it is typically used to solve for one (or sometimes a few) state at a time making it tedious for generating plenty of data, the equations are non-linear making convergence chancy in some systems, and it is mainly used for deeply bound states where none of the electrons go to far distances from the nucleus (e.g., MCHF/MCDF would not typically be used for $1s^2 2s^2 2p^2 6s$).

R-matrix

Similar to the MCHF/MCDF methods, the R-matrix method includes correlation through a superposition of antisymmetrized product states. Unlike those methods, the R-matrix method does not allow the orbitals to vary. Typically, the orbitals are determined from some external simple method (e.g. an approximation to the Hartree-Fock method) which has some resemblance to the problem of interest. The coefficients of the antisymmetrized product states are determined through a variational principle. Because the coefficients appear quadratically in the variational principle, the solution can be obtained using standard linear algebra packages. Typically, the number of product states is one or two orders of magnitude larger than that used in the MCHF/MCDF methods. This allows the R-matrix method to overcome the advantage that the MCHF/MCDF gets the most accuracy for a given number of basis functions. However, the number of product states for the R-matrix method is often one or two orders of magnitude smaller than in a perturbative calculation; the R-matrix method can compete since it finds the wave function to all orders within the restricted Hilbert space of its basis set.

The advantages of the R-matrix method are that well developed programs exist and can be used for almost any configuration of interest for almost any atom/ion, high accuracy can be achieved by using vary large basis sets, and the equations are linear which means there is no problem obtaining the solution. The difficulties of this method are that small and large corrections to the wave function are treated in the same manner and often it is impossible to have enough basis functions to achieve the desired accuracy.

Mixed Methods

Currently promising trends use a mixture of two or more methods. The most common mixture is to have a configuration interaction (CI) method (e.g. MCHF/MCDF or R-matrix) include strong effects within a restricted basis set and then use perturbation theory to include weak effects that span a wide range of basis functions. For example, use a CI method to incorporate non-relativistic correlations and use perturbation theory to include relativistic effects. Another example is to solve the for the strong correlations using a small CI calculation and then include the vast number of states that contribute to weaker correlation using perturbation theory.

The advantages of this method are that it can be more accurate than any of the other methods separately and complicated interactions can be included. The disadvantage is that often the approximations to the perturbation, $H - H_0$, are used and the perturbation might not be small enough to achieve the desired accuracy.

Accuracy of Transitions

The radiative decay and photo-excitation of atoms and ions can be computed from the wave functions of the two states. Typically, the relative error in the calculation of this process is larger than for the energy because the energy error is quadratic in the wave function error while the transition matrix element error is linear in the wave function error. Upon absorption or emission of a photon, the total angular momentum can only change by one unit and only one electron orbital can change character. For light atoms, the photon only weakly acts on the spin so the transitions that change the total spin or change the total orbital angular momentum by more than one unit are weak.

There are important trends in the error of calculated transition matrix elements. The most accurately calculated transitions are for the states that have allowed transitions between the major part of the wave function (e.g. $1s^2\ {}^1S_0$ going to $1s2p\ {}^1P_1$, the notation is ${}^{2S+1}L_J$ where S is the total spin, L is the total orbital angular momentum and J is the total angular momentum). Much less accurate are transitions between states that only occur due to correlation (e.g. $1s^2\ {}^1S_0$ going to $2s2p\ {}^1P_1$) or (in light ions) due to relativity (e.g. $1s^2\ {}^1S_0$ going to $1s2p\ {}^3P_1$). Large errors can result if two states of the same total angular momentum and parity are nearly degenerate; in this case, the precise mixing between the states might not be calculated accurately.

Especially one should beware of a transition rate of 0. There is always some type of transition connecting the states. Some wild errors in plasma simulations would have been avoided if a small value had been used in the place of 0.

ELECTRON SCATTERING

In this section, I describe some of the methods used to compute electron scattering processes. As with structure calculations, there are methods based on perturbation theory and methods that attempt to directly account for configuration interaction. Compared with structure calculations, scattering calculations are less accurate and more complicated because the basic process (a transition) depends on electron correlation. Also, the transition between two states requires the structure of the two states to be accurate; thus, if all your computer power is required to get accurate bound states, nothing will be left over to calculate the incident/scattered electron.

Perturbation Theory

When the transition between two states is weak (in other words, when the target electrons only weakly affect the incident electron), perturbation theory can be used to compute the transition rates. In principle, perturbation theory can be carried out to a high order and can give very accurate results. In practice, most calculations include only the lowest order term that allows a process. Because correlation is typically smaller for ions, perturbative results usually become more accurate with the charge of the ion.

Similar to the application of perturbation theory to structure calculations, the application to scattering involves finding a simplified Hamiltonian, H_0, from which wave functions can be calculated but near enough to the full Hamiltonian to be relevant. In some calculations, the H_0 does not contain any interaction for the incoming electron (plane wave Born), but in most applications a mean field potential is included for the incoming electron (distorted wave Born). For distorted wave Born calculations, it is sometimes ambiguous which mean field should be taken; this is especially important when the target qualitatively changes, as in ionization. Whether the incoming electron is computed in the potential for the initial state (prior form) or the final state (post form) can give quantitatively different results. This difference can serve as a gauge of accuracy.

Resonances can be included within perturbation theory. The two different paths to the final state (directly going to the final state in a single step or first going into a resonance state that decays into the final state) interfere. However, the interference often does not affect the temperature dependence of the total rates. In practice, the interference usually plays little role when the thermal electron energy is greater than the energy width of the resonance. Thus, whether or not interference should be included in a calculation depends on the process and what use will be made of the cross section.

The advantages of perturbative calculations are that the calculations are very fast, it is easy to thermally average over resonance features, and the calculations could be improved by going to higher order or by improving H_0. The difficulties include the problem of strong coupling (inadequate H_0), which mean field to use for the distorted wave, and the often inaccurate bound states used for the targets.

R-matrix

The R-matrix method for scattering is similar in idea to its use in calculating bound states. For scattering calculations, the inverse of the logarithmic derivative of the wave function (the R-matrix) at a distance r from the nucleus is found from a variational principle at a specified energy. As with the bound state calculation, the wave function in a small region near the nucleus (all electrons at a distance less than r) is represented by a superposition of antisymmetrized product states. The coefficients of the superposition are found using standard linear algebra packages. The information about the wave function at the distance r can be converted into scattering amplitudes using standard methods. The interaction of the incoming/outgoing electron with the atom/ion outside of this region is included by a numerical integration or by using perturbation theory.

The R-matrix method can be used with both relativistic or nonrelativistic quantum mechanics. Even effects like photon emission (needed for computing radiative and dielectronic recombination) can be included. The R-matrix method attempts to accurately compute the coupling of the electrons even when the coupling is strong. It can give the scattering parameters on a very fine grid of energy points but does not have a simple method for energy averaging over resonances.

The advantages of this method are that strong coupling between a limited number of channels can usually be handled routinely, calculations can be performed for almost any type of open shell targets, and the recent incorporation of pseudo-states has allowed the calculation of ionization in simple systems and faster convergence of more complicated systems. The disadvantages are that small and strong couplings are treated in the same manner, numerical artifacts that appear to be resonances (pseudo-resonances) can be present if the basis set is not properly balanced, a fine energy mesh is *required* if sharp resonances are present, and usually less accurate target states are used. The last disadvantage can be serious and arises from the fact that the target states are themselves represented by a superposition of basis states; thus for each target state and angular momentum coupling there must be (number of basis functions representing the target state) $\times$ (number of basis functions representing the incoming electron) basis functions. For excitation of complex atom/ion, there can be up to $\sim$ 100 target states forcing a very reduced representation if the calculation is to be performed at all.

Other methods

There exist many other computational techniques that can be more accurate than those described above. These other methods have not been used to generate large amounts of data for applications. However, they can be used to benchmark the more widely used methods. Some of the more important are: the convergent close-coupling (CCC) method which solves for the scattering amplitudes using a basis set technique to solve the Lippman-Schwinger equation, the time dependent close coupling (TDCC) method solves for the scattering amplitudes by directly solving the time dependent Schrödinger equation using spatial grid points, and the hyperspherical close coupling (HSCC) method solves for the wave function using hyperspherical coordinates.

Accuracy of scattering calculations

Since the transitions are due to electron correlation, the less correlation present the more accurate the calculations tend to be. Thus, the transition between two states generally becomes more accurate in an isonuclear sequence as the charge of the ion increases. Also, transitions between more compact states (e.g. $1s^2 2s^2$ going to $1s^2 2s^1 2p^1$) are more accurate than transitions between more highly excited states (e.g. $1s^2 2s^1 4s^1$ going to $1s^2 2s^1 4d^1$) because the more highly excited states interact more strongly with the incoming electron.

Transitions that mostly occur due to long range interactions are typically most accurate. The most common example is electron scattering between electric-dipole connected states. Transitions that depend on close interactions (e.g. a spin change in the atom/ion or a large angular momentum change) are usually much less accurate since the incident electron must overlap the region of the target electrons and strongly interact with them.

Transitions that are dominated by resonances are typically much less accurate for low temperature electrons than for high temperature. The reason is that the energy position of the resonances should be accurate on the scale of the temperature and thus at low temperature a high accuracy is required. A special case is a strong resonance very near threshold; sometimes calculations are not accurate enough to determine whether the state is a resonance contributing to the scattering or a bound state that does not contribute at all. A prototype of difficulties due to resonances is dielectronic recombination which is notoriously difficult for low temperature plasmas.

Electron impact ionization is a special case that deserves mention. There are always dipole allowed transitions leading to ionization and typically only the total ionization rate is desired (and not the energy distribution of the ejected electrons); both of these considerations aid the accuracy of the calculation. However, the two free electrons can strongly interact with each other which decreases accuracy. Heroic calculations of one and two electron targets are now more accurate than experiment even for neutrals. However, more complicated atoms/ions have not reached this level except for more highly charged ions. For ionization out of compact initial states, 20% accuracy can routinely be achieved for 2+ ions. Ionization from neutrals and 1+ ions or from more highly excited states can not achieve this accuracy.

FINAL REMARKS

This short paper can not begin to do reasonable job introducing atomic physics concepts and practice. The (equally brief) reference list can be used as a start in fleshing out this paper. Reference [1] gives a thorough overview of most atomic processes. Reference [2] covers many aspects of the multichannel Hartree-Fock method. Reference [3] addresses electron scattering theory with an emphasis on the R-matrix method. Reference [4] contains descriptions of several topics addressed using perturbation theory. Reference [5] contains almost everything you would want to know about atoms including chapters covering physical processes as well as chapters covering computational techniques.

ACKNOWLEDGMENTS

I appreciate the help received from M. S. Pindzola and S. D. Loch while organizing the talk and this paper.

REFERENCES

1. R. D. Cowan, *The Theory of Atomic Structure and Spectra*, University of California Press, Berkeley, 1981.
2. C. Froese Fischer, T. Brage, and P. Jönsson, *Computational atomic structure: an MCHF approach*, Institute of Physics Publishing, Bristol, 1997.
3. P. G. Burke and C. J. Joachain, *Theory of electron-atom collisions*, Plenum Press, New York, 1994.
4. *Many-body atomic physics: lectures on the application of many-body theory to atomic physics*, edited by J. J. Boyle and M. S. Pindzola, Cambridge University Press, Cambridge (1998).
5. *Atomic, Molecular, and Optical Physics Handbook*, edited by G. W. F. Drake, American Institute of Physics, Woodbury, (1996).

Developments in Soft X-Ray Spectral Modelling for Fusion: Areas of Overlap with Astrophysics

A. D. Whiteford*, N. R. Badnell*, R. Barnsley†, I. H. Coffey†, M. G. O'Mullane*, H. P. Summers* and K.-D. Zastrow**

*Department of Physics University of Strathclyde Glasgow G4 0NG UK
†Department of Physics Queens University Belfast UK
**UKAEA Fusion Culham Science Centre Abingdon OX14 3EA UK

Abstract.
This paper focuses on the diagnostic analysis in magnetically confined fusion plasmas with attention drawn both to the methodology in common with that in astrophysics and to points of departure – such as species, transport, high density and the role of neutral hydrogen. Comment is also made on the state of high quality theoretical collision data from a fusion perspective and how it can support rigorous error analysis of better spectroscopic measurements.

Keywords: Fusion, Astrophysics, X-ray, Atomic Modelling, Error Propagation
PACS: 32.30.Rj, 32.70.-n, 52.20.Fs, 95.30.Dr, 95.85.Nv, 52.70.La

1. INTRODUCTION

Quantitative x-ray spectroscopy has a long history in astrophysics. The use of the helium-like resonance region with its associated dielectronic satellite emission as a quantitative tool was first performed extensively by [1]. Since then, analysis of this spectral feature and others in the soft x-ray has been performed exhaustively in both the astrophysical and fusion communities — sometimes with different goals but with the same common underlying atomic physics and modelling requirements. It is often the case that communication between these two communities is far from ideal.

An example of a helium-like iron spectrum, recorded on the TEXTOR[1] tokamak is shown in figure 1. This same spectrum is also seen in astrophysics (especially in the flaring solar corona — albeit with small variations of detail).

An EUV/soft x-ray spectrum of tungsten inside the JET[2] tokamak is shown in figure 2. This is a spectrum with no astrophysical parallel, although, spectra of species beyond iron (i.e. $z > 26$) have been observed, e.g. nickel in the soft x-ray[22] and elements such as tantalum (very close to tungsten on the periodic table and also routinely ablated into JET) have also been observed (although not in the x-ray) and understanding them is of considerable current interest in nuclear astrophysics.

The opportunities for the fusion community differ from that of the astrophysical community. In modern fusion devices (e.g. a tokamak), the laboratory environment

[1] Tokamak EXperiment for Technology Oriented Research, see [8]
[2] Joint European Torus, see [7]

CP774, *X-Ray Diagnostics of Astrophysical Plasmas,* edited by Randall K. Smith

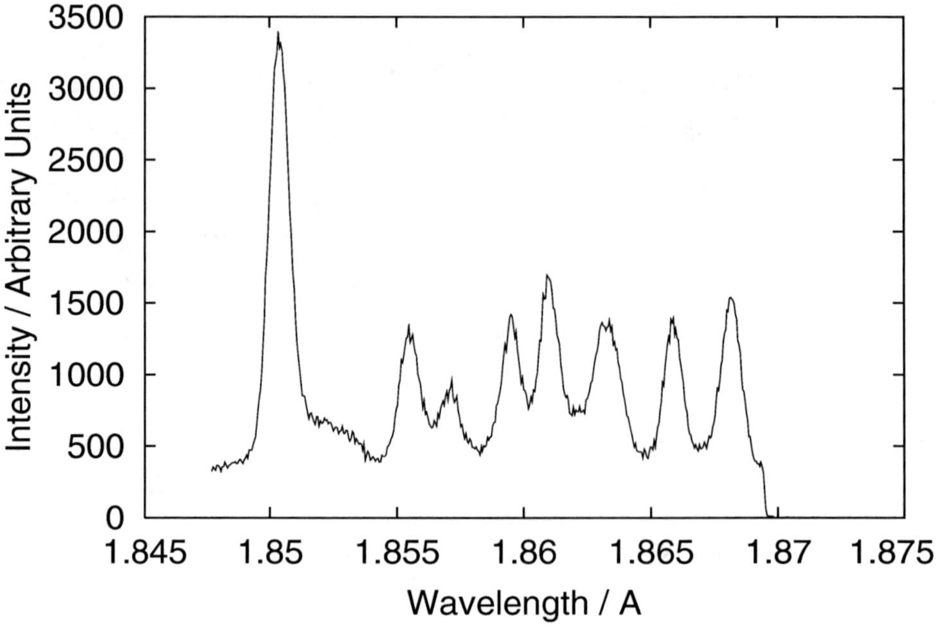

FIGURE 1. Helium-like resonance region of Fe^{24+} measured on TEXTOR in shot 15721.

for spectral measurements allows for multiple lines of sight through the plasma. Thus, for example, x-ray bolometry diagnostics can yield full tomographic reconstructions of the plasma. Also, active diagnostics play a key role in fusion analysis. Most machines have a Thomson scattering system to measure the electron temperature and density both temporally and spatially across the plasma. Although there are fewer unknowns in the modelling of the spectral emission, there is an expectation/requirement for greater accuracy. Despite the difference in applications and utilisation, the fundamental and applied atomic physics entering the modelling of soft x-ray emission are common to the two communities.

The layout of this paper is as follows: in section 2 we discuss the various methods of calculating fundamental atomic data and how they impact the modelling of emission from plasmas. In section 3 we discuss collisional–radiative modelling to provide derived atomic data suitable for direct use in plasma modelling codes and spectral analysis. In section 4 we discuss theoretical uncertainties on fundamental atomic data, comment on how to estimate them and outline the methodology used to propagate these uncertainties through to derived atomic data. Sections 2, 3 and 4 are mainly from a fusion perspective. We then finish with some short conclusions drawing attention to mutually advantageous areas of overlap between fusion and astrophysical analysis.

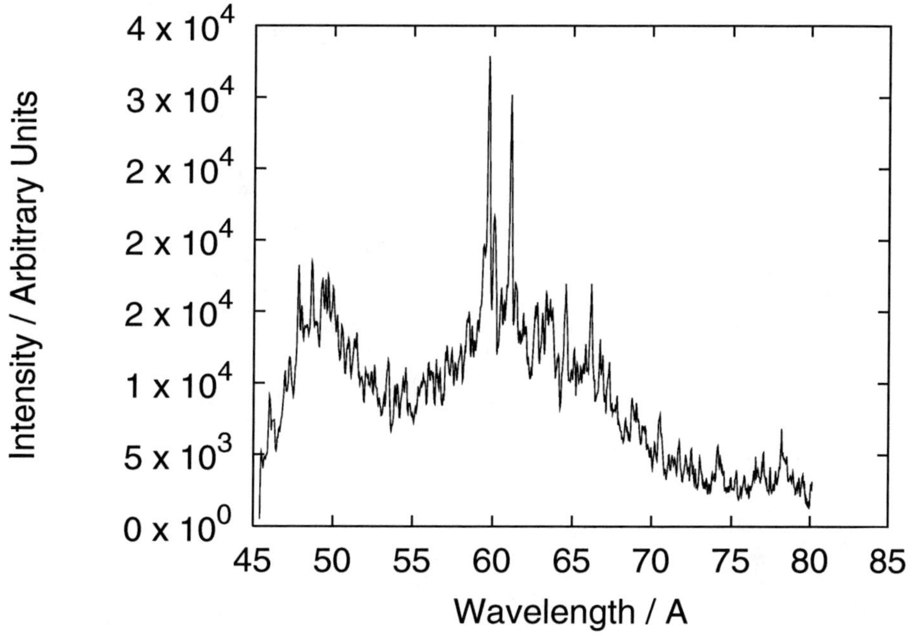

FIGURE 2. Spectrum recorded from ablation of a tungsten tile into JET using a grazing incidence spectrometer for shot 55155.

2. FUNDAMENTAL ATOMIC DATA

We distinguish three levels of fundamental atomic data, namely, baseline data, intermediate quality data and high quality data. A brief summary of methods use to calculate cross-sections is given in table 1. Key to all the methods given in table 1 is that they rely on a detailed (i.e. multi-configuration, multi-electron) atomic structure. Several codes exist to do this, such as SUPERSTRUCTURE[20], AUTOSTRUCTURE[19] and the Cowan code[21]. The possibility for quick construction of (effectively) complete atomic structures has rendered largely obsolete semi-analytical expressions such as that of [18] which are also unwarranted for precision analysis of measurements from current instrumentation. Computational advances mean that, e.g., plane wave Born calculations even for very heavy species can be performed rapidly and the quality of the atomic structure available gives a precision to the approximation which exceeds the expectations of Born.

At the high precision end of the scale, use of *R*-matrix data for the helium-like stage is now routine (see, e.g., [2], [11], [12]). However, important effects must be considered to realise the full potential of *R*-matrix. One such effect is that of radiation damping. Figure 3 shows the resonance contribution to the effective collision strength and also the reduction of this effective collision strength due to radiation damping. It can be seen that both of these effects matter — neglecting radiation damping would lead to a large overestimate of the excitation rate at low temperatures. For more details on the

TABLE 1. Summary of main cross-section determination techniques along with the major advantages and disadvantages

Type	Advantages	Disadvantages
Measurement	Real Results	Restricted energy ranges
Plane Wave Born	Quick to perform	Baseline quality results
Unitarised Born	Quick to perform	Baseline quality results
Distorted-Wave	Fairly quick	No resonances
CCC*	Valid at all energies	Slow resonance resolution
R-matrix	Rapid resonance resolution	Not valid at all energies
TDSE†	Very accurate	Very slow

* Convergent Close Coupling
† Time Dependent Schrödinger Equation

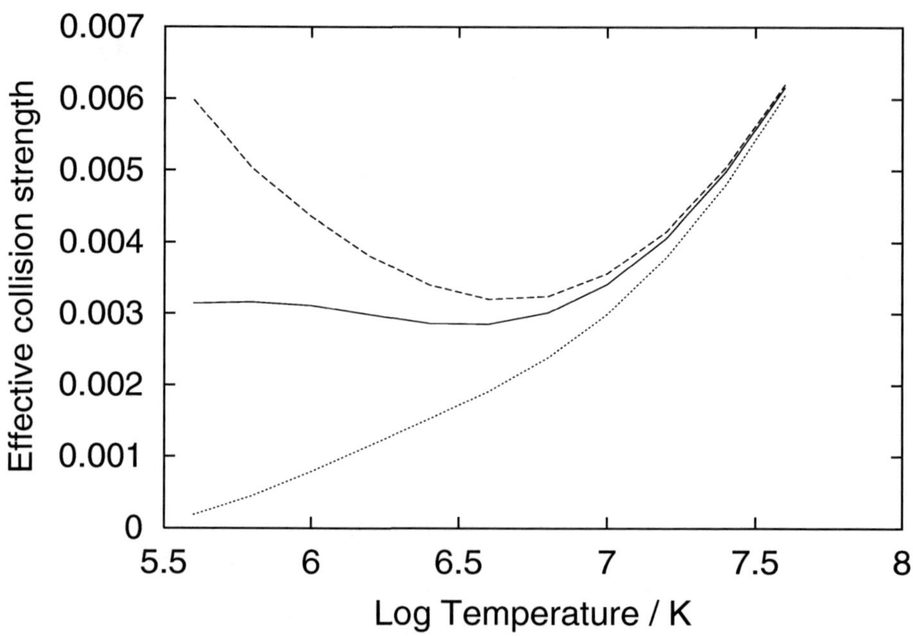

FIGURE 3. Effective collision strengths for the electron-impact excitation of the 1s2p $^1P_1 - 1s3s\,^1S_0$ transition in Ar^{16+}. The solid curve denotes results that include the radiation damping of resonances. The dashed curve denotes results that omit the radiation damping of resonances. The dotted curve denotes results for the underlying (non-resonant) background only.

background to the inclusion of radiation damping with the *R*-matrix approach see [9] and for details of the calculations themselves see [2].

For the lithium-like system, calculations on the electron impact excitation requires slightly more computational effort but are still well within the scope of moderate computer systems (see [3] for examples of calculations for argon and iron). An important effect in this case is Auger damping to intermediate resonance states outside the *R*-

TABLE 2. Contributions to the effective collision strength (divided by 10^{-4}) of the $1s^22s\,^2S_{\frac{1}{2}} - 1s2s^2\,^2S_{\frac{1}{2}}$ transition in Fe^{23+} showing contributions with Auger damping not implicitly present in the *R*-matrix method included (ADI) and such damping excluded (ADE).

Temperature	3×10^5K		3×10^6K		3×10^7K		3×10^8K	
Contribution	ADI	ADE	ADI	ADE	ADI	ADE	ADI	ADE
Background	9.48		9.49		9.51		10.0	
$1s2s2pnl$	0.92	2.92	0.05	0.47	—*	—	—	—
$1s2s3l3l'$	—	—	1.41	1.41	0.49	0.49	0.01	0.01
$1s2s3l4l'$	—	—	0.11	0.13	0.20	0.22	0.01	0.01
$1s2s3l5l'$	—	—	0.09	0.09	0.24	0.26	0.01	0.01
$1s2s3lnl'^{\dagger}$	—	0.11	—	—	0.02	0.69	—	0.01
Total	10.4	12.5	11.1	11.6	10.4	11.2	10.0	10.0

* Denotes negligible contribution

† $n \geq 6$

matrix close-coupling expansion. Table 2 shows the effects of such damping. Details of the methods used are given in [10]. The issue is also discussed in [13] where the importance of Auger damping is not only highlighted but comment is made that, with the exception of [3], all previous publications had excluded the effect.

For more complex atomic systems (principally those of iron), extensive work has been performed by the Iron Project[14] although much of the calculated data has yet to be made available publicly in a useful form for plasma modelling.

For very heavy species, fully relativistic *R*-matrix theory allows an appropriate treatment (c.f. the DARC codes[15]). A recent calculation on Ni-like xenon has been performed by [16], this ion has importance in the field of x-ray lasers and the use of high-precision data is necessary in order to accurately model measurements. The collision strength for the lasing transition is shown in figure 4.

Figure 5 outlines the number of levels which are required in a comprehensive calculation for tungsten. For the more complex ionisation stages, where *R*-matrix is not yet feasible, less accurate cross-section determination techniques are utilised (see table 1). The imperative need for modelling information on very complex species necessitates the use of simpler atomic data production techniques. The scheme used along with a discussion of targeted high-quality calculations such as those of [16] discussed above is given in section 3.

3. ATOMIC MODELLING AND KEY ATOMIC DELIVERABLES FOR FUSION

The fundamental atomic data production methods discussed in the previous section are critical for the spectroscopic analysis of both fusion and astrophysical plasmas however the completion of the collision calculation alone does not constitute a solved problem.

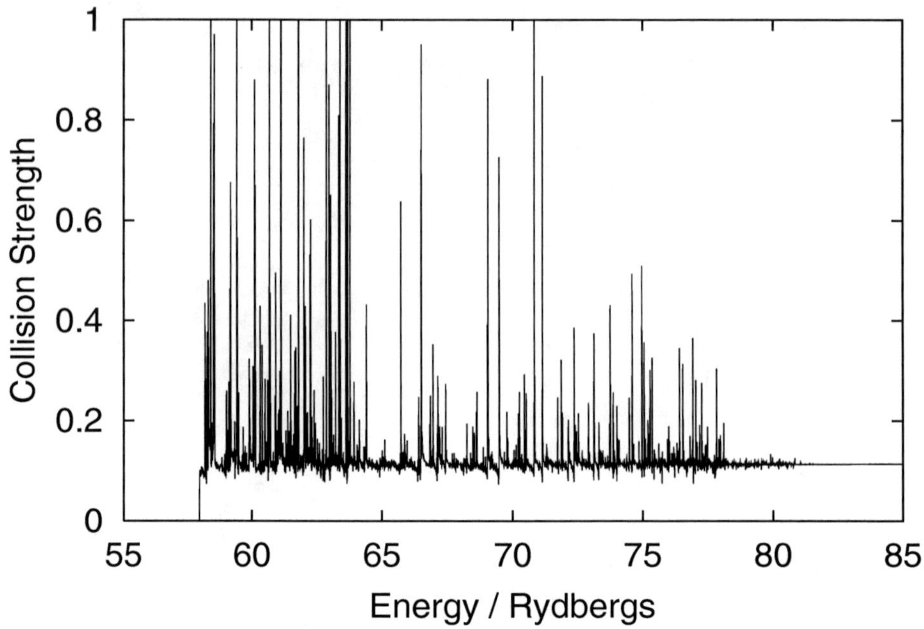

FIGURE 4. Dirac-Coloumb *R*-matrix collision strength for the $3d^{10}\,{}^1S_0 - 4d\,{}^1S_0$ transition in Xe^{26+}.

Data which originates from fundamental cross sections must be transformed into *derived atomic data* via population models. It is this latter data which is tailored to and applies to spectroscopic analysis techniques and plasma modelling. The population approach in the fusion community recognises the different plasma relaxation timescales. Specifically, we separate states of an ion into *ordinary levels* and *metastable levels*. The ordinary levels have relaxation times much faster than the timescales on which the plasma is evolving (in the fusion case, the timescales are nanoseconds versus milliseconds) so can be treated to be in quasi-equilibrium with the metastable levels. The latter may have atomic timescales similar to those of the plasma timescales. This model basis means that populations relative to metastable states can be calculated along with source terms (for transport equations) only for the metastables. These may then be archived as a function of both temperature and density. This formulation yields the key deliverables:

- Effective ionisation, recombination and power coefficients.
- Photon emissivity coefficients.

The inclusion of finite density effects along with the metastable resolution is called *generalised collisional radiative modelling*[17] in the fusion community. Such a treatment is often lacking in astrophysical analysis which assumes that the zero-density coronal picture is valid.

Turning to the quality of data required for this modelling, figure 6 shows the fractional change in ionisation potential as a function of charge in tungsten. The peaks in the

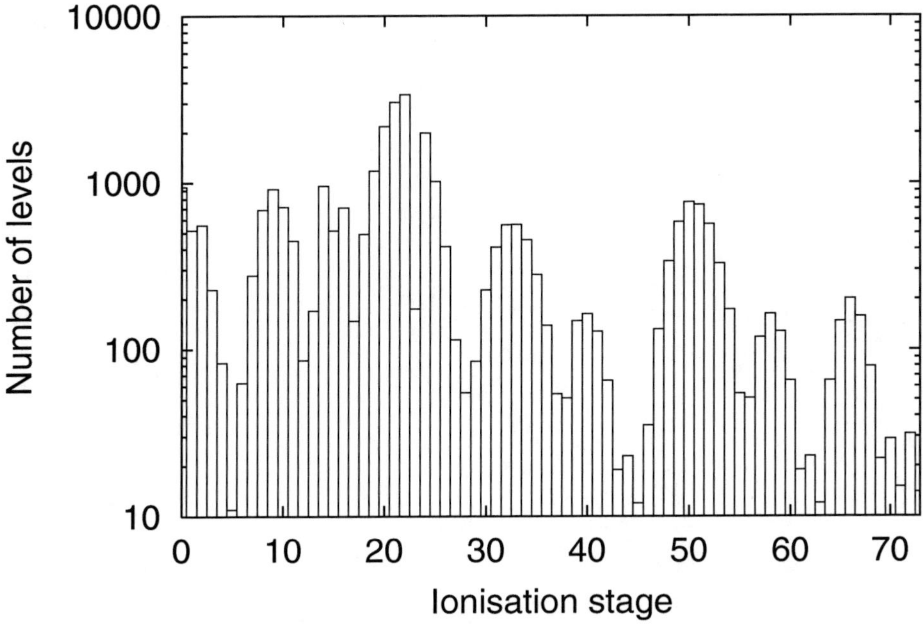

FIGURE 5. Number of levels included in tungsten excitation calculations for each ionisation stage.

curve correspond to shell boundaries and hence ions which will have extended fractional abundance in temperature space under equilibrium conditions. When interpreting spectral emission, these ions are often the ones from which spectral diagnostics arise. Hence, our approach is to use baseline cross-sections such as plane wave Born for complete coverage of a full iso-nuclear sequence (of a heavy species) and then target calculations on key ionisation stages, for more details of this approach as used for xenon see [16].

The emission from the stages not on shell boundaries for complex ions tends to have many, usually unidentifiable, lines which can even merge into a so-called *quasi-continuum*. This was exemplified in the tungsten spectrum show in figure 2. For these ions our approach is to assemble all the lines in a spectral interval (usually determined for a spectrometer detector wavelength range) so that instead of tabulating line emission as a function of temperature, density and transition we tabulate as a function of temperature, density and wavelength. We call these tabulations feature-photon-emissivity-coefficients. Such a coefficient for W^{30+} is shown in figure 7. The merging of the individual lines into a quasi-continuum also means that the atomic data used does not have to be that of the highest accuracy.

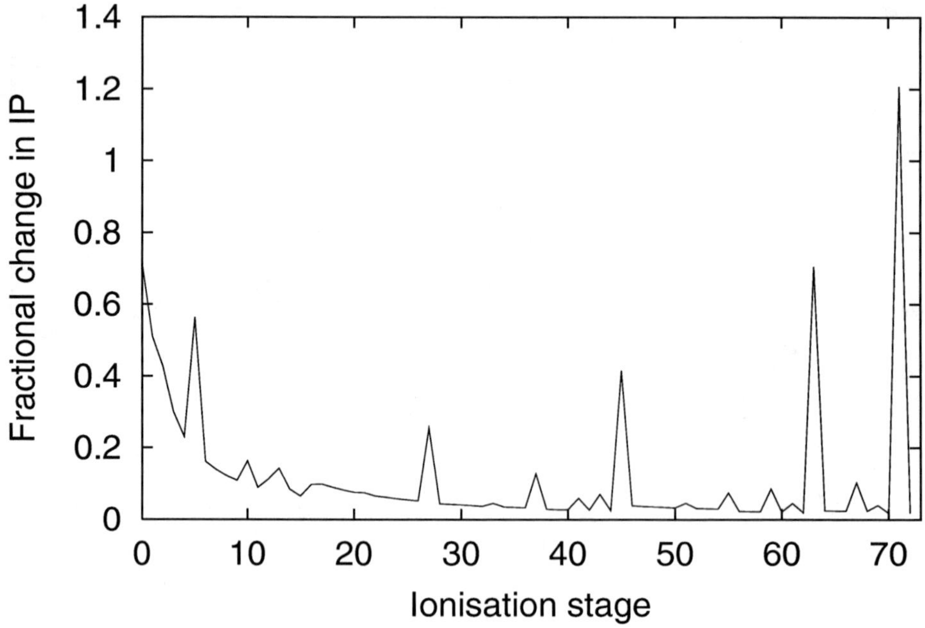

FIGURE 6. Fractional change in the ionisation potential of krypton as a function of charge state. The sharp peaks correspond to shell boundaries — see text for discussion.

4. UNCERTAINTY ESTIMATES AND ERROR PROPAGATION

The question of whether a discrepancy between a measurement and a model is due to an as yet undiscovered phenomenon, transport or inaccuracy of atomic data is one which is often asked in fusion. At issue in this confrontation is the uncertainty in the theoretical *derived* (section 3) quantities. Through the collisional-radiative and transport modelling these have a complex dependence on the uncertainties in the *fundamental* (section 2) atomic data.

Estimation of the uncertainties on theoretical cross-sections is necessary. Without such estimations it is not possible to draw firm conclusions on many results, both in astrophysical and fusion plasmas. Estimations of uncertainty in theoretical atomic data are seldom published and, if they are, then in vague generalisations rather than as tabulations on a point-by-point basis along with the calculated values. Such a practice would be almost unthinkable in the corresponding experiments to measure cross sections.

Our approach to uncertainties in the derived data is to start with a basis set of uncertainties in the fundamental atomic data[3] and then to process them through a finite-density collisional-radiative model (see section 3) using Monte-Carlo sampling. An

[3] As discussed, this basis set is never as comprehensive or precise as it should be.

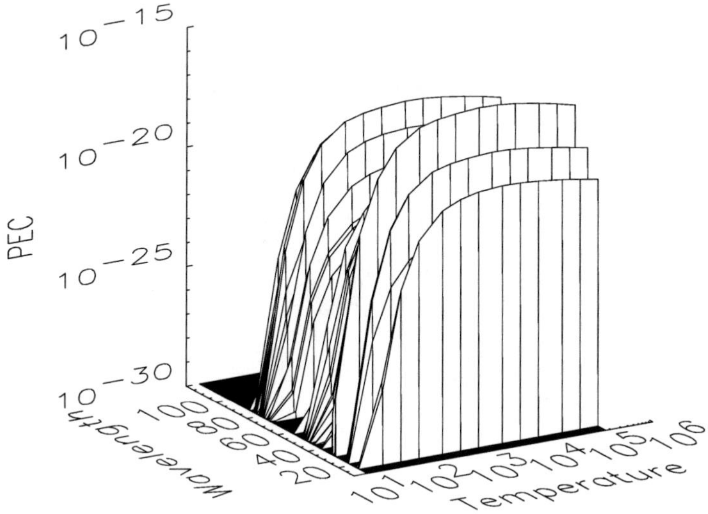

FIGURE 7. Feature-photon-emissivity coefficent for W^{30+} at an electron density of 10^{14}cm^{-3}.

implementation of this procedure[5] (extended to include dominant error analysis as well as composite error analysis) has been successfully applied to He[6] and Fe^{24+}[2]. A more detailed overview is given in [4].

Results for Fe^{24+} are shown here in illustration, where for simplicity we assumed 10% Gaussian uncertainties on all transitions (for a discussion of the effects of taking different uncertainties for different transitions see [6] and [2]). 3000 separate collisional-radiative calculations using Monte-Carlo sampling from the fundamental atomic data were performed.

Figure 8 shows the distribution of results for the population of the 1s2s $^3S_1$ level in Fe^{24+} along with a Gaussian fitted to the results. Experience has shown that in most cases, non-linearity is not an issue in the simple excitation case; Gaussian uncertainties in the collision strengths lead to Gaussian uncertainties in the populations (hence emission)[4].

Figure 9 shows the width of the fitted Gaussian (i.e. the percentage error) as a function of density. The coronal extreme where only one excitation rate matters can be clearly seen at low density along with the LTE regime where the individual excitation rates are of no importance. The finite-density collisional-radiative regime clearly shows the need for advanced error propagation such as the method described here.

---

[4] We note that non-linearity can affect other processes such as beam stopping.

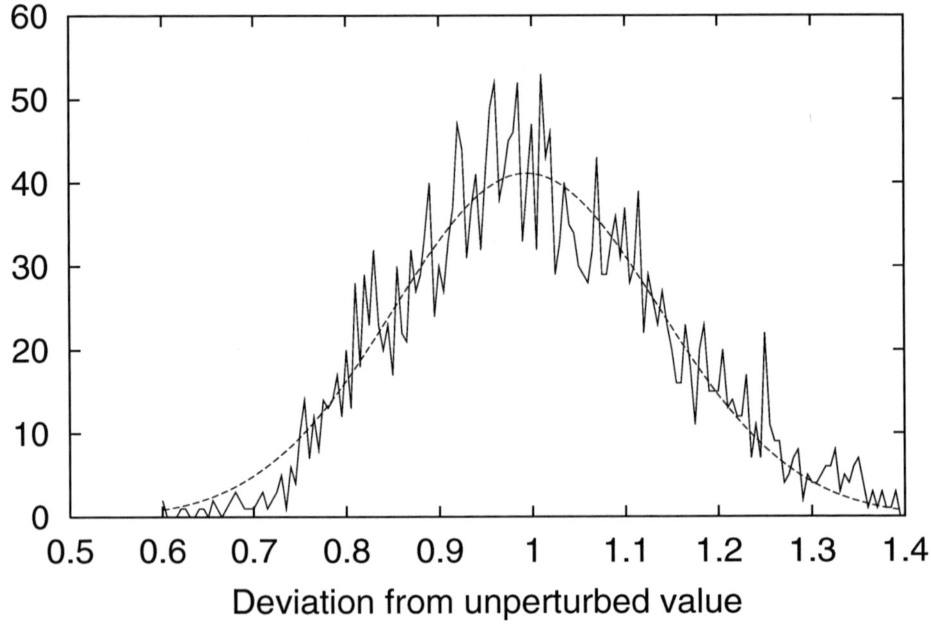

FIGURE 8. Monte-Carlo distribution of the population of the 1s2s $^3S_1$ level in Fe^{24+} assuming 10% error in all rates.

5. OVERLAPS WITH ASTROPHYSICAL ANALYSIS AND CONCLUSIONS

Shared interest in the production of electron-collisional data is a strong area of overlap between the fusion and astrophysical communities. Such data is best calculated iso-electronically as this gives the strongest check at the production phase. The two communities have different interests in which ions they require data for, however, both communities should strive for coverage along an iso-electronic sequence allowing the data to be validated in the two, often very different, scientific environments.

Finite-density effects and dynamic transport modelling are essential in the fusion domain but also impact the astrophysical regime, perhaps more than it is generally considered or implemented within the spectroscopic modelling codes available.

Shared interest in estimation of uncertainties of the fundamental atomic data is clear; it is essential in the application of models. This is not an easy task for the fundamental calculations and systematic error as well as uncertainty should always be considered. Nevertheless, the calculator generally has in his mind supplementary checks to establish the uncertainty status of their results. It is important for the user community to have these uncertainty estimates written down rather than to read a plethora of papers simply identifying discrepancies between calculations.

The area of ion-impact data and, in particular the charge exchange process, has a key

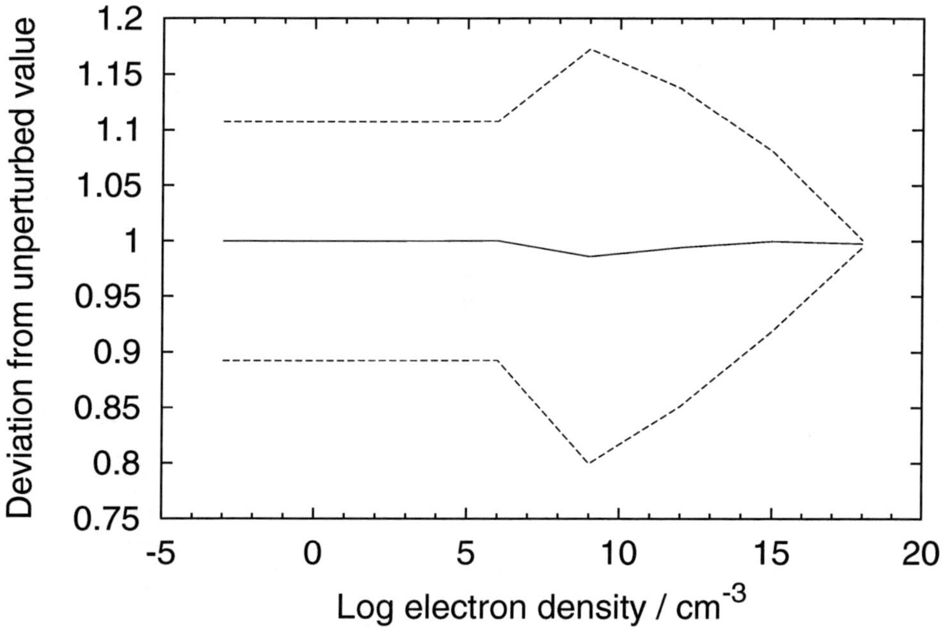

FIGURE 9. Propagated uncertainty to the population of the 1s2s $^3S_1$ level in Fe^{24+} assuming 10% error in all rates. Note the progression of the errors from the coronal, through collisional-radiative, to LTE regime.

role in the fusion community and is also being considered in astrophysics in shocks and the interaction of the solar wind with cometary atmospheres. However, there seems to be little in the way of communication (certainly much less than in the electron-impact case) on this topic. Both communities would be enhanced through comparisons of data, methods and results.

ACKNOWLEDGMENTS

ADW acknowledges an invitation and funding to attend the XDAP meeting from the Harvard-Smithsonian Center for Astrophysics. The work presented here is funded by the ADAS Project.

REFERENCES

1. A H Gabriel *Mon. Not. R. Astr. Soc.* **160** 99–119 (1972)
2. A D Whiteford, N R Badnell, C P Ballance, M G O'Mullane, H P Summers and A L Thomas *J. Phys. B: At. Mol. Opt. Phys.* **34** 3179–3191 (2001)
3. A D Whiteford, N R Badnell, C P Ballance, S D Loch, M G O'Mullane and H P Summers *J. Phys. B: At. Mol. Opt. Phys.* **35** 3729–3740 (2002)

4. H P Summers, N R Badnell, M G O'Mullane, A D Whiteford, R Bingham, B J Kellett, J Lang, K H Behringer, U Fantz, K-D Zastrow, S D Loch, M S Pindzola, D C Griffin and C P Ballance *Plasma Phys. Control. Fusion* **44** B323–B338 (2002)
5. ADAS User Manual Program: ADAS216 `http://adas.phys.strath.ac.uk`
6. Y Andrew and M G O'Mullane *Plasma Phys. Control. Fusion* **42** 301–307 (2000)
7. `http://www.jet.efda.org/`
8. `http://www.fz-juelich.de/ipp/textor_en/`
9. Robicheaux F, Gorczyca T W, Pindzola M S and Badnell N R *Phys. Rev. A* **52** 1319–33 (1995)
10. Gorczyca T W and Robicheaux F *Phys. Rev. A* **60** 1216–25 (1999)
11. F Delaheye and A K Pradhan *J. Phys. B: At. Mol. Opt. Phys.* **35** 3377–3390 (2002)
12. M A Bautista *J. Phys. B: At. Mol. Opt. Phys.* **36** 1503–1514 (2003)
13. M A Bautista, C Mendoza, T R Kallman and P Palmeri *Astronomy and Astrophysics* **403** 339–355 (2003)
14. D G Hummer, K A Berrington, W Eissner, A K Pradhan, H E Saraph and J A Tully *Astronomy and Astrophysics* **279** 298–309 (1993)
15. P H Norrington and I P Grant *J. Phys. B: At. Mol. Opt. Phys.* **20** 4869–81 (1987)
16. N R Badnell, K A Berrington, H P Summers, M G O'Mullane, A D Whiteford and C P Ballance *J. Phys. B: At. Mol. Opt. Phys.* **37** 4589–4601 (2004)
17. H P Summers and M B Hooper *Plasma Physics* **25** 1311–1344 (1983)
18. van Regemorter H *Astrophys. J.* **136** 906–15 (1962)
19. Badnell N R *J. Phys. B: At. Mol. Opt. Phys.* **30** 1–11 (1997)
20. Eissner W, Jones M and Nussbaumer H *Comput. Phys. Commun.* **4** 270–306 (1974)
21. Cowan R D *The Theory of Atomic Structure and Spectra* (Berkeley: University of California Press) (1981)
22. S Molendi, S Bianchi and G Matt *Mon. Not. R. Astr. Soc.* **343** 1 (2003)

Open Issues Regarding Ionization Balance

J. Martin Laming*, Arati Dasgupta† and Una Hwang**

*Space Science Division, Naval Research Laboratory, Code 7674L, Washington DC 20375
†Plasma Physics Division, Naval Research Laboratory, Code 6723, Washington DC 20375
**Goddard Space Flight Center, Code 662, Greenbelt MD 20771

Abstract. We discuss some or our favorite open issues in ionization balance. In particular, we are concerned with steady-state ionization balance in the presence of thermal instability, and in time dependent problems. In both cases, having accurate rate coefficients for the ionization and recombination processes is crucial to modeling the plasma state correctly, since when instabilities are involved, small errors in input physics can translate to large errors in results.

Keywords: ISM:individual(Cassiopeia A)-supernova remnants-X-rays:ISM -ISM:abundances-atomic processes-accretion, accretion disks-line: formation
PACS: 32.30.Rj, 34.80.Lx, 95.30.Dr, 97.60.Bw

INTRODUCTION

In any discussion of "Open Issues Regarding Ionization Balance", the most pressing open issue is of course where to begin. The steady state ionization balance, whether in collisional ionization equilibrium or photoionization-recombination equilibrium, forms the bedrock of much detailed spectroscopic work in many fields of astrophysics, but is arguably the area where the atomic physics is plagued with the most uncertainties. A review of dielectronic recombination rates in the literature [1] found errors of up to a factor of 2 more or less ubiquitously, both in the rates and in the resulting ionization fractions. More recently, similar problems have been identified with electron impact ionization rates (Daniel Savin, private communication 2004). Given these uncertainties it is at first sight a little surprising that any progress at all has been made in the analysis of UV and X-ray spectra from astrophysical sources, (e.g. stellar coronae), let alone the determination of variables such as electron densities, temperatures and coronal element abundance anomalies.

We cite stellar coronae this way, as opposed to other astrophysical objects for which spectroscopic observations are now available, because in this field a number of methods of data analysis and interpretation have been devised to mitigate the effects of uncertain ionization balance calculations. In particular, when observing several charge states of the same element, the constraint that all charge states must add up to unity, together with the selection of ions formed at temperatures near the peak of their ionization fractions allows procedures like the construction of a differential emission measure curve to operate in a reasonably robust fashion. However many sources, widely held in the X-ray astronomy community to be "more interesting" than stellar coronae, are not so line rich in their spectra, and in many cases cannot be considered to be in ionization equilibrium. The reduced degree of redundancy these facts entail mean that uncertainties in ionization and recombination rates can propagate much further through the data analysis and restrict

CP774, *X-Ray Diagnostics of Astrophysical Plasmas,* edited by Randall K. Smith
2005 American Institute of Physics 0-7354-0259-0/05/$22.50

scientific interpretations. It is our purpose in this paper to highlight some examples of such cases, in the hope of providing stronger motivation to atomic physicists working on the calculation of such rates, and also of interesting a wider community of X-ray astronomers in these problems. We expect that with the advent of Astro-E2 and eventually Constellation-X, the high spectral resolution observations of X-ray sources will demand renewed attention to these problems by a wider community of researchers if the scientific promise of these missions is to be fulfilled.

PHOTOIONIZATION-RECOMBINATION EQUILIBRIUM

The discovery of redshifted and broadened Fe K_α emission lines from accreting black hole candidates (BHCs) has been hailed in many quarters as the expected signal of line emission from the innermost stable orbits of the accretion disk, displaying a combination of gravitational redshift and Doppler blue and redshifting. The exciting prospect of testing strong field general relativity in this manner has recently been tempered by suggestions that these broadened and redshifted lines could also be formed in an outflow from the accretion disk, far away from the central black hole. Figure 1 (left) illustrates the basic model. Fe K_α photons formed at the inner edge of an outflow R_{inner} with Thomson optical depth of order unity are successively scattered by the plasma electrons, accumulating a redshift as they do so because of the diverging flow. (A blueshift would result from Thomson scattering in a converging flow). Monte Carlo simulations of the line profile produced [2], match very well the lines profiles observed in MCG 6-30-15 and GRO J1655-40. In particular, the outflow velocity ($\sim 0.02c$) determined in this way for MCG 6-30-15 is in very good agreement with that derived from the spectrum of the warm absorber. This treatment however neglects any consideration of the Fe charge state, which is clearly central to whether the Fe K_α photons can be produced in the correct place.

The photoionization-recombination equilibrium was computed [3] for a plasma of solar photospheric abundances outflowing with speed βc, illuminated by a Comptonization spectrum from the central source, given by $I_\nu \propto \nu^{-a} \exp(-h\nu/2k_B T)$, where $k_B T = 50$ keV, and the power law index $a = 0.25$, 0.5, or 0.75. The plasma temperature is determined by the balance between heating due to Compton scattering and photoionization by radiation from the central source, and cooling due to (assumed) optically thin radiative losses and adiabatic expansion. The results of these calculations are shown in Figure 1 (right). For winds originating close to the BHC, photoionization is sufficiently strong to ionize Fe to the heliumlike charge state or higher. In this case, photoionization of a K shell electron does not produce a K_α photon since there is no L shell electron available to radiatively decay. At an initial wind radius of about $10^{13}/L_{40}$ cm, where L_{40} is the luminosity of the central source in ergs s^{-1}, Fe starts to recombine through its L shell, before proceeding quickly to a region of thermal instability where recombination proceeds to the M shell. At this point the wind quickly becomes opaque to Fe K_α emission. The Fe K_α observations therefore may only be reproduced for winds initiated within a fairly narrow range of radii around $10^{13}/L_{40}$ cm. We emphasize that this result depends crucially on the adopted recombination rates (in which dielectronic recombination is the main source of uncertainty), with the recombination rate from heliumlike Fe

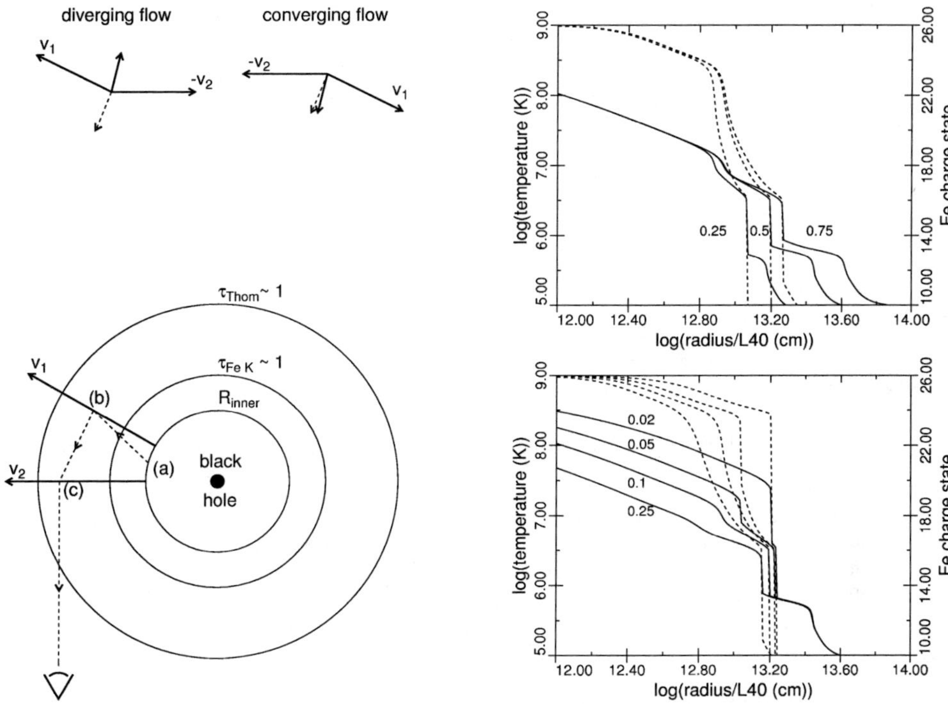

FIGURE 1. *Left* Cartoon of wind geometry in an accreting BHC. The outflow originates at the inner radius. The radii of the optical depth in the Fe K continuum and the electron scattering optical depth are of order unity. A photon emitted near the inner boundary and subsequently scattered by an electron moving with velocity v_1 is received by an electron moving with velocity v_2 as shown. The change in frequency is $\omega_2 = \omega_1 \left[1 + (v_1 - v_2) \cdot n/c\right]$, where n is a unit vector along the path of the scattered photon at the scattering point. In a diverging flow, $(v_1 - v_2) \cdot n/c < 0$ and photons are successively redshifted until they are scattered to an observer at infinity. *Right* Plot of temperature at R_{inner} against R_{inner} for an accretion wind (solid lines). the inner radius R_{inner} scales with the central source luminosity and the optical depth of the wind shell τ_T. For any $\tau_T \neq 1$, the temperature is a function of $\tau_T R_{inner}/L_{40}$. The top panel shows results for expansion speed $\beta = v/c = 0.1$ and spectral indices $\alpha = 0.25$, 0.5, and 0.75. The bottom panel shows the results for $\alpha = 0.5$ and $\beta = v/c = 0.02$, 0.05, 0.1, ad 0.25. The dashed lines show the average Fe charge state (to be read on the right hand axis).

to lithiumlike Fe being the important bottleneck. Once in the lithiumlike charge state, Fe recombines fairly rapidly through the L shell encountering another smaller bottleneck in going from neonlike to sodiumlike.

An important point about the postulated outflows from BHC accretion disks, is that while suspected on theoretical grounds from the need to transport angular momentum out from the inner regions of the flow [4], no precise mechanism or place of origin has yet been identified. Calculations of the sort described here can place valuable constraints on important parameters of such models, an example being the case of continuum driven black hole winds [5].

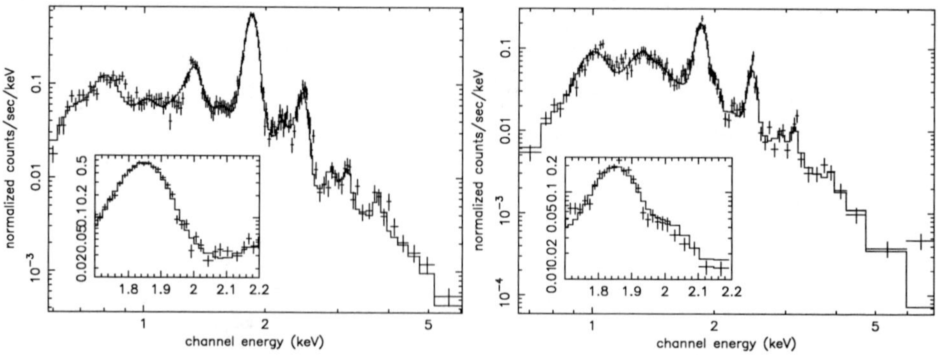

FIGURE 2. Sample spectra extracted from knot locations illustrated in Figure 4. The feature close to 2 keV energy is He- and H-like Si. At the temperature determined from the thermal bremsstrahlung continuum, Si would be ionized to the H-like or bare charge states in ionization equilibrium. The presence of the He-like charge state indicates that the plasma is not in ionization equilibrium, and this departure from equilibrium gives us a "timing" signature allowing the inference of ejecta mass coordinates. The spectrum on the right showing more H-like Si than on the left is clearly at a more advanced state of ionization ("ionization age" defined as the integral of electron density over time). The spectrum of the left has a fitted ionization age of 3×10^{10} cm^{-3}s, while that on the right has 1.8×10^{11} cm^{-3}s. Notice also how the spectrum on the left has an Fe L spectrum dominated by emission of Fe XVII and Fe XVIII at around 800 eV, while that on the right is mainly Fe XX and Fe XXI at 1 keV.

TIME DEPENDENT COLLISIONAL IONIZATION

The previous section discussed a steady state approximation to an intrinsically time varying (i.e. flowing) plasma, but where accurate atomic recombination rates are essential due to the presence of thermal instability. In this section we discuss a more complicated example, where the full time dependence of the plasma must be modeled. In April and May 2004 Chandra observed the Cassiopeia A supernova remnant (SNR) for a total of 1 Ms. We discuss the atomic data needs presented by this landmark dataset, and the possibilities arising of unique inferences from these observations into the mechanisms at work when the actual supernova exploded. Cassiopeia A is somewhat special as a core collapse SNR, in that due to extensive presupernova mass loss, the innermost layers of ejecta that would have participated in nuclear burning close to the "mass cut" (the dividing line between mass that falls back to form the compact stellar remnant and mass expelled in the explosion) are currently visible to X-ray observations. For other core-collapse remnants, whose progenitor retained more mass upon explosion, the reverse shock is generally still propagating through the slightly less interesting outer layers of ejecta.

Figures 2 and 3 show sample spectra extracted from various of the ejecta knots in Cas A, shown in images in Figure 4. Figure 2 gives examples of so-called O- or Si-rich knots [6], observed near the jet base (cluster in the NNE) and one of the more equatorial knots (E limb on the left hand side of Figure 4). One can clearly see the increase in the ratio of H- to He-like Si as the ionization age increases, and also the change in the Fe L-shell spectrum as prominent Fe XVII-XVIII emission near 800 eV gives way to Fe XX-XXI

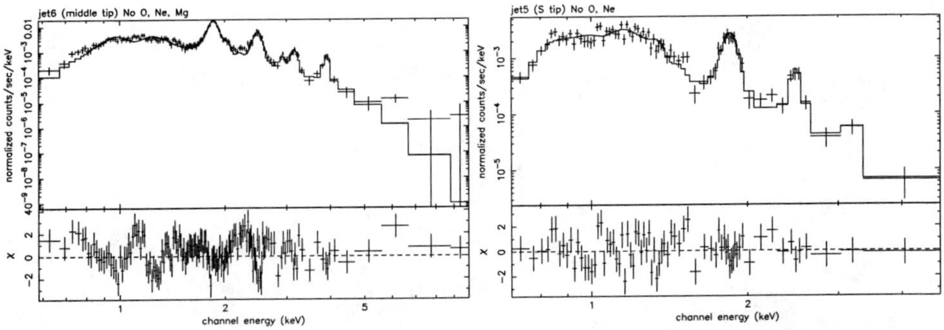

FIGURE 3. Sample spectra extracted from knots located in the jet regions illustrated in Figure 5. The He- and H-like Si features now display relative intensities consistent with ionization equilibrium at the temperature determined from the thermal bremsstrahlung spectrum. The fitted ionization age is of order 10^{13} $cm^{-3}s$. The Fe L shell spectra are also approximately consistent with ionization equilibrium, though Fe is much weaker here than in the spectra shown in Figure 2. This seems to be a generic feature of jet knots. We can't help remarking at this point that the element abundances inferred from the spectra of jet knots in Cas A, Mg, Si, S, Ar, and Ca bear more than a passing resemblance to element abundances observed in γ-ray burst afterglows. Given the its explosion energy of $\sim 2 \times 10^{51}$ ergs [6], Cas A is unlikely to have actually produced a γ-ray burst, though may have had similar jet phenomena.

emission close to 1 keV. Figure 3 shows two examples extracted from the outermost knots found in the NE jet region, shown in the right hand panel of Figure 4. The knots are estimated to have undergone reverse shock passage a year or two after explosion, and consequently have very high ionization ages, the highest ever seen in Cas A or for that matter any other SNR. The temperatures here are lower than in Figure 2, due to the greater energy density decrease due to adiabatic expansion, and so the spectra show plasma essentially in collisional ionization equilibrium.

While the ionization age in XSPEC models has been treated to date largely as an extra parameter to play with when trying to reduce a fit χ^2 to within acceptable limits, it actually has a physical interpretation. Being the time integral of the electron density since the plasma was shocked, it can be modeled using a time dependent ionization code. Figure 5 (left panel) plots the run of electron temperature against ionization age for ejecta starting out at various Lagrangian mass coordinates within the ejecta, for two different classes of models (solid and dashed lines; see [6] for more details), with the result of fits to spectra extracted from the various knots superposed. Without going into gory details, the important message of this exercise is that the degree of departure from collisional ionization equilibrium acts as a timing signature that tells us how long ago a plasma element encountered the reverse shock. This, coupled with a hydrodynamic model for the evolution of the SNR based on the observed forward shock velocity and radius, and the emission measure of forward shocked plasma, allows us to constrain the explosion model. Taking different hydrodynamic models in different directions allows an inference on the asymmetry of the explosion. Such an approach may have particular value when applied to the jets knots illustrated in Figure 4. Preliminary results are shown in the right hand panel of Figure 5. We compare the fitted temperatures and ionization ages with estimates from models. The "cavity" model is based on a simulation

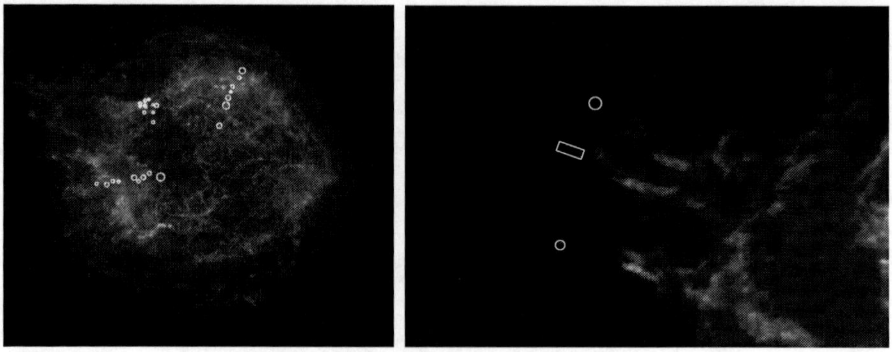

FIGURE 4. Images showing knot locations in Cas A. The left panel shows locations of O-rich knots at three locations; the "jet" base, and two more equatorial regions on either side. The right panel shows very faint knots at the outermost tip of the NE jet region.

[9] of a symmetrical explosion into an asymmetrical CSM, presumably arising from presupernova mass loss. The maximum ionization age in this model occuring in plasma at the core-envelope boundary is still at least an order of magnitude lower than than observed, strongly suggesting that the morphological features usually referred to as jets actually are a feature of the explosion and not of the CSM density distribution. We also give various possibilities for jet models. The "heavy" jet represents a model designed to match the simulations [10], in which an overdense jet propagating out from the stellar center induces the supernova explosion. The jet acceleration mechanism is not specified, but must presumably be due to radiation pressure at the jet base. Models with both O- and Si-dominated composition tend to become radiatively unstable at lower ionization ages than those inferred for the jet knots. The jets should most probably have a composition dominated by Fe (Alexei Khokhlov, private communication 2004), which would exacerbate this problem. The other possibility, much discussed in connection with γ-ray bursts, is an electron-positron dominated jet, accelerated by magnetic fields in the central regions of the supernova [11]. Such a jet becomes mass loaded on its way out through the stellar envelope,[1] and so would have a density $\leq$ that of the overlying ejecta and be composed mainly of O, like the outer ejecta. The dashed curve in Figure 4 (right) shows the trace of electron temperature against ionization age for an O dominated jet of similar density to the ambient ejecta, and as can be seen this offers the best match to the observations.

The modeling of radiative instability also plays a crucial role in analyses of the Fe-rich ejecta clumps. One important parameter in core collapse explosions is the degree to which Fe (or Ni) is mixed out into the overlying stellar envelope by Rayleigh-Taylor instabilities during the explosion. Roughly 10% of the ejected ^{56}Ni was mixed out in this manner in SN 1987A [13], and the indications from the Cas A SNR are that a similar amount of such mixing occurred during this explosion too [7]. The left hand panel of

[1] It has been argued[12] that such mass loading is necessary for jet stability.

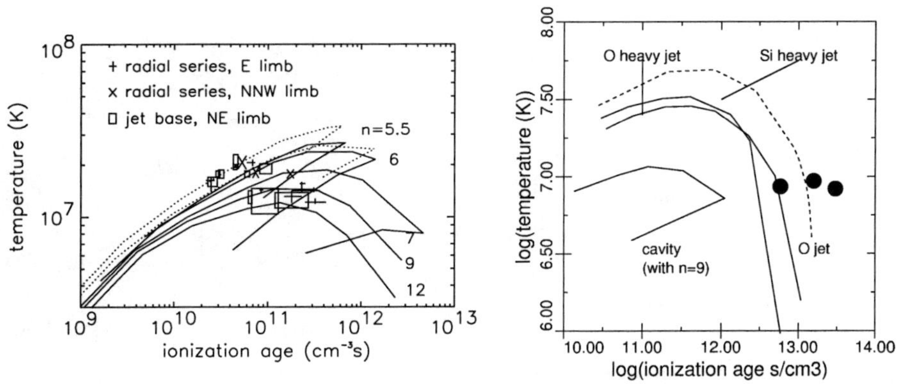

FIGURE 5. Plots of electron temperature against ionization age for different explosion models. The left hand panel shows the effect of different ejecta density profiles for explosion energies between 2 and 4×10^{51} ergs, with fitted knots superposed. The ejecta density profiles is taken to be a uniform density core with a power law envelope, following [8]. Knots from the jet base region (square symbols) imply a lower power law index, i.e. a shallower density profile, than those from the more equatorial regions. This shallower density profile suggests and higher explosion energy, and that this energy is more evenly distributed throughout the ejecta than is the case with steeper profiles [6].

Figure 6 shows the loci of electron temperature against ionization age for Fe-rich ejecta, again compared with observational results for a group of Fe-rich knots observed on the east limb, shown in the image on the right hand side of Figure 6. The precipitous drop in the temperature with ionization about 10^{12} cm^{-3}s is due to radiative instability in Fe. The fact that no Fe knots are found at ionization ages beyond that where this instability sets in is comforting, to say the least, and gives some confidence that we might not be too far wrong in our inferences. The points marked with hollow boxes (as opposed to crosses) are especially interesting, since these denote knots with very high ratios of Fe/Si. This is the result of more or less complete Si burning, and only occurs in ejecta originating from very close to the mass cut. Such knots may also harbor some of the ^{44}Ti known to have been ejected in the explosion [14, 15]. The Lagrangian mass coordinate inferred for these knots in Cas A places them at the Si-C interface in the presupernova composition distribution, reinforcing the long held notion that ^{44}Ca found in SiC X type meteoritic grains has an origin as ^{44}Ti produced in core collapse supernova [16]. This again is a comforting result suggesting that we are on the right track.

Before patting ourselves on the back since things seem to be going so well, it is worth remembering the number of assumptions and simplifications that have gone into our hydrodynamics models for the remnant evolution. Important results about shock waves in clumpy media, the plasma physics of electron heating and acceleration, approximate treatments of multi-dimensional hydrodynamic effects have all gone into the models and remain to be checked by more sophisticated numerical work. The aspect most relevant to this workshop, the atomic physics, also requires further work. So far we have used non-equilibrium ionization models in XSPEC to fit spectra extracted from ejecta knots, and then interpreted the results of these fits using simulations of time dependent postshock ionization. The radiative cooling instability in particular, and many aspects of the Fe

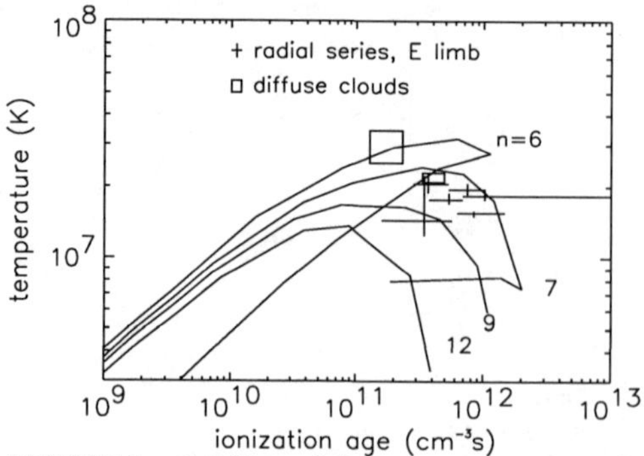

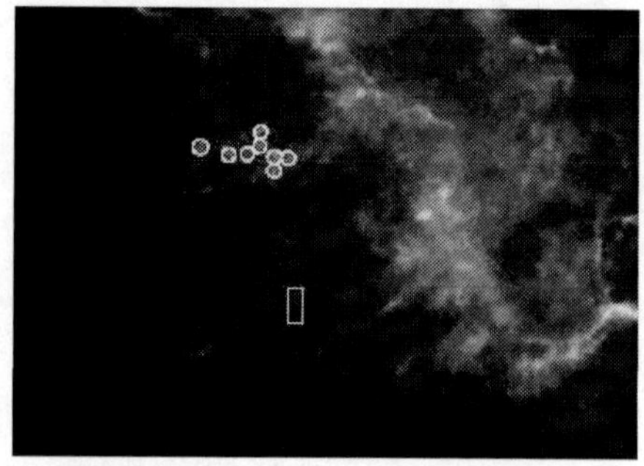

FIGURE 6. *Left* Plots of electron temperature against ionization age for varying ejecta-envelope power-law slopes, for a composition Si:Fe of 0.1:0.9 by mass. Data points from the circular regions (*right*) are plotted as crosses, with the size of the cross indicating the fit uncertainties. The diffuse Fe clouds (one is shown as a box on the right panel) are plotted as boxes, again with size indicating uncertainties.

spectra in general, are extremely sensitive to the recombinaton rate from He-like Fe to Li-like Fe, a process that is dominated by dielectronic recombination. This is illustrated in the left hand panel of Figure 7, where the ionization balance of Fe following reverse shock passage 25 years after explosion is displayed. Around 229 years following shock passage (254 years after explosion), this plasma recombines and cools catastrophically, undergoing radiative instability. It is this time that is set by the recombination rate from He-like to Li-like Fe. We conclude therefore with a brief summary of some recent work on dielectronic recombination, and on this rate in particular..

DR THEORY AND CALCULATIONS

The Maxwellian averaged DR rate coefficient α^{DR}, at a temperature T, that takes an ion from an initial state $|i\rangle$ into a final recombined state $|k\rangle$ through all intermediate autoionizing states $|j\rangle$ is given by:

$$\alpha^{DR}(i,k) = \left[\frac{4\pi R}{kT}\right]^{3/2} \frac{a_0^3}{2g_i} \sum_j F_{ijk} \exp(-\varepsilon_j/kT), \quad (1)$$

where the DR branching ratio, F_{ijk}, is given by:

$$F_{ijk} \equiv \frac{g_j A^A_{ji} A^R_{jk}}{\sum_{i'} A^A_{ji'} + \sum_{k'} A^R_{jk'}}. \quad (2)$$

In this formula, ε_j is the energy of the doubly excited state measured with respect to the initial state, g_i and g_j are the statistical weights of the initial ion state $|i\rangle$ and doubly excited state $|j\rangle$ respectively, a_0 is the Bohr radius and R is the Rydberg energy. In Eq. (2) the denominator sum goes over all possible autoionization states $|i'\rangle$ (including

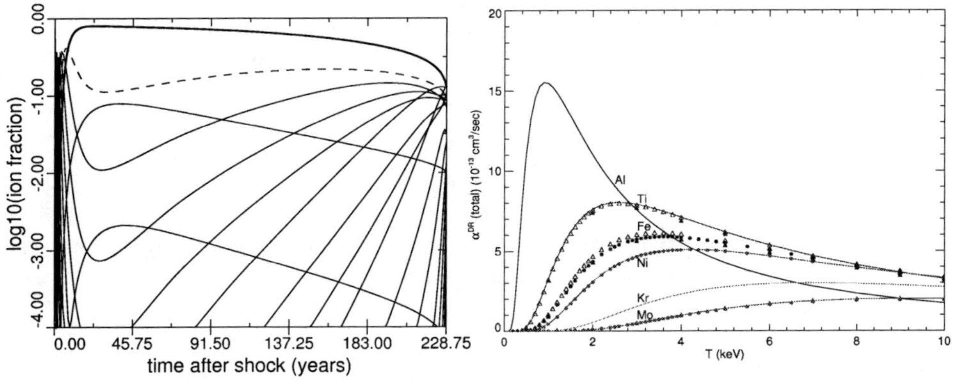

FIGURE 7. *Left* Time dependent ionization balance of Fe following reverse shock passage in Cas A 25 years after explosion. The thickest solid line (at the top) shows heliumlike Fe, and dashed line below it shows lithiumlike. Once heliumlike recombines to lithiumlike, it can be seen that the ensuing recombinations through the Fe L-shell proceed very quickly, allowing thermal instability to occur. *Right* Total dielectronic recombination rate coefficients for recombination from the ground state of He-like ions. Also shown are the scaled rate coefficients for Fe^{24+}, Ni^{26+}, and Mo^{40+} (solid circles) for comparison with calculations and other predictions. The solid and open triangles are the results of [19] and [20] respectively.

autoionizations to excited states) and and all possible final recombined states $|k'\rangle$. A^A_{ji} and A^R_{jk} are the autoionization and radiative rates from the doubly-excited autoionization state $|j\rangle$ to initial state $|i\rangle$ and final state $|k\rangle$ respectively.

In order to calculate the DR rate from an initial level i to a final level k, one has to calculate many F_{ijk}s and then sum over all resonance states j. By working at the fine-structure level, one creates an enormous amount of data. In order to specialize to atomic models that are detailed yet compact enough to require only moderate amounts of data, we have chosen to work at the configuration level. As mentioned in Refs. [17] and [18], DR data at the configuration level produce sufficient and accurate results for most detailed ionization calculations of non-coronal plasmas.

At the configuration level, a notational change is made, and i, j, and k go over into a, b, c in all the previous equations. In the present work, we include intermediate resonance states $1s2lnl'$ for recombination from the initial ground state, $1s^2$ of He-like ions. The radiative rates and autoionization rates were calculated explicitly for $n \leq 15$ and $l \leq 6$. For higher Rydberg states, we employed the usual $1/n^3$ extrapolation procedure for the autoionization rates. The doubly and singly excited levels and all the rates associated with them were judiciously lumped to develop our configuration averaged model.

Z scaling of DR data

The relations for scaling energies of singly (ε_c) and doubly (ε_b) excited states, total autoionization ($\sum_{a'} A^A_{ba'}$), total radiative ($\sum_{c'} A^R_{bc'}$) rates and DR branching ratios F_{abc}

with atomic number for ions in the fluorine and oxygen isoelectronic sequence were presented and discussed in Refs. [17] and [18]. In the configuration averaged model the DR branching ratio F_{abc} is given by:

$$F_{abc} = \frac{g_b A^A_{ba} A^R_{bc}}{\sum\limits_{a'} A^A_{ba'} + \sum\limits_{c'} A^R_{bc'}}. \tag{3}$$

where a' and c' refer to all possible initial and final recombined configurations respectively. The DR rate coefficients then can be calculated for any ion between Al and Mo using

$$\alpha^{DR}(a,c) = \left[\frac{4\pi R}{kT}\right]^{3/2} \frac{a_0^3}{2g_a} \sum_b F_{abc} \exp(-\varepsilon_b/kT). \tag{4}$$

To calculate DR rate coefficients, one needs scaling relations for ε_b and F_{abc}. The configuration averaged energy ε_b scale as:

$$\varepsilon_b = Z^2 \left(b_0^{**} + b_1^{**}/Z + b_2^{**}/Z^2\right)_b, \tag{5}$$

where Eq. (5) is in units of eV. The coefficients in these equations were determined by requiring that they reproduce the DR data that was calculated for Al, Ti, and Kr.

One can obtain an indirect Z scaling of F_{abc} from the the scaling relations of the configuration averaged A^A_{ba}, A^R_{bc}, $\sum_{a'} A^A_{ba'}$ and $\sum_{c'} A^R_{bc'}$ using Eq. (3). However a configuration averaged scaling for branching ratio F_{abc} can also be calculated directly from explicitly calculated F_{ijk}, in which case the scaling relation is given by the three-coefficients polynomial:

$$F_{abc} = Z^2 \left(b_0^F + b_1^F/Z + b_2^F/Z^2\right) \tag{6}$$

(in units of s^{-1}).

Total DR rate coefficients

Detailed calculations and scaling coefficients for DR rates of He-like ions are presented in [21]. A number of total DR rates for He-like ions are shown in Figure 7 (right). The solid curves show our total DR rates for Al^{12+}, Ti^{21+}, Ni^{27+}, Kr^{35+}, and Mo^{41+}, which were calculated using the detailed levels and then extrapolated for high Rydberg levels of the captured electron. The total configuration averaged DR rate coefficients obtained using the scaling relations are also shown as solid circles in this graph for Fe^{24+}, Ni^{26+}, and Mo^{40+}. For Ni^{26+}, the calculated and scaled results are mostly identical whereas for Mo, there is almost exact agreement for low temperature rates and the differences for higher temperatures are less than 2%. The lower values are attributed mainly to the fact that the scaled results were obtained from a lesser number of DR channels in our averaging scheme. In this figure, we also compare our results with the multiconfiguration Dirac-Fock calculation of Chen[19] and the Hartree-Fock-Slater calculation of Karim and Bhalla[20]. The excellent agreement of our calculated results with the more complex multi-configuration relativistic results of Chen for Mo is very encouraging. We also have very good agreement with Chen with our calculated results for Ti^{20+} and scaled results for Fe^{24+}. At very high temperatures, our results for both Ti^{20+} and Fe^{24+} are slightly higher than those of Chen. When we compare our rates to those of Ref.[20], we notice very good agreement for Ti^{20+} but their DR rates for Fe^{24+} are slightly higher than our scaled rates. Our rates are computed in the isolated resonance distorted wave approximation which for collisionally ionized plasmas offers a greater degree of "completeness" than is possible with more sophisticated R-matrix methods. For these applications, it is often more important to account at least approximately for all the contributing resonances than to consider only a subset of the contributing resonances, albeit with higher precision. In the lower electron temperatures characteristic of photoionized plasmas, with fewer contributing resonances, R-matrix may be sufficiently "complete" and can become the method of choice in these cases.

ACKNOWLEDGMENTS

This work has been supported by basic research funds of the Office of Naval Research, and by the Chandra Guest Investigator Program.

REFERENCES

1. D. W. Savin and J. M. Laming, *The Astrophysical Journal*, **566**, 1166–1177 (2002).
2. P. Laurent and L. Titarchuk, *The Astrophysical Journal*, submitted, (2004).
3. J. M. Laming and L. Titarchuk *The Astrophysical Journal*, **615**, L121–L124 (2004).
4. R. D. Blandford and M. C. Begelman *Monthly Notices of the RAS*, **303** L1–L5 (1999).
5. J. E. Everett and D. R. Ballantyne *The Astrophysical Journal*, **615**, L13-L16, (2004).
6. J. M. Laming and U. Hwang *The Astrophysical Journal*, **597**, 347–361 (2003).
7. U. Hwang and J. M. Laming *The Astrophysical Journal*, **597**, 362–373 (2003).
8. K. Truelove and C. F. McKee *The Astrophysical Journal Supplement*, **120**, 299–326 (1999).
9. J. M. Blondin, P. Lundqvist, and R. A. Chevalier *The Astrophysical Journal*, **472**, 257–266, (1996).
10. A. M. Khokhlov et al. *The Astrophysical Journal*, **524**, L107–L110, (1999).
11. D. Proga et al. *The Astrophysical Journal*, **599**, L5–8, (2003).
12. C. D. Matzner *Monthly Notices of the RAS*, **345**, 575–589, (2003).
13. P. A. Pinto and S. E. Woosley *Nature*, **333**, 534–537, (1988).
14. J. Vink et al. *The Astrophysical Journal*, **560**, L79–82, (2001).
15. J. Vink and J. M. Laming *The Astrophysical Journal*, **584**, 758–769, (2003).
16. D. D. Clayton et al. *The Astrophysical Journal*, **578**, L83–L86, (2002).
17. A. Dasgupta and K. G. Whitney, *P*hys. Rev. A, **42**, 2640–2654, (1990).
18. A. Dasgupta and K. G. Whitney, *A*t. Data Nucl. Data Tables, **58**, 77–99, (1994).
19. M. H. Chen, *P*hys. Rev. A, **33**, 994–999, (1986).
20. K. R. Karim and C. P. Bhalla, *P*hys. Rev. A, **39**, 3548–3554, (1989).
21. A. Dasgupta and K. G. Whitney, *P*hys. Rev. A, **69**, 022702-1 (2004).

Dr. Kate Kirby introducing a talk.

X-ray diagnostics with full radiative transfer (ALI method)

Loïc Chevallier*, Anne-Marie Dumont*, Suzy Collin*, Olivier Godet† and Anabela C. Gonçalves*

*LUTH, Observatoire de Paris, Section de Meudon, 92195 Meudon, France
†CESR, 9 av. du Colonel Roche, 31028 Toulouse Cedex 4, France

Abstract. With the advent of the present and future spatial X-ray missions, it becomes crucial to model correctly the line spectrum of X-ray emitting/absorbing media. We have built a photoionization code, TITAN, solving the transfer of a thousand lines and of the continuum with the "Accelerated Lambda Iteration" method (ALI), which is most reliable for line transfer. We give some details about this method and a justification for its use as a complement to usual approximations (e.g., escape probability or two-stream) made at present in other codes. We show that the escape probability approximation leads to a wrong estimation of the emitted X-ray line intensities, especially in the soft X-ray range. The errors can exceed one order of magnitude in the case of thick media (Thomson thickness of the order of unity). It also happens, but for different reasons, in the case of moderately thin media (Thomson thickness of 0.001 to 0.1), characteristic of the Warm Absorber in Seyfert 1 or of the X-ray emitting medium in Seyfert 2. Using TITAN, new diagnostics based on He-like ion lines of the n=2 complex are computed and some examples are presented here, concerning in particular the influence of the transfer method on the G ratio. We show also an example of the influence of the direction on the emission/absorption spectrum.

Keywords: galaxies: active – line: formation – radiative transfer – X-rays: binaries – X-rays: galaxies
PACS: 95.30.Jx – 98.54.Cm

INTRODUCTION

At the era of XMM-Newton and Chandra missions, it is crucial to use codes able to compute correctly the line spectrum of X-ray irradiated media, in order to build models for the structure and the emission of the central regions of Active Galactic Nuclei (AGN), or of X-ray binaries. All photoionization codes (like Cloudy and XSTAR), and the codes for computing the emission spectrum of accretion discs in AGN, use the "escape probability approximation" instead of solving a frequency dependent transfer in the lines. It means that the level populations, the line intensities, the contribution of the lines to the energy balance and to the ionization equilibrium, and finally to the radiation pressure, are computed in an approximate way. On the contrary, our photoionization code, TITAN, solves the transfer of a thousand lines and of the continuum with the "Accelerated Lambda Iteration" method (ALI), which is most reliable for line transfer. Since line transfer is time consuming, one could ask if it is worth performing these enormous computations. For this purpose we compare the line escape probability approximations with the ALI method in different cases relevant for AGN, keeping exactly the same method for the transfer of the continuum. We also show that computing the intensities in different directions, as TITAN does, leads to spectra strongly dependent on the inclination of

CP774, *X-Ray Diagnostics of Astrophysical Plasmas,* edited by Randall K. Smith

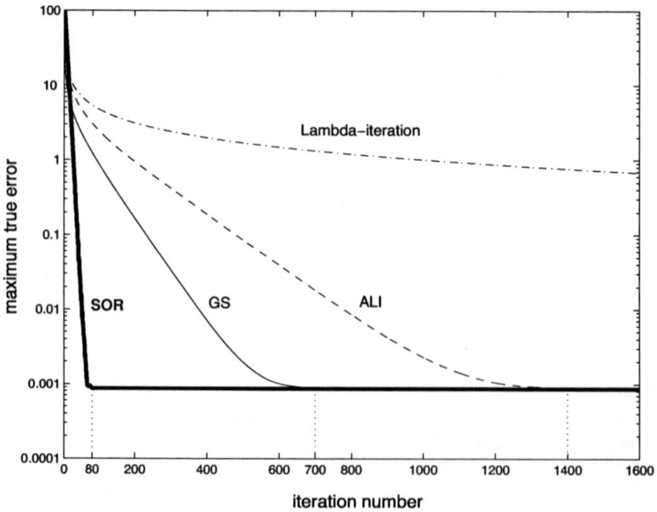

FIGURE 1. Evolution of the accuracy (maximum relative error for all depths in the medium) of several full radiative transfer algorithms (Lambda-Iteration or Jacobi, ALI, GS for Gauss-Seidel, SOR for Successive Over Relaxation) vs. the number of iterations for a very refined spatial discretization grid (20 grid-points per decade) [1, 2].

the slab or on the opening angle of the emission region.

ALI VERSUS ESCAPE PROBABILITY METHOD

TITAN is a photoionization code which includes all relevant physical processes from each level of all ions: photoionization, radiative and dielectronic recombination, ionization by high energy photons, fluorescence and Auger processes, charge exchanges, collisional ionization and 3-body recombination, radiative and collisional excitation/deexcitation. All induced processes are also included. The populations of each level are computed solving the set of ionization equations coupled with the set of statistical equations describing the excitation equilibrium, taking into account radiative and collisional ionizations from all levels and recombinations onto all levels, radiative and collisional excitations and deexcitations for all transitions. TITAN solves the ionization equilibrium of 102 ion species, the thermal equilibrium, the statistical equilibrium of all the levels of each ion (some low ionization states are treated roughly), and the transfer of the lines and of the continuum. It gives as output the ionization and temperature structures, and the reflected and outward spectra for all directions. Compton heating and Compton scattering in the lines is included through the coupling with a Monte Carlo code, NOAR [3, 4].

In TITAN, the transfer is performed with the ALI method, which is based on iterative schemes using operator splitting i.e., numerical methods better known as Jacobi's method in mathematics [5, 6, 7]. The ALI method solves accurately the full radiative

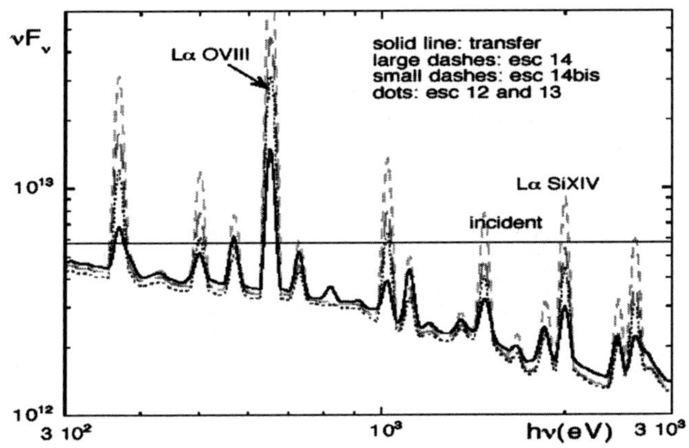

FIGURE 2. Comparison of the X-ray spectrum obtained with the most used escape probability approximations and with the ALI method: resonance lines are overestimated by up to one order of magnitude with the escape probability method. The model is a plane-parallel slab of constant density $n = 10^{12}$ cm^{-3} and total hydrogen column density 5×10^{24} cm^{-2} (Thomson thickness $\tau_{\rm th} = 4$), irradiated on one side by an isotropic incident continuum with a spectral distribution $F_\nu \propto \nu^{-1}$ extending from 0.1 eV to 100 keV. The ionization parameter ξ at the surface of the irradiated slab is equal to 10^3 [4].

transfer equation, especially its coupling with populations for spectral lines (scattering problem). This method is capable of coherently dealing with 1) any angular distribution of incident, emitted, reflected or absorbed radiative quantities like a radiative flux, 2) partial redistribution of frequencies in lines (PRD). This type of iterative algorithm is robust but rather slow to converge (e.g., Lambda–iteration as shown in Fig. 1). ALI is a fast accelerated algorithm, commonly used in stellar atmospheres modeling. Its use in TITAN gives reasonable computing times. As seen on Fig. 1, other faster than ALI algorithms (Gauss-Seidel, Successive Over Relaxation) may be useful for our code to approach typical computing times met in codes using escape probability methods.

Basically the escape probability approximation identifies the probability to escape in a single flight of a line photon emitted at an optical depth τ from the surface, with the divergence flux of the statistical equilibrium equations. Since line photons can be destroyed during their diffusion close to the point of emission, or on their way towards the surface and towards the back of the cloud, it is necessary to differentiate the probability used in the statistical equilibrium equations from those in the line emission and in the energy balance. One must also take into account photoionisations due to line photons. Several not equivalent estimates of the escape probability are used in the literature to account for these effects. The transfer computations performed with ALI have been compared to several escape approximations used in well-known photoionization codes (XSTAR and Cloudy in particular). The test models are made of a plane-parallel slab of constant density n, a given column density CD, irradiated on one side by an isotropic incident continuum $F_\nu \propto \nu^{-1}$ from 1 eV to 100 keV. The ionization parameter at the surface of the irradiated slab $\xi = 4\pi F/n$ is given, where F is the integrated incident

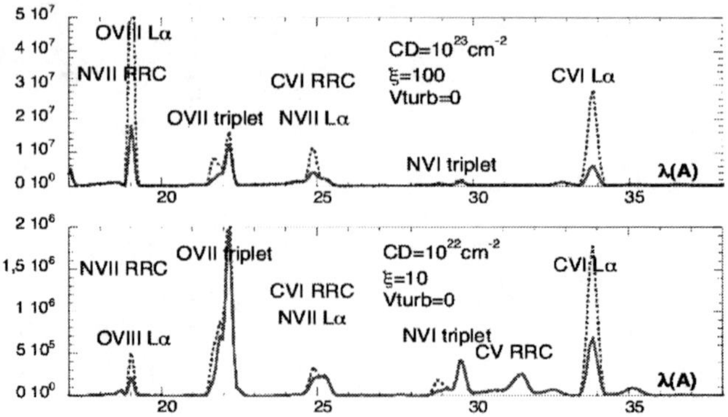

FIGURE 3. Comparison between the spectra computed with the ALI method (solid line) and the escape probability approximation used in other codes (dotted line) for two models corresponding to typical emission regions of Seyfert 2. The incident spectrum is the same as in Fig. 2. The spectral resolution is degraded to 30 [8].

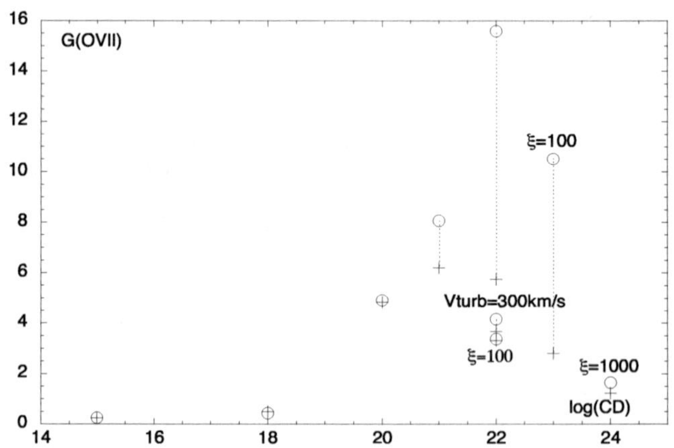

FIGURE 4. Comparison between the G ratio computed with the escape probability approximation (crosses) and with ALI (circles) for a few cases. The incident spectrum is the same as in Fig. 2. When another value is not indicated, $\xi = 10$.

flux.

For a Thomson thickness $\tau_{th} \geq 1$ (typical for the region emitting the UV-X AGN "continuum") one finds the following main results [4]:

- The escape probability computations converge very rapidly, but towards a non global energy-balanced result.
- The comparison of the most used escape probability approximations with the ALI treatment shows that resonance lines are overestimated by up to one order of mag-

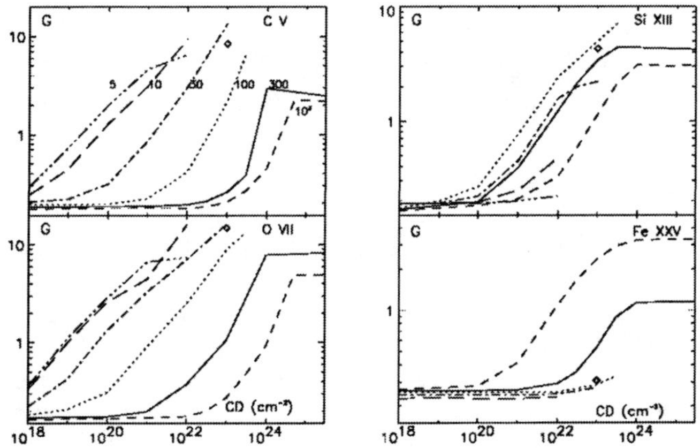

FIGURE 5. G ratio as a function of the column density for the same incident continuum as previously, for a few ions: CV, OVIII, SiXIII, FeXXV, and several values of the ionization parameter $\xi = 5, 10, 50, 100, 300$ and 1000. The incident spectrum is the same as in Fig. 2 [9].

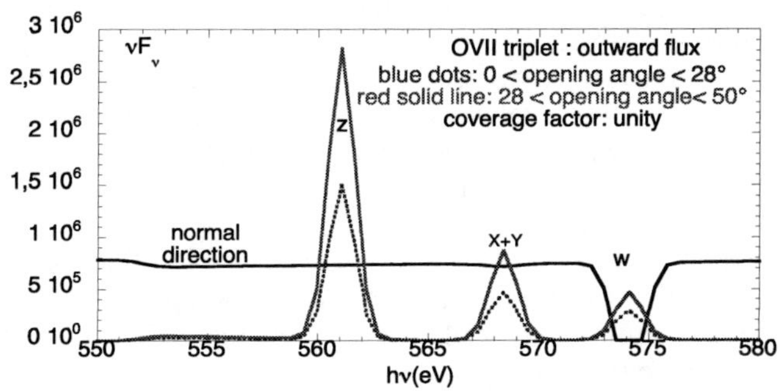

FIGURE 6. The spectrum in the region of the OVII triplet for a model with $\xi = 100, CD = 10^{22}$ cm^{-2}, and a turbulent velocity equal to 300 km s^{-1}. The resulting spectrum is the sum of the normal direction spectrum and of one or two of the other contributions. The spectral resolution is 400.

nitude with some escape approximations (cf. Fig. 2). This is due to the bad treatment of the reabsorption of line photons with escape probability approximations, particularly important in the X-ray range.

- Not only the line transfer but also the continuum transfer should be carefully handled in order to get correct reflected line intensities and a correct emission from the back side of the slab.

For a Thomson thickness $\tau_{\rm th} \sim 0.001$ to 0.1 characteristic of the Warm Absorber in Seyfert 1 or of the X-ray emitting medium in Seyfert 2 [8]:

- There are also large differences between the approximate treatments and ALI. Typically, the errors on the line fluxes and line ratios are of the order of 30% for a column density of 10^{20} cm^{-2}, and a factor of five for a column density of 10^{23} cm^{-2}, in conditions giving rise to the spectra observed in these objects.

- The reasons are subtle and the effects difficult to predict. They are almost always in the direction of an overestimation of the most intense line intensities, especially of the Lα line of H-like ions (cf. Fig. 3).

APPLICATION TO THE HELIUM-LIKE TRIPLETS

TITAN performances are illustrated here by a few examples concerning the He-like triplets. The ratio R of the forbidden line z ($1s^2\ ^1S\ - 2s\ ^3S$) and the intercombination lines y and x ($1s^2\ ^1S\ - 2p\ ^3P^o_{1,2}$), $R = z/(x+y)$ is sensitive to the electronic density n_e, while the ratio G of the forbidden plus intercombination lines over the resonant line w ($1s^2\ ^1S - 2p\ ^1P^o$), $G = (z+x+y)/w$ is sensitive to the electronic temperature. In TITAN, we use an atomic He-like description including 15 levels and a continuum, allowing to take into account all the terms for $n = 2$ and $n = 3$ and gathering the singlet and triplet terms for $n = 4$.

But G is also strongly dependent on the column density and on the transfer treatment. We have seen already on Fig. 3 the influence of the transfer treatment (ALI and the most used escape probability approximation methods) on the forbidden and resonance lines. Figure 4 illustrates these differences more quantitatively for a few examples: G can differ by up to a factor 5 between the two treatments, and is always larger with ALI than with the escape approximation. The discrepancy is smaller in the presence of a turbulent velocity and for low column densities.

Figure 5 displays G as a function of the column density for the same incident continuum as previously, for a few ions and several values of the ionization parameter ξ [9]. For typical column densities of the X-ray emitting region in Seyfert 1 and 2, the ratio differs considerably from the so-called "recombination value" (even at a very low column density, owing to radiative excitation), and it saturates at large column densities, as a given ion can only be present in a limited layer which depends on ξ. New diagnostics are derived from a wide range of models made for different values of the physical parameters (abundances, turbulent velocity, incident continuum, etc.) [9, 10].

Finally, we performed with ALI a multidirection transfer allowing to take into account the very important aspect angles, which are not considered in other photoionization codes. As an example, Fig. 6 illustrates this effect for a Warm Absorber seen in pure transmission (i.e. assuming there is only one cloud located just on the line of sight of the central source), and with two geometries of the medium giving the absorption. The resulting spectrum is the sum of the normal direction spectrum and of one or two of the other contributions. The spectrum is shown in the region of the OVII triplet. We see that the relative intensities of the permitted and forbidden lines vary with the opening angle, whereas only the resonance line is absorbed.

Other improvements are envisioned for TITAN in the near future, like a public release of this code. In particular, partial redistribution in line frequencies will be performed, and it will probably give TITAN other advantages with respect to other photoionisation codes. We intend also to add missing atomic data, but for this purpose it will be necessary to implement a new algorithm to accelerate the computation. Grids of models will be computed in order to be implemented in the reduction packages of the X-ray missions.

ACKNOWLEDGMENTS

A. C. Gonçalves acknowledges support from the *Fundação para a Ciência e a Tecnologia*, Portugal, under grant no. BPD/11641/2002.

REFERENCES

1. L. Chevallier, F. Paletou, and B. Rutily, *Astron. Astrophys.*, **411**, 221–227 (2003).
2. C. Quang, F. Paletou, and L. Chevallier, 2004, SF2A-2004: Semaine de l'Astrophysique Francaise, meeting held in Paris, France, June 14-18, 2004, Eds.: F. Combes, D. Barret, T. Contini, F. Meynadier and L. Pagani EdP-Sciences.
3. A.-M. Dumont, A. Abrassart, and S. Collin, *Astron. Astrophys.*, **357**, 823–838 (2000).
4. A.-M. Dumont, S. Collin, F. Paletou, S. Coupé, O. Godet, and D. Pelat, *Astron. Astrophys.*, **407**, 13–30 (2003).
5. C. J. Cannon, *J. Quant. Spectroscop. Radiat. Transfer*, **13**, 627–633 (1973).
6. G. L. Olson, L. H. Auer, and J. R. Buchler, *J. Quant. Spectroscop. Radiat. Transfer*, **35**, 431–442 (1986).
7. L. H. Auer, and F. Paletou, *Astron. Astrophys.*, **284**, 675–686 (1994).
8. S. Collin, A.-M. Dumont, and O. Godet, *Astron. Astrophys.*, **419**, 877–886 (2004).
9. S. Coupé, O. Godet, A.-M. Dumont, and S. Collin, *Astron. Astrophys.*, **414**, 979–992 (2004).
10. O. Godet, S. Collin, and A.-M. Dumont, *Astron. Astrophys.*, **426**, 767–776 (2004).

Drs. Jay Bookbinder and Ann Hornschemeier reviewing a presentation.

Lessons from other wavelengths

Gary J. Ferland

Physics Department, University of Kentucky, Lexington, KY 40506

Abstract I outline some lessons learned from spectroscopic observations of AGN done at lower energies. First, spectroscopy is subject to powerful selection effects. We can only detect gas that has the right density, temperature, and ionization to effectively emit the light we see. Although a temperature and density can be derived using standard plasma diagnostics, it is often the case that a range of conditions are actually present. The "Locally Optimally-emitting Cloud" (LOC) model, which uses distribution functions to describe a range of cloud parameters, was developed to treat this situation. The second lesson is the need to consider emission at all wavelengths when studying an environment. I discuss two examples – He-like ions seen in the X-ray should also emit UV lines. Although they have not yet been identified, their detections or upper limits would constrain the conditions in the gas. Another example is the very hot dust that emits the continuum near 1 μm. The gas associated with the dust is likely to have $T \approx 10^6$ K and emit in the X-rays. These are all examples of the insight that large-scale plasma simulations can bring to understand conditions in a physical environment.

INTRODUCTION

The title of this talk was given to me by the organizers – it was an interesting concept, I could talk about anything, as long as it wasn't X-rays. Despite the flexibility that comes with a job defined by what not to do, I didn't succeed – X-rays are here.

My work has focused on understanding the microphysics that determines the spectrum of an AGN. Historically, the rest-frame UV spectrum was the best studied, using large ground-based telescopes to observe $z \sim 2$ objects. The AGN phenomenology was defined by this UV spectrum, together with the optical spectrum seen in lower redshift objects. These photons have $h\nu \approx 1$ eV and detailed analysis showed that it was produced by a photoionized plasma with $kT \approx 1$ eV. Analysis of optical forbidden lines and UV intercombination lines suggested that the gas density was close to the lines' critical densities, the density where an excited level has a 50-50 chance of undergoing a radiative versus a collisional deexcitation. Davidson & Netzer (1979) offer a good review of the field at that time.

Some of these conclusions were based on selection effects introduced by atomic physics. Line-continuum reverberation studies show that the gas producing the UV spectrum exists at a wide range of densities and distances from the continuum source (Peterson 1993). X-ray line observations described at this conference show that hot ionized gas, often an outflowing wind, exists. The infrared has revealed the existence of a molecular torus, with grains nearest to the AGN being heated to their sublimation point.

It is clear that spectroscopy is subject to powerful selection effects. It is no accident that optical spectra reveal 10^4 K gas while X-rays detect much hotter gas. If a range of

CP774, *X-Ray Diagnostics of Astrophysical Plasmas,* edited by Randall K. Smith

densities and temperatures are present, as we know to be the case in an AGN, then each line will emit most efficiently near its critical density, and the spectrum is largely determined by atomic physics selection effects. The second lesson is to be aware of all emission produced by an environment. Two examples are discussed. The first is He-like ions that are seen in the X-ray. Although they have not been identified yet, these ions must emit a set of UV lines that would provide additional constraints on the properties of hot gas in AGN. The second example is the hot dust seen in innermost regions of the molecular torus. Gas associated with this dust is likely to have temperatures in the range $10^5 - 10^6$ K and so emit X-rays.

These are examples of the real purpose of large-scale numerical simulations of plasmas – they reveal the underlying physics that governs the gas properties, can quantify the selection effects that are inherent in spectroscopy, and serve as a guide to understanding the implications of the observations.

LOCALLY OPTIMALLY-EMITTING CLOUDS

Attempts at understanding conditions in the gas emitting the UV lines in quasars were among the first applications of computer simulations to astrophysics (Bahcall & Koslozski 1969). Quasars were found to have remarkably similar emission-line spectra, with far less object-to-object variation than was seen in planetary nebulae or H II regions. Photoionization calculations showed that the spectrum had a strong dependence on an "ionization parameter", the ratio of densities of ionizing photons to hydrogen nuclei. The conclusion was that an unknown process "tuned in" this special parameter. The fact that an unknown agent tuned in special cloud properties inhibited progress in using the lines as a physical probe of the AGN.

Large monitoring campaigns that measured the response of lines to continuum variations showed that the clouds occur over a broad range of radii, have a wide range in density, and that higher-ionization clouds were closer to the continuum source (Peterson 1993). At about this time, computers finally became fast enough to predict properties of clouds having a broad range of density and distance from the central source. The results of one such calculation, showing the equivalent width of one of the strongest UV lines, C IV λ1549, is shown in Figure 1. The equivalent width of a line emitted by a photoionized cloud is a measure of the cloud's reprocessing efficiency, since the cloud converts radiation produced by the central engine into emission lines.

Powerful selection effects are clearly present. This line can only be emitted when C^{+3} is present. The left-hand plot shows linear W_λ making it clear how strong a selection effect we have (spectrometers are linear detectors), while the right-hand plot shows the log of W_λ. In the right-hand plot, clouds in the lower right are too neutral, and those in the upper left are too ionized, to produce the line. The line is thermalized in high-density clouds in the upper right, and is weak relative to other emission sources in the lower left. The peak W_λ, corresponding to the peak reprocessing efficiency, corresponds to the oval contour at lower left. The nearly coincident filled circle gives the "special" parameters deduced in the original photoionization simulations.

This result suggests that a very broad range of clouds may be present with atomic physics selection effects determining the most efficient emitter. We termed this the

"Locally Optimally-emitting Cloud" (LOC) model to stress the importance of these selection effects (Baldwin et al. 1995). Different lines have different selection effects, with high-ionization lines peaking closer to the continuum source, and low-ionization lines further away, a result consistent with the structure deduced by variability studies (Korista et al. 1997; Ferguson et al. 1997). The suggestion is that the emitting region is a very chaotic environment, consistent with the picture that quasars are associated with the birth of massive galaxies (Hamann & Ferland 1999). Cloud distribution functions replace cloud parameters (Bottorff et al. 2002). The spectrum is best reproduced with a distribution function that gives equal weight to all clouds. This fully chaotic limit can be described several ways, perhaps as a fractal (Botorff & Ferland 2001) or by self-organized criticality.

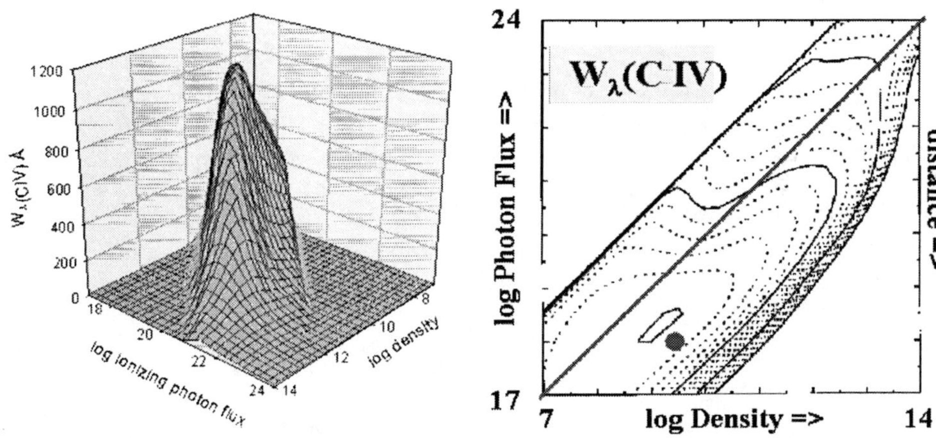

Figure 1 The predicted equivalent width W_λ of C IV λ1549, one of the strongest lines in the UV spectrum of an AGN, is shown. The axes show the cloud density (cm^{-3}) and the flux of ionizing photons striking the cloud (cm^{-2} s^{-1}), which is proportional to the inverse square of the separation between the continuum source and the cloud. The left plot shows linear W_λ (Å) and the solid contours on the right show 1 dex increments in W_λ. The diagonal line in the right plot shows constant ionization parameter. W_λ is a measure of the reprocessing efficiency of the clouds, and is sharply peaked at the oval contour in the lower left portion of the right-hand plot due to atomic physics selection effects. The filled circle shows the "special" cloud parameters deduced in the original photoionization studies. They lie very near the peak reprocessing efficiency.

UV EMISSION FROM HIGHLY CHARGED IONS

A spin-off of the large grids of emission models that we now routinely compute is that we predict emission properties of gas with an extremely broad range of density and ionization. Conventional optical / UV spectroscopy will detect gas in the lower right half of the plane shown in the right side of Figure 1. Gas in the upper left quadrant will be hot (10^5 – 10^7 K) and emit mainly in the X-rays. It is possible to detect this gas in nearby bright Seyferts with Chandra and XMM with very long exposures (Porquet & Dubau 2000) but this is not routine due to the low collecting areas of even modern X-ray observatories.

UV emission from this hot gas would be easier to detect due to the larger collecting area of ground-based telescopes. The upper left quadrant of the LOC plane contains a great deal of gas that is in an ionization state that is part of the He-like isoelectronic sequence. Table 1 lists the wavelengths of the strongest lines produced by some of these ions – these are Lα and the three lines of the 2^3P – 2^3S multiplet (the isoelectronic equivalent of He I λ10830). It is the Lα transition that has been detected in long X-ray exposures of a few AGNs.

Table 1 X-ray and UV transitions of He-like ions

Ion	λ(Lα; Å)	$\lambda(2^3P_0$-2^3S_1; Å)	$\lambda(2^3P_1$-2^3S_1; Å)	$\lambda(2^3P_2$-2^3S_1; Å)
C V	40.27	2349	2348	2339
N VI	28.79	1982	1980	1968
O VII	21.60	1698	1696	1680
Ne IX	13.45	1289	1285	1260
Mg XI	9.169	1011	1006	971.4

Figure 2 shows the predicted intensity ratio of the UV to X-ray lines for O VII. The UV lines are emitted with an intensity that is comparable to that of the X-ray line. Quantitative estimates are hampered by uncertainties in the reddening, the existence of source variability, and uncertainty in how the LOC plane is filled, but equivalent widths of 10 – 15Å are expected.

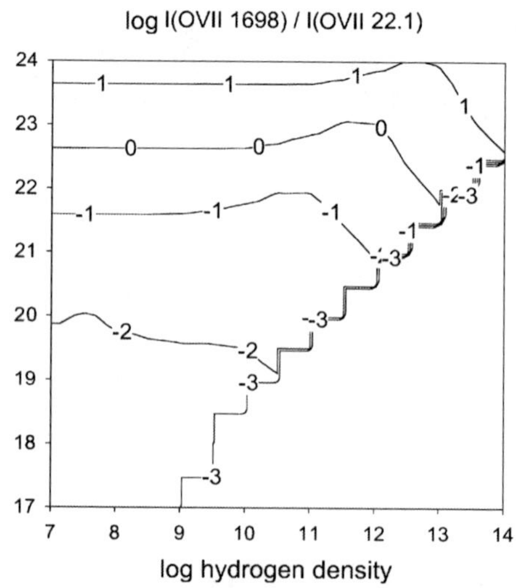

Figure 2. The log of the expected intensities of the 2^3P – 2^3S multiplet relative to Lα across the LOC plane. The UV line should have a luminosity that is comparable to the X-ray line

Although the 2^3P – 2^3S UV transitions have not been detected in AGN, they are routinely seen in stellar coronae and have been included in the analysis of physical conditions (Kahn et al. 2001). However, the AGN community has not been aware of the likely presence of these lines, so they have not been searched for in QSO or Seyfert galaxy spectra.

UV lines from He-like species are the best way to detect hot and very highly ionized gas for the broad range of sources that are too faint for high-resolution X-ray observations to be made. They would establish a new way to explore gas phases that previously could be seen in only a handful of objects. X-ray spectra that are deep enough to detect the He-like Lα transition are limited to a handful of bright nearby Seyfert galaxies. For redshifts beyond $z \approx 2$, the UV lines in Table 1 will shift into the optical, where they can then be detected out to very high redshifts. Scientific goals of this work would be to determine the mass of hot gas, see how this compares with cooler gas, and determine how these depend on luminosity or redshift. All of these would provide insight into the physics governing the properties of gas near the center of an AGN.

HOT DUST IN AGN

Infrared observations revealed the presence of very hot grains, thought to be associated with the illuminated face of the molecular torus (Sanders et al. 1989). Much of the early work on understanding this emission focused on emission properties of the grains and the heating-cooling processes that determine their temperatures.

A gaseous component must also be associated with these hot grains, and it is likely that it is hot enough to emit X-rays. Figure 3 shows results of a series of calculations in which gas with a range of densities (indicated by the lines) is placed at a variety of distances from the AGN central engine. The gas has ISM abundances and dust-to-gas ratio. Clouds close to the continuum source have hotter grains and more highly ionized gas. Since the grain temperature is the basic observation, the mean temperature of the various types and sizes of grains was derived and used as the x-axis. The plot shows the derived gas temperature, which is in the neighborhood of 10^6 K. This component may be detectable in the X-ray (Ferland et al. 2001; Ballantyne et al. 2004).

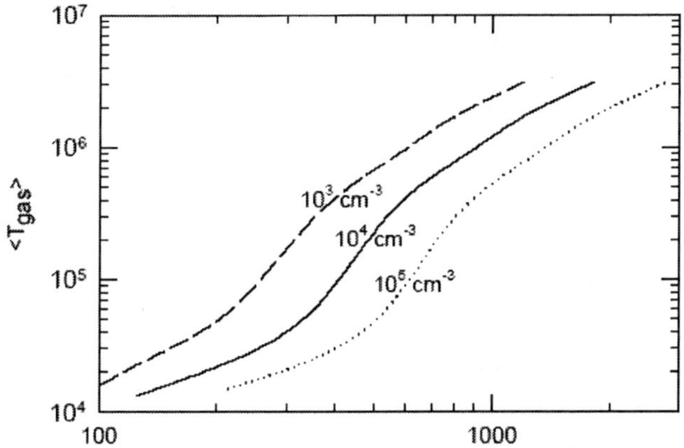

Figure 3 Results of a series of calculations in which a cloud with ISM abundances and dust to gas ratio is irradiated by an AGN continuum. The cloud densities are indicated on the plot. The clouds were located at a range of distances from the continuum source, with clouds closer to the continuum source being hotter and more highly ionized. The mean grain temperature was derived and is used as the x-axis. IR observations show that grains with temperatures near 10^3 K exist. The associated gas has a temperature around 10^6 K and emits in the X-rays.

In stellar environments grains can be fairly hot, but the associated gas has a similar temperature. This is because the incident radiation field is quite soft, having no ionizing photons. The X-ray case is different bringing in many inner shell processes that affect the conditions in the grains (Dwek & Smith 1996). The surrounding gas is heated by the hard continuum, grain photoelectrons further heat the gas, while emission from the plasma can further heat the grains. The lesson here is similar to the previous case – it is important to consider the full spectral emission of a physical component.

CONCLUSIONS

This paper has shown a few examples of how large-scale numerical simulations can identify the full observed properties of a cloud. Spectroscopy is subject to very powerful selection effects introduced by the atomic physics. Any particular ion and emission line will have a preferred temperature, density, and incident radiation field where that line is most efficiently produced. Those are the conditions that will be deduced if a broad range of temperature, density, and radiation field are present. This suggests a different approach to the analysis of the lines. Distribution functions replace cloud parameters – in our work on AGN we found that a very "flat" distribution function, which gives nearly equal weight to all possible clouds, reproduces the observed spectra the best.

An environment is likely to emit in several spectral regions, and this emission may constrain or eliminate a particular model. Examples are the X-ray emission associated with very hot dust, and UV emission from the He-like ions detected in the X-ray. Today, with large data bases and virtual observatories coming on line, a true pan-chromatic approach to understanding an object can be routine, and is often necessary to obtain an unbiased view of what is happening.

These examples underscore the real reason to do numerical simulations of an environment. We are not trying to produce a toy model of a system. Real astronomical environments are far too complex for this to succeed. Rather, the simulations can identify the full spectrum emitted by a region and show how this spectrum depends on cloud parameters (see the discussion by Ferland 2003). The simulations serve as a guide to identify the governing physics, selection effects, and how these affect our observations.

Acknowledgements I wish to thank the meeting organizers for their support and for an enjoyable meeting. All of the calculations discussed in this paper were performed with the spectral simulation code Cloudy, which is on the web at http://www.nublado.org The code has been developed with support by NSF (AST 03-07720), NASA (NAG5-12020), and the STScI theory program (AR 10316).

REFERENCES

Bahcall, J.H., & Kozlovsky, B.-Z. 1969, ApJ, 155, 1077
Baldwin, J., Ferland, G., Korista, K., Verner, D. 1995, ApJ, 455, L119
Ballantyne, D.R. Ferland, G.J. Martin, P.G. van Hoof, P.A.M. & Weingartner, J.C., 2004, Nuclear Physics B, 132, 145
Bottorff, M.C., Baldwin, J.A., Ferland, G.J., Ferguson, J.W., & Korista, K.T. 2002, ApJ, 581 932
Bottorff, M.C. & Ferland, G.J. 2001, ApJ, 549, 118

Davidson, K., & Netzer, H. 1979, Rep. Prog. in Physics 51, 715
Dwek, E., & Smith, R.K. 1996, ApJ, 459, 686
Ferguson, J. W., Korista, K. T., Baldwin, J. A., & Ferland, G. J. 1997, ApJ, 487, 122
Ferland, G.J., 2003, ARA&A, 41, 517
Ferland, G.J., Martin, P.G., van Hoof, P.A.M., & Weingartner, J.C 2001, in *X-ray spectroscopy of AGN with Chandra and XMM-Newton*, Garching, Dec 3-6, 2001, eds Th. Boller, S. Komossa, S. Kahn, H. Kunieda, MPE Report 279, p. 103 (astro-ph/0311520)
Hamann, F., & Ferland, G. J. 1999, ARAA, 37, 487
Kahn, S.M., Leutenegger, M.A., Cottam, J., Gauw, G., Vreux, J.-M., den Boffende, A.J.F., Mewe, R., & Gudel, M. A&A, 2001, A&A, 365, L312
Korista, K., Baldwin, J., Ferland, G., Verner, D. 1997, ApJS, 108, 401
Peterson, B.M. 1993, PASP, 105, 247
Porquet, D., & Dubau, J. 2000, A&AS, 143, 495
Sanders, D. B., et al. 1989, ApJ, 347, 29

.

Drs. Nancy Brickhouse and Joseph Tan watching a talk.

II: IRON L- AND K- SHELL DIAGNOSTICS

Photoionization Modeling: the K Lines and Edges of Iron

P. Palmeri*, T. R. Kallman†, C. Mendoza**, M. A. Bautista** and J. H. Krolik‡

**Astrophysique et Spectroscopie, Université de Mons–Hainaut, 20 Place du Parc, B-7000 Mons, Belgium*
†Laboratory for High Energy Astrophysics, NASA/GSFC, Code 662, Greenbelt, MD 20771, USA
***Centro de Física, Instituto Venezolano de Investigaciones Científicas (IVIC), PO Box 21827, Caracas 1020A, Venezuela*
‡Department of Physics and Astronomy, Johns Hopkins University, Homewood Campus, Baltimore, MD 21218, USA

Abstract. We calculate the efficiency of iron K line emission and iron K absorption in photoionized models using a new set of atomic data. These data are more comprehensive than those previously applied to the modeling of iron K lines and edges from photoionized gases, and allow us to systematically examine the behavior of the properties of line emission and absorption as a function of the ionization parameter, density and column density of model constant density clouds.

Keywords: atomic data, atomic processes, line formation, X-rays: spectroscopy
PACS: 95.30.Ky, 95.30.Dr, 87.64.Gb

INTRODUCTION

The K lines of iron are unique among commonly observed X-ray lines in that they can be emitted efficiently by gas over a wide range of temperatures and ionization states. Their location in an unconfused spectral region makes them good plasma diagnostics. They were first reported in the rocket observation of supernova remnant Cas A [1]. With the advent of satellite-borne telescopes they were observed from almost all classes of cosmic objects. More interests have been attracted around these lines with the observations of Seyfert galaxies and galactic black hole candidates, some of which show relativistic broadened and red-shifted lines attributed to formation within a few gravitational radii of a black hole [2].

Current improvements in resolving power and sensitivity of X-ray spectrometers onboard the Chandra and XMM-Newton spacecraft have promoted the role of Fe K lines as diagnostics, a trend that will continue with the launch of future observatories such as Astro-E2, Constellation-X, and XEUS. Such plasma diagnostics ultimately rely upon the microphysics of line formation and on the availability and the accuracy of the atomic data.

The present paper is a brief report on all the studies realized during a project dedicated to the modeling of the iron K lines and edges [3, 4, 5, 6, 7, 8]. A more detailed synthesis can be found in Kallman *et al.* [9].

CP774, *X-Ray Diagnostics of Astrophysical Plasmas,* edited by Randall K. Smith

ATOMIC DATA

Despite the solar flare line list of Seely *et al.* [10] and some EBIT and Tokamak measurements [11, 12, 13, 14, 15], the K-vacancy level structures of the highly charged Fe ions remain incomplete as can be concluded from the critical compilation of Shirai *et al.* [16]. Concerning the radiative and Auger rates, the ions Fe XVIII–Fe XXV have received much attention and some degree of data accuracy have been achieved [17, 18]. Regarding the other members of the iso-nuclear sequence, the central field calculations of Jacobs *et al.* [19] and Jacobs and Rozsnyai [20] do not meet present requirements of level-to-level data and accuracy needs of current and future X-ray missions. The situation of the inner-shell electron impact and photoionization data was even worse. Many previous close-coupling calculations [21, 22, 23, 24] omitted the Auger damping. We showed that the latter along with the radiative damping dominate the K-threshold resonance behavior and induce a smeared edge [3]. The importance of the Auger damping for low-temperature effective collision strengths has been demonstrated in Li-like systems by Whiteford *et al.* [25].

In order to remedy the incompleteness, and in some cases, the lack of atomic data, we have calculated the atomic parameters for the K-shell of all the iron ions using and comparing different state-of-the-art atomic codes, i.e. AUTOSTRUCTURE [26], HFR [27], and the Breit–Pauli R-matrix package (BPRM) [28]. While the latter was preferred for calculating the electron and photon scattering properties due to its accurate inclusion of the continuum channel coupling, the former two were chosen for their semi-empirical options which can utilize experimentally determined level energies to improve the theoretical eigenvalue separations for which the inter-system transitions rates are sensitive. Although all these programs include relativistic effects and configuration interactions, AUTOSTRUCTURE has the most complete Breit Hamiltonian. This two-body magnetic interaction plays an important role in the highly charged ions (Fe XVIII–Fe XXV) in particular for transitions with small rates and for those involving states subject to strong relativistic couplings. It was found that the energies and rates are also affected by core relaxation effects (CRE) and that these effects are more important than the higher order relativistic interactions for spectator inner-shell processes that dominate in Fe I–Fe XVII. Since, among the codes used at the time this project was initiated, HFR handled CRE more efficiently in ab initio calculations (AUTOSTRUCTURE has since been updated to treat this process efficiently), it was used for data production in cases where no experimental level energies were available for the semi-empirical corrections, i.e. in Fe I–Fe XVII.

In Fe XVIII–Fe XXV [4, 5, 8], a complete set of level energies, wavelengths, A-values, and total and partial Auger rates were computed for the K-vacancy states within the $n=2$ complex. The experimental level energies of the compilation of Shirai *et al.* [16] were used in the semi-empirical adjustment procedures. It was found that these procedures can lead to large differences in both radiative and Auger decay data of strongly mixed levels. Several experimental level energies and wavelengths were questioned, and significant discrepancies were found with previously computed decay rates that were attributed to numerical problems. The statistical accuracy of level energies and wavelengths was ranked at ± 3 eV and ± 2 mÅ, respectively, and that for A-values and partial Auger rates greater than 10^{13} s^{-1} at better than 20%. The photoabsorption cross sections across the

K edge and electron impact K-shell excitation effective collision strengths were also calculated. The target models were represented with all the fine-structure levels within the $n = 2$ complex. The effects of radiation and spectator Auger dampings were taken into account by means of an optical potential [29]. In photoabsorption, these effects cause the resonances converging to the K thresholds to display symmetric profiles of constant width that smear the edge, with important implications in the spectral analysis. In collisional excitation, they attenuate the resonances making their contributions to the effective collision strength negligible.

Regarding Fe X–Fe XVII [7], the Kα and KLL Auger widths were found to be nearly independent of both the outer-electrons and the electron occupancy keeping a constant ratio of 1.53 ± 0.06. By comparing with previous theoretical and measured wavelengths, the accuracy of these wavelengths were determined to be within 2 mÅ. Also, the good agreement found between the radiative and Auger datasets that were computed with different platforms allowed us to propose with confidence an accuracy rating of 20% for line fluorescence yields greater than 0.01. Emission and absorption spectral features were predicted finding good correlation with measurements in both laboratory and astrophysical plasmas. The photoionization cross sections across both L and K edges and the electron impact L and K shells excitation effective collision strengths were computed in LS coupling due to our computer memory limits. The targets used were modeled by all the terms within the $n = 3$ complex. As in Fe XVIII–Fe XXV, the effects of the radiative and Auger dampings were also included. Decaux *et al.* [13] have published the only experimental study of the K line spectrum of the second row, and they found that the only ion for which the K line is sufficiently unblended to derive an accurate wavelength is Fe X. We use this line to constrain our atomic structure calculations, using the semi-empirical term energy correction (TEC) in AUTOSTRUCTURE.

The ions Fe III–Fe IX have no experimental data. For Fe II, measurements have been made in the solid (e.g. [30]) and a good agreement has been found with our theoretical data [6]. Level energies, wavelengths, A-values, Auger rates and fluorescence yields have been calculated for the lowest fine-structure levels populated by photoionization of the ground state of the parent ion. Due to the complexity of the level-to-level Auger calculations, we have employed a compact formula [31] to compute the Auger widths from HFR radial integrals. We have estimated an accuracy of $\sim$10% for the transition probabilities thanks to the comparison of the Kβ/Kα branching ratios taken from different theoretical and experimental data sets in the literature. Concerning the photoabsorption cross sections, a model similar to the one used in Palmeri *et al.* [3] has been constructed in which a single $1s \rightarrow np$ Rydberg series of Lorentzian resonances converging to each K thresholds has been considered in each species. The partial photoionization cross sections connecting the ground states of the parent ions to the different K-vacancy states have been calculated by AUTOSTRUCTURE in the monoconfigurational approximation.

PHOTOIONIZED PLASMA MODELING

Into the XSTAR photoionizing code [32], we have incorporated all our published K-vacancy levels of Fe XVIII–Fe XXV and those populated by photoionization of the ground states of the parent ions in Fe I–Fe XVII according to the selection rules of Rau [33]. These are the levels whose populations are calculated explicitly, and for which we adopt the transition rates summarized in our previous. In addition we include superlevels that are associated to the photoabsorption features near the inner-shell edges using our BPRM cross sections . The populations of these superlevels are also calculated, and their effects on ionization and thermal balances are taken into account. Superlevel decays are treated as "fake" transitions, i.e., lines with wavelengths much longer than any of interest to the model results. Such fake transitions are also used to treat the decays of other levels for which we do not explicitly treat the line wavelengths such as the decays of L-vacancy levels in Fe II–Fe XVII.

For the purpose of illustrating the behavior of K lines in photoionized gas, we consider line emission and absorption in a spherical shell of gas surrounding a point source of continuum radiation. We assume cosmic element abundances [34], constant gas density, and an $F_\varepsilon \propto \varepsilon^{-1}$ power law ionizing spectrum.

We used different XSTAR model slabs with a fixed column density of 10^{17} cm^{-2} varying the ionization parameter in order to study the dependence of the K line emissivity on this parameter. It was found that the contribution of a given ion to the line emissivity is shifted down in ionization parameter to the value where the parent ion dominates the ionization distribution, e.g. Fe XIX dominates the line emission near an ionization parameter value of 100 erg s^{-1} cm^{-1}, where Fe XVIII dominates the ionization balance. In other words, ions that dominate the line emissivity are not those which dominate the ionization balance, but rather their ionization products. Thus, in situations where Fe XVII dominates the ionization balance, the Fe XVIII and Fe XIX K emission lines will appear prominently than those of Fe XVII. Similar behavior will occur for Fe X versus Fe IX. This is of interest because K-like and F-like ions have nearly filled atomic shells and therefore simpler atomic structure than ions with half-filled sub-shells. This may aid in the interpretation of spectra.

With regards to the gas density dependence, a series of model slabs were calculated with XSTAR for a fixed ionization parameter value of 100 erg s^{-1} cm^{-1} and a fixed column density value of 10^{17} cm^{-2}. Since the density acts to mix the levels of the ground term, the effect of increasing the density is to redistribute the population from the ground level to the other levels in the term. We found that the net effect summed over the lines of a given ion is small for many ions, since the line fluorescence efficiency is similar for K shell ionization of the various levels of the ground configuration. On the other hand, it was discovered that some ions show a strong density effect, and this act either to enhance or suppress the total K line emission for that ion. For instance, it was observed that the Fe XXII K line emission was enhanced at density greater than $\sim 10^{12}$ cm^{-3}, owing to the mixing of the population into the $J = 1$ level of the ground term of the parent ion, Fe XXI. This level has larger photoionization cross section than the ground level, leading to enhanced line emission. Also, collisional suppression was found to affect all the K lines at density $\geq 10^{15}$ cm^{-3}.

A family of XSTAR model slabs with an ionization parameter of 100 erg s^{-1} cm^{-1} and a gas density of 10^{12} cm^{-3} were considered for several column densities. For columns $\geq 10^{23}$ cm^{-2}, a net decrease in the K line emissivity was found, particularly for Fe XVIII–Fe XXII. The effect is primarily due to the change in the ionization balance of the slab; at high columns the average ionization of the slab decreases owing to the shielding of the deeper regions. Evidence of this was observed through the increase of the K lines emitted by Fe VI–Fe XVII as those emitted by Fe XVIII–Fe XXII decrease for column density values ranging from $\sim 10^{22}$ to $\sim 10^{23}$ cm^{-2}.

The K line emission spectrum was studied for a series of thin (column density of 10^{17} cm^{-2}) slabs of gas density of 10^{12} cm^{-3} and of various ionization parameter values ranging from 0.01 to $\sim$ 1260 erg s^{-1} cm^{-1}. It has a high degree of complexity, particularly for Fe XI–Fe XXIV. The Fe I–Fe X show a shift in the centroid of the Kα feature with ionization stage, in the sense that the centroid shifts to lower energies as the ionization stage increases from neutral, and then back to higher energies for Fe XI–Fe XVII. This can be used as an observational diagnostic for relatively bright objects viewed with high-resolution instruments. The behavior of the line profile can be crudely divided into three ionization parameter ranges, depending on which ions dominate the ionization balance. These are distinct in their character and thus provide potential diagnostics of the ionization parameter in observed spectra. At low ionization parameter ($<$ 10 erg s^{-1} cm^{-1}), the line profile has a characteristic shape consisting of the Kα_1 and Kα_2 components. The separation of these two components is approximately constant at $\sim$13–15 eV, and the ratio varies slowly with ionization stage, from 2:1 for Fe II to 1.7:1 for Fe IX. The energies of the Kα components moves down with increasing ionization, by $\sim$5 eV, across the Fe II–Fe IX ions, because of the interaction between the $2p$ and the $3d$ orbitals. The Kβ energy increases monotonically with increasing ionization stage. The character of the profile changes qualitatively at ionization parameter $\geq$ 10 erg s^{-1} cm^{-1}, where the Fe XI–Fe XVII ions cause the line centroid to move up in energy by a significant amount. The line shape consists of a blend of narrow components with total width 20–30 eV, which may still appear as a single line if observed by an instrument with moderate spectral resolution or statistics. If represented by a single broad line component, these blends would have an apparent width of $\sim$1000–1500 km s^{-1}. The profile changes qualitatively again at ionization parameter $\geq$ 100 erg s^{-1} cm^{-1}, where the Fe XVIII–Fe XXVI ions dominate. Here distinct contributions to the line from various ionization stages are separated in energy by $\sim$50–75 eV and are less likely to be interpreted as part of a single feature.

The same kind of analysis was done for the absorption spectrum in the iron K edge region. For that purpose, a family of thick (column of 3×10^{23} cm^{-2}) slabs were calculated with XSTAR for an ionization parameter range of 0.01–1000 erg s^{-1} cm^{-1}. We found that the K shell opacity is dominated by the strongly damped resonances at energies below the K ionization threshold. These conspire to smear and weaken the change in opacity at the edge and also present numerous narrow features in the absorption spectrum. A good illustration of this effect is given in Figure 1. In this figure, a simulated spectrum of a 100 mcrab source with an ionization parameter of 100 erg s^{-1} cm^{-1} and a column density of 3×10^{23} cm^{-2} is shown for an exposure time of 10^5 s using the XRS instrument onboard Astro-E2.

CONCLUSIONS

We calculated the efficiency of iron K line emission and iron K absorption in photoionized models using a new set of atomic data. These data are more comprehensive than

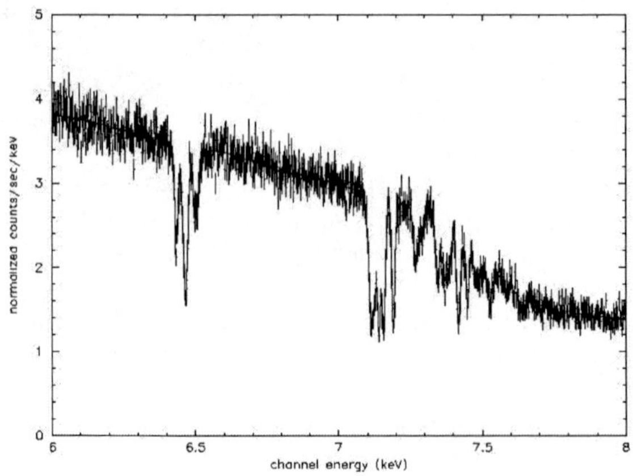

FIGURE 1. Simulated spectrum of a 100 mcrab source with an ionization parameter of 100 erg s^{-1} cm^{-1} and a column density of 3×10^{23} cm^{-2} for an exposure time of 10^5 s using the XRS instrument onboard Astro-E2.

those previously applied to the modeling of the iron K lines and edges from photoionized gases and allowed us to examine the diagnostic properties of these spectral features. The level of complexity of the iron K spectrum will be accessible to future instruments like the XRS microcalorimeter onboard Astro-E2 and the high-resolution spectra registered with this new generation of observatories will constrain the ionization of the gas, for example, and likely other properties such as gas flows, relativistic effects, and multiple components.

All of the atomic data used in these calculations are available online through electronic tables associated with the earlier papers [3, 4, 5, 6, 7, 8], and as part of the TOPbase electronic database [35]. The results reported in this paper are incorporated into the publicly available photoionization and spectrum synthesis code XSTAR and the associated XSPEC tables [36].

ACKNOWLEDGMENTS

This work was supported by grant NRA-00-01-ATP-025 through the NASA Astrophysics Theory Program. MB and CM acknowledge partial support from FONACIT, Venezuela, under Project S1-20011000912. PP is funded by a return grant of the Belgian Federal Scientific Policy.

REFERENCES

1. P. J. Serlemitsos, E. A. Boldt, S. S. Holt, R. Ramaty, and A. F. Brisken, *Astrophys. J.*, **184**, L1–L5 (1973).
2. Y. Tanaka, K. Nandra, A. C. Fabian, H. Inoue, C. Otani, T. Dotani, K. Hayashida, K. Iwasawa, T. Kii, H. Kunieda, F. Makino, and M. Matsuoka, *Nature*, **375**, 659–661 (1995).
3. P. Palmeri, C. Mendoza, T. R. Kallman, and M. A. Bautista, *Astrophys. J.*, **577**, L119–L122 (2002).
4. M. A. Bautista, C. Mendoza, T. R. Kallman, and P. Palmeri, *Astron. Astrophys.*, **403**, 339–355 (2003).
5. P. Palmeri, C. Mendoza, T. R. Kallman, and M. A. Bautista, *Astron. Astrophys.*, **403**, 1175–1184 (2003).
6. P. Palmeri, C. Mendoza, T. R. Kallman, M. A. Bautista, M. Meléndez, *Astron. Astrophys.*, **410**, 359–364 (2003).
7. C. Mendoza, T. R. Kallman, M. A. Bautista, and P. Palmeri, *Astron. Astrophys.*, **414**, 377–388 (2004).
8. M. A. Bautista, C. Mendoza, T. R. Kallman, and P. Palmeri, *Astron. Astrophys.*, **418**, 1171–1178 (2004).
9. T. R. Kallman, P. Palmeri, M. A. Bautista, C. Mendoza, and J. H. Krolik, *Astrophys. J. Suppl. Ser.*, **155**, 675–701 (2004).
10. J. F. Seely, U. Feldman, and U. I. Safronova, *Astrophys. J.*, **304**, 838–848 (1986).
11. P. Beiersdorfer, M. Bitter, S. von Goeler, and K. W. Hill, *Phys. Rev. A*, **40**, 150–157 (1989).
12. P. Beiersdorfer, T. Phillips, V. L. Jacobs, K. W. Hill, M. Bitter, S. von Goeler, and S. M. Kahn, *Astrophys. J.*, **409**, 846–859 (1993).
13. V. Decaux and P. Beiersdorfer, *Phys. Scr.*, **T47**, 80 (1993).
14. V. Decaux, P. Beiersdorfer, A. Osterheld, M. Chen, and S. M. Kahn, *Astrophys. J.*, **443**, 464–468 (1995).
15. V. Decaux, P. Beiersdorfer, S. M. Kahn, and V. L. Jacobs, *Astrophys. J.*, **482**, 1076–1084 (1997).
16. T. Shirai, J. Sugar, A. Musgrove, and W. L. Wiese, *J. Phys. Chem. Ref. Data*, Monograph 8 (2000).
17. M. H. Chen, *At. Data Nucl. Data Tables*, **34**, 301–356 (1986).
18. T. Kato, U. I. Safronova, A. S. Shlyaptseva, M. Cornille, J. Dubau, and J. Nilsen, *At. Data Nucl. Data Tables*, **67**, 225–329 (1997).
19. V. L. Jacobs, J. Davis, B. .F. Rozsnyai, F. Balazs, and J. W. Cooper, *Phys. Rev. A*, **21**, 1917–1926 (1980).
20. V. L. Jacobs and B. .F. Rozsnyai, *Phys. Rev. A*, **34**, 216–226 (1986).
21. K. A. Berrington, L. Quigley, and H. L Zhang, *J. Phys. B*, **30**, 5409–5417 (1997).
22. D. W. Donnelly, K. L. Bell, M. P. Scott, and F. P. Keenan, *Astrophys. J.*, **531**, 1168–1172 (2000).
23. K. A. Berrington and C. P. Ballance, *J. Phys. B*, **34**, 2697–2705 (2001).
24. C. P. Ballance, N. R. Badnell, and K. A. Berrington, *J. Phys. B*, **34**, 3287–3300 (2001).
25. A. D. Whiteford, N. R. Badnell, C. P. Ballance, S. D. Loch, M. G. O'Mullane, and H. P. Summers, *J. Phys. B*, **35**, 3729–3740 (2002).
26. N. R. Badnell, *J. Phys. B*, **30**, 1–11 (1997).
27. R. D. Cowan, *The Theory of Atomic Structure and Spectra*, University of California Press, Berkeley, 1981.
28. P. G. Burke and K. A. Berrington, *Atomic and Molecular Processes: an R-matrix Approach*, IOP Publishing Ltd, Bristol, 1993.
29. T. W. Gorczyca and N. R. Badnell, *J. Phys. B*, **33**, 2511–2523 (2000).
30. G. Holzer, M. Fritsch, and M. Deutsch, *Phys. Rev. A*, **56**, 4554–4568 (1997).
31. P.Palmeri, P. Quinet, N. Zitane, N. Vaeck, *J. Phys. B*, **34**, 4125–4139 (2001).
32. T. R. Kallman and M. A. Bautista, Astrophys. J. Suppl. Ser., **133**, 221–253 (2001).
33. A. R. P. Rau, in *Electron and Photon Interactions with Atoms*, ed. H. Kleinpoppen and M. R. C. McDowell, Plenum, New York, 1976, 141.
34. N. Grevesse and A. Sauval, *Space Sci. Rev.*, **85**, 161–174 (1998).
35. URL `http://heasarc.gsfc.nasa.gov/topbase/topbase.html`
36. URL `http://heasarc.gsfc.nasa.gov/docs/software/xstar/xstar.html`

Dr. Martin Elvis displaying a satellite canoli from the reception.

Laboratory Studies of the Fe K-shell Emission

P. Beiersdorfer

Lawrence Livermore National Laboratory, Livermore, CA 94550

Abstract.
An overview is given of measurements of the Fe K-shell spectrum from low-density laboratory sources. The measurements include wavelengths, electron-impact collision cross sections, innershell ionization phenomena, dielectronic recombination resonance strengths, charge exchange recombination, electron beam polarization effects, resonance excitation, and radiative cascade contributions. K-shell spectra have now also been obtained with microcalorimeters, including microcalorimeter arrays that are twins of the *ASTRO-E* and *ASTRO-E2* missions, which illustrate typical resolving powers and spectral capabilities.

INTRODUCTION

The importance of the Fe K-shell emission for astrophysics has been evident for many years. Data have typically been obtained with rather low-resolution solid-state detectors such as those used on *ASCA* or presently on *Chandra*. Next year's anticipated launch of *ASTRO-E2* is expected to provide Fe K-shell spectra with high-resolution ($E/\Delta E \approx 1000$). This will enable spectroscopy of the Fe K lines and their use as diagnostics of such source parameters as temperature, ionization equilibrium, red shift, and potentially electron density.

In order to analyze Fe K-shell data, spectral models must be both accurate and complete. Laboratory data provide tests for both. The Fe K-shell spectrum has been studied in low-density laboratory sources in much detail, i.e., on tokamaks and the Livermore electron beam ion traps. Laboratory data now include accurate line lists and identifications, electron-ion collision cross sections, innershell ionization phenomena and cross sections, dielectronic recombination resonance strengths, spectra produced by charge exchange recombination, electron beam polarization effects, resonance excitation, and radiative cascade contributions.

A review of laboratory X-ray astrophysics measurements has recently been given [1]. This review covered many measurements of the Fe K-shell emission, including measurements of non-equilibrium phenomena in ionizing plasmas, such as those found in young supernova remnants, as well as in recombining plasma, such as those found in photo-ionized sources. In the following, we will provide an overview of laboratory measurements complementary to this review.

WAVELENGTH MEASUREMENTS

Line lists of the Fe K lines have been produced by detailed spectral measurements from various sources. Measurements using low-density sources include the Sun [2], the PLT

CP774, *X-Ray Diagnostics of Astrophysical Plasmas,* edited by Randall K. Smith

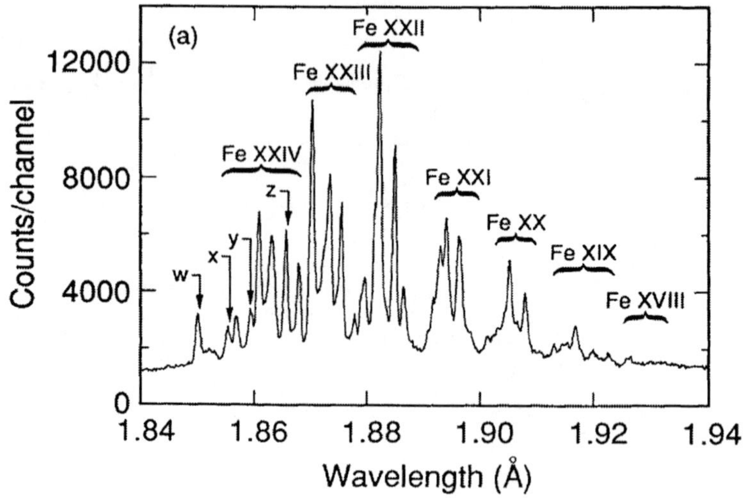

FIGURE 1. Tokamak spectrum of the Fe K-shell emission. From [3].

tokamak [3], and the Livermore EBIT-II electron beam ion trap [4].

A spectrum of the $2 \rightarrow 1$ Fe K-shell emission is shown in Fig. 1. The spectrum was recorded with a crystal spectrometer on the PLT tokamak with ≤ 1.0 keV electron temperature. At these temperatures, the spectrum is dominated by lines from Fe XXII and Fe XXIII, and lines from charge states as low as Fe XVII and as high as Fe XXV are seen.

Essentially all strong lines in the spectrum have been identified. However, there is some uncertainty concerning blends. Moreover, the emission from the intermediate charge states of iron is sensitive to the electron density for $n_e \geq 10^{11}$ cm^{-3}, which complicates the analysis [3, 5, 6].

Lines have also been studied for the higher members of the Fe K-shell Rydberg series. As an example we show in Fig. 2 crystal-spectrometer data of the $3 \rightarrow 1$ Fe K emission recorded on the TFTR tokamak.

LINE STRENGTHS

Excitation processes of the Fe K-shell emission comprise a complex mix, as described in more detail by Jacobs *et al.* in these proceedings. A typical modeling spectrum of the Fe $3 \rightarrow 1$ transitions is shown in Fig. 2. The model spectrum was produced with the HULLAC package of atomic codes [8] and generally reproduces well the observed emission.

A notable difference between the tokamak spectrum and the HULLAC model can be seen in the size of the $1s3p\ ^3P_1 \rightarrow 1s^2\ ^1S_0$ intercombination line in Fe XXV (labeled **1** in Fig. 2) relative to the $1s3p\ ^1P_1 \rightarrow 1s^2\ ^1S_0$ resonance line (labeled **2** in Fig. 2). The

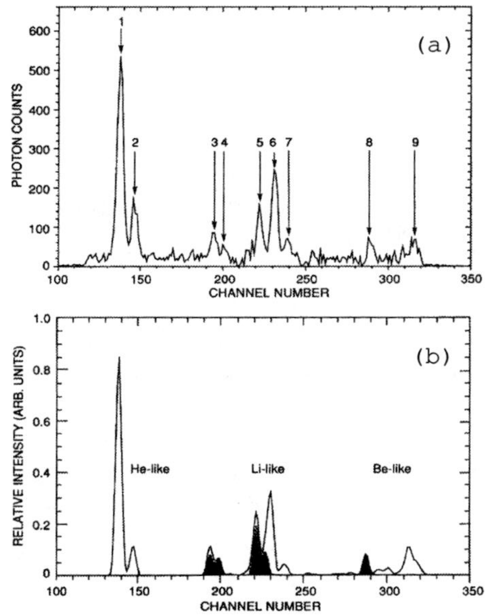

FIGURE 2. Spectrum of the Fe 3 → 1 K-shell emission: (a) Tokamak data; (b) collisional radiative modelling calculations. From [7].

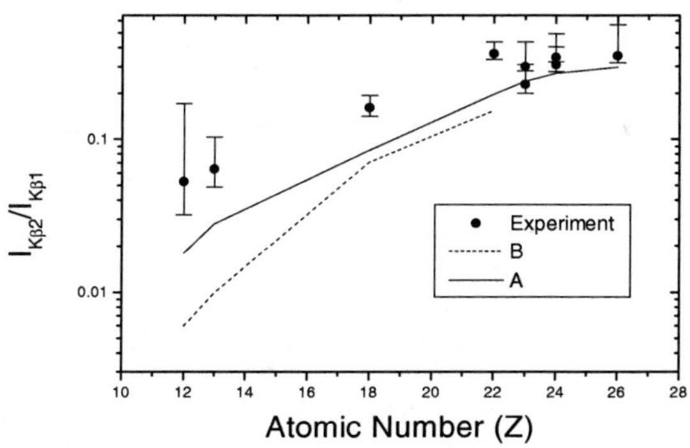

FIGURE 3. Comparison of the ratio of the intensities of the 3 → 1 intercombination and resonance lines in various heliumlike ions with two predictions. From [7].

HULLAC model predicts a ratio that is about a factor of two less than observed [7, 9].

This ratio was investigated further using the ion traps at Livermore [10]. The measurements found a systematic underestimation of this ratio by modeling calculations, as illustrated in Fig. 3. In part this discrepancy was attributed to inaccuracies in the

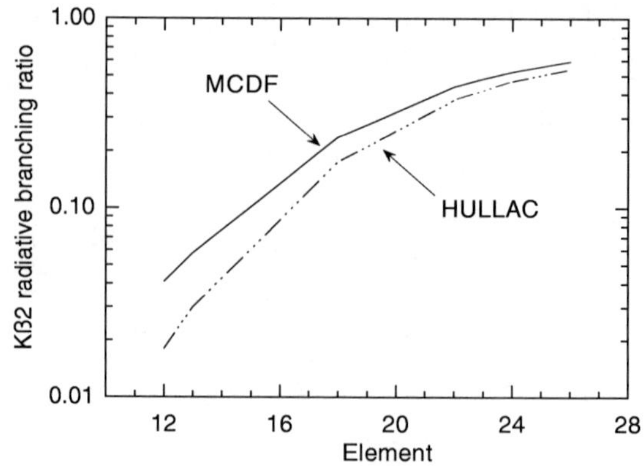

FIGURE 4. Comparison of the radiative branching ratio affecting the $3 \rightarrow 1$ intercombination line emanating from the $1s3p\ ^3P_1$ upper level in various heliumlike ions. From [10].

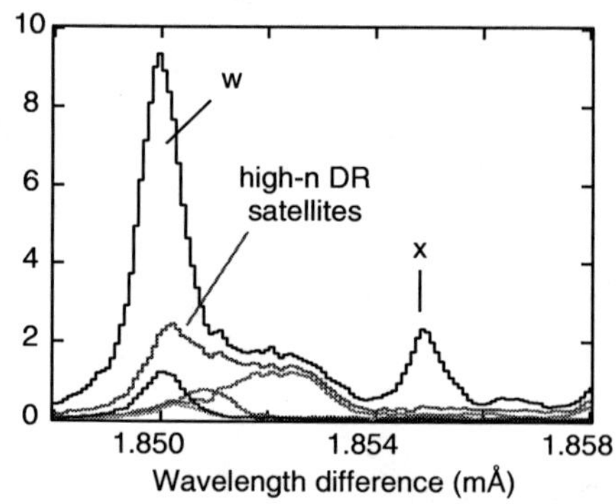

FIGURE 5. Contributions from high-n satellites ($n \geq 3$) to the Fe XXV resonance line w. From [13].

calculation of the radiative branching ratio that affects the strength of the intercombination line. The radiative branching ratio calculated by HULLAC is systematically smaller than those obtained with a multi-configuration Dirac-Fock (MCDF) calculation [10], as shown in Fig. 4. However, even correcting for the branching ratio, calculations do not yet match the laboratory ratios, as shown by the solid line in Fig. 3.

DIELECTRONIC RECOMBINATION

The K-shell Fe spectrum contains many lines produced exclusively by dielectronic recombination [11]. While many lines are well resolved, others blend with some of the strongest spectral lines. For example, dielectronic satellite transitions overlap with the Fe XXV $1s2p\ ^1P_1 \rightarrow 1s^2\ ^1S_0$ resonance line w and can significantly change its apparent intensity, especially at low temperature [12].

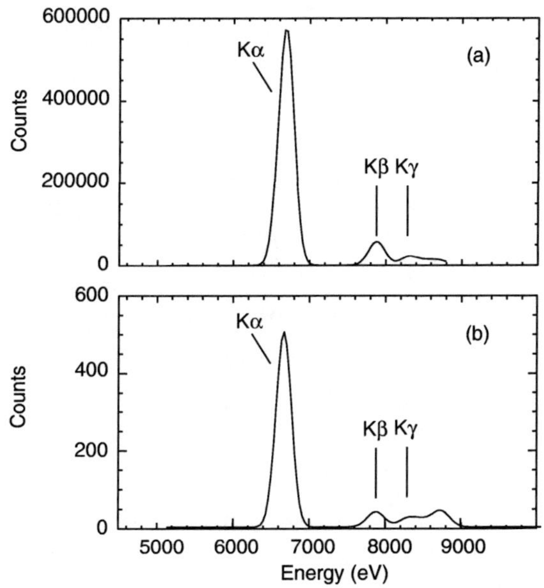

FIGURE 6. Fe XXV K-shell emission: (a) formed by electron-impact collisions; (b) formed by charge exchange of neutral gases with Fe^{25+} ions. From [15].

Using the EBIT-II electron beam ion trap at Livermore the emission from dielectronic satellite lines $1s2pn\ell \rightarrow 1s^2n\ell$ with $n \geq 3$ in the vicinity of line w were measured using a high-resolution crystal spectrometer [13], as shown in Fig. 5. The measured resonance strengths agree with calculations at the 10–20 % level [1].

CHARGE EXCHANGE

Charge exchange between neutral gases and highly charged ions is a very effective process. Charge exchange cross sections are on the order of 10^{-15} cm^2, which is a value many orders of magnitude larger than those for electron-impact excitation — Fe K-shell excitation cross sections are on the order of 10^{-22} cm^2 [14]. As a result, charge exchange may dominate the line formation processes in those cases where neutral species are present, even at a very low fraction.

K-shell X-ray production via charge exchange has been studied using the Livermore electron beam ion traps [16, 17]. A recent spectrum [15] of the K-shell emission from Fe XXV produced in the reaction

$$Fe^{25+} + A \rightarrow Fe^{24+*} + A^{+} \tag{1}$$

is shown in Fig. 6. Here A represents a neutral species, N_2 in the present case. The emission is compared with the emission by electron-impact excitation. The two spectra, taken with a low-resolution solid-state detector, look very similar. However, the centroid

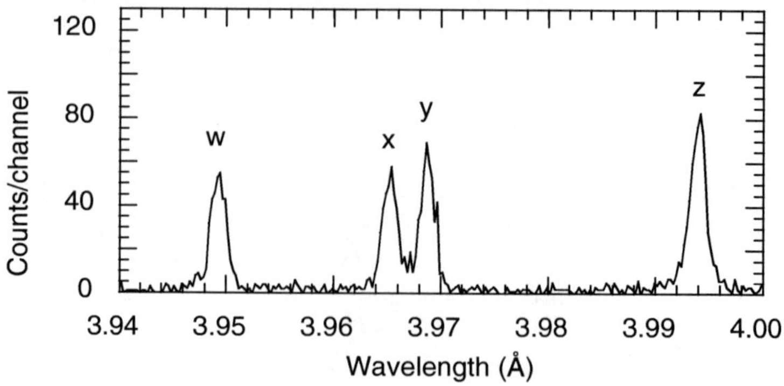

FIGURE 7. Spectrum of the Ar XVII K-shell emission produced by charge exchange in the reaction $Ar^{17+} + H \rightarrow Ar^{16+*} + H^{+}$. From [18].

of the $2 \rightarrow 1$ emission produced by charge exchange is shifted by more than 20 eV from that produced by electron impact excitation, indicating that the spectral details have changed markedly [15].

In order to investigate the spectral details of charge exchange produced line emission, high-resolution crystal spectrometer measurements were made of the K-shell spectrum of Ar XVII on the NSTX tokamak [18]. As intimated by the Fe XXV measurements, the relative intensity of the resonance line *w*, intercombination lines *x* and *y*, and forbidden line *z* in the Ar XVII is very different than expected from spectra recorded from collisional plasmas. In particular, the resonance line *w* is weaker than the other three lines. This is reminiscent of the K-shell Fe XXV excited by radiative recombination, i.e., by the capture of a free electron by an Fe^{25+} ion [1].

MICROCALORIMETER MEASUREMENTS

The use of microcalorimeters for X-ray spectroscopy opens a new era of astrophysics. The first microcalorimeter measurement of the Fe K emission was carried out on the Livermore EBIT-II electron beam ion trap using a Livermore microcalorimeter [19]. That measurement achieved a resolution of about 20–25 eV. The best spectrum of the measurement series, selected for the highest apparent resolution, is shown in Fig. 8(a). Upon relocation to the Smithsonian Astrophysical Observatory and further development, the instrument was used to record an Fe XXV spectrum on the electron beam ion trap at NIST [22, 21] (see Fig. 8(c)). A resolution of 6 eV was now attributed to the instrument [22]. Precision microcalorimeter instruments have been developed at the Goddard Space Flight Center as part of the *ASTRO-E* and *ASTRO-E2* programs. A measurement of the Fe XXV K lines performed with a twin detector of the *ASTRO-E* mission is shown in Fig. 8(b). This measurement was carried out on the Livermore EBIT-I electron beam ion trap [20]. A resolution of 11 eV was attributed to this spectrum. A unique combination of long-term gain stability and high effective area enabled the high statistics measurement shown.

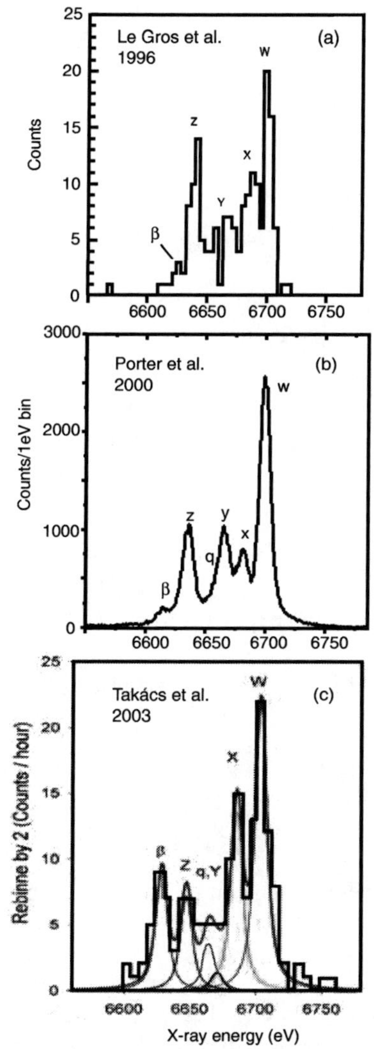

FIGURE 8. Spectrum of the Fe K-shell emission recorded with a microcalorimeter: (a) single-pixel Livermore calorimeter (from [19]); (b) 32-pixel GSFC calorimeter (from [20]); (c) single-pixel SAO calorimeter (reprinted from [21], "Astrophysics and spectroscopy with microcalorimeters on an electron beam ion trap", with permission from Elsevier.)

Microcalorimeters represent new X-ray instrumentation, and the data they produce have to measure up to decades of data produced by crystal spectrometers (which generally still have higher resolving powers than calorimeters). In this context it is interesting to note that the intercombination line x is larger than y in (a) and (c). The effect is grossly exaggerated in (c), where x is the second strongest line in the spectrum. Such a large ratio

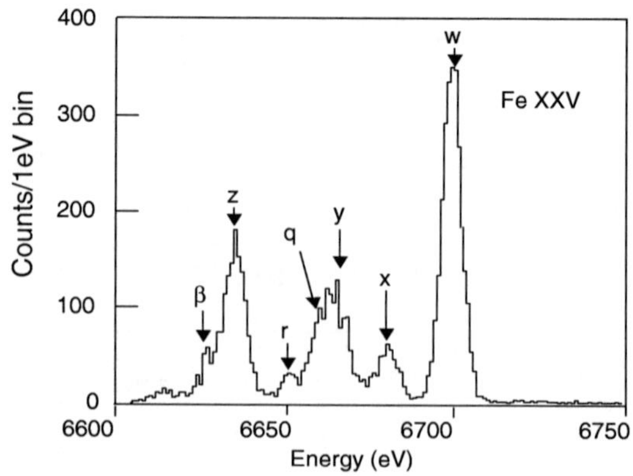

FIGURE 9. Microcalorimeter spectrum of the Fe K-shell emission from EBIT-I using the twin XRS from *ASTRO-E2*. From [25].

of x to w cannot be consistently explained by atomic physics, especially when considering that line y is nearly absent in this spectrum. By contrast, the intensity ratio of x to w and x to y in spectrum (b) is as expected from crystal spectrometer measurements, e.g. [14, 23].

It is also interesting to note that the resonance line β in Fe XXIII, situated on the low-energy side of the Fe XXV forbidden line z, is large in the measurement carried out on the NIST electron beam ion trap (cf. Fig. 8(c)). In fact, it is the third largest line in the spectrum. From the spectrum and the fits provided by Takács et al. [21] we derive an abundance ratio of 0.40 : 0.20 : 1.00 for Fe XXIII : Fe XXIV : Fe XXV. In other words, there is more Fe XXIII and Fe XXV than Fe XXIV. Such a hollow abundance ratio is atypical for an electron beam ion trap. By contrast, β is small in the measurements carried on the Livermore electron beam ion traps (cf. Fig. 8(a) and (b)). This is consistent with a charge balance that peaks at Fe XXV. Indeed, from the spectrum in (b) we derive an abundance ratio of 0.07 : 0.22 : 1.00 for Fe XXIII : Fe XXIV : Fe XXV. Such a charge balance is sensible for ion trap plasmas and in accordance with abundance measurements carried out with crystal spectrometers [24].

As illustrated by our discussion, the *ASTRO-E* microcalorimeter generally matches the expectations set by the multitude of crystal spectrometer measurements and has been found to have high spectral fidelity.

The *ASTRO-E* calorimeter has now been superseded by the *ASTRO-E2* calorimeter. Its twin X-ray Spectrometer (XRS) was installed at the Livermore EBIT-I electron beam ion trap in October 2003. A spectrum of the Fe K emission is shown in Fig. 9. The *ASTRO-E2* microcalorimeter has an improved resolution compared to the *ASTRO-E* calorimeter. This is evident when comparing the spectrum in Fig. 9 with that in Fig. 8(b). The instrument has a resolution of 6 eV [25]. Although 6 eV is the same resolution assigned to the spectrum in Fig. 8(c) by Silver *et al.*, a comparison shows that the *ASTRO-E2* microcalorimeter clearly resolves lines in a superior fashion. In Fig. 10 we show the K-shell spectrum of Ni recorded with the *ASTRO-E2* XRS on EBIT-I [26].

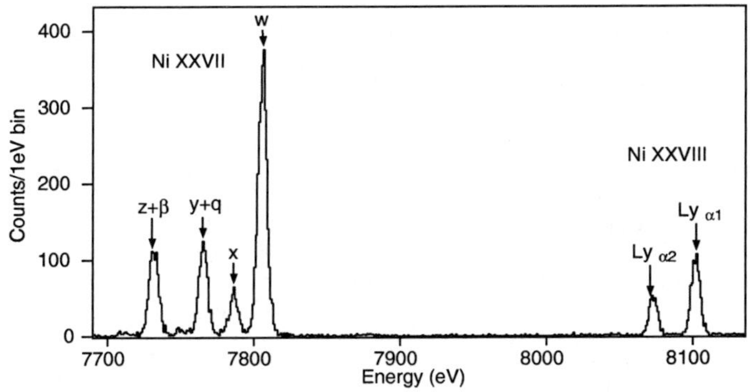

FIGURE 10. Microcalorimeter spectrum of the Ni K-shell emission from EBIT-I using the twin XRS from *ASTRO-E2*. From [26].

Like its predecessor, the *ASTRO-E2* microcalorimeter has high spectral fidelity and matches laboratory measurements carried out with proven x-ray instrumentation. In particular, the ratios of *x* and *y*, *x* and *w*, etc. agree well with those recorded with crystal spectrometers, as does the inferred abundance ratio of Fe XXIII, Fe XXIV and Fe XXV. This bodes very well for the capabilities of the *ASTRO-E2* mission.

CONCLUSIONS

The Fe K spectrum has been studied in great detail, and much laboratory data are available for anchoring spectral modeling codes. Nevertheless, it still offers surprises, and new results are expected to come forth in the near future, especially pertaining to line formation by charge exchange and microcalorimeter measurements.

ACKNOWLEDGMENTS

The author wishes to thank F. Scott Porter and Bradford Wargelin for making results available before publication. This work was performed under the auspices of the U.S. DOE by the University of California Lawrence Livermore National Laboratory under contract W-7405-Eng-48 and support by NASA's Astronomy and Physics Research and Analysis Program under contract S-06553-G.

REFERENCES

1. Beiersdorfer, P., *Ann. Rev. Astron. Astrophys.*, **412**, 343 (2003).
2. Seely, J. F., Feldman, U., and Safronova, U. I., *Astrophys. J.*, **304**, 838 (1986).
3. Beiersdorfer, P., Phillips, T., Jacobs, V. L., Hill, K. W., Bitter, M., von Goeler, S., and Kahn, S. M., *Astrophys. J.*, **409**, 846 (1993).
4. Decaux, V., Jacobs, V. L., Beiersdorfer, P., Liedahl, D. A., and Kahn, S. M., *Phys. Rev. A*, **68**, 012509 (2003).

5. J. R. Lemen, K. J. H. Phillips, R. D. Cowan, and I. P. Grant, *Astron. Astrophys.*, **135**, 313 (1984).
6. Jacobs, V. L., Doschek, G. A., Seely, J. F., and Cowan, R. D., *Phys. Rev. A*, **39**, 2411 (1989).
7. Smith, A. J., Bitter, M., Hsuan, H., Hill, K. W., von Goeler, S., Timberlake, J., Beiersdorfer, P., and Osterheld, A., *Phys. Rev. A*, **47**, 3073 (1993).
8. Bar-Shalom, A., Klapisch, M., and Oreg, J., *J. Quant. Spectrosc. Rad. Transfer*, **71**, 179 (2001).
9. Beiersdorfer, P., Osterheld, A. L., Phillips, T. W., Bitter, M., Hill, K. W., and von Goeler, S., *Phys. Rev. E*, **52**, 1980 (1995).
10. A.J. Smith, P. Beiersdorfer, K.J. Reed, A.L. Osterheld, V. Decaux, K. Widmann, and M.H. Chen, *Phys. Rev. A*, **62**, 012704 (2000).
11. Gabriel, A. H., *Mon. Not. R. Astron. Soc.*, **160**, 99 (1972).
12. Bitter, M., von Goeler, S., Hill, K. W., Horton, R., Johnson, D., Roney, W., Sauthoff, N., Silver, E., and Stodiek, W., *Phys. Rev. Lett.*, **47**, 921 (1981).
13. Beiersdorfer, P., Schneider, M. B., Bitter, M., and von Goeler, S., *Rev. Sci. Instrum.*, **63**, 5029 (1992).
14. Wong, K. L., Beiersdorfer, P., Reed, K. J., and Vogel, D. A., *Phys. Rev. A*, **51**, 1214 (1995).
15. B. J. Wargelin et al., *in preparation.*
16. Beiersdorfer, P., Olson, R. E., Brown, G. V., Chen, H., Harris, C. L., Neill, P. A., Schweikhard, L., Utter, S. B., and Widmann, K., *Phys. Rev. Lett.*, **85**, 5090 (2000).
17. Beiersdorfer, P., Lisse, C. M., Olson, R. E., Brown, G. V., and Chen, H., *Astrophys. J. (Lett.)*, **549**, L147 (2001).
18. Beiersdorfer, P., Bitter, M., Marion, M., and Olson, R. E., *in preparation.*
19. Le Gros, M., Silver, E., Beiersdorfer, P., López-Urrutia, J. C., Widmann, K., and Kahn, S. M., *in Electron Beam Ion Trap Annual Report 1995, University of California Lawrence Livermore National Laboratory Report UCRL-ID-124429 (Livermore, CA), p.22* (1996).
20. Porter, F. S., Audley, M. D., Beiersdorfer, P., Boyce, K. R., Brekosky, R. P., Brown, G. V., Gendreau, K. C., Gygax, J., Kahn, S., Kelley, R. L., Stahle, C. K., and Szymkowiak, A. E., *Proc. SPIE*, **4140**, 407 (2000).
21. Takács, E., Silver, E., Laming, J. M., Gillaspy, J. D., Schnopper, H., Brickhouse, N., Barbera, M., Mantraga, M., Ratliff, L. P., Tawara, H., Makónyi, K., Madden, N., Landis, D., Beeman, J., and Jaller, E. E., *Nucl. Instrum. Methods Phys. Res. B*, **205**, 144 (2003).
22. Silver, E., Schnopper, H., Bandler, S., Brickhouse, N., Murray, S., Barbera, M., Takács, E., Gillaspy, J. D., Porto, J. V., Kink, I., Ratliff, P. R., Hudson, L., Laming, J. M., Madden, N., Landis, D., Beeman, J., Haller, E. E., and Schuch, R., *in Photonic, Electronic & Atomic Collisions, ed. by J. Burgdörfer and J. S Cohen and S. Datz and C. R. Vane (Rinton Press, Princeton), p. 312* (2003).
23. Bitter, M., von Goeler, S., Horton, R., Goldman, M., Hill, K., Sauthoff, N. R., and Stodiek, W., *Phys. Rev. Lett.*, **42**, 304 (1979).
24. Wong, K., Beiersdorfer, P., Marrs, R. E., Penetrante, B. M., Reed, K. J., Scofield, J. H., Vogel, D. A., and Zasadzinski, R., *Nucl. Instrum. Methods*, **B72**, 234 (1992).
25. F. S. Porter et al., *in preparation.*
26. Porter, F. S., Brown, G. V., Boyce, K. R., Kelley, R. L., Kilbourne, C. A., Beiersdorfer, P., Chen, H., Terracol, S., Kahn, S. M., and Szymkowiak, A. E., *Rev. Sci. Instrum.*, **75**, 3772 (2004).

Compton scattering in fluorescent Fe K line profiles

A. Różańska* and J. Madej†

*Copernicus Astronomical Center, Bartycka 18, 00-716 Warsaw, Poland (agata@camk.edu.pl)
†Astronomical Observatory, University of Warsaw, Al. Ujazdowskie 4, 00-478 Warsaw, Poland (jm@astrouw.edu.pl)

Abstract. We discuss the recent progress in modeling of X-ray irradiated atmospheres, where radiative transfer of both external radiation and self radiation of affected gas are computed. We present model atmospheres in both radiative and hydrostatic equilibrium, which include precise treatment of multiple Compton scattering process. Outgoing continuum spectra of our models demonstrate the effects of Compton redistribution of hard X-rays to lower energies. These spectra include also detailed profiles of fluorescent iron K_α lines, which are broadened and distorted by Compton scattering. Model spectra exhibit distinct Compton shoulders from the red side of the line with equivalent widths of the order of 50 eV.

Keywords: radiative transfer – X-rays – scattering: atmospheres – X-rays: stars – galaxies:quasars
PACS: 32.30;32.70;97.80;98.54

INTRODUCTION

We investigated the influence of external X-ray irradiation on the soft X-ray spectrum emitted by the plane parallel atmosphere parameterized by effective temperature and constant gravity. We took into account effects of Compton scattering on free electrons and the generation of fluorescent Fe K_α line. Both effects, the redistribution of external irradiation, and the emission in Fe fluorescence line were observed by X-ray satellite experiments in many classes of astrophysical objects. The latter include galactic black hole candidates (GBHC), active galactic nuclei (AGN), and various classes of interacting binary stars.

In many of these objects both effects are generated in atmospheres of accretion disks. We believe that the stellar effective temperatures and surface gravity assumed here can represent a single ring on an accretion disk in AGN. In near future we shall expand our research on the formation of fluorescent lines into atmospheres where energy generation via viscosity is included in the equation of radiative equilibrium [1], and non-LTE equation of state is taken into account [1], [2].

THE EQUATION OF TRANSFER

In this paper we aim to construct emission profiles of the fluorescent K_α line of iron and put them into a model atmosphere, in order to compute the equivalent widths and demonstrate the influence of Compton scattering on the line. Full non-LTE treatment of

CP774, *X-Ray Diagnostics of Astrophysical Plasmas,* edited by Randall K. Smith

various sublevels in iron ions and the construction of series of fluorescent Fe K_β and Fe L lines is beyond the scope of our paper.

In case of a static atmosphere in plane parallel geometry the equation of radiative transfer is

$$\mu \frac{\partial I_\nu}{\rho \partial z} = j_\nu - (\kappa_\nu + \sigma_\nu) I_\nu , \tag{1}$$

where ρ and z denote density of gas and geometrical depth, respectively. Eq. 1 contains the emission coefficient j_ν, which is the sum of thermal and scattering emission of both intrinsic and external radiation plus the fluorescent line emission coefficient, E_{fl}, taken for 1 gram of matter [3]:

$$j_\nu = j_\nu^{th} + j_\nu^{sc} + E_{fl}^{\alpha} \varphi_\nu^{\alpha} . \tag{2}$$

Variable φ_ν^α denotes the joint profile of the fluorescent line doublet K_{α_1} and K_{α_2} normalized to unity. For Fe K_α doublet, the integrated emissivity is:

$$E_{fl}^{\alpha} = 0.34\, h\nu_0 \int\limits_{\nu_{tr}}^{\infty} \frac{\alpha_\nu}{h\nu} (J_\nu + U_\nu)\, d\nu , \tag{3}$$

where ν_0 corresponds to the central energy of the line equal to 6.400 keV. The threshold frequency ν_{tr} is equivalent to 7.11 keV, and α_ν denotes absorption coefficient from K shell of low ionized iron. Natural widths of the K_α components were taken from [4]. Integrated emissivities of K_{α_1} and K_{α_2} lines are in the ratio 1:2. The coefficient 0.34 denotes the fluorescent yield for the K_α transition of neutral iron. Variable U_ν denotes the mean intensity of external illumination I_ν^{ext}, directly penetrating the atmosphere to the actual point. Monochromatic optical depth scale is defined as $d\tau_\nu = -(\kappa_\nu + \sigma_\nu)\rho dz$.

$$U_\nu(\tau_\nu) = \frac{1}{4\pi} \int\limits_{\Omega} I_\nu^{ext}(\omega) \exp(-\tau_\nu/\mu_\omega)\, d\omega . \tag{4}$$

The last term in the Eq. 2 represents the extreme non-LTE contribution of a fluorescent line to the emission coefficient.

Thermal emissivity j_ν^{th} in our code contains numerous contributions from bound-free transitions of hydrogen, both ions of helium, and all the existing ions of iron, and free-free transitions from all positive ions. In case of LTE the emissivity $j_\nu^{th} = \kappa_\nu B_\nu$ for each transition.

Scattering emissivity j_ν^{sc} results from Compton scattering of photons which could have different frequency and the direction of propagation before scattering. The exact expression of j_ν^{sc} was extensively discussed in our earlier papers (e.g. [5], [6]), here we briefly present its final form. Compton scattering emissivity j_ν^{sc} is expressed by auxiliary functions Φ_1 and Φ_2, which are defined by:

$$\Phi_1(\nu,\nu') = \left(1 + \frac{c^2}{2h\nu'^3} J_{\nu'}\right) \Phi(\nu,\nu') , \tag{5}$$

$$\Phi_2(\nu,\nu') = \left[1 + \frac{c^2}{2h\nu^3}(J_\nu + U_\nu)\right] \left(\frac{\nu}{\nu'}\right)^3 \times \times \exp\left(-\frac{h(\nu-\nu')}{kT}\right) \Phi(\nu,\nu') , \tag{6}$$

where $\Phi(\nu,\nu')$ denotes the zeroth angular moment (angular average) of real angle-dependent Compton scattering redistribution function ϕ, taken from [7]:

$$\Phi(\nu,\nu') = \oint_{\omega'} \phi(\nu,\nu',\vec{n}\cdot\vec{n}') \frac{d\omega'}{4\pi}. \tag{7}$$

Both functions Φ_1 and Φ_2 include stimulated scattering correction factors, and fulfill strictly the detailed balancing condition in thermodynamic equilibrium.

The equation of transfer with all the relevant emission coefficients is given by:

$$\begin{aligned}\mu \frac{\partial I_\nu}{\partial \tau_\nu} &= I_\nu - \varepsilon_\nu B_\nu - (1-\varepsilon_\nu) J_\nu - \frac{E^\alpha_{fl}\varphi^\alpha_\nu}{\kappa_\nu + \sigma_\nu} + (1-\varepsilon_\nu) J_\nu \int_0^\infty \Phi_1(\nu,\nu')\, d\nu' \\ &- (1-\varepsilon_\nu) \int_0^\infty (J_{\nu'} + U_{\nu'}) \Phi_2(\nu,\nu')\, d\nu',\end{aligned} \tag{8}$$

where the dimensionless monochromatic absorption $\varepsilon_\nu = \kappa_\nu/(\kappa_\nu + \sigma_\nu)$.

Our equation of transfer works correctly in cases of both large and small energy exchange between X-ray photon and free electron at the time of scattering. The above equation makes no reference to the diffusion approximation of Compton scattering (Kompaneets equation). We can obtain accurate numerical solutions also in cases when the initial photon energy before or after scattering exceeds the electron rest mass, 511 keV.

More details on the following transformations of the equation of transfer, and discussion of both upper and lower boundary conditions were presented in [5].

The equilibrium conditions and linearization procedure

Static model atmospheres are subject to the constraints of radiative and hydrostatic equilibrium. We accept the equation of hydrostatic equilibrium in the following form (see [5]):

$$\frac{dP_g}{d\tau} = \frac{g}{(\kappa+\sigma)_{std}} - \frac{4\pi}{c} \int_0^\infty \eta_\nu (H_\nu - V_\nu)\, d\nu, \tag{9}$$

where variable V_ν denotes the flux of external irradiation

$$V_\nu(\tau_\nu) = \frac{1}{4\pi} \int_\Omega I^{ext}_\nu(\omega)\, \mu_\omega \exp(-\tau_\nu/\mu_\omega)\, d\omega. \tag{10}$$

The condition of radiative equilibrium at τ_ν requires that

$$\int_0^\infty H_\nu(\tau_\nu)\, d\nu - \int_0^\infty V_\nu(\tau_\nu)\, d\nu = \frac{\sigma_R T^4_{\rm eff}}{4\pi}. \tag{11}$$

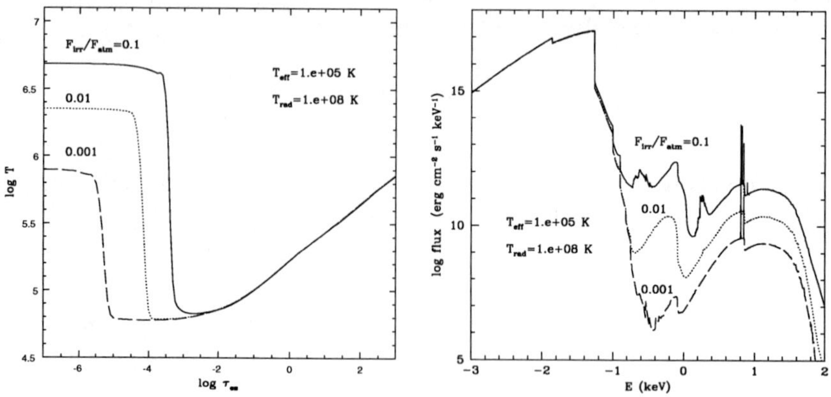

FIGURE 1. Left panel: run of temperature T vs. Thomson scattering optical depth τ_{es} for the series of models with $T_{irr} = 10^8$ K. Right panel: energy spectra of outgoing radiation for the same series of irradiated models.

The equation of radiative equilibrium, Eq. 11, is expanded in the Taylor series of the first order, with respect to unknown temperature stratification ([5]; [8]; [9]). The actual version of our code uses linearization of four functions: Planck function, Compton redistribution functions, Φ_1 and Φ_2, and the profile of fluorescent line φ_ν,

RESULTS

We assumed that the spectrum of the external irradiation is given by a diluted blackbody of an arbitrary temperature T_{irr}

$$F_{irr} = w\,\sigma_R T_{irr}^4, \tag{12}$$

where w denotes the usual dilution factor [3]. We attempted to obtain as big F_{irr} as possible, changing the value of w. The scheme of full radiative transfer calculations of an atmosphere in hydrostatic equilibrium does not solve the problem of thermal instabilities due to X-ray illumination. Therefore, if X-ray flux is too high and eventual instabilities occur then we are not able to continue calculations. The present version of the code allowed us to reach $F_{irr} = 0.1F_{atm}$, where F_{atm} denotes the bolometric flux emitted by a star, $F_{atm} = \sigma_R T_{\rm eff}^4$.

We have computed two series of models corresponding to an atmosphere of $T_{\rm eff} = 10^5$ K and $\log g = 8.0$ (a hot white dwarf). The model consists of hydrogen, helium and iron with solar number abundances. We assumed irradiation by very hard X-rays with $T_{irr} = 10^8$ K and 10^9 K, and various dilution factors.

Fig. 1 (left panel) presents run of temperature T vs. Thomson scattering optical depth τ_{es} for a series of models with $T_{irr} = 10^8$ K computed with account of Compton scattering. One can see that external hard X-rays can heat the outermost atmosphere by 1.5 orders of magnitude. The heated layer always has very well defined boundaries. Right panel displays the energy spectra of outgoing radiation from illuminated atmospheres.

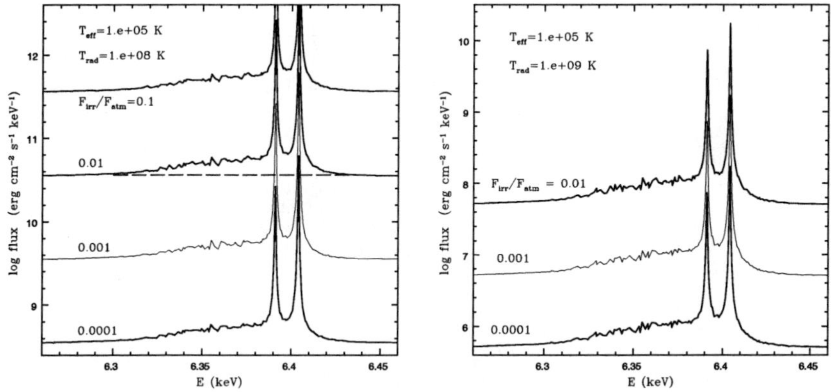

FIGURE 2. The series of fluorescent iron K_α line profiles (flux spectra) computed either for $T_{irr} = 10^8$ K (left panel) or 10^9 K (right panel), and for various dilution factors w rescaled to F_{irr}/F_{atm}. All K_α lines present Compton shoulders and the main doublet line structure. corresponding here to neutral iron.

One can note, that irradiating hard X-ray flux is entirely shifted to EUV and soft X-rays by Compton scattering on relatively cold thermal electrons.

Figs. 2-3 show our theoretical profiles of the iron K_α fluorescent line, which is affected by Compton scattering in the irradiated atmosphere. We present both flux (Fig. 2) and intensity spectra (Fig. 3). The line profile in the outgoing radiation consists of the sharp emission doublet peak and a clearly visible bump from the low-energy side of the peak. The latter is the so-called Compton shoulder of the fluorescent line. and results from a single or multiple Compton downscattering of line radiation.

The equivalent width of the Compton shoulder itself amounts to 50 eV in our models. Note, that we kept central energies of both components of the emission doublet, K_{α_1} and K_{α_2}, and other line parameters to the values corresponding to neutral iron.

CONCLUSIONS

In this paper we demonstrate the impact of Compton scattering on free electrons on the shapes and equivalent widths of the iron fluorescent line K_α. The line originates in an atmosphere of hot star with the effective temperature $T_{\rm eff} = 10^5$ K irradiated by external X-rays of thermal spectral distribution. The characteristic temperature of diluted external radiation is either $T_{rad} = 10^8$ or 10^9 K.

An atmosphere irradiated by hard X-rays develops an outermost very hot zone of temperature T approaching few times 10^6 K. The transition region between the hot zone and deeper colder layers is very narrow, which means that the temperature changes dramatically by about two orders of magnitude there.

Our code reproduces Compton redistribution of very hard external X-rays to lower photon energies both in an outgoing continuum radiation and spectral lines.

We demonstrate by theoretical model calculations that the fluorescent emission line

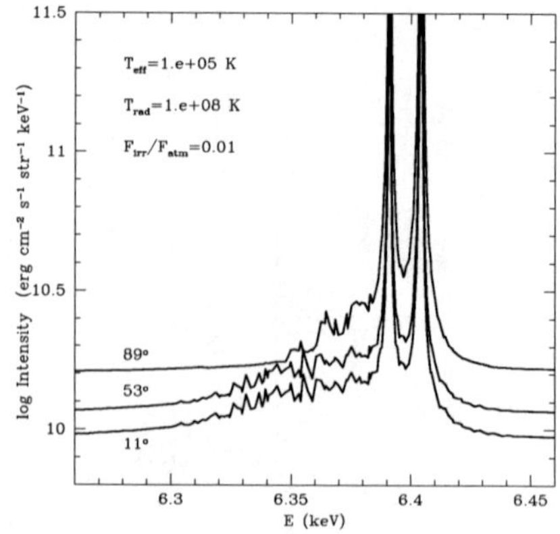

FIGURE 3. Intensity spectra for a sample iron K_α line. Line profiles in the Figure are labeled by the value of angle between a given direction and the vertical axis. The appearance of the Compton shoulder clearly depends on that angle.

emerging from the atmosphere exhibits both narrow emission peak and broad Compton shoulder from the red side of the peak. Equivalent widths of our theoretical Compton shoulder are of the order of 50 eV.

Resolved Compton shoulder of iron K_α line was recently reported by [10] in Circinus Galaxy, [11] in X-ray spectra of NGC 3783, and [12] in case of GX 301-2.

ACKNOWLEDGMENTS

We acknowledge support of the grants No. 2 P03D 021 22 and No. 2P03D 003 22 from the Polish Committee for Scientific Research.

REFERENCES

1. Hubeny, I., Blaes, O., Krolik, J.H., Agol, E., *ApJ*, **559**, 680-702, 2001
2. Różańska, A., Dumont A.-M., Czerny B., Collin S., *MNRAS*, **332**, 799-813, 2002
3. Mihalas, D., *Stellar Atmospheres*, W.H. Freeman and Co., San Francisco 1978
4. Krause, M. O.; Oliver, J. H. *Journal of Physical and Chemical Reference Data*, **8**, 329-338, 1979
5. Madej, J., Różańska, A., *MNRAS*, **347**, 1266-1272, 2004
6. Madej, J., Joss, P.C., Różańska, A., *ApJ*, **602**, 904-912, 2004
7. Guilbert, P.W., *MNRAS*, **197**, 451-460, 1981
8. Madej, J., *ApJ*, **376**, 161-176, 1991
9. Madej, J., Różańska, A., *A&A*, **363**, 1055-1064, 2000
10. Bianchi, S., Matt, G., Fiore, F., Iwasawa, K., Nicastro, G., *A&A*, **396**, 793-799, 2002
11. Kaspi, S., Brandt, W.N., George, I.M. et al., *ApJ*, **574**, 643-662, 2002
12. Watanabe, S., Sako, M., Ishida, M. et al., *ApJ*, **597**, L37-L40, 2003

Resonant Auger Destruction and Iron Kα Spectra in Compact X-ray Sources

Duane A. Liedahl

Department of Physics and Advanced Technologies, Lawrence Livermore National Laboratory, 7000 East Ave, L-473, Livermore, CA 94550, USA

Abstract. We examine the effects of resonant Auger destruction in modifying the intensities and flux distributions of Kα spectra from iron L-shell ions. Applications include X-ray irradiated stellar winds in X-ray binaries and accretion disk atmospheres. Using detailed atomic models, we find that resonant Auger destruction is selective, in that only a subset of the emitted Kα lines is highly attenuated. We also show that that the local excitation conditions can have a dramatic effect on the Kα emissivity spectrum.

Keywords: X-ray spectroscopy, accretion disks, stellar winds
PACS: 95.30.Dr, 95.30.Jx, 97.60.Lf, 97.80.Jp

INTRODUCTION

In high-temperature, highly ionized plasmas, iron K lines, arising typically from *1s-np* transitions, are driven by collisional excitation of H-like and He-like ions. Except under transient ionizing conditions [1] [2], iron K lines from lower charge states contribute relatively subtle modifications to the spectrum [3] [4]. By contrast, in accretion-powered X-ray sources, such as active galactic nuclei (AGN) and X-ray binaries (XRBs), reprocessing of a hard X-ray continuum in relatively cool matter (~10^5 – 10^6 K) can generate intense iron K radiation from more neutral iron species [5].

The reprocessing mechanism, for any species with more than two bound electrons, begins with photoionization of a *1s* electron by a photon of energy ε, sending an element A in charge state i-1 to charge state i, where i is in a quasi-bound state that is coupled to the continuum, denoted by the double asterisk below.

$$\varepsilon + A_{i-1} \rightarrow A_i^{**} + e^- \quad \text{(K-shell photoionization)}$$

$$A_i^{**} \rightarrow \begin{cases} A_{i+1} + e^- & \text{(autoionization)} \\ A_i^{*} + \varepsilon_K & \text{(K photon emission)} \end{cases}$$

Autoionization – referred to as the *Auger effect* for *1s* vacancy states – is a possible decay route for the *1s*-hole state. The reaction products are an Auger electron, i.e., an electron with a kinetic energy that is characteristic of the atomic energy level structure, and an ion in charge state $i+1$, where, for the problem at hand, we assume

CP774, *X-Ray Diagnostics of Astrophysical Plasmas*, edited by Randall K. Smith

that $i+1$ is left in its ground state. Spontaneous radiative decay, the other possible decay route, can stabilize the ion against autoionization. Stabilizing transitions are typically of the parity-changing type *np-1s,* where $n = 2$ for L-shell ions (Kα), and where $n = 2$ or 3 for M-shell ions (Kα or Kβ, respectively). Although the final atomic state is stabilized by a K transition, the final state may be excited, which we denote by the single asterisk.

The two-step process – inner-shell photoionization followed by K line emission – is called *fluorescence*. The *K fluorescence yield,* which we denote by Y_K, is the quotient of the rate at which K lines are generated and the rate at which K-shell holes are produced [6]. Once a K-shell hole is produced, Y_K is, for our purposes, determined by the competition between radiative decay and Auger decay. Since electric dipole radiative rates scale as Z^4 for $\Delta n > 0$ transitions, and since Auger rates are roughly constant with Z, Y_K increases rapidly with Z [7]. Coupled with the fact that iron has the highest cosmic abundance among intermediate-Z elements, iron K fluorescence lines tend to be the most commonly observed in astrophysics.

Historically, we have had to satisfy ourselves with the fact that spectrometers have, at best, resolved the iron Kα lines into three features: the Fe XXVI Lyα doublet at 6.97 keV, the Fe XXV blend, near 6.7 keV, and a blend of emission from a composite of lines from Fe II-Fe XVIII near 6.4 keV [8]. Readers may note the lack of reference to the ions Fe XIX-XXIV, the subjects of this paper.

For convenience, we refer to the 6.4 keV blend as arising in "near-neutral" material. The blending of lines from near-neutral material [9] [10] can be understood as follows: proceeding through the M shell, successive changes in the screening of the atomic potential experienced by *2p* electrons are small compared to the line energy. Even as we move into the L shell, the successive stripping of M-shell electrons dictates the energy separations. For example, comparing the Kα lines (*1s-2p*) from Fe XVII (Ne-like) and Fe XVIII (F-like), the upper configurations are $1s2s^2 2p^6 3s$ and $1s2s^2 2p^6$, respectively. The small difference in line energy results from the small screening of the nuclear potential by the *3s* electron in the Fe XVII *1s*-hole state. In fact, the Fe XVII and Fe XVIII Kα centroids are separated by only 5 eV. Thus even Kα lines from Fe XVII and Fe XVIII can be thought of as part of the 6.4 keV blend and, in a practical sense, as arising in near-neutral material. By contrast, the Kα energy centroids for Fe XVIII and Fe XIX are separated by 34 eV. This is easily understood by noting that the upper configuration for the Fe XIX lines is $1s2s^2 2p^5$, where the absence of a *2p* electron has a relatively large effect on the differential screening when compared to the neighboring Fe XVIII. Thus O-like Kα lines are the first to break away from the pack of blended lines from the near-neutral ions. Energy separations of roughly this magnitude persist through the L-shell sequence.

Part of the motivation for this work is the scheduled deployment in 2005 of the XRS calorimeter aboard *Astro-E2*, with a resolution of approximately 6 eV in the iron Kα spectral band [11], which is capable of resolving much of the Kα structure contributed by the L-shell ions. The lack of spectroscopic data with high resolution in this band has undoubtedly played a part in dissuading extensive calculations of iron K spectra. This situation is improving, and detailed calculations of the atomic physics processes associated with iron K spectral formation, using state-of-the-art computer codes, are now becoming available [12] [13].

We have calculated several atomic models in order to search for plasma diagnostics involving iron K spectra. Thus far, the search has focused on the L-shell species, which is a follow-on to similar work on Si Kα spectra (Liedahl, et al., in prep.), which was motivated by an observation of the high-mass X-ray binary Vela X-1 with the gratings aboard *Chandra*. We find that the part of the Si Kα line complex formerly attributed to neutral or near-neutral species [14] is actually made up of features corresponding to Si III – Si XII. The Kα spectra of L-shell ions provide unique diagnostic information on plasmas that exist over intermediate ranges of ionization. It is likely that observations with *Astro-E2* will show that, for some sources, the iron L shell contributes to the overall iron K complex.

Fluorescence from a given ion entails a conversion of part of the radiation field above the corresponding photoelectric edge into discrete emission and Auger electrons. Thus the problem of calculating the fluorescence spectrum from an astrophysical source requires us to consider continuum radiation transfer. Having generated K lines, it makes sense to ask whether or not these lines can escape the fluorescing medium and make their way to an observer. In this paper, we focus on the contributions from discrete opacity to the suppression of photon escape, which involves a process known as resonant Auger destruction, discussed below.

Calculations of atomic structure, radiative transition rates, and autoionization rates and were performed using the HULLAC package [15]. When working with L-shell ions, the stepwise set of processes – photoionization, then autoionization – couples three charge states [16]. In more complex cases involving iron ions of low charge, Auger cascades can couple as many as ten charge states [17]. We will not be concerned with calculations of the charge state distribution here, since we need only consider ``per-atom" processes.

We assembled six atomic models, one for each of the ions Fe XIX – Fe XXIV. Each model consists of three charge states. For example, in calculating the spectrum of ion *i*, the model also includes charge states *i-1* and *i+1*, where *i-1* serves as a source, through photoionization, of excited states in *i*, and where *i+1* serves as a sink, through autoionization, for these excited states. The treatment of level population kinetics is somewhat involved. Owing to space limitations, a full discussion will be presented elsewhere.

RESONANT AUGER DESTRUCTION

As with any other photon, discrete opacity can, in principle, suppress emission of fluorescence lines. There is a special circumstance in effect in this case, however: the upper level is autoionizing, which implies a large destruction probability per scatter. The following two-step reaction illustrates the process.

$$\varepsilon_K + A_i^* \rightarrow A_i^{**} \quad \text{(resonant absorption)}$$

$$A_i^{**} \rightarrow A_{i+1} + e^- \quad \text{(Auger decay - photon destruction)}$$

By picking out three of the key words from the above two reactions, one can form the name of the process – *resonant Auger destruction* [18] [19], which we abbreviate RAD. Since the probability of Auger decay is usually high, multiple scattering leading to escape, whether by a random walk in space, or by a migration in frequency to the wings of the line profile, is extremely unlikely.

Clearly, the effect of RAD in attenuating a given line depends on the optical depth in the ion A_i*. Consider two examples in order to make a distinction: Fe XXI (an L-shell ion) and Fe XV (an M-shell ion). For Fe XXI, subsequent to the creation of a *1s* hole, we have the following set of reactions:

$$1s2s^2 2p^3 \rightarrow 1s^2 2s^2 2p^2 + \varepsilon_K$$
$$\varepsilon_K + 1s^2 2s^2 2p^2 \rightarrow 1s2s^2 2p^3$$

The second reaction is the inverse of the first, except that it involves a distinct ion, since the K photon will travel, on average, a distance corresponding to unity line optical depth from its creation site. The probability of resonant absorption in this case is relatively high, since most of the Fe XXI population will reside in levels belonging to the ground configuration. For Fe XV, we have

$$1s2s^2 2p^6 3s^2 3p \rightarrow 1s^2 2s^2 2p^5 3s^2 3p + \varepsilon_K$$
$$\varepsilon_K + 1s^2 2s^2 2p^5 3s^2 3p \rightarrow 1s2s^2 2p^6 3s^2 3p$$

The probability of resonant absorption in this case is relatively low, since the lifetime of $1s^2 2s^2 2p^5 3s^2 3p$ is exceedingly short, and only a miniscule fraction of the Fe XV population will lie in this configuration. Therefore, as a rule of thumb, RAD does not affect Kα emission from M-shell ions. Note that in the previous example, if Kβ were produced, RAD could efficiently suppress it.

Suppose we adopt a simple picture of a fluorescing zone of an accretion disk in which the zone is represented by a semi-infinite slab that is irradiated from above. Then, based on the preceding discussion, we might expect that only an upper layer of the zone corresponding roughly to unity line optical depth will contribute to Kα fluorescence from L-shell ions. Below that depth, escape is prohibited by resonant Auger destruction. Compare this to the case of fluorescence from M-shell ions, where the fluorescing region for near-neutrals has an extent corresponding to roughly one continuum optical depth. Crudely speaking, therefore, the ratio of the emergent Kα line flux from L-shell ions to the Kα line flux from near-neutrals is of order $\sigma_{cont}/\sigma_{line}$, which is of order 10^{-3}. This provides a plausible basis for simply setting to zero the Kα line emissivities for L-shell ions, since such low flux levels would imply unobservably small equivalent widths. However, this argument is incomplete, as we discuss below.

Selectivity

In assessing the effect of RAD, the essential ingredient, apart from the atomic physics, is the line optical depth. For a given transition between an upper level u and a lower level l, the line optical depth, in terms of the hydrogen column density N_H, the elemental abundance A_z, and the ionic fraction F_{ion}, can be written

$$\tau_{lu} = N_H A_z F_{ion}\, p_l \frac{\pi e^2}{mc} f_{lu}\ \phi(\nu)$$

Where f_{lu} is the absorption oscillator strength, $\phi(\nu)$ is the line profile function, and p_l is the fractional population density of the ion. Therefore, we must solve for the level populations p_l for the lower levels of each Kα transition. To justify this additional effort, we show in Fig. 1 that Kα transitions often terminate on excited states. Under many conditions, for example, in a stellar wind in a HMXB, we expect that the level populations for excited states will usually be much less than for ground, and that plasmas that are optically thick to lines corresponding to ground state transitions need not be optically thick for others. Thus RAD selectively affects the Kα spectrum.

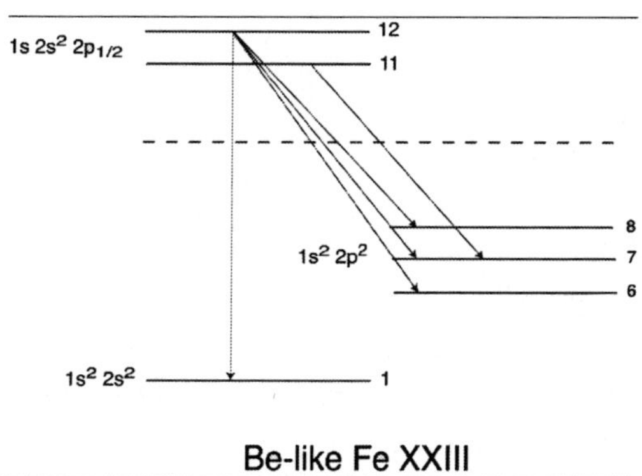

FIGURE 1. Partial Grotrian diagram for Be-like Fe XXIII, showing that Kα transitions often terminate on excited levels. The numerals indicate the level index for the model ion. The dashed line indicates the first ionization limit.

Effect of Varying Level Population Distributions

To model iron Kα spectra in a disk environment, we add a diluted blackbody radiation field to represent the effect of an underlying accretion disk in driving populations among low-lying excited levels. Once we determine the set p_l, we can calculate the line optical depths. Implicit in this discussion is the presence of a hard X-ray field that determines the charge state distribution but does not substantially affect the bound-state level populations. For consistency, we must also calculate the populations for levels in the adjacent ion, the pre-ionization charge state [20]. We note that L-shell ions immersed in plasmas such that the electron densities are near or above the critical densities for n=2 intrashell collisional transitions will be similarly affected. For the purpose of illustration, we work with the Be-like/Li-like pair, and determine the Fe XXIV spectrum. We calculate two cases. First, the spectrum is calculated in the limit of low density and no blackbody field. The result is shown in the left panel of Fig. 2. Second, a blackbody with kT_{rad} = 80 eV and a dilution factor of 0.5 is added. The result is shown in the right panel of Fig. 2. Here, we are illustrating only the effects of varying the population kinetics – neither example accounts for RAD. The spectrum with the EUV field is characterized by higher energy lines and by brighter Kα emission.

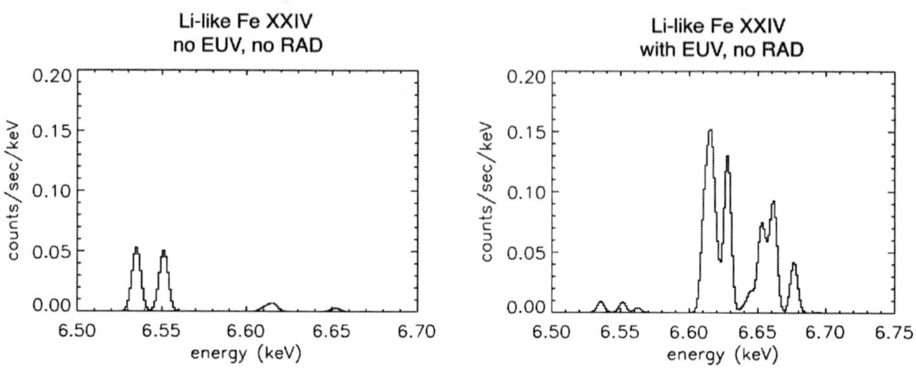

FIGURE 2. The left panel shows the Fe XXIV spectrum appropriate to a low-UV environment, in which nearly all charge states exist in their ground levels. The right panel shows the spectrum in the case that the Be-like level population distribution is modified by the presence of a diluted blackbody radiation field. Both examples are "zero-D," i.e., no RAD is applied.

The drastic change induced by application of the Planckian field can be understood by referring to Fig. 3. First, if we assume that nearly all the Be-like population is in the ground state, then the only Li-like autoionizing level produced by K-shell photoionization is $1s2s^2$, which may decay radiatively, as shown by the wavy lines. The line fluorescence yield of each line, i.e., the radiative branching ratio including the total autoionization rate as a sink, is relatively low, as indicated in the figure. When the level populations are re-calculated in the presence of an EUV field, the ground state becomes depleted, while low-lying levels in $1s^2 2s2p$ gain population. Subsequent K-shell photoionization from this latter configuration populates Li-like $1s2s2p$, which decays by $1s$-$2p$, with a much higher yield.

Next, for a range of ionic column densities appropriate to accretion disk atmospheres [21], we apply RAD to these two emissivity spectra. Using the Voigt function, assuming that the line is the only source of opacity, we calculate an escape probability for each line that is angle-averaged and frequency-averaged. The resulting spectra for the "EUV on" case are shown in Fig. 4, starting with a column density of zero, which represents the spectrum when RAD is neglected. The four panels illustrate the selectivity of the process, and show that, even with moderately large absorbing column densities, RAD does not entirely quench the Kα spectrum.

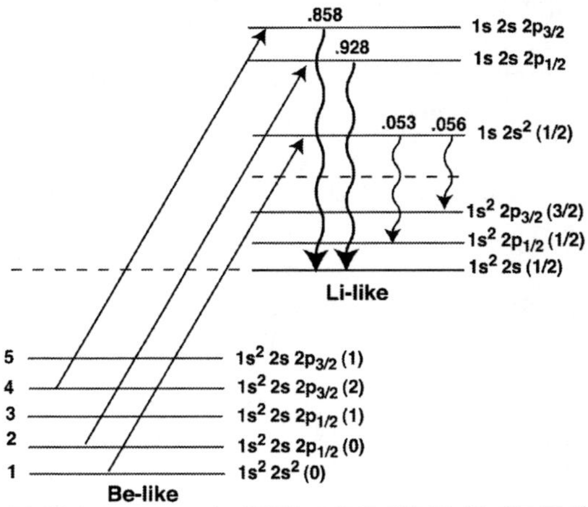

FIGURE 3. Schematic of Be-like/Li-like iron showing effect on Fe XXIV Kα spectrum of enhancing populations of low-lying Be-like levels. With the Be-like population concentrated in the ground level, only the $1s2s^2$ level is driven by K-shell photoionization, which can decay radiatively as shown, though with low line fluorescence yields of about 0.05, shown in the diagram. When Be-like configurations with a *2p* orbital become substantially populated, influx to Li-like *1s2s2p* increases, which has levels with large line fluorescent yields of about 0.9.

A similar calculation was performed for the case of no EUV field. To summarize the results for Fe XXIV, we plot in Fig. 5 (upper left panel) effective fluorescent yields (the apparent fluorescence yield after accounting for RAD) against ionic column density for each of the two cases (EUV on, EUV off). As noted above, the line emissivity for the zero-D case is brighter in the presence of the EUV field than without it. This corresponds to the leftmost side of the plot, where it is seen that the yield is about five times larger than for the "no-UV" case. As the absorbing column density is increased, the effective yield drops, as expected. Note, however, that across this range of column densities, the effective yield is larger than one would calculate assuming that the Be-like population resides exclusively in the ground level. The remaining three panels in Fig. 5 show that Fe XXIV is not unique; the same qualitative behavior is found for Fe XXI, Fe XXII, and Fe XXIII. (Note that, perhaps against convention, that we are assigning the fluorescence yields to the post-ionization charge state, rather than the pre-ionization charge state.) We believe that this casts doubt on the common practice of zeroing out Kα lines from iron L-shell ions.

CONCLUSIONS

After a re-examination of resonant Auger destruction, where details of the level population distribution are treated explicitly, we find that the effective fluorescent yields for several iron L-shell ions remain high over the range of column densities expected in accretion disk atmospheres. We suggest that models of X-ray irradiated disks can and should be constrained by comparing model predictions of L-shell Kα emission with inferences drawn from observational data. The apparent absence of Kα emission from L-shell ions in accretion disk sources, if verified, for example, by *Astro-E2* observations, may point to problems with the assumptions and/or methodologies adopted for model calculations.

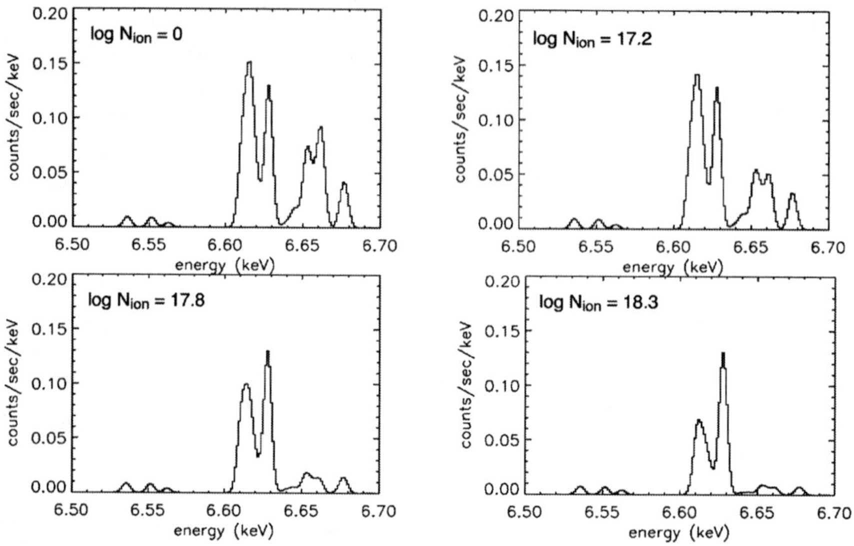

FIGURE 4. Fe XXIV spectra plotted for four ionic column densities, showing the modifications engendered by resonant Auger destruction using an escape probability method. The top-left panel shows the Fe XXIV spectrum appropriate to a zero-D calculation. The remaining three show the selective action of RAD. Model spectra are convolved with a 6 eV FWHM gaussian resolution kernel, representative of the *Astro-E2* XRS.

By contrast, we have clear evidence for Kα emission from L-shell ions in HMXBs. However, these detections are not of iron emission, but rather, silicon and sulfur emission, and we are not aware of any reports of this emission from L-shell iron. While this apparent discrepancy has not been studied in detail, it does appear to pose an interesting puzzle. The problem will be better defined by results from *Astro-E2*. In HMXBs, the RAD effect may turn out to be a useful diagnostic based on analyses of line ratios or "line" shapes, recalling that most ionic "lines" are blends. Clearly, having observational access to a class of lines from several elements that may respond differently to RAD, and that respond differently to local excitation conditions, offers an exciting opportunity to gain insight as to the composition, physical state, and dynamics of accretion flows in these systems. On the other hand, the application of specific diagnostics based on RAD in relativistic accretion disks, where the detail is blurred into a single, though possibly structured, line complex, is not likely. There, it is the energy distribution and total equivalent width that matter most. Nevertheless, these gross attributes depend on detailed modeling calculations.

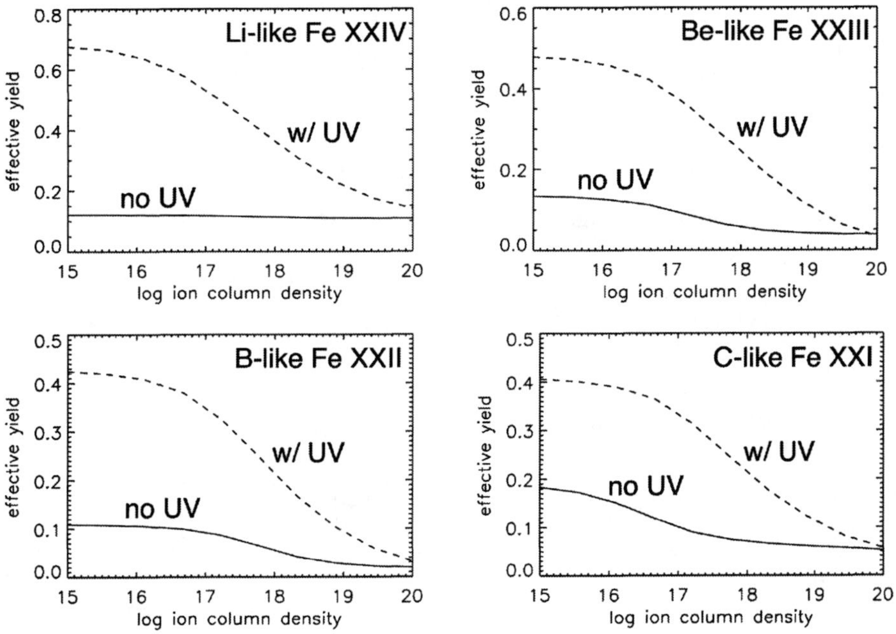

FIGURE 5. Effective fluorescent yields plotted against ionic column density for Fe XXI – Fe XXIV. Each plot shows two cases: (dotted line) level populations of low-lying states driven by 80 eV half-diluted blackbody and (dashed line) no perturbing radiation field.

We acknowledge the inadequacy of the escape probability treatment in assessing spectral modifications to the zero-D emissivity spectrum. The underlying assumption in using an escape probability formalism is that a photon, once resonantly absorbed, is destroyed via the Auger effect. This is not a horrible approximation, since the destruction probabilities are, on average, rather high. But the distribution in line fluorescent yields is broad (see Fig. 3, for example), and multiple scattering leading to escape cannot be dismissed outright. Moreover, we have ignored line splitting (see Fig. 1), where the photon energy is converted to one corresponding to a different energy level separation, such that the lower level is less populated than the lower level corresponding to the pre-absorption line, thereby increasing the probability of escape. Finally, we assumed a homogeneous slab in calculating spectral modifications for different column densities, whereas a more reliable calculation would account for density, temperature, and ionization gradients, and ultimately, the effects of velocity fields. These aspects of the problem will be taken up in a future paper, using the Monte Carlo radiation transport capability of the COMPASS code [22].

ACKNOWLEDGMENTS

The author thanks Mau Chen, Tim Kallman, and Chris Mauche for useful discussions. This work was performed under the auspices of the U.S. Department of Energy by the University of California Lawrence Livermore National Laboratory under contract No. W-7405-Eng-48.

REFERENCES

1. Mewe, R., and Schrijver, J., *Astron. Astrophys.*, **87**, 261 (1980).
2. Doschek, G.A., and Tanaka, K., *Ap. J.*, **323**, 799 (1987).
3. Doschek, G.A., Kreplin, R.W., and Feldman, U., *Ap. J.*, **233**, L157 (1979).
4. Swartz, D.A., and Sulkanen, M.E., *Ap. J.*, **417**, 487 (1993).
5. George, I.M, and Fabian, A.C., *M.N.R.A.S.*, **249**, 352 (1991).
6. Bambynek, W., et al., *Rev. Mod. Phys.*, **44**, 716.
7. Cowan, R.D., *The Theory of Atomic Structure*, Berkeley, Los Angeles, London: University of California Press, 1981, p. 589.
8. Ebisawa, K., et al., *P.A.S.J.*, **48**, 425 (1996).
9. House, L.L., *Ap. J. Suppl.*, **155**, 21 (1969).
10. Kaastra, J.S., and Mewe, R., *Astron. Astrophys. Suppl.*, **97**, 443 (1993).
11. Stahle, C.K., et al., *Nucl. Inst. and Meth. A*, **520**, 466 (2004).
12. Palmeri, P., Mendoza, C., Kallman, T.R., and Bautista, M.A., *Astron. Astrophys.*, **403**, 1175 (2003).
13. Kallman, T.R., Palmeri, P., Bautista, M., Mendoza, C., and Krolik, J.H., astro-ph/0405210v1 (2004).
14. Sako, M., Liedahl, D.A., Kahn, S.M., and Paerels, F., *Ap. J.*, **525**, 921 (1999).
15. Bar-Shalom, A., Klapisch, M., and Oreg, J., *J. Quant. Spectrosc. Radiat. Transfer*, **71**, 169, (2001).
16. Weisheit, J.C., *Ap. J.*, **190**, 735 (1974).
17. Kaastra, J.S., and Mewe, R., *Astron. Astrophys. Suppl.*, **97**, 443 (1993).
18. Ross, R.R., Fabian, A.C., and Brandt, W.N., *M.N.R.A.S.*, **278**, 1082 (1996).
19. Band, D.L., Klein, R.I., Castor, J.I., and Nash, J.K., *Ap. J.*, **362**, 90 (1990).
20. Jacobs, V.L., Doschek, G.A., Seely, J.F., and Cowan, R.D., *Phys. Rev. A*, **39**, 2411 (1989).
21. Nayakshin, S., and Kallman, T.R., *Ap. J.*, **546**, 406 (2001).
22. Mauche, C.W., Liedahl, D.A., Mathiesen, B.F., Jimenez-Garate, M.A., and Raymond, J.C., *Ap. J.*, **606**, 168 (2004).

Laboratory studies of X-ray emission from Fe L-shell transitions and their diagnostic utility

G.V. Brown*, P. Beiersdorfer*, H. Chen*, J. H. Scofield*, K. R. Boyce,†, R. L. Kelley†, C. A. Kilbourne†, F. S. Porter†, N. S. Brickhouse**, M. F. Gu‡, S. M. Kahn‡ and A. E. Szymkowiak§

**Lawrence Livermore National Laboratory, 7000 East Avenue, Livermore, CA 94550*
†NASA/Goddard Space Flight Center, Greenbelt, MD, 20771
***Harvard-Smithsonian Center for Astrophysics, 60 Garden Street, Cambridge, MA 02138*
‡Department of Physics, Stanford University, Stanford, CA 94305
§Department of Physics, Yale University, New Haven, CT 06520

Abstract. Celestial objects are often home to complex, dynamic, intriguing environments. High-resolution x-ray spectra from these sources measured by satellites such as the *Chandra*, *XMM-Newton*, the *Solar Maximum Mission*, and the soon-to-be-launched *Astro-E2* provide a means for understanding the physics governing these sources. Especially rich is the x-ray emission from L-shell transitions in highly charged iron ions. This emission is the source of a variety of diagnostics whose utility lies in the accuracy of the atomic data employed to model the x-ray spectra. The atomic data used to describe these diagnostics are generally provided by large theoretical calculations and benchmarked by laboratory data. In this paper we discuss laboratory measurements of Fe L-shell x-ray emission including wavelengths, relative and absolute excitation cross sections, and line ratios that provide diagnostics of temperature and density.

INTRODUCTION

Beginning in the mid 1960s, spectrometers carried by sounding rockets and orbiting satellites have measured high-resolution spectra from the Sun covering the 10–20 Å bandwidth [1, 2, 3, 4, 5, 6, 7]. From these measurements it was realized that the strong emission from Fe L-shell ions contains great diagnostic potential. With the successful launches of the *Chandra* and *XMM-Newton* X-ray observatories, high-resolution spectra of Fe L-shell emission are now routinely measured from a variety of extra-solar sources with the hope of taking full advantage of the diagnostics associated with the Fe L-shell ions. Reliable interpretation of spectra from solar and extra-solar sources requires accurate and complete databases of transition wavelengths and line strengths. Many improvements in spectral modeling packages used by the astrophysics community have been made such as the inclusion of large sets of recently available experimental data where available, and by utilizing experimentally benchmarked atomic theories to provide data when experimental results do not exist. In some cases, the inclusion of experimental data has made reliable interpretation of high-resolution spectra possible for the first time. Fe L-shell experimental data has been produced by several experimental facilities including storage rings (see Savin et al. these proceedings), tokamaks [8, 9], and electron beam ion traps [10]. Here we give a brief review of the laboratory astrophysics studies being conducted in the study of Fe L-shell x-ray emission at the LLNL EBIT facility.

CP774, *X-Ray Diagnostics of Astrophysical Plasmas*, edited by Randall K. Smith

WAVELENGTHS AND LINE IDENTIFICATION

Accurate line identification of observed emission is the first step towards understanding a spectrum and its source. In the past, the identification of Fe L-shell x-ray lines has proceeded by comparing calculated wavelengths and line intensities to experimental data from solar observations [4, 5, 6, 11] and emission from tokamak and laser produced plasmas [9, 12]. Problems arise because these experimental sources all emit x-rays from several different Fe ions simultaneously. This limits the number of lines that can be identified and the accuracy of the wavelengths that can be measured because lines from different charge states blend even when measured with high-resolution spectrometers. In spite of these challenges, several databases of line intensities and wavelengths have been compiled [13, 14, 15]. For these databases, calculations are used to provide wavelengths and line intensities where no experimental value exists.

In anticipation of the launch of the high-resolution instruments on board *Chandra* and *XMM-Newton* and to address problems found in the interpretation of *ASCA* data, all the line emission from Fe XVII–XXIV was measured using the LLNL electron beam ion traps (EBITs) [16, 17]. Because the electron beam on an EBIT is mono-energetic, we are able to produce and trap a single ionic species of Fe and measure its x-ray emission. This eliminates the spectral confusion among lines from different charge states. The wavelength of each transition is determined by comparing to well-known wavelengths of x-ray emission from hydrogenic and helium-like ions. Figure 1 shows the Rydberg series of helium-like O VII and two iron spectra. One iron spectrum was measured above the ionization potential of Fe^{16+} and one below. Above the ionization potential, the Fe XVIII line F1 is produced [18]. Thus, by changing the electron beam energy from above and below ionization potentials of different iron ions, line emission can be positively identified with a specific charge state. Once the charge state and wavelength are known, identification of each x-ray line was completed by comparing measured spectrum to a spectrum calculated by HULLAC. The wavelengths measured at the LLNL EBIT have been implemented in the Atomic Plasma Emission Code (APEC) making reliable line identifications of many lines measured by *Chandra* and *XMM-Newton* possible, and are being used to benchmark more recent calculations.

The wavelengths and line identifications provided by our experiments not only make possible the diagnostic utility of Fe L-shell transitions, but also increase the reliability of the diagnostics of other ionic species. For example, emission from helium-like neon, magnesium, and hydrogenic oxygen and neon also fall in the same wavelength band as the Fe L-shell lines. Without properly accounting for each iron line, the temperature and density diagnostics of each of these ions is precluded. Figure 2 shows a *Chandra* spectrum of Capella emphasizing the wavelength region around the Ne IX triplet. This spectrum shows that the Fe L-shell emission must be accounted for when using the Ne IX emission as a diagnostic.

THE RELATIVE INTENSITY OF THE FE XVII LINES 3C AND 3D

Two of the more distinct lines found in spectra of Fe L-shell emission are the resonance and intercombination lines from Fe XVII, known as 3C and 3D respectively. These lines

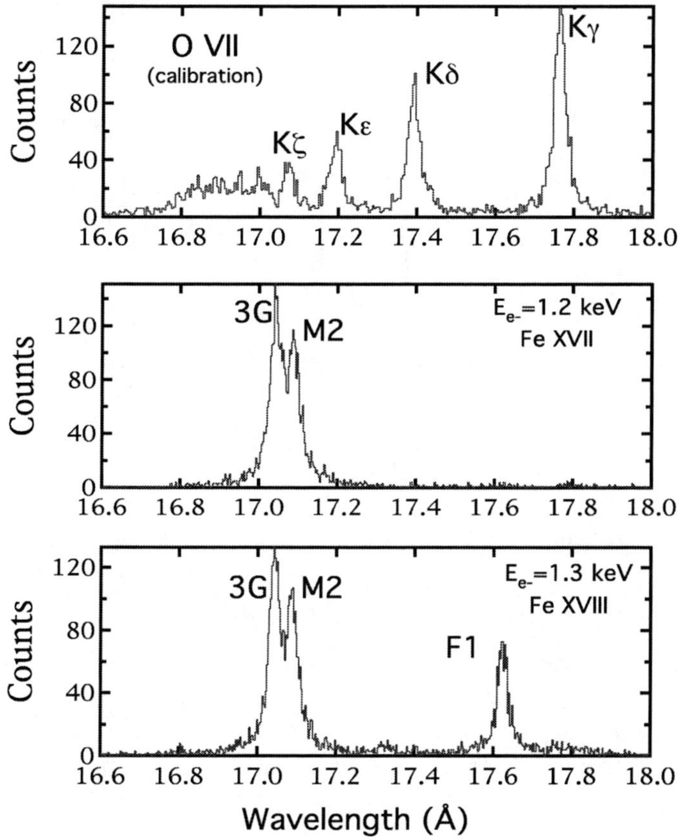

FIGURE 1. (top) Spectrum of helium-like O^{6+} x-ray emission used for calibration. (middle) Fe XVII spectrum taken at an electron beam energy of 1.2 keV. (bottom) Fe XVIII spectrum measured at an electron beam energy of 1.3 keV. The wavelength scale is determined by the line emission from O^{6+} and the lines are identified of the Fe L-shell emission are determined by comparing spectrum measured at different beam energies. These spectra demonstrate the utility of a mono-energetic electron beam in the identification of x-ray line emission. This figure is from [18].

are a result of the transitions $2p^5 3d_{3/2}\ ^1P_1$ and $2p^5 3d_{5/2}\ ^3D_1$ to the $2p^6\ ^1S_0$ ground state, and are located at 15.01 and 15.25 Å, respectively. They have been observed in observations of the Solar corona [1, 7, 3] and from other extra-solar sources [20, 21]. Owing to their prevalence, these lines have been a topic of studies for several decades. Comparison of the observed relative intensity I_{3C}/I_{3D} and the relative intensity calculated

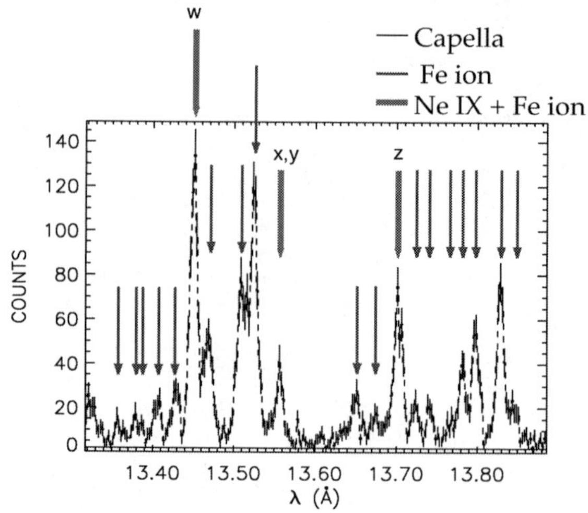

FIGURE 2. Spectrum of Capella depicting the region around the helium-like Ne^{8+} triplet. Several emission lines from Fe L-shell ions are present in this wavelength region and some blend with the lines w, x, y, and z of Ne^{8+}. Without accounting for all the iron emission, the accuracy of the temperature and density diagnostics associated with Ne IX is limited. This data can be found in [19].

by modern atomic codes shows that the calculation generally predict values significantly higher than observed. To account for the difference, some concluded that the emission from 3C is resonantly scattered out of the line of sight, reducing the relative intensity of I_{3C}/I_{3D} compared to the optically thin case. The reduced ratio I_{3C}/I_{3D} was then used to infer the optical depth of line 3C and the column density of the source [22]. However, because of the large variation among theoretical values for the optically thin ratio, correctly estimating the optical depth of 3C or the column density of the source was not possible.

To benchmark the relative intensity of I_{3C}/I_{3D} in the optically thin limit, it was measured at the LLNL EBIT [16]. Our results show the ratio to be 3 while most modern atomic codes predict values that are more than 20 % larger. Using our measured ratio and the methods given in Waljeski et al. [22], we derived the optical depth of line 3C as a function of I_{3C}/I_{3D}. Figure 3 shows a plot of τ(3C) versus I_{3C}/I_{3D} and compares the curve derived from our experimental ratio with different calculated optically thin ratios.

Although the measured value of $I_{3C}/I_{3D} = 3.04 \pm 0.12$ is significantly lower than many of the calculations, it is still not as low as some values measured in the corona of the Sun. To address the cause for these low ratios, we used the LLNL EBIT to measure the line ratio under conditions where a significant amount of Na-like Fe^{15+} was present at the same time as neon-like Fe^{16+} [23]. Figure 4 shows three spectra measured at the LLNL EBIT each with different relative abundances of Fe^{15+}/Fe^{16+}. In the top spectrum, the relative abundance is about 15%, in the middle spectrum 50 %, and in the bottom spectrum no Fe^{15+} is present. Our results show the presence of several Fe^{15+} x-ray lines in this wavelength band and that one of these lines coincides with the line 3D.

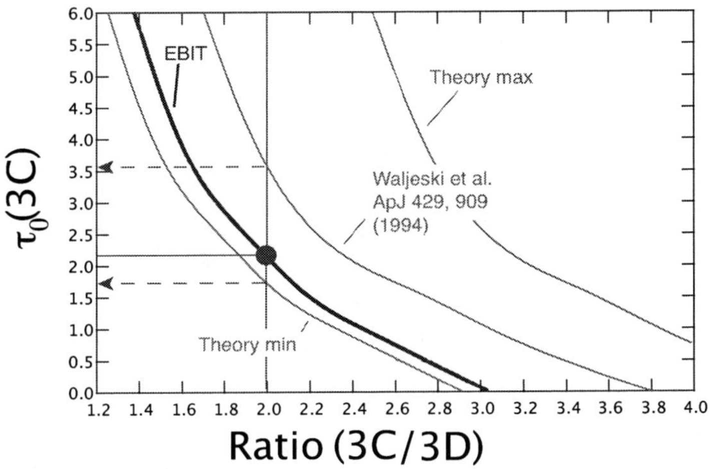

FIGURE 3. Plot of optical depth of the resonance line 3C versus intensity ratio 3C/3D. This figure compares the curve obtained using the experimental ratio to those obtained from different calculations.

Thus, when a significant amount of Fe^{15+} is present the ratio of I_{3C}/I_{3D} is reduced. The influence of Fe^{15+} innershell satellites was also pointed out by Behar, Cottam & Kahn [20] in their analysis of the high resolution spectrum of Capella provided by *Chandra*, and these line were tentatively identified by Phillips et al. in their analysis of solar spectra [7]. Also, the lower ratio measured from tokamak measurements provided by Beiersdorfer et al. [8] is consistent with our result.

The fact that an Fe^{15+} innershell satellite line increases the apparent strength of line 3D means I_{3C}/I_{3D} depends on the relative abundance of Fe^{15+} to Fe^{16+} and is, therefore, a function of temperature. Using the abundance versus temperature tables of Arnaud & Raymond [24], we plot the ratio I_{3C}/I_{3D} as a function of temperature in figure 5. Using the observed values of I_{3C}/I_{3D}, we also infer the temperature of Capella and of flaring and non-flaring active regions of the Sun.

EXCITATION CROSS SECTIONS

Although our laboratory measurements have provided a benchmark for the optically thin ratio of 3C/3D and demonstrated that the ratio can be reduced by an Fe $^{15+}$ innershell satellite that coincides with line 3D, the measurements of the relative intensity cannot point out if calculations predict the incorrect cross section for 3C, 3D, or both. To uncover the source of the problem, we have used the X-ray microcalorimeter built by the Goddard Space Flight Center/University of Wisconson X-ray microcalorimeter group and implemented at the LLNL EBIT facility [25, 26, 27] to measure the absolute cross section of both lines 3C and 3D [28]. Our method is the same as described in [29]. In brief, we take advantage of the fact the NASA/GSFC calorimeters have a

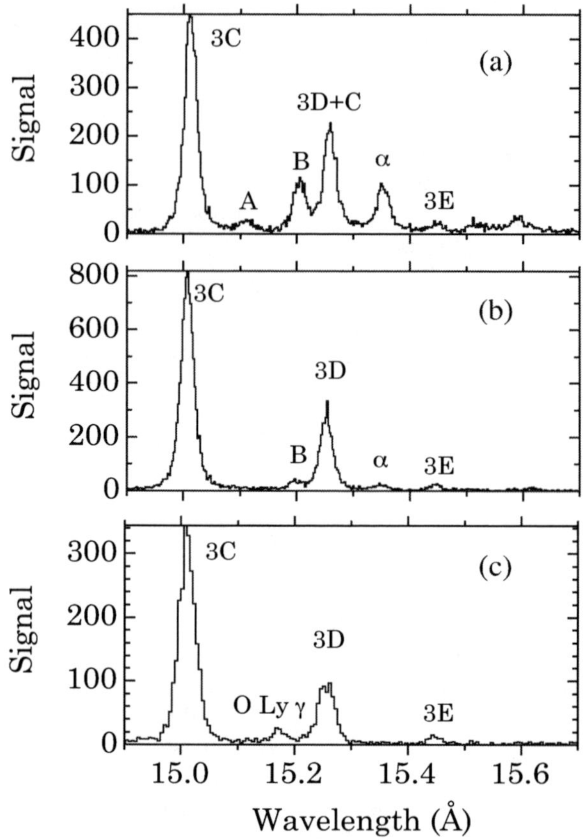

FIGURE 4. Comparison of Fe XVII spectra measured with different relative abundances of Fe^{15+} to Fe^{16+}. (a) $Fe^{15+}/Fe^{16+} \approx 1$. (b) $Fe^{15+}/Fe^{16+} \approx 0.15$. (c) $Fe^{15+}/Fe^{16+} = 0$. Lines 3C, 3D, and 3E are from Fe^{16+}, lines A, B, and C are from Fe^{15+}, and line α is from Fe^{14+}. This figure is from [23].

large bandwidth, long-time gain stability, a large collecting area, and relatively high resolution to simultaneously measure the emission from direct excitation and radiative recombination. Once both DE and RR emission are measured, we normalize our entire spectrum to the well-known cross sections of RR. In the case of Fe XVII, our results show that modern atomic theories predict cross sections for the resonance line 3C to be too large, while correctly predicting the cross section for the intercombination line 3D. This result not only provides the reason modern theories predict I_{3C}/I_{3D} to be too large, but also demonstrates the reason calculations underestimate the relative intensity of the $2p-3s$ Fe XVII lines located around 17 Å relative to line 3C [30, 31]. Our result thus reduces the source of two long standing puzzles in the study of Fe XVII x-ray emission to a single emission line.

We have also measured excitation cross sections of several other Fe L-shell transitions both at single electron impact energies [29, 32] and as a function of electron impact

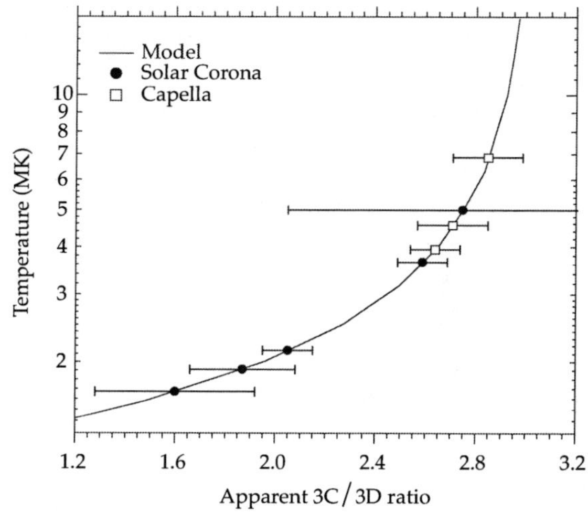

FIGURE 5. Correlation of the apparent I_{3C}/I_{3D} ratio with the electron temperature. This curve is from [23].

energy [33, 34]. Figure 6 shows the results of one of these measurements. In this case, the Fe XXIV line Li3 was measured as a function of electron energy by sweeping the beam energy linearly from an energy below the threshold for direct excitation to above threshold and recording the spectrum as function of energy. Once complete, the energy function is then absolutely calibrated by the cross section measured with the calorimeter at one or more single-electron beam energies. We note that this excitation function includes all relevant atomic processes that contribute to the line strength as a function of electron impact energy in collisional plasmas, including dielectronic recombination and resonance excitation. Therefore, this function can be integrated over any electron energy distribution and provide the proper rate coefficient and line strength.

DENSITY DIAGNOSTICS

In addition to temperature and opacity diagnostics, Fe L-shell transitions can provide a diagnostic of electron density. One example is the magnetic dipole line M2 in Fe XVII located at 17.096 Å. This transitions has a long lifetime and, in the case of high density plasmas can be collisionally de-excited. An example case where the absence of line M2 provides a lower limit on the electron density is given by Mauche et al. 2001 [35]. Recent studies also show a less well-known density diagnostic from boron-like Fe XXII [36]. In this case, in contrast to a long lived state being collisionally de-populated, a virtual ground state is populated collisionally from the true ground state. Once populated, collisional excitation from metastable ground state can significantly enhance the line strength of some x-ray lines. Figure 7 shows this density dependence in boron-like Fe XXII. Plotted is the ratio $R = (I_{3d_{5/2}\to 2p_{3/2}} + I_{3d_{3/2}\to 2p_{3/2}})/I_{3d_{3/2}\to 2p_{1/2}}$

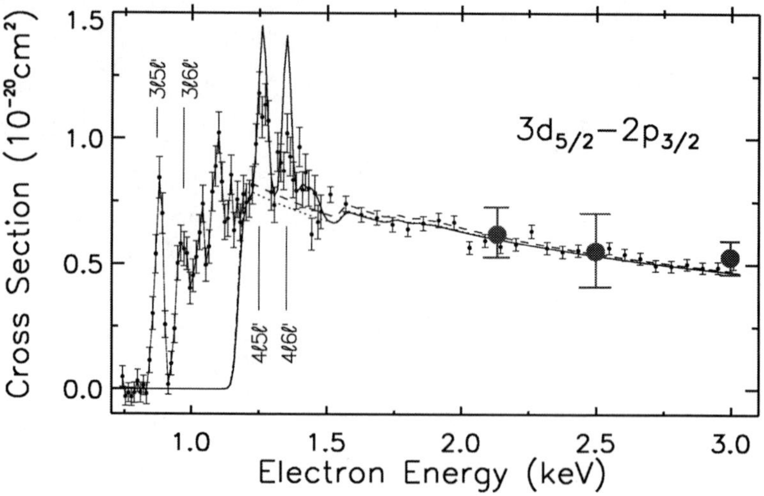

FIGURE 6. Excitation function of the Li-like Fe XXIV x-ray transitions. By measuring the line strength as a function of electron impact energy and by measuring its absolute cross sections, the excitation function of this line has been measured and absolutely calibrated in the laboratory. The large dots represent the absolute cross section measurements. This figure is from [34].

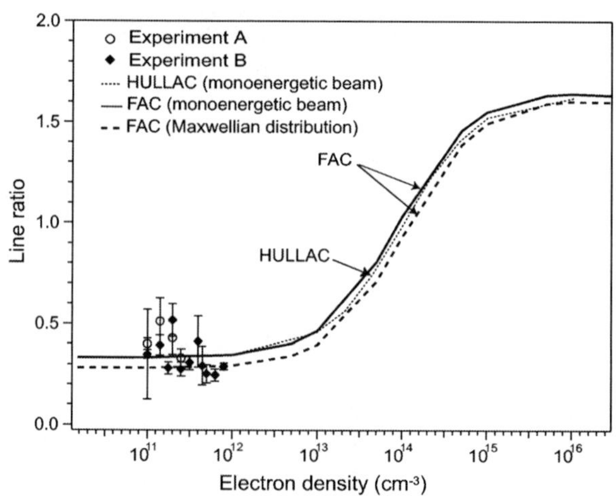

FIGURE 7. Ratio of $R = (I_{3d_{5/2}\to 2p_{3/2}} + I_{3d_{3/2}\to 2p_{3/2}})/I_{3d_{3/2}\to 2p_{1/2}}$ versus density. Two independent measurements are presented, labelled experiment A and experiment B. The experimental results are compared to calulcations from HULLAC and FAC. This figure is from [36].

versus density. These results are compared to calculations using HULLAC and FAC. The comparison shows good agreement between both theories and experiment. For a more detailed description of this result see Chen et al. [36].

SUMMARY

Laboratory astrophysics experiments in the study of Fe L-shell emission have provided accurate, complete sets of wavelengths, and several relative line intensities and absolute cross sections necessary for proper interpretation of astrophysical x-ray spectra. Future work includes measurement of the absolute excitation functions of all the significant Fe L-shell x-ray transitions, and extending our wavelength survey to below 10 Å. We have also extended our measurements to L-shell transitions from other ions where similar diagnostics exist. These include nickel [37], argon, sulfur [38, 39], and silicon. Experimental results such as the ones presented here provide quantitative error bars that can be used to estimate the accuracy of the physical parameters inferred from high-resolution spectra of astrophysical sources.

ACKNOWLEDGMENTS

Work at LLNL was completed under the auspices of the U.S. D.o.E by the University of California Lawrence Livermore National Laboratory under contract W-7405-Eng-48 and supported by NASA's Astronomy and Physics Research and Analysis Program under work order S-06553-G.

REFERENCES

1. Blake, R. L., Chubb, T. A., Friedman, H., and Unizicker, A. E., *Astrophys. J.*, **142**, 1 (1965).
2. Freeman, F. F., and Jones, B. B., *Solar Physics*, **15**, 288 (1970).
3. Parkinson, J. H., *Astron. Astrophys.*, **24**, 215 (1973).
4. Parkinson, J. H., *Solar Physics*, **42**, 183 (1975).
5. Hutcheon, R. H., Pye, F. P., and Evans, K. D., *Mon. Not. R. astr. Soc.*, **175**, 489 (1976).
6. McKenzie, D. L., Landecker, P. B., Broussard, R. M., Rugge, H. R., and Young, R. M., *Astrophys. J.*, **241**, 409 (1980).
7. Phillips, K. J. H., Leibacher, J. W., Wolfson, C. J., Parkinson, J. H., Kent, B. J., Mason, H. E., Acton, L. W., Culhane, J. L., and Gabriel, A. H., *Astrophys. J.*, **256**, 774–787 (1982).
8. Beiersdorfer, P., von Goeler, S., Bitter, M., and Thorn, D. B., *Phys. Rev. A*, **64**, 032705 (2001).
9. Von Goeler, S., Bitter, M., Cohen, S., Eames, D., Hill, K., Hillis, D., Hulse, R., Lenner, G., Manos, D., Roney, P., Roney, W., Sauthoff, N., Sesnic, S., Stodiek, W., Tenney, F., and Timberlake, J., "X-Ray Spectroscopy on Tokamaks," in *Diagnostics for Fusion Reactor Conditions*, edited by P. Scott., Commission of the European Communities, Belgium, 1982, p. 109.
10. Beiersdorfer, P., *Astron. Astrophys. Review*, **41**, 343–390 (2003).
11. McKenzie, D. L., and Landecker, P. B., *Astrophys. J.*, **254**, 309–317 (1982).
12. Boiko, V. A., Faenov, A. Y., and Pikuz, S. A., *J. Quant. Spectrosc. Radiat. Transfer*, **19**, 11–50 (1978).
13. Kelly, R. R., *J. Phys. Chem. Ref. Data*, **16**, 861–873 (1987), supplement Series.
14. Mewe, R., Gronenschild, E. H. B. M., and van den Oord, G. H. J., *Astron. Astrophys. Supp.*, **62**, 197–254 (1985).
15. Mewe, R., and Gronenschild, E. H. B. M., *Astron. Astrophys. Supp.*, **45**, 11–52 (1981).
16. Brown, G. V., Beiersdorfer, P., Kahn, S. M., Liedahl, D. A., and Widmann, K., *Astrophys. J.*, **502**, 1015–1026 (1998).
17. Brown, G. V., Beiersdorfer, P., Liedahl, D. A., Widmann, K., and amd E. J. Clothiaux, S. M. K., *Astrophys. J. Supp.*, **140**, 589–607 (2002).
18. Drake, J. J., Swartz, D. A., Beiersdorfer, P., Brown, G. V., and Kahn, S. M., *Astrophys. J.*, **521**, 839–843 (1999).
19. Ness, J., Brickhouse, N. S., Drake, J., and Huenmoerder, D. P., *Astrophys. J.*, **598**, 1277–1289 (2003).

20. Behar, E., Cottam, J., and Kahn, S. M., *Astrophys. J.*, **548**, 966–975 (2001).
21. Xu, H., Kahn, S. M., Peterson, J. R., Behar, E., Paerls, F. B. S., Mushotzky, R. F., Jernigan, J. G., and Makishima, K., *Astrophys. J.*, **579**, 600 (2002).
22. Waljeski, K., Moses, D., Dere, K., Saba, J. L. R., Web, D. F., and Zarro, D. M., *Astrophys. J.*, **429**, 909–923 (1994).
23. Brown, G. V., Beiersdorfer, P., Chen, H., Chen, M. H., and Reed, K. J., *Astrophys. J. Lett.*, **557**, L75–L78 (2001).
24. Arnaud, M., and Raymond, J., *Astrophys. J.*, **398**, 394–406 (1992).
25. Porter, F. S., Beiersdorfer, P., Boyce, K. R., Brown, G. V., Chen, H., Kelley, R. L., and Kilbourne, C. A., *Rev. Sci. Instrum.* (2004), in press.
26. Porter, F. S., Beiersdorfer, P., Boyce, K. R., Brown, G. V., Chen, H., Gendreau, K. C., Gygax, J., Kahn, S. M., Kelley, R. L., Liedahl, D. A., Stahl, C. K., Szymkowiak, A. E., and Widmann, K., p. 184 (2001).
27. Porter, F. S., Audley, M. D., Beiersdorfer, P., Boyce, K. R., Brekosky, R. P., Brown, G. V., Gendreau, K. C., Gygax, J., Kahn, S. M., Kelley, R. L., Stahle, C. K., and Szymkowiak, A. E., "Laboratory Astrophysics using a Spare XRS Microcalorimeter," in *Proceedings of the 45th annual SPIE meeting on Optical Science and Technology*, SPIE Press, 2000, p. 4140.
28. Brown, G. V., Beiersdorfer, P., Chen, H., Scofield, K., Boyce, K. R., Kelley, R. L., Kilbourne, C. A., Porter, F. S., Szymkowiak, A. E., and Kahn, S. M. (2004).
29. Chen, H., Beiersdorfer, P., Brown, G. V., Gendreau, K. C., Boyce, K. R., Kelley, R. L., Porter, F. S., k. Stahle, C., Szymkowiak, A. E., Kahn, S. M., and Scofield, J., *Astrophys. J. Lett.*, **567**, L169–L172 (2002).
30. Beiersdorfer, P., Bitter, M., and von Goeler, S., *Astrophys. J.*, **610**, 616–623 (2004).
31. Beiersdorfer, P., Behar, E., Boyce, K. R., Brown, G. V., Chen, H., gendreau, K. C., Gu, M.-F., Gygax, J., Kahn, S. M., Kelley, R. L., Porter, F. S., Stahle, C. K., and Szymkowiak, A. E., *Astrophys. J. Lett.*, **576**, L169–172 (2002).
32. Chen, H., Beiersdorfer, P., Scofield, J. H., Brown, G. V., Boyce, K. R., Kelley, R. L., Kilbourne, C. A., Porter, F. S., Gu, M. F., and Kahn, S. M., *Astrophys. J.* (2004), accepted.
33. Gu, M. F., Beiersdorfer, P., Brown, G. V., Kahn, S. M., Liedahl, D. A., Reed, K. J., and Savin, D. W., *Astrophys. J.*, **563** (2001), dec.
34. Gu, M. F., Kahn, S. M., Savin, D. W., Behar, E., Beiersdorfer, P., Brown, G. V., Liedahl, D. A., and Reed, K. J., *Astrophys. J.*, **518**, 1002 (1999).
35. Mauche, C. W., Liedahl, D. A., and Fournier, K. B., *Astrophys. J.*, **560**, 992–996 (2001).
36. Chen, H., Beiersdorfer, P., Heeter, L. A., Liedahl, D. A., Naranio-Rivera, K. L., Trabert, E., Gu, M. F., and Lepson, J. K., *Astrophys. J.*, **611**, 598 (2004).
37. Gu, M. F., Beiersdorfer, P., Brown, G. V., Chen H., Boyce, K. R., Kilbourne, C. K., Porter, F. S., Kahn, and S. M., *Astrophys. J. Lett.*, **607**, L143 (2004).
38. Lepson, J. K., Beiersdorfer, P., Behar. E., and Kahn, S. M., *Astrophys. J.*, submitted (2005).
39. Lepson, J. K., Beiersdorfer, P., Brown, G. V., Kahn, S. M., Liedahl, D. A., Mauche, C. W., and Utter, S. B. *Proceedings of the NASA Laboratory Astrophysics Workshop*, (2002).

Atomic Processes, Theories, and Data for X-Ray Astronomy

Anil K. Pradhan

The Ohio State University

Abstract. Physical processes underlying X-ray spectral formation may be influenced considerably by particle energy distribution in the source. It is found that different electron distributions — Maxwellian or Gaussian — in the plasma yield significantly different X-ray line ratios. Owing to energetically uneven occurrence of dense resonances in electron impact excitation cross sections for transitions in Fe XVII, line ratios with Maxwellian and Gaussian averaged rates may differ by up to a factor of two, thereby explaining outstanding discrepancies between two independent laboratory measurements using Electron-Beam-Ion-Traps, as well as with astrophysical observations. The Close Coupling R-matrix theory which includes resonances via an ab initio wavefunction expansion, vs. several variants of the Distorted Wave theory which does not, is discussed. The R-matrix method also enables a unified and self-consistent theoretical treatment of photoionization and (e + ion) recombination (inverse) processes, that includes both the radiative and dielectronic recombination and yields total and level-specific rate coefficients valid at all temperatures. Finally, atomic databases from the Opacity Project and Iron Project at the Center de Donnees Astronomiques de Strasbourg and the Ohio Supercomputer Center are described, including a novel on-line database to compute stellar opacities for an arbitrary mixture of elements.

INTRODUCTION

Recent advances in X-ray spectroscopy afford unique insight into most high-temperature astrophysical sources. X-ray spectra are relatively clean, and high-resolution instruments aboard *Chandra*, *XMM-Newton*, and other space observatories enable studies of a variety of objects, such as stars, active galactic nuclei, supernovae, and afterglows from from gamma-ray bursts. However, high-precision theoretical analysis of observations is faced with serious problems in (i) the understanding of some important underlying physical processes, (ii) theoretical methods to take account of those processes, and (iii) accuracy of atomic data. We address specific aspects of these issues.

This paper is divided into three separate but related topics: (i) the role of electron distribution functions and resonances in atomic processes, (ii) theoretical methods to compute cross sections for atomic processes — electron impact excitation, photoionization, and (e + ion) recombination, and (iii) large-scale databases from the Opacity Project [1] and the Iron Project [2].

ELECTRON DISTRIBUTION IN PLASMA SOURCES

The presence of resonances in the cross sections of atomic processes is well established. In particular, photoionization, electron impact excitation, and (e + ion) recombination generally entail infinite series of resonances that may dominate the effective cross sec-

CP774, *X-Ray Diagnostics of Astrophysical Plasmas,* edited by Randall K. Smith

tion and the rate coefficient (see, for example, the review by Nahar and Pradhan [3]). Generally, forbidden and intercombination transitions are most susceptible to resonant contributions since the background (non-resonant) cross section is small. The rate coefficients of these weak transitions are therefore enhanced relatively more than those for the strong dipole allowed transitions with larger background cross section (dominated by the Coulomb potential). The energy variation of the cross section due to resonances translates into temperature variation of the rate coefficient — cross section convolved over an electron distribution function.

However, resonances are distributed unevenly in energy, with widely different strengths depending on the nature of target excitations and associated threshold levels. Therefore, if the electron distribution in the source needs to be specified or known in order to obtain the correct rate coefficient, which is thus source-specific. A corollary is that the rate (and hence the line intensity) is independent of the distribution function if, and only if, resonances play no role in the excitation process.

It is reasonable to assume astrophysical sources generally have a maxwellian distribution since electron-electron collisions thermalize rapidly. On the other hand, laboratory sources may have quite different electron distributions. For example, very elaborate measurements are being made on electron-beam-ion-traps (EBIT) that are believed to have a gaussian distribution centered around a beam energy of width of tens of eV. Such a distribution would preferentially sample resonances that are strong or dense, and lead to an enhanced gaussian averaged rate coefficient in at those energies, as opposed to other energies where resonances are weak or sparse.

Electron Impact Excitation of Fe XVII

Ne-like and He-like ions are prominent constituents characterstic of high temperature environments. Particular attention has been paid to Fe XVII which has strong soft X-ray lines in the 15-17 Årange. Experimental measurements for intensity ratios of some of these lines, excited primarily by electron impact, have been carried out on electron-beam-ion-traps (EBITs) and tokamaks [4, 5, 6]. However, there are long-standing discrepancies among theoretical, experimental, and observational data for Fe XVII. We have carried out extensive collisional and radiative calculations using the BPRM method for electron impact excitation, photoionization, unified recombination, and transition probabilities [7, 8, 9, 11, 10]. It was shown that the influence of resonances is crucial in obtaining accurate excitation rates [7, 8]. Employing these datasets, a large 89-level collisional-radiative model has been constructed to compute line intensities to compare with measured and observed line ratios.

Of particular interest are three prominent X-ray transitions to the ground level $1s^22s^22p^6\ {}^1S_0$ from excited levels: 3C (λ 15.014Å) $1s^22s^22p^5[1/2]3d_{3/2}\ {}^1P^o_1$, 3D λ 15.265Å$1s^22s^22p^5[3/2]3d_{5/2}\ {}^3D^o_1$, and 3E λ 15.456Å: $1s^22s^22p^5[3/2]3d_{5/2}\ {}^3P^o_1$. The present line ratios for these three benchmark diagnostic lines 3C, 3D, and 3E at 15.014, 15.265, and 15.456 Årespectively, are in excellent agreement with two independent measurements on EBITs at the Livermore National Laboratory (LLNL) [4], and at the National Institute for Standards and Technology (NIST) [5]. The ratio 3F/3C is especially susceptible to resonance effects since the collision strengths for intercombination transitions (spin change) have much larger resonance enhancements than the allowed transitions. The computed line ratio is $\sim$0.7 at the temperature of maximum abundance of Fe XVII in collisional (coronal) equilibrium, in nearly perfect agreement with the measured value in EBIT and tokamak plasmas [6]. The theoretical curve using the BPRM collision strengths is quite different from earlier results using the DW values neglecting resonances. This difference between the close coupling BPRM and the DW is highly pronounced at low temperatures found in photoionized plasmas.

Electron Distribution Function and Rate Coefficients

Although the agreement between BPRM calculations and the two EBIT experiments, is very good for the 3C, 3D, and 3E transitions, there is a factor of two difference between the two EBIT experiments [5] and [4] for the *combined* line intensity ratio '3s/3d', which represents the ratio of the sum of all transitions to the $(2p^5)$3s levels, to the $(2p^5)$3d levels of Fe XVII. We introduce the electron distribution functions (EDF) as a new physical parameter which is shown to have significant influence by up to a factor of two on 3s/3d ratios and may play a key role in the solution of this 3s/3d problem in Fe XVII. Based on large-scale relativistic close coupling calculations using the Breit-Pauli *R*-matrix method, revealing the precise effect of resonances in collisional excitation, we constructed extended collisional-radiative models using cross sections convolved with both the gaussian and the maxwellian EDF. A direct physical consequence is that the departure from a maxwellian should manifest itself in, and be used as a diagnostics of, particle distributions in plasmas. Comparison with the EBIT measurements indicates the solution, implying (a) the electron distributions and plasma conditions in the two EBITs are different, and (b) observed 3s/3d X-ray line ratios in general are 'source-specific', contrary to recent interpretations.

Fig. 1 shows various results from theory, experiments, and observations.

Fig. 1 shows that the NIST data agree better with the primary MA calculations. However, the EDF inside both the NIST and the LLNL EBITs is expected to be different from a Maxwellian. The two NIST points could possibly correspond to the minima in the fundamental oscillations in the 3s/3d line ratio, with a narrower EDF than the LLNL EBIT [15]. Thus the present results show that a much more detailed investigation of the fundamental nature of the measurements is required, such as whether the energy values are determined better than beam width. At present, the two sets of experimental data do not fully trace the theoretically predicted oscillations. While this result may definitively solve the 3s/3d problem, and reconcile the two sets of EBIT measurements, assuming that both are of claimed accuracy, further work is needed to ascertain the source of these differences between the experiments. Accurate measurements of beam widths are crucial, but to our knowledge have not been accurately determined or reported. The EDF, in turn, bears on physical effects such as space charge potential, beam currents, polarization corrections etc. [15]. Nevertheless, the theoretical sets of 3s/3d ratios clearly demostrate that they are sensitive to the plasma conditions, i. e. they are source-specific (c.f. [6]). In order to interpret additional experimental information on these issues, the

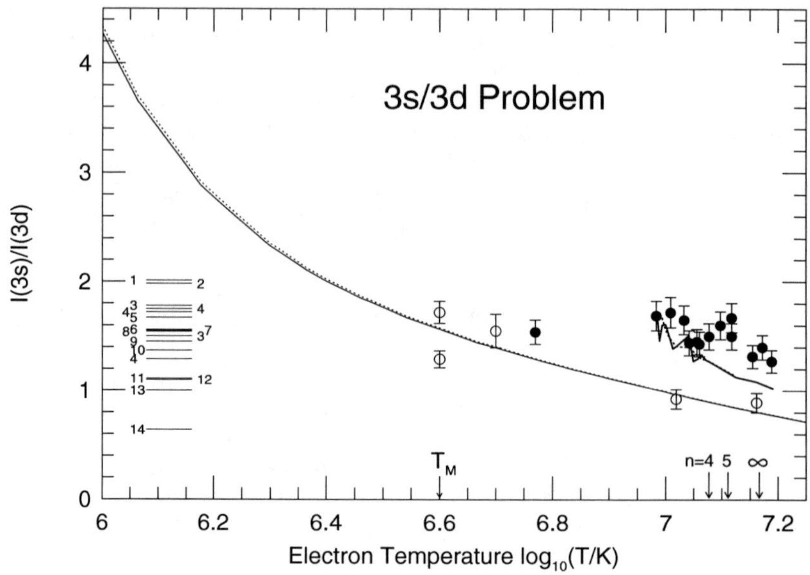

FIGURE 1. Theoretical 3s/3d line ratios in Fe XVII computed with Maxwellian and Gaussian averaged collision strengths (MA and GA respectively), and comparisons with experiments and observations. Intensity ratios using MA collision strengths in a 89-level collisional-radiative model are the monotonically decreasing curves, calculated at two electron densities 10^{13} cm^{-3} (solid line) and 10^9 cm^{-3} (dotted line); dependence upon electron density is negligible. The solid 'oscillating' curve using GA collision strengths, with FWHM = 30 eV, lie up to a factor of two above the MA curves between log (T/K) = 70 - 7.2. The oscillations are due to the distribution, strength, and the density of resonances in the collision strengths in different energy regions as sampled by the gaussian electron distribution function, and could explain the energy (temperature) dependence of 3s/3d ratios in EBIT experiments. The LLNL EBIT experimental values [4] are shown as filled circles with error bars, and the NIST EBIT experimental values [5] as the two open circles. The observed coronal 3s/3d ratios from the Sun and the solar-type star Capella [12, 13] are respectively shown with open and filled circles at or above T_M, log (T/K) = 6.6 - 6.8. Also shown for comparison are the very scattered 3s/3d ratios from the long history of various observations of solar, stellar, and disk coronae (corresponding references are in Ref. [14]).

theoretical results will also need to be extended to other EDF to elicit a more complete description of resulting variations in line intensities.

THE R-MATRIX CLOSE COUPLING METHOD AND THE DISTORTED WAVE METHOD

From a theoretical point of view resonances are a manifestion of the coupling among different levels of the target ion. The total wavefunction expansion for the (e + ion) system is a quantum superposition of components associated with all important pathways, or channels, for excitation between any two levels of the ion forming a line. Resonances arise when a quasi-bound state of the (e + ion) system forms temporarily, before decaying (autoionizing) into the continuum with a free electron and the ion. The resonance is therefore represented by the coupling between a continuum-type wavefunction (open channel), and a bound-state type wavefunction (closed channel). Since there are an infinite number of Rydberg bound (e + ion) levels, converging on to each level of the target ion, there are also several infinite series of resonances associated all levels of the target ion included in the total wavefunction expansion.

The thoeretical problem amounts to including these coupling effects to take full account of the complexities of the infinite series of resonance structures. The Close Coupling (CC) approximation in atomic collision theory includes resonances in an ab initio manner for all resonances along the infinite Rydberg series converging on to various levels of the target ion. The powerful R-matrix (RM) method, developed by P.G. Burke and collaborators and extended during the Opacity Project by M.J. Seaton and collaborators, implements the CC approximation [1, 2, 16]. Historically, the CC method was developed to account for strong coupling effects among the levels of an atom that are not considered in simpler approximations such as the distorted wave (DW) approximation. The CC or the RM method has been used extensively in the calculation of photoioinization and electron impact excitation cross sections, especially in large-scale calculations for astrophysically abundant ions under the Opacity Project [1] and the Iron Project [2], as well an extension of the Iron Project, called the RmaX Project (e.g. [8], aimed at X-ray astronomy.

The Distorted Wave (DW) approximation, on the other hand, includes coupling only between the two intial and final levels involved in forming the line, i.e. indirect couplings to levels that lead to formation of resonances are not included in the wavefunction expansion. Nonetheless, since the calculations are much simpler, the DW approach widely employed. Especially for highly charged ions it is assumed that the coupling among levels is generally weak and the DW approximation should yield reasonable accuracy. Also, a limited account of resonances may be taken in the so called 'isolated resonance approximation' by explicitly including indirect channels.

The CC wavefunction expansion for a (N+1)-electron (e + ion) system is described as:

$$\Psi_E(e+ion) = A\sum_i^N \chi_i(ion)\theta_i + \sum_j c_j\Phi_j(e+ion), \quad (1)$$

where χ_i is the target ion or core wavefunction in a specific state $S_iL_i\pi_i$ or level $J_i\pi_i$, θ_i is the wavefunction of the interacting (N+1)th electron in a channel labeled as $S_iL_i(J_i)\pi_i\ k_i^2\ell_i(SL\pi\ or\ J\pi)$; k_i^2 is the incident kinetic energy. Φ_j is the correlation functions of (e+ion) system that compensates the orthogonality condition and short range correlation interactions. The complex resonant structures in photoionization, recombination, and in electron impact excitation are included through channel couplings. The Breit-Pauli R-matrix (BPRM) method is the relativistic version of the R-matrix method, including relativistic effects and fine structure in intermediate coupling [2, 16].

The most powerful feature of the R-matrix method is that scattering or photoionization cross sections may be computed at an arbitrarily large number of energies. This enables a precise delineation of resonance profiles. As we shall show later, a complete description of resonance phenomena are essential for collision and radiative cross sections.

Much of the atomic data for applications in fusion and astrophysical X-ray sources is computed using the DW approximation, which omits the summation in the first term on the RHS of Eq. (1) from the total wavefunction expansion. Neglecting channel coupling may not be of adequate accuracy for all highly charged ions, since strong coupling effects and resonances appear in the cross sections for many transitions, as exemplified in the previous section on Fe XVII.

PHOTOIONIZATION AND RECOMBINATION: SELF-CONSISTENT AND UNIFIED METHOD

The need for self-consistency in atomic parameters for photoionization and (e + ion) recombination is manifest under photoionization equilibrium characterized by the following equation

$$\int_{\nu_0}^{\infty} \frac{4\pi J_\nu}{h\nu} N(X^z)\sigma_{PI}(\nu,X^z)d\nu = \sum_j N_e N(X^{z+1})\alpha_R(X_j^z;T), \tag{2}$$

where $\alpha_R(X_j^z;T)$ is the total electron-ion recombination rate coefficient of the recombined ion of charge z, X_j^z, to state j at electron temperature T, C_I is the rate coefficient for electron impact ionization, and σ_{PI} is the photoionization cross section evaluated at photon frequency ν and convoluted with the isotropic radiation density J_ν of the source; N_e, $N(X^{z+1})$, and $N(X^z)$ are the densities for the free electrons, and the recombining and recombined ions respectively.

Resonances are of particular importance in photoionization and (e + ion) recombination — inverse processes related by detailed balance. Unlike past theoretical treatments, mainly based on variations of the DW approximation and the isolated resonance approximation, the CC approximation enables a self-consistent and unified treatment of both processes.

Resonant and non-resonant phenomena are always unified in nature. For example, in experimental measurements resonances appear as integral part of the cross sections. Therefore a unified theoretical treatment is preferable to the heretofore separate methods for non-resonant recombination, called radiative recombination (RR), and resonant recombination called dielectronic recombination(DR). Owing to simplicity, these are generally treated separately using: (A) background photoionization cross sections (for example, calculated in the central field approximation [17]) that yield the RR rate, and (B) using a DW-type approximation to compute auotoionization and radiative rates for isolated resonances independently to obtain DR rates [18, 20].

In recent years a self-consistent and unified method has been developed by Nahar and Pradhan ([3]) employing the R-matrix method. Although the R-matrix calculations are very extensive, self-consistent and unified sets of photoionization and recombination data are now available for over 50 atoms and ions. The unified (e + ion) recombination cross sections have been benchmarked against available experimental data for several ions, and the results agree to within 10-20% with the measured values (which are naturally unified!). The detailed comparison between theory and experiment for Fe XVII and (e.g. [9]) shows that there are practically no significant shortcomings in the R-matrix method for (e + ion) recombination (c.f. [20]). Moreover, both the astrophysical and laboratory plasma sources are affected by the cut-offs in the high lying members of the Rydberg series of resonances converging on to excited levels of the target (recombining) ion. Finally, we note that large-scale BPRM calculations are in progress for Fe XVII in order to provide a complete set of photoionization and recombination data for fusion and astrophysical applications. These calculations include 60 levels of the target ion Fe XVIII in the wavefunction expansion (Eq. 1), and all resonance structures among them. Both the total and the level-specific photoionization and recombination cross sections will be obtained up to sufficiently high energies to yield rate coefficients valid at all high temperatures for practical applications. In contrast, previous calculations and experiments considered only the very low energy range including the first 3 levels of Fe XVIII ([9, 19, 20].

A detailed discussion of photoionization and recombination is given in the review article by Nahar and Pradhan ([3], astro-ph/0310624). Following is the list of atoms and ions for which self-consistent sets of total and level-specific photoionization cross sections and unified (e + ion) recombination rate coefficients have so far been obtained for over 50 ions.
Carbon: C I, C II, C III, C IV, C V, C VI
Nitrogen: N I, N II, N II, N IV, N V, N VI, N VI
Oxygen: O I, O II, O III, O IV, O V, O VI, O VII, O VII
Si: Si I, Si II, Si IX
S: S II, S III, S XI
Ar: Ar V, Ar XIII Ca: Ca VII, Ca XV Fe: Fe I, Fe II, Fe III, Fe IV, Fe V, Fe XIII, Fe XVII, Fe XXI, Fe XXIV, Fe XXV, Fe XXVI
Ni: Ni II, Ni XXVI
Carbon-like: F IV, Ne V, Na VI, Mg VII, Al VIII
Li-He-H like: C IV-VI, O VI-VIII, Ne VIII-XI, Fe XXIV-XXVI, Ni XXVI-XXVIII

The above mentioned review article and data are also available from (www.astronomy.ohio-state.edu/∼pradhan), or from Sultana Nahar (nahar@astronomy.ohio-state.edu).

TRANSIENT PLASMAS

Modeling the spectra of He-like ions under transient conditions is be of special interest in certain astrophysical situations, such as in accretion disc X-ray flares. He-like ions provide the most useful X-ray diagnostics since they remain in the plasma across the widest temperature range of any ionic species. Furthermore, He-like ions have a very rich spectrum, consisting of a set of 4 primary $n = 2 \rightarrow 1$ transitions, as well as dielectronic satellite lines that appear at temperatures lower than that of maximum abundance in collisional equilibrium. Fig. 2 shows a numerical simulation of Fe XXV spectra under time-dependent conditions in a flare less than one hour duration with both radiative and collisional ionization present, the so called 'hybrid' case as opposed to purely photoionization or coronal equilibrium. A new code HELINE ([21, 22]) solves the time-dependent ionization balance and collisional-radiative equations to obtain instantaneous line intensities. Salient features of the simulated spectra include the dominance of dielectronic satellites at early times during the flare, and a 'spectral inversion' of the main line spectrum at late times when the forbidden z-line is stronger than the allowed w-line.

THE OPACITY PROJECT AND IRON PROJECT DATABASES

Electronically accessible web based databases haver been constructed for the OP/IP/RmaX data by C. Mendoza and the OP/IP team. They are accessible from two websites, one in Europe at the Center de Donnees Astronomiques de Strasbourg (CDS) France (http://cdsweb.u-strasbf.fr/topbase/op.html), and the other at the Ohio Supercomputer Center (OSC) in the the US (http://osc.osc.edu/hpc/opacities/). The latest database TIPTOPbase included the OP/IP, is an upgraded version of TOPbase which includes only the OP data. The TIPbase facility includes the fine-structure data :radiative transition probabilities and electron impact excitation cross sections and rate coefficients for fine structure transitions. Links are also provided to publications from the OP and IP collaborations.

In addition, a large amount of data has been computed but not yet input into TIPTOP-base, for example, for the low ionization stages of Fe I-IV. Information on additional data may be obtained from Sultana Nahar (nahar@astronomy.ohio-state.edu).

Both the CDS and the OSC websites enable on-line computations of plasma opacities and radiative accelerations for an arbitrary mixture of elements using new data and codes by M.J. Seaton and collaborators (M.J. Seaton, MNRAS, submitted, astro-ph/0411010).

ACKNOWLEDGMENTS

The extensive work on Fe XVII and electron distribution functions was carried by Guo Xin Chen, now a postdoctoral fellow at the Harvard-Smithsonian Institute for Atomic and Molecular Physics (ITAMP); results on X-ray spectroscopy of transient plasmas were obtained by Justin Oelgoetz; the self-consistent and unified method for R-matrix calculations for photoionization and recombination was developed in collaboration with Sultana Nahar and Hong Lin Zhang. This work was supported by a grant from the NASA Astrophysical Theory Program. The computational work was carried out mainly on the Cray SV1 at the Ohio Supercomputer Center in Columbus Ohio.

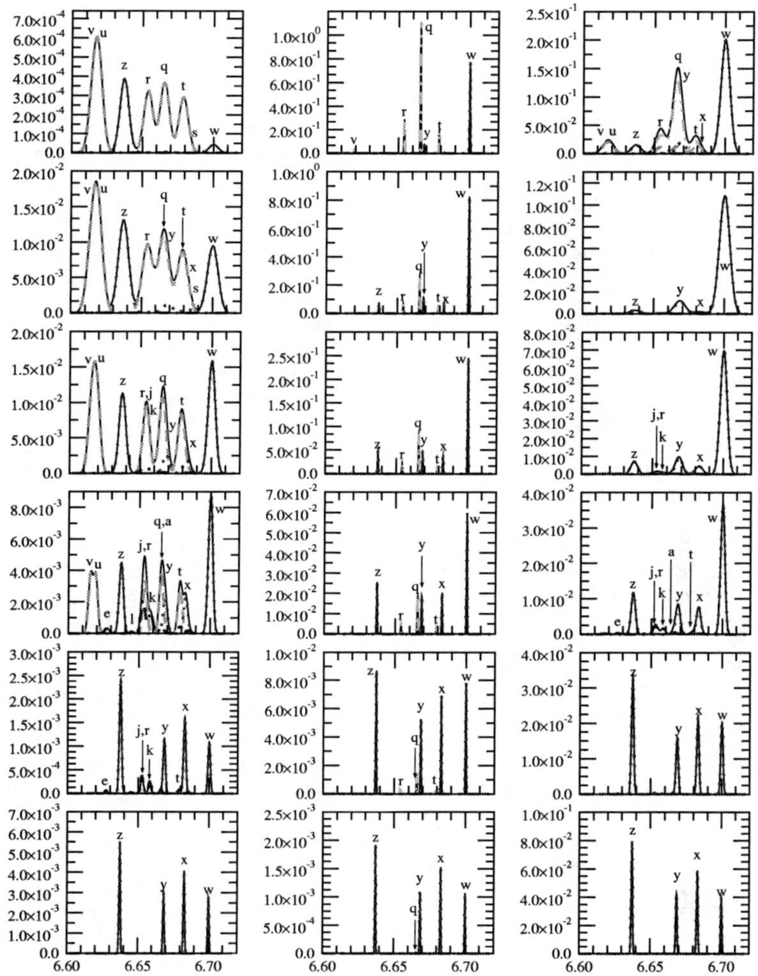

FIGURE 2. Evolution of transient spectra at various times throughout the time-dependent runs, with $T_e(t)$ and ionization parameter U(t) as in a flaring plasma with $\lesssim$ 2400 second duration (relative intensities). The LHS panels correspond to pure collisional case, the middle panels to photoionized case, and the RHS panels to the hybrid case. Of particular note are the enhanced satellite and z lines at early times in the collisional case (signature of collisional ionization), the enhanced resonance (w) and inner shell photoexcitation (q,r & t) lines at early times in the photoionized case and hybrid cases (signature of ongoing photoionization), and the enhanced non-dipole (x and z) lines at late times in all cases (signature of recombination). All intensities are given in the same set of arbitrary, relative units.

REFERENCES

1. M.J. Seaton, Y. Yu, D. Mihalas, A.K. Pradhan, *Mon. Not. R. Astr. Soc.* **266**, 805 (1994); *The Opacity Project Vols. 1 & 2*, compiled by the Opacity Project TeamL Institute of Physics, London UK 1995 and 1996 — (OP)

2. D.G. Hummer, K.A. Berrington, W. Eissner, A.K. Pradhan, H.E. Saraph, & J.A. Tully, *Astro. Astrophys.* **279**, 298 (1993) — (IP)
3. *Self-Consistent R-matrix Approach To Photoionization And Unified Electron-Ion Recombination*, Sultana N. Nahar and Anil K. Pradhan, Review in Radiation Processes In Physics and Chemistry 70, 323-344, Elsevier (2004), astro-ph/0310624.
4. G.V. Brown, P. Beiersdorfer, D.A. Liedahl, and K. Widmann, *Astrophys. J.* , **502**, 1015 (1998); P. Beiersdorfer, E. Behar, K.R. Boyce, G.V. Brown, H. Chen, K.C. Gendreau, M.F. Gu, J. Gygax, S.M. Kahn, R.L. Kelley, F.S. Porter, C.K. Stahle, and A.E. Szymkowiak, *Astrophys. J. Lett.* **576**, L169 (2002)
5. J.M. Laming, I. Kink, E. Takács, J.D. Gillaspy, E.H. Silver, H.W. Schnopper, S.R. Bandler, N.S. Brickhouse, S.S. Murray, M. Barbera, A.K. Bhatia, G.A. Doschek, N. Madden, D. Landis, J. Beeman, and E.E. Haller, *Astrophys. J. Lett.* **545**, L161 (2000)
6. P. Beiersdorfer, M. Bitter, S. von Goeler, and K.W. Hill, *Astrophys. J.* **610**, 616 (2004)
7. G.X. Chen and A.K. Pradhan, *Phys. Rev. Letters* **89**, 013202 (2002)
8. G.X. Chen, A.K. Pradhan, and W.B. Eissner, *J. Phys. B: At. Mol. Opt. Phys.* **36**, 453 (2003)
9. A.K. Pradhan, S.N. Nahar, H.L. Zhang, *Astrophys. J. Lett.* **549**, L265 (2001)
10. S.N. Nahar, W. Eissner, G.X. Chen, and A.K. Pradhan, *Astro. Astrophys.* **408**, 789 (2003)
11. H.L. Zhang, S.N. Nahar, and A.K. Pradhan, *Phys. Rev. A* **64**, 032719 (2001)
12. C.R. Canizares, D.P. Huenemoerder, D.S. Davis, D. Dewey, K.A. Fla nagan, J. Houck, T.H. Markert, H.L. Marshall, M.L. Schattenburg, N.S. Schulz, and M. Wise, *Astrophys. J. Lett.* **539**, L41 (2000)
13. M. Audard, E. Behar, M. Güdel, A.J.J. Raassen, D. Porquet, R. Mewe, C.R. Foley, and G.E. Bromage, *Astro. Astrophys.* **365**, L329 (2001)
14. D.L. McKenzie, P.B. Landecker, R.M. Broussard, H.R. Rugge, R.M. Young, U. Feldma n, and G.A. Doschek, *Astrophys. J.* **241**, 409 (1980); J.H. Parkinson, *Sol. Phys.* **42**, 183 (1975); Fawcett, B. J. Kent, 774 (1 982); D.P. Huenemoerder, C.R. Canizares, and N.S. Schulz, *Astrophys. J.* **559**, 1135 (2001) ; B. Stelzer, V. Burwitz, M. Audard, M. Güdel, J.-U. Ness, N. Grosso, R. Neuh äuser, J.H.M.M. Schmitt, P. Predehl, and B. Aschenbach, *Astro. Astrophys.* **392**, 585 (2002); J.-U. Ness, J.H.M.M. Schmitt, M.Audard, M. Güdel, and R. Mewe, *Astro. Astrophys.* **407** , 347 (2003); R. Mewe, A.J.J. Raassen, J.J. Drake, J.S. Kaastra, R.L.J. van der Meer, and D. Porquer, *Astro. Astrophys.* **368**, 888 (2001); E. Behar, J. Cottam, and S.M. Kahn, *Astrophys. J.* **548**, 966 (2001); A.C. Brinkman, C.J.T. Gunsing, J.S. Kaastra, R.L.J. van der Meer, R. Mewe, F. Pa erels, A.J.J. Raassen, J.J. van Rooijen, H. Bräuninger, W. Burkert, V. Burwitz, G. Hartner, P. Predehl, J.-U. Ness,, J.H.M.M. Schmitt, J.J. Drake, O. Johnson. M. Juda, V. Kashyap, S.S. Murray, D. Pease, P. Ratzlaff, and B.J. Wargelin, *Astrophys. J. Lett.* **530**, L111 (2000); A.J.J. Raassen, R. Mewe, M. Audard, M. Güdel, E. Behar, J.S. Kaastra, R.L.J. van der Meer, C.R. Foley, and J.-U. Ness, *Astro. Astrophys.* **389**, 228 (2002); K.J.H. Phillips, C.J. Greer, A.K. Bhatia, I.H. Coffey, and F.P. Keenan, *Astro. Astrophys.* **324**, 381 (1997); S.M. Kahn, M.A. Leutenegger, J. Cottam, G. Rauw, J.-M. Vreux, A.J.F. den Boggend e, R. Mewe, and M. Güdel, *Astro. Astrophys.* **365**, L312 (2001); H. Xu, S.M. Kahn, J.R. Peterson, E. Behar, F.B.S. Paerels, R.F. Mushotzky, J.G. Jernigan, A.C. Brinkman, and K. Makishima, *Astrophys. J.* **579**, 600 (2002)
15. Private comunication; see also J.D. Gillaspy *et al.*, Proceedings of the 14th APS Topical Conference on Atomic Processes in Plasmas (in press, 2004)
16. K.A. Berrington, W. Eissner, and P.H. Norrington, *Comput. Phys. Commun.* **92**, 290 (1995)
17. R.F. Reilman and S.T. Manson, *Astrophys. J. Suppl.* **40**, 815 (1979)
18. Badnell, N.R., *J. Phys. B: At. Mol. Opt. Phys.* **19**, 3827 (1986)
19. Savin, D.W. *et al.*, *Astrophys. J. Suppl.* bf 123, 687 (1999)
20. T.W. Gorczyca, N.R. Badnell, and D.W. Savin, *Phys. Rev. A* **65**, 062707 (2002)
21. J. Oelgoetz and Anil K. Pradhan, *Mon. Not. R. Astr. Soc.* **327**, L42
22. J. Oelgoetz and Anil K. Pradhan, *Mon. Not. R. Astr. Soc.* **354**, 1093 (2004)

Fe XVII Emission from Hot, Collisional Plasmas

P. Beiersdorfer*, M. Bitter†, S. von Goeler† and K. W. Hill†

*Lawrence Livermore National Laboratory, Livermore, CA 94550
†Princeton Plasma Physics Laboratory, Princeton, NJ 08543

Abstract.
The ratios of the Fe XVII $3s \rightarrow 2p$ transitions to that of the dominant $3d \rightarrow 2p$ transition measured in high-temperature tokamak plasmas are compared to solar and astrophysical observations. Good agreement is found, demonstrating that the collisional line formation processes active in opacity-free, low-density, high-temperature laboratory plasmas are a good description of those found in astrophysical plasmas.

The intensity ratios of the $3s \rightarrow 2p$ and $3d \rightarrow 2p$ lines in Fe XVII observed from the Sun and various astrophysical sources have not been reproduced by spectral modelling codes. For example, the ratio of the sum of the three $3s \rightarrow 2p$ lines commonly denoted $3F$, $3G$, and $3H$ (or $M2$) to the strongest $3d \rightarrow 2p$ line denoted $3C$ is predicted by the APEC model to be $I_{3F+3G+M2}/I_{3C} \approx 1.33$ at the temperature of maximum abundance for Fe XVII. The ratios typically measured in the Sun [1, 2, 3, 4, 5] or stellar atmospheres [6, 7, 8, 9, 10, 11, 12, 13] range from 2 to 3 with an average of $I_{3F+3G+M2}/I_{3C} \approx 2.4$, as illustrated in Fig. 1.

The reason for the discrepancy has been the object of much speculation. Measurements using a microcalorimeter and the NIST electron beam ion trap had resulted in excellent agreement between measurement and collisional-radiative models based on electron-impact excitation of selected levels [14]. The agreement with APEC was also excellent, as shown in Fig. 1. Because of the large discrepancy between their measurements and astrophysical and solar observations the authors speculated that processes not found in their measurement must play a role in astrophysical plasmas.

Using a microcalorimeter and the Livermore EBIT-II electron beam ion trap a different conclusion was obtained. As illustrated in Fig. 1, the Livermore microcalorimeter measurements showed no agreement with APEC. On the contrary, they showed good agreement with solar and astrophysical observations [15]. Additional measurements were conducted with a grazing-incidence grating spectrometer and with a flat-crystal spectrometer. These had been tested and proven reliable in many other experiments [16]. The grating and crystal spectrometer measurements confirmed the Livermore microcalorimeter results. Another set of measurements have been made using a second-generation microcalorimeter and the Livermore EBIT-I electron beam ion trap on Ni XIX, i.e., using a different calorimeter, a different electron beam ion trap, and a different element [17]. The results again confirm the Livermore calorimeter data of Fe XVII.

In order to test for collisional effects possibly not addressed in the Livermore and NIST electron beam ion trap measurements, we have analyzed the Fe XVII emission from tokamak plasmas, including the PLT, DITE, and JET tokamaks [18]. Like in the

CP774, *X-Ray Diagnostics of Astrophysical Plasmas,* edited by Randall K. Smith

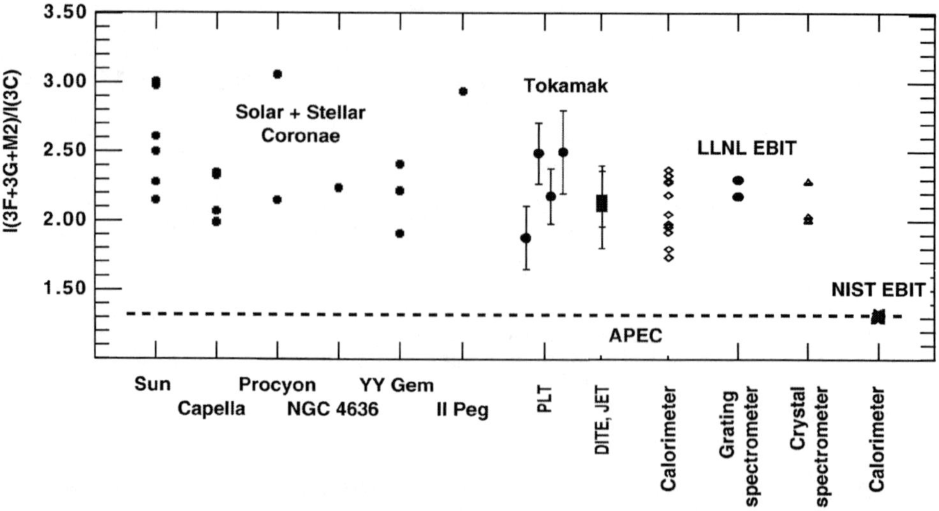

FIGURE 1. Comparison of the ratios of the sum of the three $3s \rightarrow 2p$ lines commonly denoted $3F$, $3G$, and $3H$ (or $M2$) to the strongest $3d \rightarrow 2p$ line denoted $3C$ measured in the laboratory and inferred from observations of various astrophysical sources and the Sun. APEC calculations shown as a dashed line are for an electron temperature of 432 eV.

Sun and stellar coronae, the Fe XVII emission from tokamak plasmas arises predominantly from those temperature regions where Fe XVII reaches its maximum abundance. Neighboring ionization states may also contribute, either by recombination or by inner-shell ionization [19]. Tokamak plasmas are optically thin in the soft X-ray regime, and resonant scattering will not affect the line ratios.

We found that the tokamak data reproduced those from astrophysical sources, as illustrated in Fig. 1. We conclude that the collisional processes dominating in hot tokamak plasma are a good description of Fe XVII excitation. Moreover, the tokamak data also agree rather well with the ratios obtained in the Livermore electron beam ion traps for electron beam energies where indirect line formation processes involving neighboring ions do not play a role. These processes, however, appear to contribute at some level, as the tokamak ratios are on average slightly higher than the LLNL ion trap data, in line with earlier determinations of the role of indirect line formation processes affecting neonlike ions [19]. Non-collisional processes need not be invoked to explain the Fe XVII emission. This means that the problems describing the astrophysical line ratios must lie with the collisional atomic data directly involving the Fe XVII ion.

Several high-resolution spectrometers have newly been installed on the National Spherical Torus Experiments (NSTX) tokamak [20]. These will observe the Fe XVII under various plasma conditions in upcoming experiments with the hope that further light will be shed on the Fe XVII problem.

ACKNOWLEDGMENTS

This work was performed under the auspices of the U.S. DOE by the University of California Lawrence Livermore National Laboratory under contract W-7405-Eng-48 and by the Princeton Plasma Physics Laboratory under contract DE-AC02-76CHO3073.

REFERENCES

1. R. J. Hutcheon, J. P. P., and Evans, K. D., *Mon. Not. R. Astron. Soc.*, **175**, 489 (1976).
2. Parkinson, J. H., *Astron. Astrophys.*, **24**, 215 (1973).
3. McKenzie, D. L., Landecker, P. B., Broussard, R. M., Rugge, H. R., Young, R. M., Feldman, U., , and Doschek, G. A., *Astrophys. J.*, **241**, 409 (1980).
4. Parkinson, J. H., *Solar Phys.*, **42**, 183 (1975).
5. Rugge, H. R., and McKenzie, D. L., *Astrophys. J.*, **297**, 338 (1985).
6. Behar, E., Cottam, J., and Kahn, S. M., *Astrophys. J.*, **548**, 966 (2001).
7. Brinkman, A. C., Gunsing, C. J. T., Kaastra, J. S., van der Meer, R. L. J., Mewe, R., Paerels, F., Raassen, A. J. J., van Rooijen, J. J., Bräuninger, H., Burkert, W., Burwitz, V., Hartner, G., Predehl, P., Ness, J.-U., Schmitt, J. H. M. M., Drake, J. J., Johnson, O., Juda, M., Kashyap, V., Murray, S. S., Pease, D., Ratzlaff, P., and Wargelin, B. J., *Astrophys. J. (Lett.)*, **530**, L111 (2000).
8. Mewe, R., a. J. J. Raassen, Drake, J. J., Kaastra, J. S., van der Meer, R. L. J., and Porquet, D., *Astron. Astrophys.*, **368**, 888 (2001).
9. Canizares, C. R., Huenemoerder, D. P., Davis, D. S., Dewey, D., Flanagan, K. A., Houck, J., Markert, T. H., Marshall, H. L., Schattenburg, M. L., Schulz, N. S., Wise, M., Drake, J. J., and Brickhouse, N. S., *Astrophys. J.*, **539**, L41 (2000).
10. Raassen, A. J. J., Mewe, R., Audard, M., Güdel, M., Behar, E., Kaastra, J. S., van der Meer, R. L. J., Foley, C. R., and Ness, J.-U., *Astron. Astrophys.*, **389**, 228 (2002).
11. Xu, H., Kahn, S. M., Peterson, J. R., Behar, E., Paerels, F. B. S., Mushotzky, R. F., Jernigan, J. G., and Makishima, K., *Astrophys. J.*, **579**, 600 (2002).
12. Stelzer, B., Burwitz, V., Audard, M., Güdel, M., Ness, J.-U., Grosso, N., Neuhäuser, R., Schmitt, J., Predehl, P., and Aschenbach, B., *Astron. Astrophys.*, **392**, 585 (2002).
13. Huenemoerder, D. P., Canizares, C. R., and Schulz, N. S., *Astrophys. J.*, **559**, 1135 (2001).
14. Laming, J. M., Kink, I., Takacs, E., Porto, J. V., Gillaspy, J. D., Silver, E. H., Schopper, H. W., Bandler, S. R., Brickhouse, N. S., Murray, S. S., Barbera, M., Bhatia, A. K., Doschek, G. A., Madden, N., Landis, D., Beeman, J., and Haller, E. E., *Astrophys. J.*, **545**, L161 (2000).
15. Beiersdorfer, P., Behar, E., Boyce, K. R., Brown, G. V., Chen, H., Gendreau, K. C., Gu, M.-F., Gygax, J., Kahn, S. M., Kelley, R. L., Porter, F. S., Stahle, C. K., and Szymkowiak, A. E., *Astrophys. J. (Lett.)*, **576**, L169 (2002).
16. Beiersdorfer, P., *Ann. Rev. Astron. Astrophys.*, **412**, 343 (2003).
17. Gu, M.-F., Beiersdorfer, P., Brown, G. V., Chen, H., Boyce, K. R., Kelley, R. L., Kilbourne, C. A., Porter, F. S., and Kahn, S. M., *Astrophys. J. (Lett.)*, **607**, L143 (2004).
18. Beiersdorfer, P., Bitter, M., von Goeler, S., and Hill, K. W., *Astrophys. J.*, **610**, 616 (2004).
19. Beiersdorfer, P., Osterheld, A. L., Chen, M. H., Henderson, J. R., Knapp, D. A., Levine, M. A., Marrs, R. E., Reed, K. J., Schneider, M. B., and Vogel, D. A., *Phys. Rev. Lett.*, **65**, 1995 (1990).
20. Beiersdorfer, P., Bitter, M., May, M. J., and Roquemore, L., *Rev. Sci. Instrum.*, **74**, 1974 (2003).

Drs. Duncan Galloway, Jazmina Lazendic, Cara Rakowski, and Parviz Ghavamian

Fe L-Shell Density Diagnostics in Theory and Practice

Christopher W. Mauche*, Duane A. Liedahl* and Kevin B. Fournier*

*Lawrence Livermore National Laboratory, L-473, 7000 East Avenue, Livermore, CA 94550

Abstract.
We provide a discussion of the density and photoexcitation sensitivity of the X-ray spectra of Fe L-shell ions (Fe XVII–Fe XXIV) calculated with the Livermore X-ray Spectral Synthesizer, a suite of IDL codes that calculates spectral models of highly charged ions based primarily on HULLAC atomic data. These models are applicable to collisionally ionized laboratory or cosmic plasmas with electron temperatures $T_e \approx 2$–45 MK (0.2–4 keV) and electron densities $n_e \gtrsim 10^{11}$ cm^{-3}. Potentially useful density diagnostics are identified for Fe XVII and Fe XIX–Fe XXIII, with the most straightforward being the Fe XVII I(17.10 Å)/I(17.05 Å) line ratio and the Fe XXII I(11.92 Å)/I(11.77 Å) line ratio. Applying these models to the *Chandra* X-ray Observatory High Energy Transmission Grating spectrum of the intermediate polar EX Hya, we find that the strength of all the Fe L-shell lines are consistent with electron densities $n_e \gtrsim 1\times 10^{14}$ cm^{-3}. Specifically, from the observed Fe XVII I(17.10 Å)/I(17.05 Å) line ratio, we infer an electron density $n_e \gtrsim 2\times 10^{14}$ cm^{-3} at the $3\,\sigma$ level, while from the observed Fe XXII I(11.92 Å)/I(11.77 Å) line ratio, we infer $n_e = 1.0^{+2.0}_{-0.5}\times 10^{14}$ cm^{-3} at the $1\,\sigma$ level and $n_e \gtrsim 2\times 10^{13}$ cm^{-3} at the $3\,\sigma$ level.

Keywords: atomic processes — binaries: close — stars: individual (EX Hydrae) — stars: magnetic fields — X-rays: binaries
PACS: 32.30.Rj, 97.80.Gm

INTRODUCTION

The standard density diagnostic of high-temperature plasmas is the intensity ratio $R \equiv f/i$ of the forbidden to intercombination lines of He-like ions, which falls from its low-density value R_0 to zero at a critical density (where $R/R_0 = 0.5$) that increases from $n_e \approx 3\times 10^8$ cm^{-3} for C to $n_e \approx 3\times 10^{17}$ cm^{-3} for Fe [16, 4, 35, 36] (see left panel of Fig 1). Unfortunately, this diagnostic is compromised in ultraviolet-bright sources like O stars and cataclysmic variables, wherein photoexcitation competes with collisional excitation to depopulate the upper level of the forbidden line, causing the R line ratio to appear to be in the high-density limit regardless of the density; for a plasma illuminated by a 30 kK blackbody, this is true for all elements through Mg [26] (see right panel of Fig 1). In an attempt to circumvent this problem, we have undertaken an investigation of potential density diagnostics of lines of Fe L-shell ions (Fe XVII–Fe XXIV). These diagnostics are applicable to collisionally ionized plasmas with electron temperatures $T_e \approx 2$–45 MK (0.2–4 keV) and electron densities $n_e \gtrsim 10^{11}$ cm^{-3}. Given the high densities, these diagnostics are particularly suitable to magnetic cataclysmic variables (polars and intermediate polars [40]), in which the density of the X-ray–emitting plasma is increased by (1) the magnetic channeling of the mass lost by the secondary onto small

CP774, *X-Ray Diagnostics of Astrophysical Plasmas,* edited by Randall K. Smith

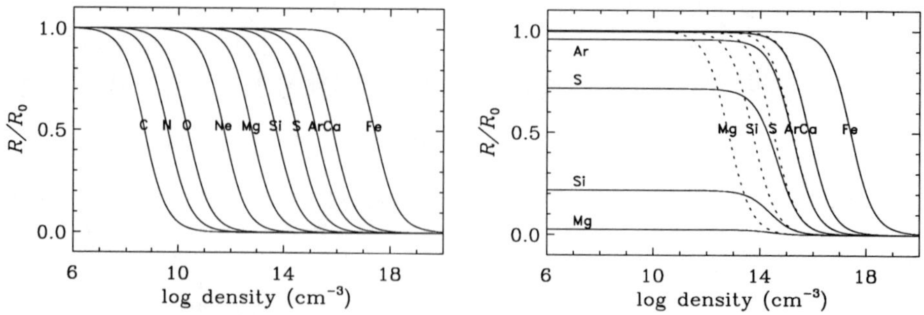

FIGURE 1. *Left panel*: He-like $R \equiv z/(x+y) = f/i$ line ratios of the abundant elements as a function of electron density. For C, N, O, ..., Fe, the plasma temperatures $T_e = 0.46, 0.57, 0.86, \ldots, 36.1$ MK, the temperature of the peak ionization fraction for each ion [29]. *Right panel*: Similar to the left panel (*dotted curves*), but accounts for photoexcitation by a $T_{bb} = 30$ kK blackbody (*full curves*). In both panels the line ratios are scaled to the low-density limit R_0 for ease of comparison.

spots near the white dwarf magnetic poles; (2) the factor-of-four density jump across the standoff shock, which heats the plasma to a temperature $T_s = 3GM_{wd}\mu m_H/8kR_{wd} \approx 200$ MK; and (3) the settling nature of the post-shock flow, wherein the pressure is roughly constant and the density scales inversely with the decreasing temperature [14]. For a mass-accretion rate $\dot{M} = 10^{15}$ g s^{-1} (hence luminosity $L = GM_{wd}\dot{M}/R_{wd} \approx 6 \times 10^{31}$ erg s^{-1}), a free-fall velocity $v_{ff} = (2GM_{wd}/R_{wd})^{1/2} \approx 3600$ km s^{-1}, and a relative spot size $f = 0.1$, the density of the accretion flow immediately behind the shock $n = \dot{M}/4\pi f R_{wd}^2 \mu m_H (v_{ff}/4) \approx 10^{13}$ cm^{-3}.

In two previous publications, we explored the density and photoexcitation sensitivity of the Fe XVII I(17.10 Å)/I(17.05 Å) line ratio [27] and the Fe XXII I(11.92 Å)/I(11.77 Å) line ratio [28] and applied them to the *Chandra* X-ray Observatory High-Energy Transmission Grating Spectrometer (HETGS) Medium Energy Grating (MEG) spectrum of the intermediate polar EX Hya [17, 19, 15, 10, 18]. We summarize the results of those investigations below, but first provide a general look at the *Chandra* HETG spectrum of EX Hya and the density sensitivity of the X-ray spectra of all Fe L-shell ions.

CHANDRA SPECTRUM OF EX HYA

We begin by comparing the *Chandra* MEG spectrum of EX Hya with that of the coronally active RS CVn binary HR 1099 [1, 30, 31]. As one of the targets of the *Chandra* Emission Line Project (`http://asc.harvard.edu/elp/ELP.html`) [5], HR 1099 was observed extensively by *Chandra*, and its X-ray spectrum is one of the benchmarks for models of collisionally ionized plasmas. Figure 2 shows the MEG count spectra of EX Hya (*gray histogram*) and HR 1099 (*black histogram*) in the 10.5–17.5 Å bandpass, which contains the strongest lines of Fe L-shell ions. The spectrum of EX Hya differs from that of HR 1099 in (1) the absence of the Fe XVII 17.10 Å line, (2) the strength of the Fe XVII 17.05 Å line, (3) the absence of the Ne IX 13.70 Å forbidden

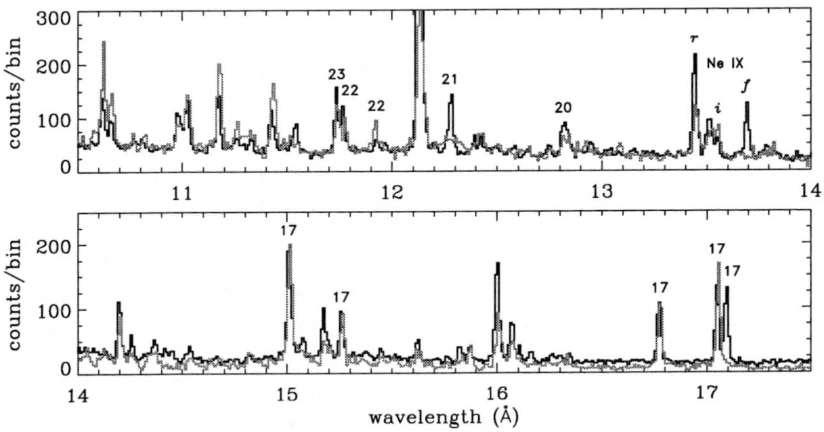

FIGURE 2. *Chandra* MEG count spectra (combining ±1st orders and binned to 0.005 Å) of EX Hya (*gray histogram*) and HR1099 (*black histogram*). The HR 1099 data is scaled by a factor of 0.25 (0.34) in the upper (lower) panel, and the Ne IX and Fe L-shell lines discussed in the text are labeled.

line, (4) the weakness of the Fe XX 12.83 Å line, (5) the absence of the Fe XXI 12.26 Å line, and (6) the strength of the Fe XXII 11.92 Å line. As shown by Fig 1, the absence of the Ne IX forbidden line is indicative of electron densities $n_e \gtrsim 1 \times 10^{13}$ cm^{-3} or of photoexcitation by a $T_{bb} \gtrsim 30$ kK blackbody. The later possibility cannot be excluded in EX Hya, whose far ultraviolet spectrum has been modeled as a $T_{bb} \approx 25$ kK white dwarf [17] and as a $T_{bb} \approx 20$ kK white dwarf with a $T_{bb} \approx 37$ kK hot spot [25]. As we will see below, the Fe XVII I(17.10 Å)/I(17.05 Å) line ratio and the Fe XXII I(11.92 Å)/I(11.77 Å) line ratio are less sensitive to photoexcitation, and indeed the strength of all the Fe L-shell lines are consistent with electron densities $n_e \gtrsim 1 \times 10^{14}$ cm^{-3}.

LXSS FE L-SHELL MODEL SPECTRA

We constructed model spectra of Fe L-shell ions using the Livermore X-ray Spectral Synthesizer (LXSS), a suite of IDL codes that calculates spectral models of highly charged ions based on Hebrew University/Lawrence Livermore Atomic Code (HULLAC) atomic data. HULLAC calculates atomic wavefunctions, level energies, radiative transition rates, and oscillator strengths according to the fully relativistic, multiconfiguration, parametric potential method [21, 22]. Electron impact excitation rate coefficients are computed quasi-relativistically in the distorted wave approximation [2] assuming a Maxwellian velocity distribution. Table 1 lists the number of energy levels, radiative transition rates for E1, E2, M1, and M2 decays, and electron impact excitation rate coefficients for each of our Fe L-shell models. Using these data, LXSS calculates the level populations for a given temperature and density assuming collisional-radiative equilibrium. The line intensities are then simply the product of the level populations and the radiative transition rates. In the sections below on Fe XVII and Fe XXII, we account for

TABLE 1. LXSS Fe L-shell Models

Ion	Levels	Collisional Rates	Radiative Rates
Fe XVII	281	33,887	49,882
Fe XVIII	456	93,583	141,229
Fe XIX	605	164,496	240,948
Fe XX	609	165,350	257,765
Fe XXI	591	153,953	227,743
Fe XXII	228	24,084	37,300
Fe XXIII	116	6,478	8,798
Fe XXIV	76	1,704	4,100

photoexcitation by including in the LXSS population kinetics calculation the photoexcitation rates $(\pi e^2/m_e c) f_{ij} F_\nu(T)$, where $F_\nu(T)$ is the continuum spectral energy distribution and f_{ij} are the oscillator strengths of the various transitions. For simplicity, we assume that $F_\nu(T) = (4\pi/h\nu)B_\nu(T_{\rm bb})$ (i.e., the radiation field is that of a blackbody of temperature $T_{\rm bb}$) and the dilution factor of the radiation field is equal to $\frac{1}{2}$ (i.e., the X-ray emitting plasma is in close proximity to the source of the photoexcitation continuum).

The resulting Fe L-shell ion spectra are shown in Figs 3–6. In each case, the atomic model was calculated at the temperature of the peak ionization fraction for each ion [29] for electron densities $n_{\rm e} = 10^{10}$–10^{18} cm^{-3}, and the spectra were binned to 0.02 Å, which is approximately equal to the FWHM of the MEG spectra. The upper and lower panels of each figure show the model spectra at the extremes of the density grid: $n_{\rm e} = 10^{10}$ cm^{-3} in black and $n_{\rm e} = 10^{18}$ cm^{-3} in gray. Note that each panel encompasses a different 4 Å bandpass, allowing them to be compared directly to Fig 2. In each case, the spectrum is scaled to the brightest line in the $n_{\rm e} = 10^{10}$ cm^{-3} model and by a factor of $(10^{10}\ {\rm cm}^{-3})/n_{\rm e}$. The middle panels of each figure show as a function of electron density the relative line strengths for lines brighter than 0.25 photons/bin. Each of these models is discussed in turn.

Fe XVII: The X-ray spectrum of Ne-like Fe XVII is dominated at low densities by the three $3d$–$2p$ lines near 15 Å and the three $3s$–$2p$ lines near 17 Å. As the density increases, the 17.10 Å line disappears and the 17.05 Å line brightens. As shown below, the I(17.10 Å)/I(17.05 Å) line ratio observed in EX Hya is consistent with $n_{\rm e} \gtrsim 2 \times 10^{14}$ cm^{-3} at $T_{\rm e} \approx 4.1$ MK.

Fe XVIII: The X-ray spectrum of F-like Fe XVIII is dominated at low densities by the $3d$–$2p$ line at 14.20 Å. As the density increases, the three brightest lines of this ion all dim, providing no useful density diagnostic.

Fe XIX: The X-ray spectrum of O-like Fe XIX is dominated at low densities by a tight clump of $3d$–$2p$ lines near 13.5 Å. As the density increases, the 13.52 Å and 13.50 Å lines dim and the 13.48 Å line brightens, providing a potential density diagnostic, although these lines are blended with the (typically brighter) He α lines of Ne IX.

Fe XX: The X-ray spectrum of N-like Fe XX is dominated at low densities by a broad clump of $3d$–$2p$ lines near 12.8 Å. As the density increases, the 12.80 Å and 12.82 Å lines dim and the 12.86 Å, 12.96 Å, and 12.90 Å lines brighten, providing a potential

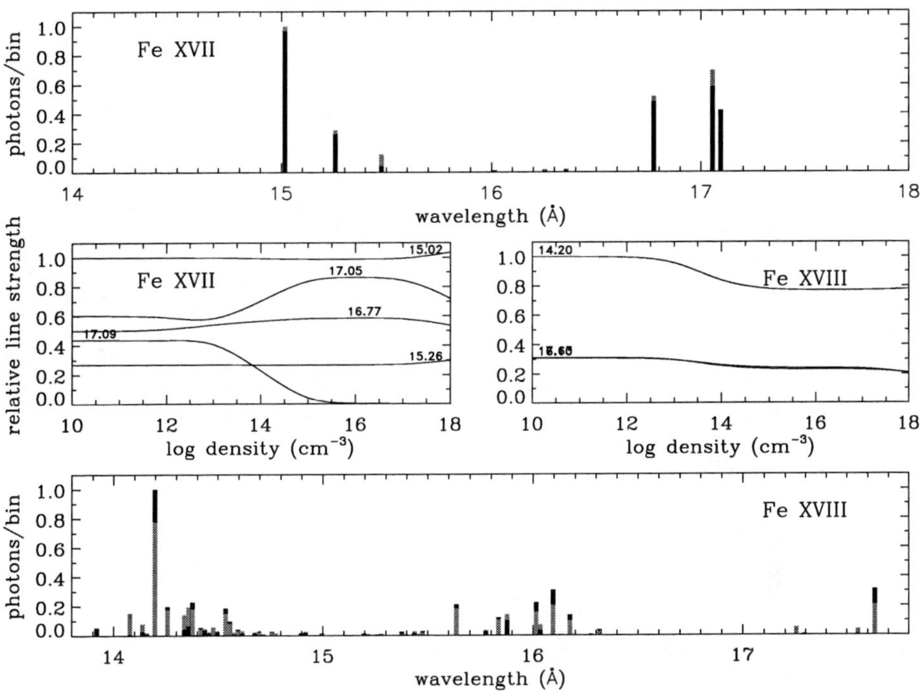

FIGURE 3. LXSS models of Fe XVII at $T_e = 4.1$ MK and Fe XVIII at $T_e = 6.6$ MK. Upper and lower panels show the model spectra binned to 0.02 Å for $n_e = 10^{10}$ cm^{-3} (*black bars*) and $n_e = 10^{18}$ cm^{-3} (*gray bars*); middle panels show as a function of electron density the relative line strengths for lines brighter than 0.25 photons/bin.

density diagnostic. The weakness of the 12.8 Å line in EX Hya is consistent with $n_e \gtrsim 3 \times 10^{14}$ cm^{-3} at $T_e \approx 9.5$ MK.

Fe XXI: The relatively simple X-ray spectrum of C-like Fe XXI is dominated at low densities by a single $3d$–$2p$ line at 12.26 Å. As the density increases, this line dims and the 12.38 Å line brightens, providing a potential density diagnostic. The absence of the 12.26 Å line in EX Hya is consistent with $n_e \gtrsim 3 \times 10^{13}$ cm^{-3} at $T_e \approx 10.5$ MK.

Fe XXII: The X-ray spectrum of B-like Fe XXII is dominated at low densities by the $3d_{3/2}$–$2p_{1/2}$ line at 11.77 Å. As the density increases, this line dims and the $3d_{5/2}$–$2p_{3/2}$ line at 11.92 Å line brightens, providing a clear density diagnostic. As shown below, the I(11.92 Å)/I(11.77 Å) line ratio observed in EX Hya is consistent $n_e \approx 1 \times 10^{14}$ cm^{-3} at $T_e \approx 12.3$ MK.

Fe XXIII: The X-ray spectrum of Be-like Fe XXIII consists of a few strong $3d$–$2p$ lines between 11.0–12.2 Å. As the densities increases, the 11.74 Å line dims and the 11.44 Å line brightens, but they do not become comparable in strength until $n_e \approx 10^{18}$ cm^{-3}.

Fe XXIV: The X-ray spectrum of Li-like Fe XXIV consists of a few strong $3s$–$2p$, $3d$–$2p$, and $3p$–$2s$ lines between 10.6–11.4 Å. There is little density dependence to its X-ray

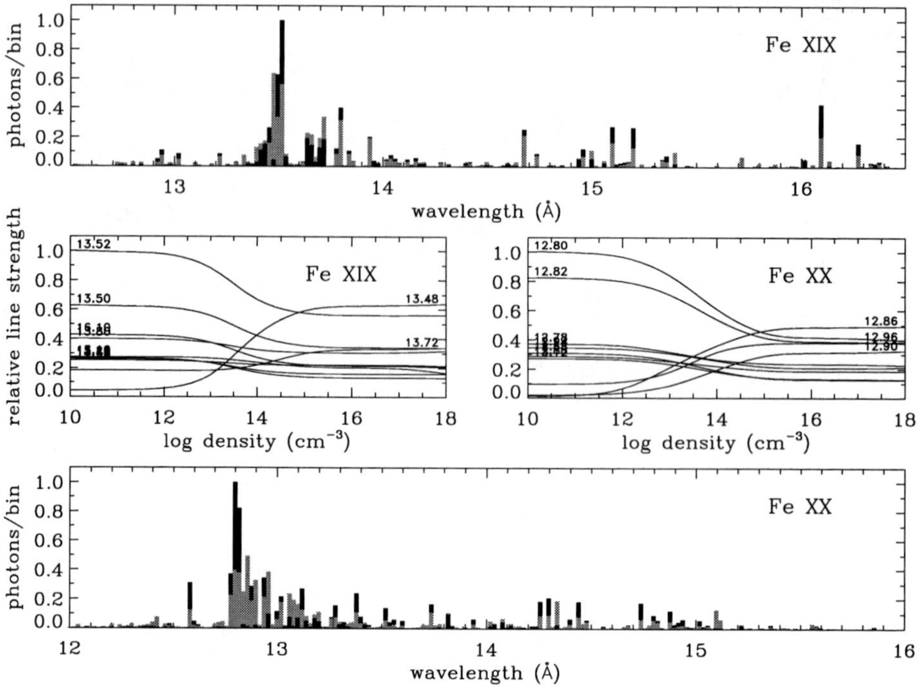

FIGURE 4. Similar to Fig 2, for Fe XIX at $T_e = 7.9$ MK and Fe XX at $T_e = 9.5$ MK.

spectrum.

Based on the foregoing, it is clear that the Fe XVII $I(17.10\ \text{Å})/I(17.05\ \text{Å})$ line ratio and the Fe XXII $I(11.92\ \text{Å})/I(11.77\ \text{Å})$ line ratio provide the most straightforward density diagnostics of Fe L-shell ions. Each of these is discussed in more detail below.

FE XVII

Because of the persistence of Ne-like Fe XVII over a broad temperature range ($T_e \approx 2$–12 MK) and the large collision strengths for $2p \rightarrow nd$ transitions, the $2p^6$–$2p^5 3l$ ($l = s$, d) lines of Fe XVII at 15–17 Å are prominent in the X-ray spectra of high-temperature plasmas in tokamaks [23, 34, 3], the Sun [37, 34, 38], and both late- and early-type stars [6, 1, 20, 39]. The lowest-lying configurations of Fe XVII are the $2p^6$ ground state, the $2p^5 3s$ manifold (4 levels), the $2p^5 3p$ manifold (10 levels), and the $2p^5 3d$ manifold (12 levels) (Fig 7). Because electron impact excitation from the ground state is strong for transitions into the $2p^5 nd$ configurations, but weak for transitions into the $2p^5 3s$ configuration, the population flux into the lowest-lying configuration is dominated by radiative cascades originating on higher-lying energy levels, rather than direct excitation from the ground state. The four levels of the $2p^5 3s$ configuration have, in increasing

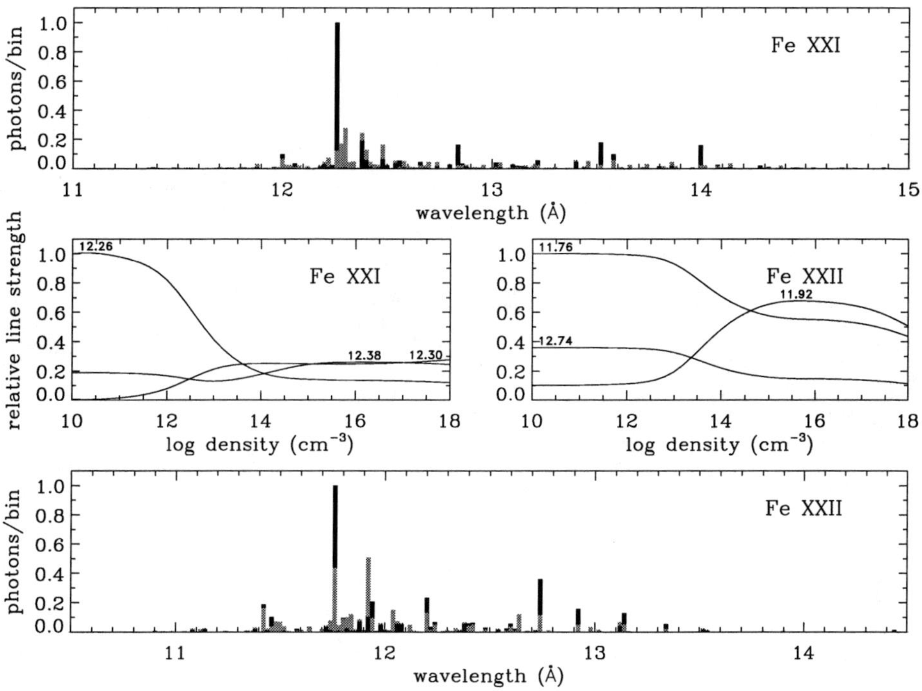

FIGURE 5. Similar to Fig 2, for Fe XXI at $T_e = 10.5$ MK and Fe XXII at $T_e = 12.3$ MK.

energy order, total angular momenta $J = 2, 1, 0$, and 1. The $2p^6$–$2p^5 3s$ ($J = 0$) transition is strictly forbidden, but the remaining three levels $J = 2$, 1, and 1 decay to ground, producing lines at 17.10, 17.05, and 16.78 Å, respectively. The 17.10 Å line is produced by an M2 transition, but it is nevertheless bright because the upper level is populated efficiently by radiative cascades, and its radiative branching ratio to ground is 1.0. Since the radiative decay rate of this transition is slow compared to the other $2p$–$3l$ lines, collisional depopulation sets in at lower densities. Thus, the intensity ratio of the 17.10 Å line to any of the other $2p$–$3l$ lines provides a density diagnostic, as was first pointed out by Klapisch et al. [23].

To investigate quantitatively the density and photoexcitation dependence of the Fe XVII I(17.10 Å)/I(17.05 Å) line ratio, we calculated LXSS models for this ion for $T_e = 4.1$ MK, $n_e = 10^{10}$–10^{18} cm^{-3}, and $T_{bb} = 0$–60 kK (for additional details, see [27]). Figure 8 shows the LXSS model I(17.10 Å)/I(17.05 Å) line ratio, the *Chandra* MEG spectrum of EX Hya in the neighborhood of the Fe XVII $2p$–$3s$ lines, and the value and 1, 2, and 3σ error envelopes of the measured I(17.10 Å)/I(17.05 Å) line ratio. The calculation shows that the blackbody temperature must be $T_{bb} \gtrsim 60$ kK to drive the I(17.10 Å)/I(17.05 Å) line ratio into the high-density limit, while the figure shows that, in EX Hya, $n_e \gtrsim 2 \times 10^{14}$ cm^{-3} or $T_{bb} \gtrsim 50$ kK at the 3σ level.

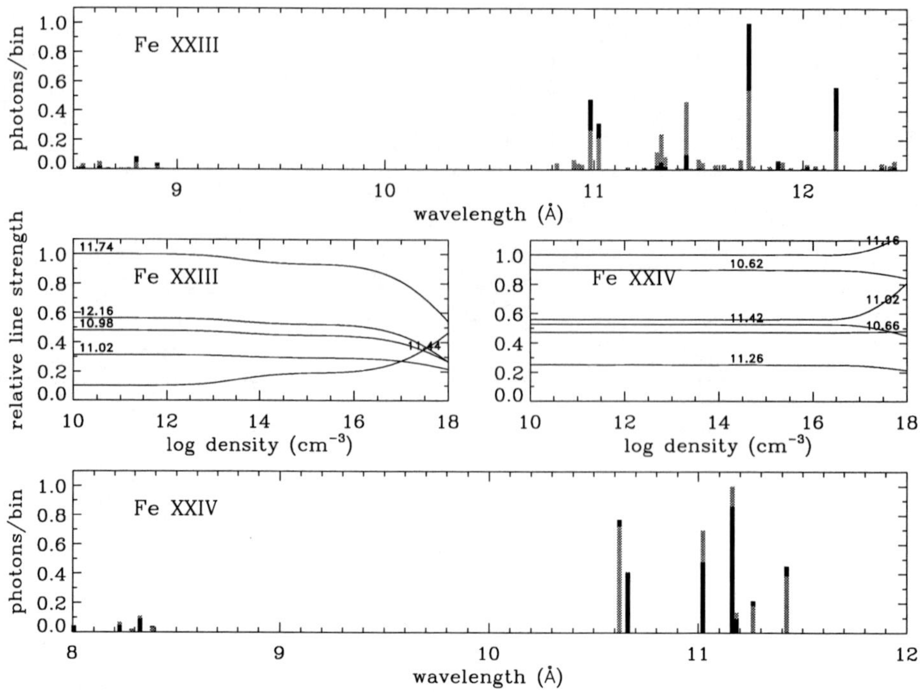

FIGURE 6. Similar to Fig 2, for Fe XXIII at $T_e = 14.1$ MK and Fe XXIV at $T_e = 17.8$ MK.

FE XXII

The density sensitivity of the X-ray lines of B-like Fe XXII has been discussed extensively in studies of the X-ray spectra of solar and laboratory high-temperature plasmas [9, 24, 8, 33, 11, 32, 41, 7]. At low densities, the Fe XXII electron population is primarily in the $2s^22p_{1/2}$ ground state, and collisional excitations are predominantly from the ground state into the $2s2p^2$ manifold (8 levels) and the $2s^23d_{3/2}$ level, both of which decay primarily to ground, producing lines in the extreme ultraviolet (EUV) and at 11.77 Å, respectively (Fig 7). However, these levels also decay to the $2s^22p_{3/2}$ first-excited level with approximately 15% probability, producing lines in the EUV and at 11.92 Å; when *that* level decays to ground, it produces a line in the ultraviolet. As the density increases, electron population builds up in the $2s^22p_{3/2}$ first-excited level because the M1 transition to ground is slow. At high densities, the $2s^22p_{3/2}$ first-excited level is fed primarily by radiative decays from the $2s2p^2$ manifold. Collisional excitations out of the first-excited level are primarily into the $2s^23d_{5/2}$ level, which decays primarily back to the first-excited level, producing a line at 11.92 Å. Consequently, the Fe XXII 11.92 Å line is relatively strong in the X-ray spectra of high-density plasmas.

To investigate quantitatively the density and photoexcitation dependence of the

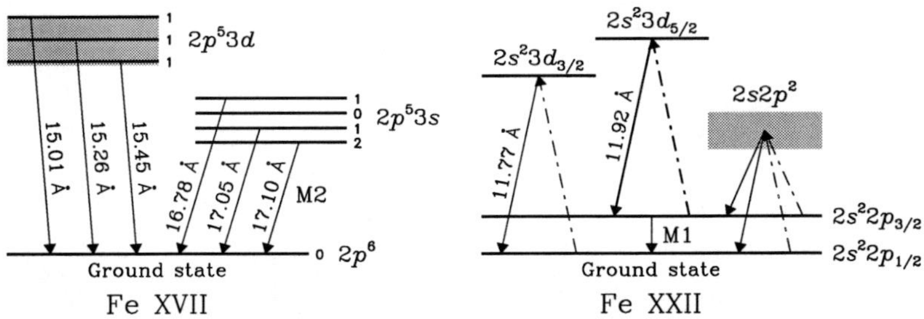

FIGURE 7. Simplified Grotrian diagrams for Fe XVII and Fe XXII. Collisional channels are shown with dot-dashed lines and radiative channels with solid lines; X-ray transitions are labeled with the wavelength.

Fe XXII $I(11.92\ \text{Å})/I(11.77\ \text{Å})$ line ratio, we calculated LXSS models for this ion for $T_e = 12.8$ MK, $n_e = 10^{10}$–10^{18} cm^{-3}, and $T_{bb} = 0$–100 kK. Motivated by the detailed study of Fournier et al. [13] of the density-sensitive Fe L-shell lines in the EUV, in these models we made two modifications to the collisional excitation data used in LXSS. First, we replaced, for all transitions between and among the $2s^22p$ and $2s2p^2$ levels of Fe XXII, the electron impact excitation rate coefficients computed with HULLAC with those of Zhang & Pradhan [42] computed with the relativistic R-matrix method. Second, we added proton excitations for transitions among the levels of the $2s^22p$ and $2s2p^2$ configuration using the proton impact excitation rate coefficients of Foster, Keenan, & Reid [12]. For additional details, see [28].

Figure 9 shows the resulting LXSS model $I(11.92\ \text{Å})/I(11.77\ \text{Å})$ line ratio, the *Chandra* MEG spectrum of EX Hya in the neighborhood of the Fe XXII $2p$–$3d$ lines, and the value and 1, 2, and 3σ error envelops of the measured $I(11.92\ \text{Å})/I(11.77\ \text{Å})$ line ratio. The figure shows that the $I(11.92\ \text{Å})/I(11.77\ \text{Å})$ line ratio is relatively insensitive to photoexcitation, and that, in EX Hya, $n_e = 1.0^{+2.0}_{-0.5} \times 10^{14}$ cm^{-3} at the 1σ level and $n_e \gtrsim 2 \times 10^{13}$ cm^{-3} or $T_{bb} \gtrsim 100$ kK at the 3σ level.

SUMMARY

We have provided a discussion of the density and photoexcitation sensitivity of the X-ray spectra of collisionally ionized Fe L-shell ions. Potentially useful density diagnostics are identified for Fe XVII and Fe XIX–Fe XXIII, with the most straightforward being the Fe XVII $I(17.10\ \text{Å})/I(17.05\ \text{Å})$ line ratio and the Fe XXII $I(11.92\ \text{Å})/I(11.77\ \text{Å})$ line ratio. Compared to the He-like R density diagnostics, the Fe L-shell density diagnostics are less sensitive to photoexcitation, hence are particularly valuable for sources like O stars and cataclysmic variables that are bright in the ultraviolet. Applying these models to the *Chandra* HETG spectrum of the intermediate polar EX Hya, we have shown that the observed Fe XVII $I(17.10\ \text{Å})/I(17.05\ \text{Å})$ line ratio is consistent with an electron density $n_e \gtrsim 2 \times 10^{14}$ cm^{-3} (or a blackbody photoexcitation temperature $T_{bb} \gtrsim 50$ kK) at the 3σ level, while from the observed Fe XXII $I(11.92\ \text{Å})/I(11.77\ \text{Å})$ line

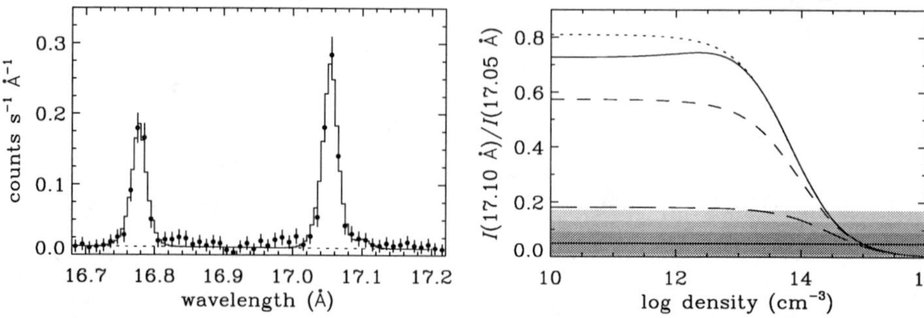

FIGURE 8. *Left panel*: Detail of the *Chandra* MEG spectrum of EX Hya in the neighborhood of the Fe XVII $2p$–$3s$ lines. Data (combining ± 1st orders and binned to 0.01 Å) are shown by the filled circles with error bars and the model fit is shown by the histogram. *Right panel*: LXSS model Fe XVII I(17.10 Å)/I(17.05 Å) line ratio as a function of electron density for $T_e = 4.1$ MK and $T_{bb} = 10, 30, 40$, and 50 kK (*solid, dotted, short dashed, and long dashed curves, respectively*). Horizontal line and the dark, medium, and light shaded stripes indicate respectively the value and the 68%, 90%, and 99% confidence error envelopes of the line ratio measured in EX Hya.

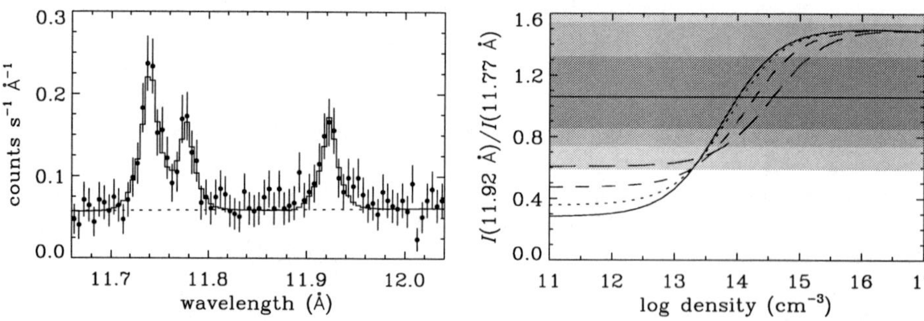

FIGURE 9. *Left panel*: Detail of the *Chandra* MEG spectrum of EX Hya in the neighborhood of the Fe XXII $2p$–$3d$ lines. Data (combining ± 1st orders and binned to 0.005 Å) are shown by the filled circles with error bars and the model fit is shown by the histogram. *Right panel*: LXSS model Fe XXII I(11.92 Å)/I(11.77 Å) line ratio as a function of electron density for $T_e = 12.8$ MK and $T_{bb} = 0, 60, 80$, and 100 kK (*solid, dotted, short dashed, and long dashed curves, respectively*). Horizontal line and the dark, medium, and light shaded stripes indicate respectively the value and the 68%, 90%, and 99% confidence error envelopes of the line ratio measured in EX Hya.

ratio, we infer $n_e = 1.0^{+2.0}_{-0.5} \times 10^{14}$ cm^{-3} at the $1\,\sigma$ level and $n_e \gtrsim 2 \times 10^{13}$ cm^{-3} (or $T_{bb} \gtrsim 100$ kK) at the $3\,\sigma$ level. The high densities are assured in EX Hya, whose far ultraviolet spectrum has been modeled as a $T_{bb} \approx 25$ kK white dwarf [17] and as a $T_{bb} \approx 20$ kK white dwarf with a $T_{bb} \approx 37$ kK hot spot [25]. While these results are particular to EX Hya, the Fe L-shell density diagnostics are applicable to collisionally ionized laboratory (e.g., tokamak) or cosmic (e.g., cataclysmic variable) plasmas with electron temperatures $T_e \approx 2$–45 MK and electron densities $n_e \gtrsim 10^{11}$ cm^{-3}.

ACKNOWLEDGMENTS

Support for this work was provided in part by NASA through *Chandra* Award Number DD0-1004B issued by the *Chandra* X-Ray Observatory Center, which is operated by SAO for and on behalf of NASA under contract NAS8-39073. This work was performed under the auspices of the U.S. Department of Energy by University of California Lawrence Livermore National Laboratory under contract No. W-7405-Eng-48.

REFERENCES

1. Ayres, T. R., et al. 2001, ApJ, 549, 554
2. Bar-Shalom, A., Klapisch, M., & Oreg, J. 1988, Phys. Rev. A, 38, 1773
3. Beiersdorfer, P., von Goeler, S., Bitter, M., & Thorn, D. B. 2001, Phys. Rev. A, 64, 032705
4. Blumenthal, G. R., Drake, G. W. F., & Tucker, W. H. 1972, ApJ, 172, 205
5. Brickhouse, N. S., & Drake, J. J. 2000, RevMexAA (Serie de Conferencias), 9, 24
6. Canizares, C. R., et al. 2000, ApJ, 539, L41
7. Chen, H., et al. 2004, ApJ, 611, 598
8. Doschek, G. A., Feldman, U., & Cowan, R. D. 1981, ApJ, 245, 315
9. Doschek, G. A., Meekins, J. F., & Cowan, R. D. 1973, Sol. Phys., 29, 125
10. Eisenbart, S., Beuermann, K., Reinsch, K., & Gänsicke, B. T. 2002, A&A, 382, 984
11. Fawcett, B. C., Jordan, C., Lemen, J. R., & Phillips, K. J. H. 1987, MNRAS, 225, 1013
12. Foster, V. J., Keenan, F. P., & Reid, R. H. G. 1997, At. Data Nucl. Data Tables, 67, 99
13. Fournier, K. B., et al. 2001, ApJ, 561, 1144
14. Frank, J., King, A., & Raine, D. 1992, Accretion Power in Astrophysics (Cambridge: CUP)
15. Fujimoto, R., & Ishida, M. 1997, ApJ, 474, 774
16. Gabriel, A. H., & Jordan, C. 1969, MNRAS, 145, 241
17. Greeley, B. W., Blair, W. P., Long, K. S., & Knigge, C. 1997, ApJ, 488, 419
18. Hoogerwerf, R., Brickhouse, N. S., & Mauche, C. W. 2004, ApJ, 610, 411
19. Hurwitz, M., Sirk, M., Bowyer, S., & Ko, Y.-K. 1997, ApJ, 477, 390
20. Kahn, S. M., et al. 2001, A&A, 365, L312
21. Klapisch, M. 1971, Computer Phys. Comm., 2, 239
22. Klapisch, M., Schwob, J., Fraenkel, B., & Oreg, J. 1977, J. Opt. Soc. Am., 67, 148
23. Klapisch, M., et al. 1978, Phys. Letters, 69A, 34
24. Mason, H. E., & Storey, P. J. 1980, MNRAS, 191, 631
25. Mauche, C. W. 1999, ApJ, 520, 822.
26. Mauche, C. W. 2002, in The Physics of Cataclysmic Variables and Related Objects, ed. B. T. Gänsicke, K. Beuermann, & K. Reinsch (San Francisco: ASP), 113
27. Mauche, C. W., Liedahl, D. A., & Fournier, K. B. 2001, ApJ, 560, 992
28. Mauche, C. W., Liedahl, D. A., & Fournier, K. B. 2003, ApJ, 588, L101
29. Mazzotta, P., Mazzitelli, G., Colafrancesco, S., & Vittorio, N. 1998, A&AS, 133, 403
30. Ness, J.-U., et al. 2002, A&Ap, 394, 911
31. Osten, R. A., et al. 2004, ApJS, 153, 317
32. Phillips, K. J. H., Bhatia, A. K., Mason, H. E., & Zarro, D. M. 1996, ApJ, 466, 549
33. Phillips, K. J. H., et al. 1982, ApJ, 256, 774
34. Phillips, K. J. H., et al. 1997, A&A, 324, 381
35. Porquet, D., Mewe, R., Dubau, J., Raassen, A. J. J., & Kaastra, J. S. 2001, A&A, 376, 1113
36. Porquet, D. 2005, this volume
37. Rugge, H. R., & McKenzie, D. L. 1985, ApJ, 297, 338
38. Saba, J. L. R., Schmelz, J. T., Bhatia, A. K., & Strong, K. T. 1999, ApJ, 510, 1064
39. Waldron, W. L., & Cassinelli, J. P. 2001, ApJ, 548, L45
40. Warner, B. 1995, Cataclysmic Variable Stars (Cambridge: CUP)
41. Wargelin, B. J., Beiersdorfer, P., Liedahl, D. A., Kahn, S. M., & von Goeler, S. 1998, ApJ, 496, 1031
42. Zhang, H. L., & Pradhan, A. K., 1997, A&AS, 123, 575

Drs. Mario Jimenez-Garate and Q. Daniel Wang discussing results during a break.

Recent Theoretical Developments on Atomic Data for Iron L-shell Ions

Ming Feng Gu

382 Via Pueblo Mall, Varian, Department of Physics, Stanford University, CA 94305

Abstract. Emission lines from L-shell ions of iron dominates the soft X-ray spectral band for hot collisionally-ionized plasmas with temperatures near 10^6 K. These lines also contribute significantly in the spectra of photoionized plasmas at lower temperatures. To accurately model such spectra, a large set of atomic data is needed, including wavelengths, radiative transition rates, collisional excitation, radiative and dielectronic recombination rates, and collisional ionization. We review the recent theoretical efforts in producing such atomic parameters and assess their accuracy by comparing them with the experimental measurements wherever available. We also compare the close-coupling and distorted-wave theoretical approaches in treating the collisional excitation of Fe L-shell ions.

INTRODUCTION

Due to the high abundance of iron in the cosmic environments and the complex structure of its L-shell ions, the X-ray lines resulting from the radiative transitions between $n \geq 3$ and $n = 2$ levels of Fe XVII–XXIV are the most prominent discrete features in the X-ray spectra of a wide variety of astrophysical sources, including stellar coronae, elliptical galaxies, and low temperature clusters of galaxies. The observations of solar corona [1], the recent high resolution X-ray spectrum of Capella [2, 3] and II Pegasi [4] obtained with *Chandra*, and that of the elliptical galaxy NGC 4636 obtained with *XMM Newton* [5] have demonstrated the richness of iron L-shell emission and the unique diagnostic opportunities these lines provide.

However, modeling the X-ray emission from Fe L-shell ions requires large amounts of accurate atomic data which have not been available until recently. The widely used plasma spectral synthesis codes, such as MEKA [6] and RS [7], have a very limited set of emission lines from Fe XVII-XXIV, and the atomic data used for these lines are generally of poor quality. The atomic data deficiencies of these codes have even been demonstrated with relatively low spectral resolution instruments such as *ASCA*. Liedahl et al. [8] later improved the MEKA model by including Fe L-shell atomic data calculated with the Hebrew University Lawrence Livermore Atomic Code (HULLAC) developed by Bar-Shalom et al. [35]. These data also form the foundation of the newer plasma codes, such as the astrophysical plasma emission code, or APEC [9]. Recently, Gu [10] developed a new distorted-wave atomic code, the Flexible Atomic Code (FAC), which can generate large sets of atomic data for complex ions. In this paper, we describe some of the data recently calculated with FAC, which includes wavelengths, radiative transition rates, collisional excitation and ionization cross sections, and radiative and dielectronic recombination rate coefficients. We assess the accuracy of these data by

CP774, *X-Ray Diagnostics of Astrophysical Plasmas,* edited by Randall K. Smith

comparing theoretical results with the experimental measurements wherever available. We also compare the close-coupling and distorted-wave theoretical approaches in treating the collisional excitation of Fe L-shell ions in some examples.

WAVELENGTHS OF L-SHELL TRANSITIONS IN IRON IONS

The most widely adopted theoretical method for calculating level energies and transition wavelengths is based on configuration interaction (CI) technique, in which the total wavefunction of an atomic state is assumed to be a summation over uncorrelated Slater determinants. Unfortunately, the accuracy of the transition energies obtained with this method is typically a few eV for the L-shell ions of Fe. In wavelength units, this amounts to $\sim$ 50 mÅ for the $3 \rightarrow 2$ transitions located at 10–18 Å. In contrast, the grating spectrometers on board *Chandra* can measure wavelength accurately to within a few mÅ. In the crowded L-shell spectral region, there is significant line blending that need to be taken into account in theoretical models.

Many of the strong and relatively isolated lines of Fe ions have been identified and measured in both solar spectra [11] and laboratory experiments. The resulting wavelengths have been compiled and tabulated in several atomic databases, e.g., the spectroscopic database at the National Institute of Standard and Technology, and the line table of Kelly [12]. A recent experimental effort at the Lawrence Livermore National Laboratory's electron beam ion traps have resulted in more accurate wavelength measurements for most iron L-shell lines [13, 14]. These new laboratory wavelengths have proven to be critical in analyzing the high resolution spectra from *Chandra* and *XMM-Newton* observatories [3].

However, there are still numerous weak or blended line features in Fe ions whose wavelengths are not well known. Gu [15] has recently carried out combined configuration interaction and many-body perturbation theory calculations of the $3 \rightarrow 2$ transition energies, which have accuracies in the sub-eV range. In this method, the configuration space is divided into two sub-spaces, M and N. M consists of configurations included in the CI expansion, whose correlation is treated to all orders, while N contains the rest of configurations whose correlation is accounted for with 2nd-order MBPT. The standard multi-reference Rayleigh-Schrödinger perturbation theory provides order-by-order expressions for the level energies

$$E_k = E_k^{CI} + E_k^2 + \cdots, \tag{1}$$

where E_k^{CI} is the CI eigenvalues, and E_k^2 is the 2nd-order correction due to correlations in space N [16]

$$E_k^2 = \sum_{n \in N} \sum_{i,j \in M} b_{ki} b_{kj} \frac{\langle \Phi_i | H_{BCD} | \Phi_n \rangle \langle \Phi_n | H_{BCD} | \Phi_j \rangle}{E_j^0 - E_n^0}, \tag{2}$$

where E_j^0 and E_n^0 are the zero-th order energies of the configuration state functions, i.e., the sum of the one-electron orbital energies. H_{BCD} is the standard Breit-Coulomb Dirac Hamiltonian.

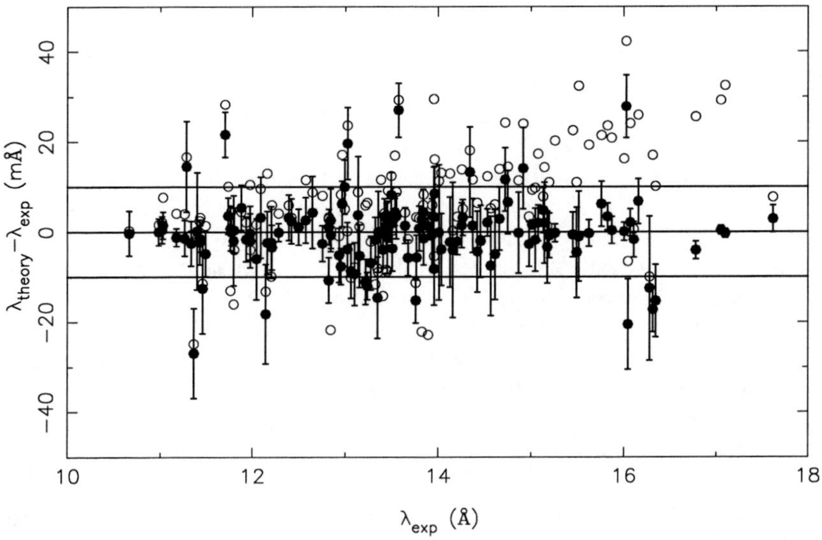

FIGURE 1. Comparison of theoretical and experimental wavelengths of $2l \rightarrow 3l'$ transitions of L-shell Fe ions. Filled circles with experimental error bars are for the combined CI and MBPT results; open circles are for the present CI calculations. The horizontal lines in the upper panel indicate the ± 10 mÅ region. Experimental wavelengths are from Brown et al. [13, 14].

In the present calculation of $1s^2 2l^q$ energies, the M space is chosen to be the $1s^2 2l^q$ configurations, and the N space consists of all single and double excitations from M. For the calculation of $1s^2 2l^{q-1} 3l'$ energies, the M space consists of $1s^2 2l^q$ and $1s^2 2l^{q-1} 3l'$ configurations, and N space contains all single and double excitations from them. The N space contains not only bound states, but also continuum. A finite basis set method is used to evaluate the summation in Equation 2 to convergence. The basis set is derived by imposing the boundary condition $Q(r_b)/P(r_b) = b$ for the single-electron radial wavefunctions, where Q and P are the small and large components of the Dirac spinor, r_b is the boundary radius, which is chosen to be large enough to contain the $1s$, $2l$, and $3l$ wavefunction amplitudes, and b is an arbitrary constant chosen to be $\kappa/2r_b c$, where κ is the relativistic angular momentum quantum number of the one-electron orbital, and c is the speed of light. With this choice, the boundary condition reduces to $P'(r_b)/P(r_b) = 0$ in the non-relativistic limit.

In Figure 1, we compare the present theoretical wavelengths with the experimental measurements of Brown et al. [13, 14] for L-shell Fe ions. It is clear that the MBPT method improves the wavelength accuracy significantly.

TOTAL RECOMBINATION RATE COEFFICIENTS

Recombination processes, including radiative recombination (RR) and dielectronic recombination (DR), are important in determining the ionization equilibrium of hot plasmas.

The commonly used set of RR rate coefficients in modern spectral modeling codes is due to the calculations and fits of Péquignot et al. [17] for C, N, O and Ne ions, and those of Shull and Van Steenberg [18] for Mg, Si, Ar, Ca, Fe, and Ni ions. Data for K and L-shell Fe ions have been revised by Arnaud and Raymond [19], which are based on the photoionization cross sections of Clark et al. [20]. The fitting formulae for these data are usually only valid in a restricted temperature range, and for many ions, interpolations or extrapolations were used to derive the rate coefficients when detailed photoionization calculations are not available. Verner and Ferland [21] presented RR rate coefficients of bare, H-like, He-like, and Ne-like ions for elements with $Z \leq 30$ in a wide temperature range, using the analytic fits to the photoionization cross sections of Verner et al. [22] for the ground-state shells, and those of Clark et al. [20] for the excited-state shells. Gu [23] carried out systematic calculations of total RR rate coefficients for bare to F-like isoelectronic sequences of Mg, Si, S, Ar, Ca, Fe, and Ni using detailed photoionization cross sections obtained with FAC.

Gu [24] also carried out both $\Delta N = 0$ and $\Delta N > 0$ total DR rate coefficients of the H-like through Ne-like isoelectronic sequences of Mg, Si, S, Ar, Ca, Fe, and Ni. The theoretical DR rate coefficients are found to agree well with the storage ring measurements for Fe ions [25, 26, 27].

Using these new RR and DR rate coefficients, we calculated the coronal ionization equilibria of L-shell ions, and found large differences as compared with the ionization balance tables of Mazzotta et al. [28], as shown in Figure 2 for iron ions.

L-SHELL LINE EMISSION IN COLLISIONAL PLASMAS

There are several long standing discrepancies between the observations and theories on the intensities of some strong iron L-shell lines in collisional plasmas, which have limited their diagnostic utility. These discrepancies are most frequently observed in Fe XVII lines because the closed electronic shell structure of Fe XVII makes it the most abundant ion in a wide temperature range at ionization equilibrium. The brightest $3 \rightarrow 2$ lines of Fe XVII arise from the $3d \rightarrow 2p$ and $3s \rightarrow 2p$ transitions, which are located at 15 and 17 Å approximately. The first problem is that the observed intensity of the 15.01 Å line arising from the $2p^5 3d\,^1P_1 \rightarrow 2p^6\,^1S_0$ transition is significantly smaller than the theoretical predictions as compared with its nearby $2p^5 3d\,^3D_1 \rightarrow 2p^6\,^1S_0$ line at 15.26 Å. This has led to the speculation that the opacity effects may have affected this line ratio [29]. However, recent experimental measurements using the electron beam ion trap (EBIT-II) at the Lawrence Livermore National Laboratory (LLNL) have shown that the existing theoretical models overestimate the 15.01 Å to the 15.26 Å line ratio [13]. In addition, the observed ratio is further reduced by line blending with one or more Fe XVI lines, depending on the spectral resolution [30, 3].

Another frequently encountered problem with the Fe XVII lines is that the overall intensities of $3s \rightarrow 2p$ lines at 17 Å are consistently larger than the calculations when compared with the $3d \rightarrow 2p$ lines at 15 Å. Several theoretical explanations have been offered to address this discrepancy. For example, Feldman [31] suggests that the $2p$ inner shell ionization of Fe XVI might enhance the $3s \rightarrow 2p$ intensities. Loulergue and

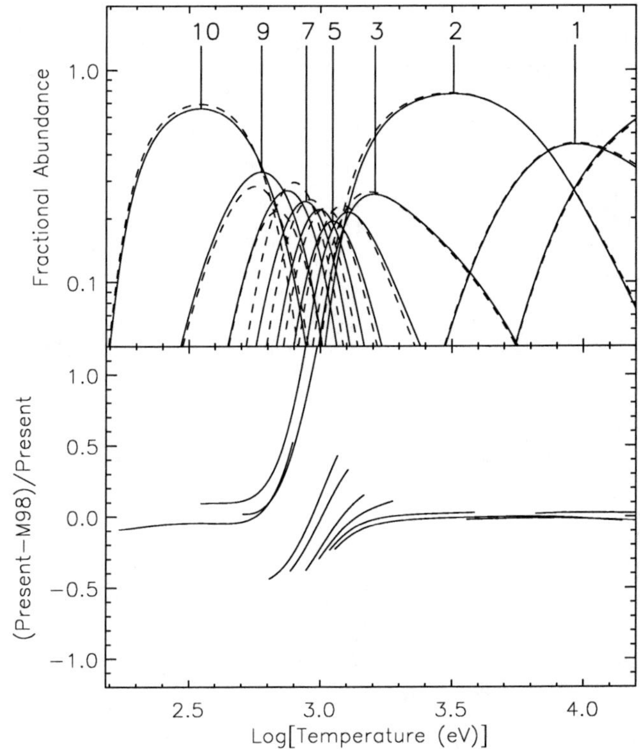

FIGURE 2. Comparison of coronal ionization equilibrium of Fe ions. Top: Fractional abundances of Fe XVII–XXVII. Solid lines are the present results, and dashed lines are those of Mazzotta et al. [28]. Numbers on the curves indicate the number of electrons of respective ions; Bottom: Relative differences of fractional abundances between the present calculations and those of Mazzotta et al. [28].

Nussbaumer [32] and Smith et al. [33] point out that resonance excitation (RE) can make significant contribution to the population of 3*s* levels. Finally, the contribution to the level population from DR of Fe XVIII has also been investigated [29]. All these processes preferentially increase the intensities of $3s \rightarrow 2p$ lines while have little effects on $3d \rightarrow 2p$ ones. However, residual discrepancies persist even after the consideration of these processes. Recently, in an effort to improve the atomic models, Doron and Behar [34] have constructed a three-ion collisional radiative model that includes atomic states of Fe XVIII, XVII, and XVI using HULLAC. This revised model uses the calculation of Chen and Reed [36] for the RE rate coefficients of Fe XVII and includes a limited set of DR channels which are believed to be most important at the relevant temperatures. The resulting $3s \rightarrow 2p$ to $3d \rightarrow 2p$ line ratio, although a vast improvement over the earlier models, is still smaller than most astrophysical observations, except at very low temperatures.

Gu [10] used a multi-ion model similar in spirit to that of Doron and Behar [34] and

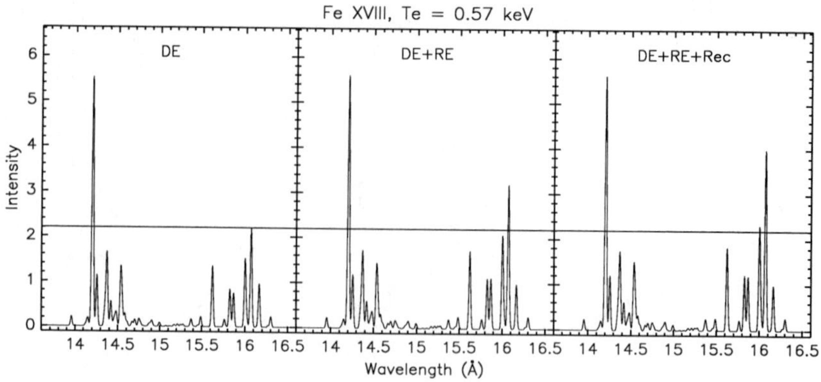

FIGURE 3. Spectra of Fe XVIII at a temperature of 570 eV. The left panel includes only direct collisional excitation, the middle panel includes resonance excitation, and the right panel also includes the recombination, mostly DR, contributions.

Doron et al. [37] on more complex L-shell ions of iron. The line formation processes due to RE and DR are found to be important for Fe XVII–XX. In some cases, the contributions from RE and DR can enhance the line intensities by a factor of 2. In Figure 3, we show the theoretical spectrum of XVIII at the temperature of peak abundance for the ion. The lines most affected by RE and DR are those due to $3s \rightarrow 2p$ transitions.

COMPARISON OF CLOSE-COUPLING AND DISTORTED-WAVE METHODS FOR COLLISIONAL EXCITATION

The applicability of distorted-wave (DW) method to treat collisional excitation for a particular ion is sometimes non-trivial, and is source of some debates. It is well known that the DW method only calculates the direct excitation without resonance contributions in its simplest form. The resonances can be added using the independent-process isolated-resonance (IPID) approximation, ignoring the interference between resonances and the background. In this section, we compare the $\Delta n = 0$ collision strengths within the $n = 2$ complex of B-like Fe XXII calculated with two methods. One is the DW plus IPID approximations, and the other is the more sophisticated relativistic R-matrix theory, both implemented in FAC [38]. We find that the DW plus IPID approximation can describe the collisional excitations very well except for those involving two-electron transitions, which are also heavily affected by configuration interaction.

The electron impact excitation of B-like Fe XXII has been studied extensively in the past. Zhang and Sampson [39] calculated all $\Delta n = 0$ collision strengths using a relativistic distorted-wave (RDW) approximation. The non-relativistic R-matrix method coupled with a term coupling coefficients (TCC) transformation was applied to the B-like iso-electronic sequence including Fe XXII [40]. Zhang and Pradhan [41, 42] used a relativistic R-matrix code employing the Breit-Pauli (BP) Hamiltonian to calculate the collision strengths of Fe XXII within the $n = 2$ and 3 complexes. A fully relativistic

Dirac R-matrix Atomic Code (DARC) was also used to study the collisional excitation within the $n = 2$ complex [43].

These previous studies have established the importance of both relativistic effects and the near-threshold resonances for this ion, and have shown that the R-matrix collision strengths in the non-resonant energy regions do not always agree with the RDW results even for some strong dipole-allowed transitions. For example, both TCC and DARC calculations show that the background collision strength of $2s^2 2p\ (^2P_{1/2})$–$2s2p^2\ (^2D_{3/2})$ is about a factor of 2 smaller than the RDW results at energies below the highest thresholds of $n = 2$ states. The DARC calculation also indicates that the collision strengths of dipole transitions at high energies are significantly smaller than the RDW results, even though the convergence of the partial-wave summation has been reportedly verified. The close-coupling collision strengths of the intercombination transition $2s^2 2p$ $(^2P_{1/2})$–$2s2p^2\ (^4P_{1/2})$ were also shown to be quite different from the RDW calculation at the resonance-free energy region.

The theoretical basis of Dirac R-matrix method was described in Chang [44], and a numerical implementation, DARC, was developed in Norrington and Grant [45]. The present work is a reimplementation of the same theory. As in the non-relativistic case [46], the configuration space is partitioned into two regions separated by the R-matrix boundary r_0. r_0 is chosen such that the exchange between the incident electron and the target electrons are negligible for $r > r_0$. In the inner region, the total scattering wavefunction is expanded on a basis set

$$\Psi_k = \sum_{ij} c_{ijk} \mathscr{A} [\Phi_i, \phi_{ij}] + \sum_m d_{mk} \theta_m, \tag{3}$$

where Φ_i is the wavefunction of the target states, ϕ_{ij} is the R-matrix basis for the continuum electron, θ_m is the $(N+1)$-electron correlation wavefunctions introduced to compensate for the orthogonality constraints imposed on ϕ_{ij}, and $\mathscr{A}$ represents the angular coupling and asymmetrization between the target and incident electrons. The coefficients c_{ijk} and d_{mk} are derived by diagonalizing the $(N+1)$-electron Hamiltionian matrix, which is the standard Dirac-Coulomb Hamiltonian.

The R-matrix radial basis functions satisfy the boundary condition

$$\frac{\mathscr{Q}_i(r_0)}{\mathscr{P}_i(r_0)} = \frac{b+\kappa}{2r_0 c}, \tag{4}$$

where $\mathscr{P}_i$ and $\mathscr{Q}_i$ are the large and small components of the Dirac spinor for the basis wavefunction, c is the speed of light, κ is the relativistic angular momentum quantum number, and b is an arbitrary constant usually chosen to be 0. The multi-channel scattering wavefunctions then satisfy the following boundary condition

$$P_i(r_0) = \sum_j R_{ij} \left[2r_0 c Q_j(r_0) - (b + \kappa_j) P_j(r_0) \right], \tag{5}$$

where R_{ij} is the R-matrix defined by

$$R_{ij} = \frac{1}{2r_0} \sum_k \frac{w_{ik}(r_0) w_{jk}(r_0)}{E_k^{N+1} - E}, \tag{6}$$

where w_{ik} are the surface amplitudes of Ψ_k in channel i, E_k^{N+1} are the R-matrix poles, and E is the total energy of the scattering system.

In the external region, the exchange between the scattered electron and the target electrons are neglected, and the channel wavefunctions satisfy a system of coupled differential equations, which is easier to solve than the usual differential-integral equations. The reactance K-matrix is defined through the standard asymptotic form of the channel wavefunctions as described in Young and Norrington [47]. The matching of the external and internal solutions at r_0 gives the K-matrix, from which the S-matrix and collision strengths are deduced.

A relativistic DW method is also used to calculate the same set of collision strengths, with the identical target description. The computational method for the background DW collision strengths is similar to that of Zhang et al. [48]. The near-threshold resonances are also calculated in the independent-process, isolated resonance approximation, in which the excitation from state a to b is treated as a two-step process, dielectronic capture forming a doubly excited state d which is followed by autoionization into state b. The resonance contribution to the collision strength is expressed as

$$\Omega_{ab} = \sum_d \pi g_d A^a_{da} \frac{A^a_{db}}{\sum_i A^a_{di} + \sum_k A^r_{dk}} \delta(E - E_{ad}), \tag{7}$$

where g_d is the statistical weight of the doubly excited state d, A^a is the autoionization rate, A^r is the radiative decay rate, and E_{ad} is the resonance energy. The resonance profile is approximated with a δ-function.

In Figure 4, we show the collision strengths of transitions between levels 1–2, 1–3, 1–6, and 1–11 of B-like Fe XXII. It is seen that previous R-matrix results for the background values do not agree with the present R-matrix and RDW calculations for 1–3, 1–6, and 1–11 transitions. The present RDW results, however, agrees well with the DW calculations of Zhang and Sampson [39], and also the present R-matrix calculations. The IPID approximation is also shown to be reasonably accurate in describing the near threshold resonances.

CONCLUSIONS

A variety of atomic processes and associated data calculated with the Flexible Atomic Code are presented, including a combined CI and MBPT calculation of transition energies and wavelengths of L-shell iron ions, the calculation of total RR and DR rate coefficients in the DW approximation, the various line formation processes, such as direct excitation, resonance excitation, and recombination, in the collisionally ionized plasmas, and a comparative study of relativistic R-matrix and DW approaches to the collisional excitation of Fe XXII.

Some of the work presented here were supported by NASA through Chandra Postdoctoral Fellowship Award Number PF01-10014 issued by the Chandra X-ray Observatory Center, which is operated by Smithsonian Astrophysical Observatory for and on behalf of NASA under contract NAS8-39073. We also gladly acknowledge the support from NASA grant NAG5-5419.

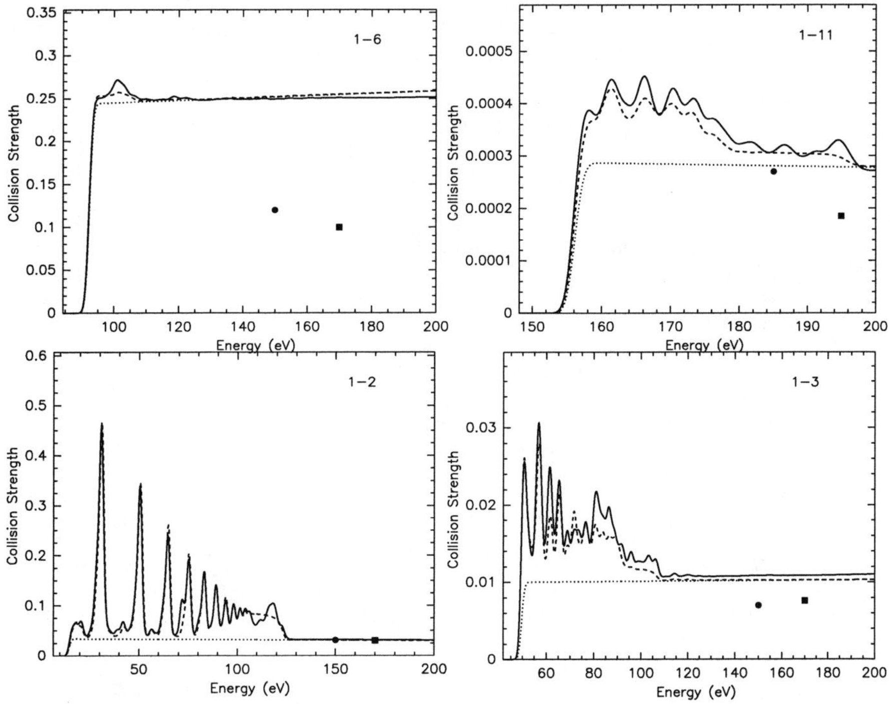

FIGURE 4. The collision strength of the transitions 1–2, 1–3, 1–6, and 1–11 of Fe XXII in the resonance energy region. The resonances have been convolved with a Gaussian function of FWHM=2.35 eV. The solid line is the present R-matrix result, the dotted line is the present RDW background, and the dashed line is the isolated-resonance DW result. The filled circle is the DARC results of Ait-Tahar et al. [43], and filled square is the non-relativistic R-matrix result of Zhang and Pradhan [40] for the background collision strengths.

REFERENCES

1. Rugge, H. R., and McKenzie, D. L., *ApJ*, **297**, 338 (1985).
2. Canizares, C. R., Huenemoerder, D. P., Davis, D. S., Flanagan, D. D. K. A., Houck, J., Markert, T. H., Marshall, H. L., Schattenburg, M. L., Schulz, N. S., Wise, M., Drake, J. J., and Brickhouse, N. S., *ApJ*, **539**, L41 (2000).
3. Behar, E., Cottam, J., and Kahn, S. M., *ApJ*, **548**, 966 (2001).
4. Huenemoerder, D. P., Canizares, C. R., and Schulz, N. S., *ApJ*, **559**, 1135 (2001).
5. Xu, H., Kahn, S. M., Peterson, J. R., Behar, E., Paerels, F. B. S., Moshutzky, R. F., Jernigan, J. G., and Makishima, K., *ApJ*, **579**, 600 (2002).
6. Mewe, R., Gronemschild, E. H. B. M., and van den Oord, G. H. J., *A & AS*, **62**, 197 (1985).
7. Raymond, J. C., and Smith, B. W., *ApJS*, **35**, 419 (1977).
8. Liedahl, D. A., Osterheld, A. L., and Goldstein, W. H., *ApJ*, **438**, L115 (1995).
9. Smith, R. K., Brickhouse, N. S., Liedahl, D. A., and Raymond, J. C., *ApJ*, **556**, 91 (2001).
10. Gu, M. F., *ApJ*, **582**, 1241 (2003).
11. Phillips, K. J. H., Mewe, R., Harra-Murnion, L. K., Kaastra, J. S., Beiersdorfer, P., Brown, G. V., and Liedahl, D. A., *A & A*, **138**, 381 (1999).
12. Kelly, R. R., *J. Phys. Chem. Ref. Data*, **16**, 861 (1987).

13. Brown, G. V., Beiersdorfer, P., Kahn, S. M., Liedahl, D. A., and Widmann, K., *ApJ*, **502**, 1015 (1998).
14. Brown, G. V., Beiersdorfer, P., Liedahl, D. A., Widmann, K., Kahn, S. M., and Clothiaux, E. J., *ApJS*, **140**, 589 (2002).
15. Gu, M. F., *At. Data Nucl. Data Tables*, p. in press (2004).
16. Vilkas, M. J., Ishikawa, Y., and Koc, K., *Phys. Rev. A*, **60**, 2808 (1999).
17. Péquignot, D., Petitjean, P., and Boisson, C., *A & A*, **251**, 680 (1991).
18. Shull, J. M., and Van Steenberg, M., *ApJS*, **48**, 95 (1982).
19. Arnaud, M., and Raymond, J., *ApJ*, **398**, 394 (1992).
20. Clark, R. E. H., Cowan, R. D., and Bobrowicz, F. W., *At. Data Nucl. Data Tables*, **34**, 415 (1986).
21. Verner, D. A., and Ferland, G. J., *ApJS*, **103**, 467 (1996).
22. Verner, D. A., Ferland, G. J., Korista, K. T., and Yakovlev, D. G., *ApJ*, **465**, 487 (1996).
23. Gu, M. F., *ApJ*, **589**, 1085 (2003).
24. Gu, M. F., *ApJ*, **590**, 1131 (2003).
25. Savin, D. W., Kahn, S. M., Linkemann, J., Saghiri, A. A., Schmitt, M., Grieser, M., Repnow, R., Schwalm, D., Wolf, A., Bartsch, T., Brandau, C., Hoffknecht, A., Müller, A., Schippers, S., Chen, M. H., and Badnell, N. R., *ApJS*, **123**, 687 (1999).
26. Savin, D. W., Behar, E., Kahn, S. M., gwinner, G., Saghiri, A., Schmitt, M., Grieser, M., Repnow, R., Schwalm, D., Wolf, A., Bartsch, T., , Müller, A., Schippers, S., Badnell, N. R., Chen, M., and Gorczyca, T. W., *ApJS*, **138**, 337 (2002).
27. Savin, D. W., Kahn, S. M., Linkemann, J., Saghiri, A., Schmitt, M., Grieser, M., Repnow, R., Schwalm, D., Wolf, A., Bartsch, T., , Müller, A., Schippers, S., Chen, M., Badnell, N. R., Gorczyca, T. W., and Zatsarinny, O., *ApJ*, **576**, 1098 (2002).
28. Mazzotta, P., Mazzitelli, G., Colafrancesco, S., and Vittorio, N., *A & AS*, **133**, 403 (1998).
29. Saba, J. L., Schmelz, J. Y., Bhatia, A. K., and Strong, K. T., *ApJ*, **510**, 1064 (1999).
30. Brown, G. V., Beiersdorfer, P., Chen, H., Chen, M. H., and Reed, K. J., *ApJ*, **557**, L75 (2001).
31. Feldman, U., *Comments At. Mol. Phys.*, **31**, 11 (1995).
32. Loulergue, M., and Nussbaumer, H., *AA*, **45**, 125 (1975).
33. Smith, B. W., Rymond, J. C., Mann, J. B., and Cowan, R. D., *ApJ*, **298**, 898 (1985).
34. Doron, R., and Behar, E., *ApJ*, **574**, 518 (2002).
35. Bar-Shalom, A., Klapisch, M., and Oreg, J., *J. Quant. Spect. Rad. Trans.*, **71**, 169 (2001).
36. Chen, M. H., and Reed, K. J., *Phys. Rev. A*, **40**, 2292 (1989).
37. Doron, R., Behar, E., Fraenkel, M., Mandelbaum, P., Schwob, J. L., Zigler, A., Faenov, A. Y., and Pikuz, T. A., *Phys. Rev. A*, **58**, 1859 (1998).
38. Gu, M. F., *Phys. Rev. A*, **70**, 062704 (2004).
39. Zhang, H. L., and Sampson, D. H., *At. Data Nucl. Data Tables*, **56**, 41 (1994).
40. Zhang, H. L., and Pradhan, A. K., *Phys. Rev. A*, **50**, 3105 (1994).
41. Zhang, H. L., and Pradhan, A. K., *J. Phys. B*, **28**, L285 (1995).
42. Zhang, H. L., and Pradhan, A. K., *A & AS*, **123**, 575 (1997).
43. Ait-Tahar, S., Grant, I. P., and Norrington, P. H., *Phys. Rev. A*, **54**, 3984 (1996).
44. Chang, J. J., *J. Phys. B*, **10**, 3335 (1975).
45. Norrington, P. H., and Grant, I. P., *J. Phys. B*, **20**, 4869 (1987).
46. Burke, P. G., Hibbert, A., and Robb, W. D., *J. Phys. B*, **4**, 153 (1971).
47. Young, I. G., and Norrington, P. H., *Comp. Phys. Comm.*, **83**, 215 (1994).
48. Zhang, H. L., Sampson, D. H., and Mohanty, A. K., *Phys. Rev. A*, **40**, 616 (1989).

Comparison of Fe XVIII and Fe XIX Line Emissions with Spectral Models

P. Desai*, N.S. Brickhouse*, J. J. Drake*, R. J. Edgar*, R. Hoogerwerf*, V. Kashyap*, B. J. Wargelin*, R. K. Smith†, D. P. Huenemoerder** and D. A. Liedahl‡

*Harvard-Smithsonian Center for Astrophysics, 60 Garden Street, Cambridge, MA 02138
†NASA Goddard Space Flight Center/JHU, Code 662, Greenbelt, MD 20771
**MIT Center for Space Research, 70 Vassar Street, Cambridge, MA 02139
‡Lawrence Livermore National Laboratory, P.O. Box 808, L-41, Livermore, CA 94551

Abstract. We discuss here the observations of Fe XVIII and XIX emission lines and compare the X-ray, EUV and FUV lines with the spectral codes widely used today (e.g. FAC and APEC). We assess the relative accuracy of these spectral models and try to identify the critical atomic data and processes. Capella with a narrow enhancement in its emission measure distribution at 6 MK provides a unique opportunity to test the Fe XVIII and Fe XIX model emissivities which peak from 6 to 8 MK. We use the summed spectra from Chandra HETG/ACIS-S and LETG/HRC-S, as well as contemporaneous EUVE and FUSE observations, to measure line ratios for comparison with predictions.

Keywords: Capella; atomic processes; X-rays; stars; coronae
PACS: 32.30.Rj

MOTIVATION

In recent years spectral codes have undergone major improvements; however their accuracy and completeness for diagnostic analysis have not been fully assessed. The Emission Line Project observations of Capella (Fig. 1) offer a unique opportunity to compare model agreement among FUV, EUV and X-ray lines. Multiwavelength comparisons of Fe XVIII and Fe XIX should be particularly useful since their strong n= 2 $\rightarrow$ 2 lines (not present in Fe XVII for which problems with line ratios are long standing) are almost entirely produced by direct collisional excitation, and hence should be useful for comparison with the X-ray lines. The relative flux calibration for the Chandra gratings is believed accurate to $\sim$ 15%, and thus these data can provide good constraints.

The Capella spectra (Fig. 1) are unusually steady over time and thus we do not expect the diagnostics to be affected by varying plasma conditions. Figure 2 shows multi-wavelength comparisons between measured and predicted line fluxes, including only the strongest of the X-ray lines, for APEC V1.3 (Smith et al. 2001), CHIANTI V4.2 (Young et al. 2003), SPEX V1.0 (approx MEKAL, Mewe et al. 1985), and FAC (Gu 2003). Predictions use the emission measure distribution shape from Brickhouse et al.(2000) normalized to current fluxes.

CP774, *X-Ray Diagnostics of Astrophysical Plasmas*, edited by Randall K. Smith

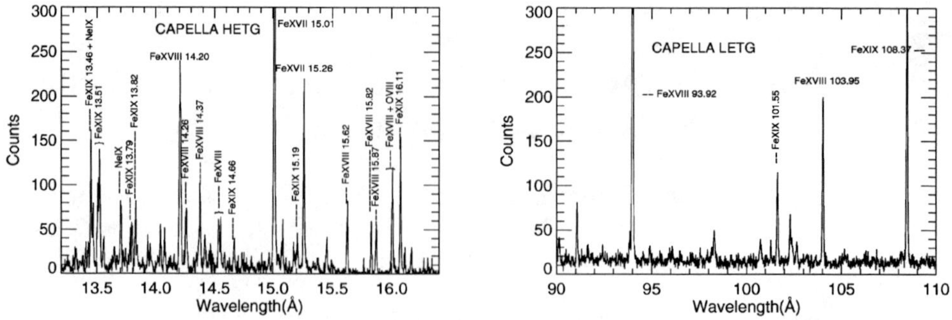

FIGURE 1. Combined HETG/ACIS-S observations of Capella provide high signal-to-noise ratio in the region of Fe XVIII and XIX lines (top left). The combined LETG/HRC-S spectrum (top right) between 90 and 110 Å shows strong Fe XVIII and XIX EUV lines. The $\lambda 93.9$ and $\lambda 108.4$ resonance lines are formed by direct excitation from the ground and thus we expect them to have accurate emissivities. Corrections for interstellar absorption are less than 8%; observed branching ratios with other EUV lines agree well with predictions.

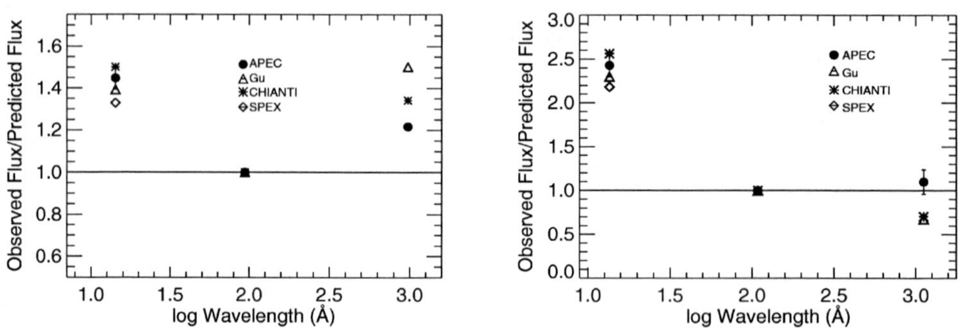

FIGURE 2. (Left) Observed to predicted ratios of strong Fe XVIII lines in the X-ray, EUV and FUV regions, ie.$\lambda 14.2$, $\lambda 93.9$, $\lambda 974.4$, normalized to the $\lambda 93.9$ resonance line. (Right) Observed to predicted ratios of strong Fe XIX lines in the X-ray, EUV and FUV regions, ie. $\lambda 13.5$, $\lambda 108.4$, $\lambda 1118.1$, normalized to the $\lambda 108.4$ resonance line. Shown for comparison are the ratios obtained using APEC, CHIANTI, SPEX, and FAC spectral codes. Error bars are only shown for APEC comparisons. FUV line fluxes from FUSE are from Young et al.(2001).

Conclusion

There is a striking discrepancy between the EUV and X-ray lines which is not explained by calibration or line blending. Even the X-ray resonance lines $\lambda 14.2$ and $\lambda 13.5$

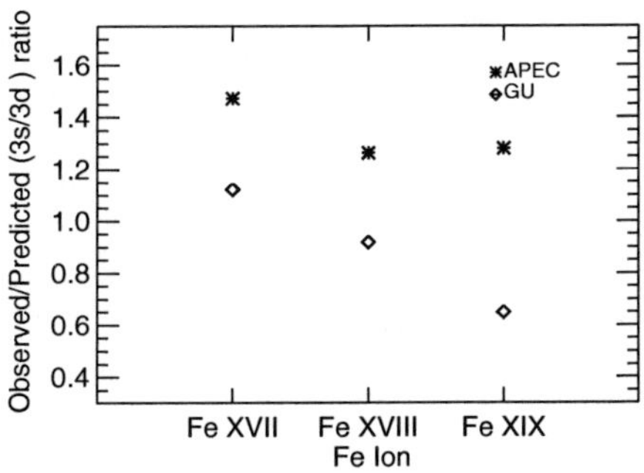

FIGURE 3. Observed to predicted (3s→2p/3d→2p) line ratios of Fe XVII, Fe XVIII and Fe XIX lines. The predicted models are from APEC and FAC at 6 million degrees. The 3s lines included for Fe XVII are λ16.78, λ17.05 and λ17.1. The 3d lines included are λ15.01 and λ15.26. For Fe XVIII the 3s line included are λ15.62, λ15.82, λ15.87. The 3d lines are λ14.20, λ14.26, λ14.37, λ14.53 and λ14.57. For Fe XIX the 3s lines are λ14.66 and λ15.07. The 3d lines are λ13.46, λ13.49, λ13.51, λ13.79. The Fe XVII line fluxes have been taken from Canizares et al.(2000). It should be noted that the models used do not include weak ($>$ 10%) blends.

are under-predicted relative to their EUV counterparts (Fig 2).

New calculations from FAC show significant improvement for lines originating from the 3s level, indicating the importance of DR in line formation. However the 3d→2p line predictions from APEC and FAC are fairly similar. Figure 3 shows the general pattern for the (3s→2p/3d→2p) line ratio, which is useful for low resolution X-ray spectra, however individual lines show a more complex pattern of behavior (Desai et al. 2005) and there are many discrepancies larger than 50%.

REFERENCES

1. Brickhouse, N. S., et al. 2000, ApJ, 530, 387
2. Canizares, C. R., et al. 2000, ApJ, 539, L41
3. Desai, P., et al 2005, ApJ Letters, in prep.
4. Gu, M.F., 2003, ApJ, 582, 1241
5. Mewe, R., Kaastra, J., & Liedahl, D. A. 1985, Legacy, 6, 16
6. Smith, R. K., et al. 2001, ApJ, 556, L91
7. Young, P., et al. 2001, ApJ, 555, L121
8. Young, P. R., Del Zanna, G., Landi, E., Dere, K. P., Mason, H. E., & Landini, M. 2003, ApJS, 144, 135

III: HELIUM-LIKE TRANSITIONS

Accurate wavelengths for X-ray spectroscopy and the NIST hydrogen-like ion database

S. A. Kotochigova[*,†], K. P. Kirby[**,‡], N. S. Brickhouse[‡], P. J. Mohr[†] and I. I. Tupitsyn[§]

[*]*Department of Physics, Temple University, Philadelphia, PA 19122*
[†]*National Institute of Standards and Technology, 100 Bureau Drive, stop 8423, Gaithersburg, Maryland 20899*
[**]*Institute for Theoretical Atomic and Molecular Physics, 60 Garden St., Cambridge, MA 02138*
[‡]*Harvard-Smithsonian Center for Astrophysics, 60 Garden St., Cambridge, MA 02138*
[§]*St. Petersburg State University, Petrodvorets, 198504 Russia*

Abstract. We have developed an *ab initio* multi-configuration Dirac-Fock-Sturm method for the precise calculation of X-ray emission spectra, including energies, transition wavelengths and transition probabilities. The calculations are based on non-orthogonal basis sets, generated by solving the Dirac-Fock and Dirac-Fock-Sturm equations. Inclusion of Sturm functions into the basis set provides an efficient description of correlation effects in highly charged ions and fast convergence of the configuration interaction procedure.

A second part of our study is devoted to developing a theoretical procedure and creating an interactive database to generate energies and transition frequencies for hydrogen-like ions. This procedure is highly accurate and based on current knowledge of the relevant theory, which includes relativistic, quantum electrodynamic, recoil, and nuclear size effects.

Keywords: atomic data, X-ray emission spectra, Dirac-Fock theory, Sturmian functions, highly-charged ions of iron, hydrogen-like ions
PACS: 32.30.Rj

INTRODUCTION

This paper describes two different theoretical methods that we have developed to provide reliable and accurate atomic data needed for X-ray spectroscopy and astrophysical modeling. The first method is based on Multi-configuration Dirac-Fock-Sturm (MDFS) theory (see, for example, Refs. [1, 2] and references therein). It has unique features designed for efficient description of correlation effects in highly charged ions. The second method allows us to calculate highly accurate energies and transition frequencies of hydrogen-like ions, where correlation effects do not exist. Numerous relativistic and quantum electrodynamic (QED) corrections to the Dirac energy, together with nuclear size effects, have been included to obtain precise transition frequencies. This data will be available through a Physics Laboratory interactive web site at the National Institute of Standards and Technology.

We applied the MDFS method to calculate L-shell emission spectra of Fe XVIII up to Fe XXII. This work is part of the Emission Line Project at the Smithsonian Astrophysical Observatory to improve atomic databases for X-ray diagnostics. High quality observations from Chandra and XMM-Newton contain a wealth of emission lines which, if properly identified, can be used for determining plasma properties. The

CP774, *X-Ray Diagnostics of Astrophysical Plasmas,* edited by Randall K. Smith

L-shell emission spectra of iron have been relatively well studied by both theoretical and laboratory work (See for example Refs.[3, 4]). The accuracy obtained by these studies is very impressive. The best values of the transition wavelengths are determined with an accuracy of about 10^{-3} Å. At the same time the information provided is far from complete. For example, there are weak lines of Fe XIX between 13 Å and 14 Å, which have to be known better in order to understand their contribution to blends in the observed spectra. In addition, the resolution of laboratory measurements for Fe XX up to Fe XXII is not sufficiently high to avoid line blending in some spectral regions.

Our goal is to obtain information that can help to identify weak and blended lines in X-ray spectra of iron. The high accuracy that we expect to obtain is due to our computational approach, which is a combination of a relativistic MDFS and second order many-body perturbation theory (MBPT2) and includes an advanced treatment of correlation effects. In this presentation we focus on the calculation of energies and transition wavelengths only. The results for other properties such as oscillator strengths, photoionization and autoionization cross-sections are forthcoming.

MDFS THEORY

In our configuration interaction (CI) approach, the total wavefunction of an atom or ion is written as a linear combination of Slater determinants constructed from four component Dirac spinors and numerical Sturm functions:

$$\Psi(x_1,x_2,...,x_N) = \sum_{\alpha} C_\alpha \det{}_\alpha\left(\phi(x_i)\right), \tag{1}$$

where N is the number of electrons and α counts the Slater determinants. The radial parts of the Dirac spinors or Sturm functions, $\phi(x)$, are obtained numerically by solving self-consistently either the Dirac-Fock or Dirac-Fock-Sturm equations. The coefficients C_α are obtained by solving a generalized eigenvalue problem by applying the iterative Davidson eigensolver. Higher-order correlation effects from highly excited orbitals, which are not included in the CI, are treated by second-order perturbation theory.

Dirac-Fock basis set

Dirac-Fock functions are chosen to describe the valence electrons and Sturm-Dirac functions represent the virtual orbitals. For example, Fe XIX has $1s^2$ $2s^2$ $2p^4$ electrons in the ground configuration and $1s^2$ $2s^2$ $2p^3$ 3s or 3p or 3d electrons in low-lying excited states of L-shell transitions. In our model all these valence orbitals are occupied. The valence orbitals are obtained by solving the Dirac-Fock (DF) equations. Higher-excited orbitals, such as 4s, 4p, 4d, and etc, if they are not occupied, play the role of virtual orbitals.

The use of DF functions as virtual orbitals has been shown to be ineffective. The radius of the excited DF orbitals grows rapidly with the level of excitation and becomes much larger than the radius of the valence orbitals. As a result the overlap between these orbitals is small, leading to slow convergence with an increase in the number of

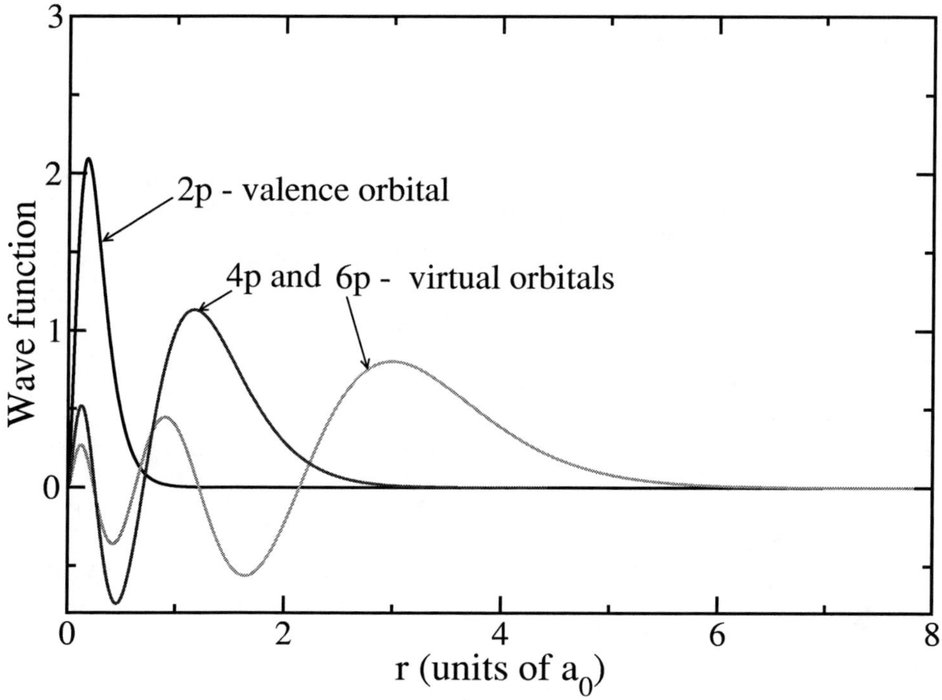

FIGURE 1. Dirac-Fock p-orbitals of Fe XIX.

orbitals in the CI procedure. Figure 1 shows the Dirac-Fock p orbitals of Fe XIX as an example. The 2p valence orbital is seen to have minimal overlap with the 4p and 6p virtual orbitals. The overlap between valence and virtual orbitals decreases as the principal quantum number n increases.

Sturm basis set

An alternative is to use Sturm functions obtained by solving the Dirac-Fock-Sturm equations, which are approximately given by

$$[H_{DF} - \varepsilon_0]\phi_j = \mu_j W_l(r)\phi_j \tag{2}$$

and solved for a fixed energy ε_0 equal to the energy of one of the valence electrons and fixed l. Here, H_{DF} is the Dirac-Fock part of the equation. The parameter μ plays the role of the eigenvalue of the Sturm operator. This operator is Hermitian and its eigenfunctions form a complete, discrete set of functions. We thus avoid the use of extended continuum functions. The l-dependent function $W_l(r)$ is a weight function with which the Sturm functions are orthogonalized:

$$\int_0^\infty drW_l(r)P_{nl}(r)P_{n'l}(r) = 0, \quad n \neq n'. \tag{3}$$

In Refs [5, 6] W_l was chosen to have the form $1/r$, which works well for simple systems. However, this form has singularities at r=0 and slow convergence for $r \to \infty$. We have found a better form for $W(r)$:

$$W_l(r) = -\frac{1-\exp(-[\alpha_l r]^2)}{[\alpha_l r]^2}. \tag{4}$$

This form approaches a constant for $r < 1/\alpha_l$ and decreases to zero more rapidly than $1/r$ as $r \to \infty$.

Figure 2 shows the Dirac-Fock 2p valence orbital and virtual 4p and 6p Sturm orbitals. Sturm functions are calculated using the ε_0 of the 2p valence orbital and are localized in the same spatial region as this reference function. They differ in the number of nodes. Near the core the functions are nearly identical and have a similar asymptotic behavior as the corresponding valence orbital. All these properties of Sturm functions help to construct efficient excited configurations that lead to a compact and rapidly convergent CI wave function.

COMPUTATIONAL MODEL

The Hamiltonian for the electrons contains three electron-electron operators. These are the Coulomb, magnetic, and retardation interactions. The basis sets are determined by including the Coulomb operator only. The magnetic and retardation contributions are then added for the CI calculation. Nuclear finite-size corrections are taken into account by using a Fermi charge distribution with parameters estimated from semi-empirical formulas using the Fe mass number.

We applied the MDFS+MBPT2 method to calculate wavelengths of L-shell transitions in Fe XVIII up to Fe XX. These spectra are sufficiently well understood from combined experimental and theoretical studies. Once we are able to reproduce and even improve upon this data, we will be able to make predictions for spectral regions where the assignment is not certain.

The scale of the calculations is determined by the size of the Hamiltonian matrix. For all three cases of Fe XVIII, Fe XIX, and Fe XX we have used the same number of orbitals (46 relativistic orbitals) to construct the basis sets. However, the number of electrons in the 2p-shell differs and determines the size of the H-matrix. In the basis set not all orbitals are treated equally. Only orbitals from $1s^2$ up to 5d are used to construct configurations that appear in the CI. Higher excited orbitals, up to n = 8, are treated by second-order perturbation theory. For the Fe XIX $2p^3 3s$ calculation, for example, we have about 60 000 determinants in the Hamiltonian matrix and CI calculation. About 10^7 determinants are included using perturbation theory.

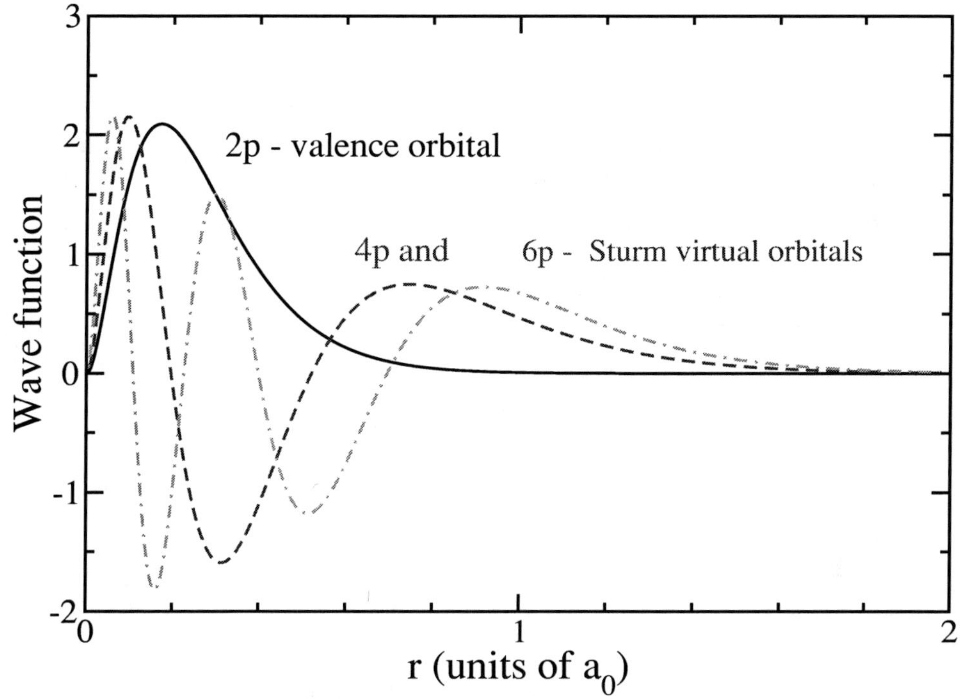

FIGURE 2. Dirac-Fock and Sturmian *p*-orbitals for Fe XIX.

RESULTS AND COMPARISON

As a first step we have compared our calculations with results of an EBIT experiment and calculations with the HULLAC code, given in Ref. [3]. This is the most recent and complete study of the L-shell emission spectra of Fe XVIII to Fe XX. We show our results for the medium-sized calculation of Fe XIX 3s to 2p transitions. Figure 3 shows a number of lines with wavelengths ranging from 14.6 Å to 15.2 Å, which we are able to compare to the EBIT data and HULLAC calculations of Ref. [3]. The labels on the horizontal axis of Fig.3 correspond to a labeling of lines from the EBIT experiment. The markers connected by a full line show the wavelengths of our calculation. The markers connected by a dotted line correspond to HULLAC data and the error bars show the experimental results. Clearly, our calculations of transition wave lengths are consistently lower than HULLAC by a few hundredths of an Angstrom. We are in better agreement with experiment than HULLAC for most of the lines, with the exception of the *O*9 line. Table 1 lists the wavelengths as well as the quantum labels of 3s $\rightarrow$ 2p transitions in Fe XIX obtained in our calculation and by Ref. [3].

Figure 4 shows a similar graph for the Fe XX spectrum. The line labels correspond to the notation of Ref. [3]. For the lines that have been assigned in the EBIT experiment the agreement between the three sources is similar to that for Fe XIX. The exception

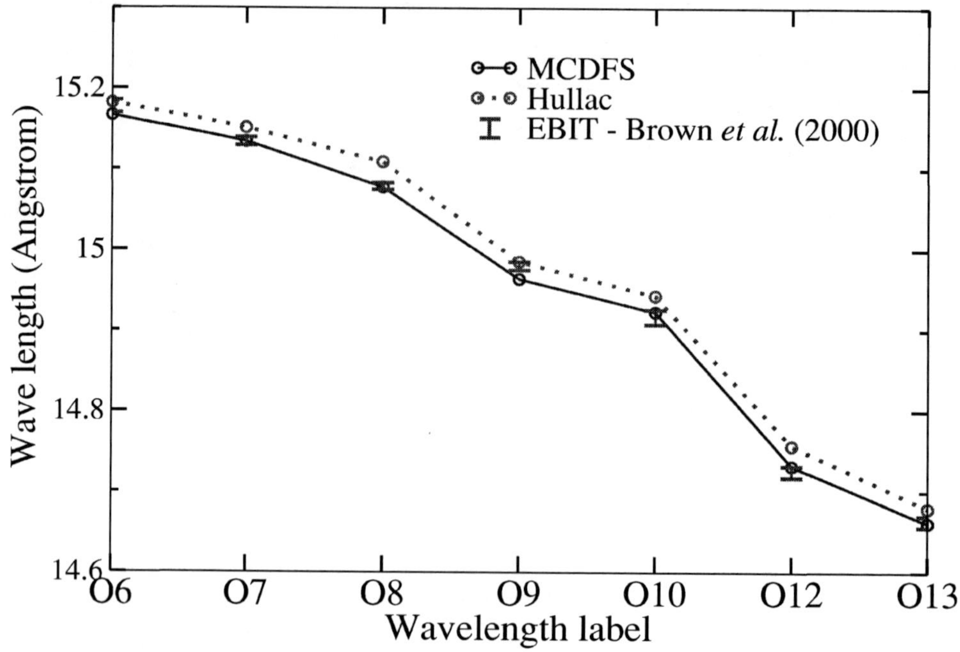

FIGURE 3. Wavelengths of $1s^22s^22p^33s \rightarrow 1s^22s^22p^4$ transitions in Fe XIX from our calculation and Ref. [3].

TABLE 1. Wavelengths of $1s^22s^22p^33s \rightarrow 1s^22s^22p^4$ transitions in Fe XIX from our calculation and Ref. [3]. Columns 2 and 3 give the total angular momentum of the upper and lower state of the transition.

Line label	Upper J	Lower J	MDFS	HULLAC	EBIT
	2	1	15.283		
*O*6	1	1	15.167	15.182	15.177(8)
*O*7	1	0	15.134	15.151	15.134(5)
*O*8	2	2	15.078	15.109	15.079(4)
*O*9	1	2	14.964	14.985	14.980(5)
	2	1	14.928	14.948	
	1	1	14.923	14.943	14.917(9)
	1	0	14.892		
	2	1	14.813		
*O*12	2	2	14.732	14.757	14.725(7)
	1	2	14.727		
	0	1	14.668		
*O*13	3	2	14.662	14.680	14.664(7)

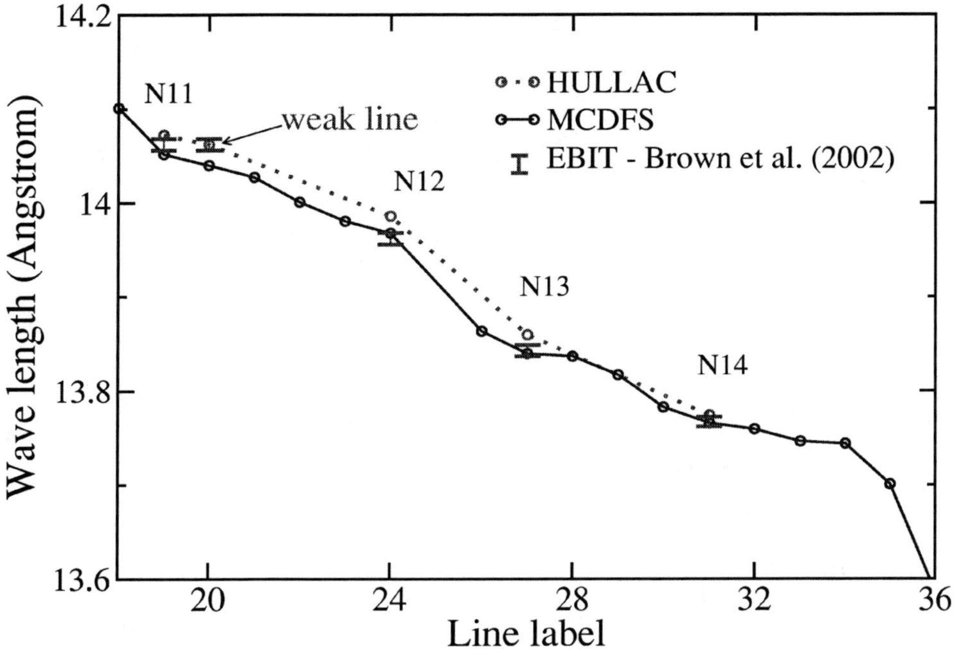

FIGURE 4. Wavelengths of $1s^2 2s^2 2p^2 3s \rightarrow 1s^2 2s^2 2p^3$ transitions in Fe XX. Our calculations are compared with results of Ref. [3]

is the line between $N11$ and $N12$, which is very weak in the experimental spectrum. In addition, we note that our calculation shows many more 3s $\rightarrow$ 2p transitions than are identified in Ref. [3]. To assign other observed lines with the help of our calculations we have to know the oscillator strength of each transition. This information will be forthcoming.

Figure 5 shows the Fe XVIII spectrum between 15.7 Å and 16.1 Å. Again we can see the same sort of agreement between the three sources. The two theoretical calculations differ significantly from experiment for the $F7$ line. This line, however, is reported as weak in the experimentally observed spectrum [3] and could have been assigned incorrectly.

The comparison of the MDFS calculations with data from Ref. [3] indicates that we reached 10^{-3} Å agreement with the experimental EBIT wavelengths for the three iron ions Fe XVIII, Fe XIX, and Fe XX. In our calculation we do not include QED corrections. An estimate shows that these corrections are expected to be on the order of 10^{-3}-10^{-4} Å and therefore might be relevant. We plan to include the major QED corrections in our calculations in the near future. The systematic blue-shift of our wavelengths from HULLAC's data can be explained by the fact that the total energy of the ground states in our calculation is lower and according to the variational principal

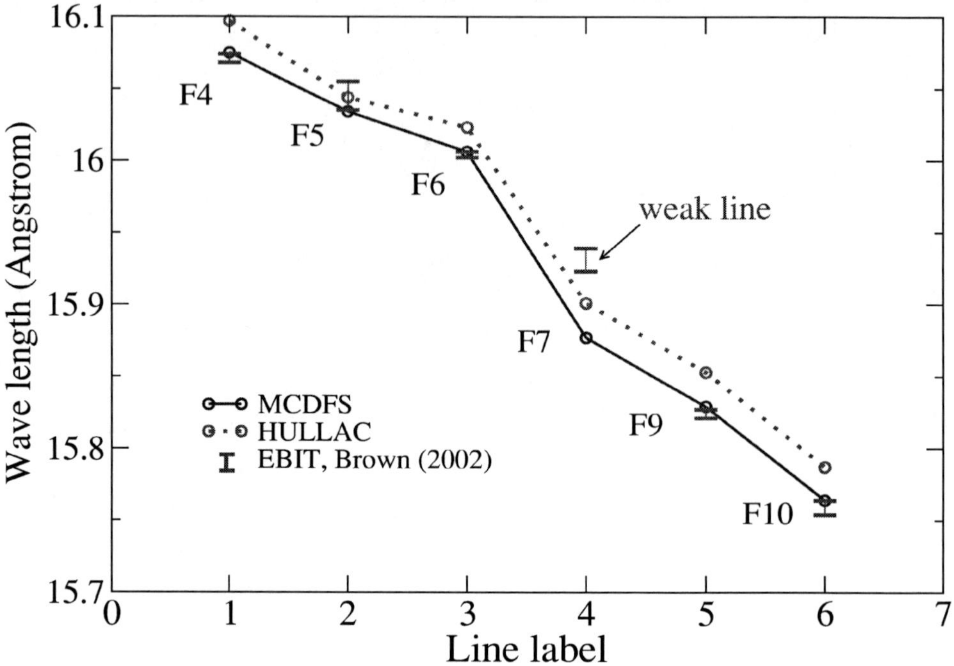

FIGURE 5. Wavelengths of $1s^22s^22p^43s \rightarrow 1s^22s^22p^5$ transitions in Fe XVIII. Our calculations are compared with results of Ref. [3]

is more precise.

ENERGY LEVELS OF HYDROGEN-LIKE IONS

In this section, we describe the NIST project to create an interactive database, which provides theoretical values of energy levels and transition frequencies for hydrogen-like ions. The first phase of this project includes hydrogen-like ions from He^+ to Ne^{+9} and is planned to be completed in 2005. The values are based on the most recent knowledge of the relevant theoretical contributions to the Dirac energy of an electron, including relativistic, quantum electrodynamic, recoil, and nuclear size effects. Contributions that are known only for low principal quantum numbers n have been extrapolated to highly-excited states using techniques developed in our previous study of hydrogen and deuterium energy levels (see Refs. [7, 8]). The total list of contributions to the relativistic Dirac energy is given in Table 2. Detailed descriptions of these contributions will be given in a forthcoming publication. The present study is limited to levels with principal quantum numbers $n = 1$ to 20 and all allowed orbital angular momenta ℓ and total angular momenta j.

TABLE 2. List of relativistic and QED contributions used in calculations of hydrogen-like ion energy levels

Relativistic Recoil	Self Energy
Vacuum Polarization	Two-photon Corrections
Three-photon Corrections	Finite Nuclear Size
Radiative-Recoil Correction	Nuclear Size Correction
Nuclear Polarization	to Self Energy and
Nuclear Self Energy	Vacuum Polarization

TABLE 3. Atomic and nuclear parameters obtained from the literature

Z	Atomic mass (a.m.u.)	Ionization energy (cm^{-1})	Nuclear radius (fm)
1	1.007 825 032 07(10)	109 678.7717	0.852 1(70)
2	4.002 603 254 15(6)	637 219.638 4	1.675 8(27)
3	7.016 004 548(84)	1 641 227.159 2	2.395 2(510)
4	9.012 182 201(426)	3 219 335.722 4	2.518 0(116)
5	11.009 930 540 6(448)	5 411 855.820 6	2.405 9(293)
6	12	8 308 360.968 9	2.470 4(23)
7	14.003 074 004 78(62)	11 985 917.103 2	2.552 0(85)
8	15.994 914 619 56(16)	16 484 719.293 5	2.702 8(76)
9	18.998 403 224(74)	21 904 891.239 8	2.897 6(27)
10	19.992 440 175 42(192)	28 323 307.502 2	3.004 6(38)

Fundamental constants used in the theoretical expressions are taken from the 2002 least squares adjustment published on the Fundamental Physical Constants NIST web site. The nuclear parameters needed for the evaluation of the hydrogenic energy levels are the mass and charge radius. To obtain the nuclear mass, we take the atomic mass from the 2003 update of the atomic mass evaluation of Refs. [15, 16]. The atomic mass data are also available in an electronic version at www−csnsm.in2p3.fr/amdc/web/masseval.html, the Web site of the Atomic Mass Data Center (AMDC) Centre de Spectrométrie Nucléaire et de Spectrométrie de Masse (CSNSM), Orsay, France. The values of the atomic masses are given in the first column of Table 3.

A calculation of the relative masses of various nuclei requires the total ionization energies, or binding energies E_{b} (the sum of the individual ionization energies). These are calculated by summing the ionization energies of each stage of ionization of a given atom, and the results are given in the second column of Table 3.

The nuclear root-mean-square (rms) charge radii are given in the last column of Table 3. They are taken from the compilation of Angeli [17].

Uncertainties in the energies and transition frequencies are estimated. They take into account uncertainties in the theoretical calculations and in the fundamental constants.

As an example of the available data on hydrogen-like ions Fig. 6 shows a comparison of our theoretical values for the Lamb shift and experimental data. References to the experimental data are given in the caption of Fig. 6.

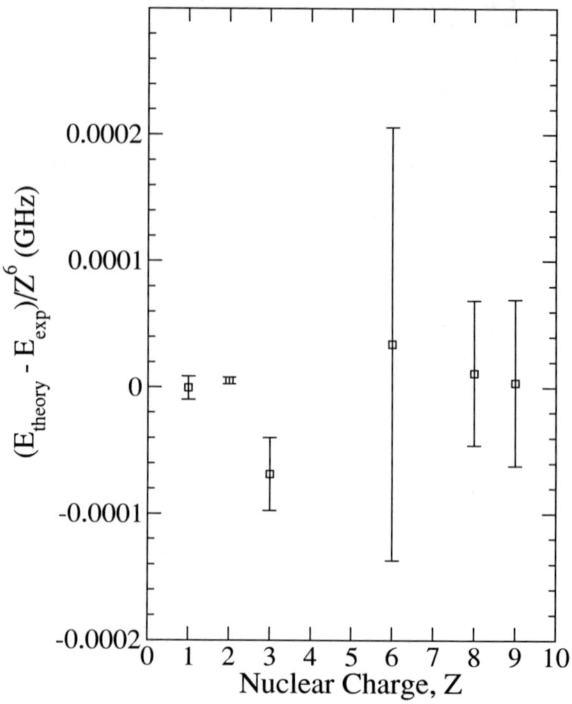

FIGURE 6. The Lamb shift difference between our theoretical values and experimental data for hydrogen and hydrogen-like ions. The experimental data are taken from Refs.: H - [9], He^{+} - [10], Li^{+2} - [11], C^{+5} - [12], O^{+7} - [13], and F^{+8} - [14]. The error bars primarily show experimental uncertainties.

ACKNOWLEDGMENTS

We acknowledge helpful discussions with Tim Kallman and Patrick Palmeri.

REFERENCES

1. S. A. Kotochigova and I. I. Tupitsyn, *J. Phys. B: At. Mol. Phys.*, **20**, 4759-4772 (1987).
2. S. Kotochigova, E. Tiesinga, and I. Tupitsyn, in "New Trends in Quantum Systems in Chemistry and Physics" **1** (Kluwer Academic Publ., The Netherlands), (2001), pp. 219-242.
3. G. V. Brown, P. Beiersdorfer, D. A. Liedahl, K. Widmann, S. M. Kahn, and E. J. Clothiaux, *ApJ Supp. Ser.*, **140**, 589-607 (2002).
4. T. Shirai, Y. Funatake, K. Mori, J. Sugar, W. L. Wiese, and Y. Nakai, *J. Phys. Chem. Ref. Data*, **19**, 127-257 (1990).
5. V. A. Fock, in *Principles of Quantum Mechanics*, (1976)
6. R. Szmykowski, *J.Phys.B*, **30**, 825-861 (1997)
7. S. Kotochigova, P. J. Mohr, and B. N. Taylor, *Can. J. Phys.*, **80**, 1373-1382 (2002).
8. S. Kotochigova, P. J. Mohr, and B. N. Taylor, at http://physics.nist.gov/PhysRefData/HDEL (2004)
9. S. R. Lundeen and F. M. Pipkin, *Metrologia*, **22**, 9-54 (1986).
10. A. van Wijngaarden, F. Holuj, and G. W. F. Drake, *Phys. Rev. A* **63** 012505 (2001)

11. M. Leventhal, *Phys. Rev. A*, **11**, 427-441 (1975).
12. H. W. Kugel, M. Leventhal, and D. E. Murnich, *Phys. Rev. A*, **6**, 1306-1321 (1972).
13. B. Curnutte, C. L. Cocke, and R. D. Dubois, in "Proceedings of Int. Conf. Fast Ion Beam Spectr.", Laval (1981).
14. H. W. Kugel, M. Leventhal, D. E. Murnich, C. K. N. Patel, and O. R. Wood, *Phys. Rev. Lett.*, **35**, 647-650 (1975).
15. A. H. Wapstra, G. Audi, and C. Thibault, *Nucl. Phys. A*, **729**, 129-336 (2003).
16. G. Audi, A. H. Wapstra, and C. Thibault, *Nucl. Phys. A*, **729**, 337-676 (2003).
17. I. Angeli, *Heavy Ion Physics*, **8**, 23-29 (1998).

Drs. Jonathan Gelbord, Agata Rozanska, and Tahir Yaqoob at a break in the workshop.

Dielectronic Satellites Near the $1s^2 - 1snp$ ($n > 2$) Lines of He-like Ions in Solar X-ray Spectra

E. Landi*, K. J. H. Phillips†, J. Sylwester and B. Sylwester**, B. R. Dennis‡ and J. Dubau§

*Artep, Inc., Ellicott City, MD 21042 and Naval Research Laboratory, Washington DC, USA
†NRC Senior Research Associate, Code 682, NASA GSFC, Greenbelt, MD 20771, U.S.A.
**Space Research Centre, Polish Academy of Sciences, 51-622, Kopernika 11, Wrocław, Poland
‡Code 682, NASA Goddard Space Flight Center, Greenbelt, MD 20771, U.S.A.
§Observatoire de Paris, Section de Meudon, France

Abstract. We discuss the importance of dielectronic satellites with transitions $1s^22l - 1s2l3p$ on the long-wavelength side of He-like ion lines $1s^2 - 1snp$ ($n > 2$). Their intensity ratios with the He-like ion lines have an inverse temperature dependence, making them useful as diagnostics. This is examined in the case of Si and Fe satellites.

Keywords: X-ray spectra – X-ray diagnostics – Satellite lines
PACS: 32.30.Rj

INTRODUCTION

The importance of dielectronic satellites as temperature diagnostics in solar and tokamak soft X-ray spectra is well known, but the satellites in question most generally used are those near the $1s^2 - 1s2p$ lines of He-like ions. These suffer from spectral confusion with other He-like ion lines to the long-wavelength side of the resonance line, notably the forbidden line ($1s^2\,{}^1S_0 - 1s2s\,{}^3S_1$). Relatively little attention has been paid to satellites on the long-wavelength side of $1s^2 - 1snp$ lines of He-like ions in solar spectra, yet their temperature dependence (approximately T^{-1}) is identical to the $1s^2 - 1s2p$ satellites and (owing to the absence of an equivalent forbidden line) there is no spectral confusion. This paper identifies recent solar flare observations of these satellites and their importance for temperature information. RESIK, a high-resolution spectrometer flying on the Russian *CORONAS-F* spacecraft, has made observations of Si XII and S XIV satellites near the (He-like) Si XIII and S XV $1s^2 - 1snp$, n =3, 4, and 5. We find that line ratios of Si and temperatures from *GOES* spacecraft data confirm the T^{-1} dependence. In addition, we discuss spectra from *RHESSI*, which observes solar flare thermal spectra with broad-band resolution ($\simeq$1 keV FWHM) at wavelengths less than 4 Å (photon energies > 3 keV); they include line features at 6.7 keV due to Fe XXV $1s^2 - 1s2p$ lines and satellites and at ~ 8 keV which for $T \sim 20$ MK is made up of Fe XXV $1s^2 - 1snp$ ($n > 2$) lines and Fe XXIV satellites of the type discussed here.

CP774, *X-Ray Diagnostics of Astrophysical Plasmas*, edited by Randall K. Smith

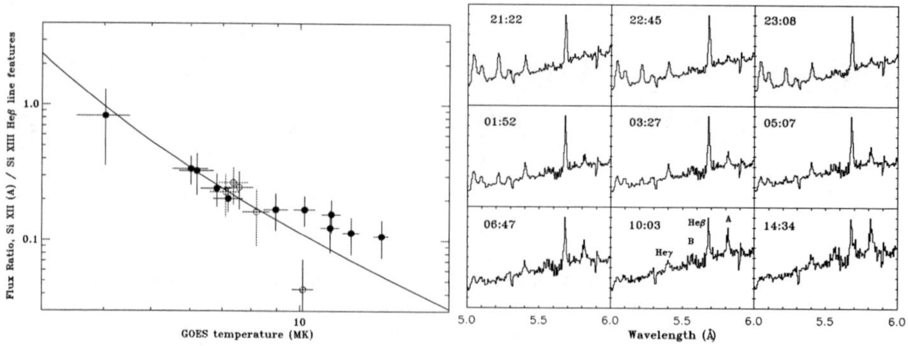

FIGURE 1. **Left**: Measurements of the ratio of the Si XII satellite feature A to the Si XIII Heβ (5.681Å) line vs. *GOES* temperature during long-duration flares on 2002 July 23 (open circles) and 2002 July 26/27 (filled circles) compared with the predicted variation of ratio with temperature from CHIANTI (Dere et al. 1997) and Dubau data. **Right**: RESIK spectra of the 5 to 6 Å region during the solar flare on 2002 July 26/27, showing the Si XIII $1s^2 - 1s3p$ (5.681 Å Heβ) and $1s^2 - 1s4p$ (5.384 Å Heγ) and Si XII satellites at 5.818 Å (A) and 5.565 Å (B). As the flare cools, the satellites become relatively more intense.

SI SATELLITES

The RESIK instrument on *CORONAS-F* which operated as a crystal spectrometer from 2001 till May 2003 made observations of Si XIII lines at 5.681 Å (Heβ, $1s^{2\,1}S_0 - 1s3p^1P_1$), 5.384 Å (Heγ, $1s^{2\,1}S_0 - 1s4p^1P_1$), and 5.253 Å (Heδ, $1s^{2\,1}S_0 - 1s5p^1P_1$). To the long-wavelength side of each of these lines is a line feature made of several dielectronic lines, with transitions like $1s^22l - 1s2l3p$. The most intense are those with transitions $1s^22p^2P_{3/2} - 1s2p3p^2D_{5/2}$ and $1s^22p^2P_{1/2} - 1s2p3p^2D_{3/2}$, equivalent to satellites j and k near the Heα line [2]. The principal mode of excitation of these satellites is dielectronic recombination of the He-like Si ion followed by stabilizing transitions, though inner-shell excitation is important for some satellites. Intensity factors have been calculated by one of us (J.D.), allowing the theoretical intensity ratio of satellite/He-like ion line to be found as a function of T (see Figure 1, left). The T^{-1} temperature dependence is evident.

Observations were made of these lines with RESIK during long-duration flares on 2002 July 26/27 and 2002 July 23 (see [3] for further details). Figure 1, right shows how the intensities of the Si XII satellite features A and B (wavelengths 5.681 Å and 5.565 Å) near the Si XIII Heβ and Heγ lines vary over a 17-hour period during the July 26/27 flare, with the flare slowly decaying and the temperature declining. These observations have been analyzed to give line intensities at various times during each flare. Approximate values of temperature were derived for these times using the ratio of the two X-ray channels of the *GOES* satellites, from standard software available on *SolarSoft*. The measured intensity ratios are plotted against the *GOES* temperatures in Figure 1 (left). The close correspondence illustrates the accuracy of the atomic calculations and the temperature dependence.

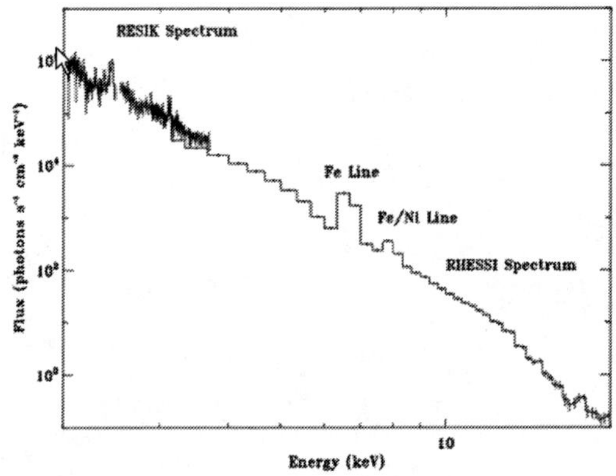

FIGURE 2. *RHESSI* spectrum taken during the flare of 2003 April 26 (03:00 UT) (histogram) and RESIK spectrum plotted against photon energy (range 2-20 keV), showing the Fe line feature at ~ 6.7 keV and Fe/Ni line feature at ~ 8 keV. For the temperature of this spectrum, 18.6 MK, the dominant lines in the Fe/Ni features are Fe XXIV satellites of the type described in the text.

FE SATELLITES

The *RHESSI* mission, looking at solar flares in the 3 keV to 17 GeV energy band, observes line emission at 6.7 keV (1.85Å Å, so-called Fe line feature: see [4]) and 8.1 keV (1.55 Å, Fe/Ni line feature) with broad-band resolution. The Fe line feature is due to Fe XXV lines and dielectronic satellites; the Fe/Ni line feature, at $T < 40$ MK, is due to Fe XXV $1s^2 1snp$ ($n = 3$, 4 etc.) lines and, at $T < 25$ MK, to Fe XXIV satellites $1s^2 2p\,^2P_{1/2} - 1s2p3p\,^2D_{3/2}$ and $1s^2 2p\,^2P_{1/2} - 1s2p4p\,^2D_{3/2}$. The recent addition of these satellites to the CHIANTI code (Dere et al. 1997) has greatly improved the agreement between the observed and theoretical intensity ratio R of the 6.7 keV to the 8.1 keV line feature. The combined *RHESSI* and RESIK spectrum, during the flare of 2003 April 26 at 03:00 UT, shows the (free-free and free-bound) continuum and line emission, with the Fe and Fe/Ni line features marked. The observed and calculated values of the ratio R are both about 10.

REFERENCES

1. Dere, K. P., A&AS, **125**, 149 (1997)
2. Gabriel, A. H. MNRAS, **160**, 99 (1972)
3. Phillips, K. J. H., Sylwester, J., Sylwester, B., & Landi, E. ApJ, **589**, L113 (2003)
4. Phillips, K. J. H. ApJ, **605**, 921 (2004)

The XDAP workshop in session.

The He-like triplet ratios as powerful plasma diagnostic tools

Delphine Porquet

Max-Planck-Institut für extraterrestrische Physik, Postfach 1312, 85741 Garching, Germany

Abstract. With the advent of the new generation of X-ray satellites (Chandra and XMM-Newton), for the first time X-ray spectroscopy with unprecedented spectral resolution is now available for extra-solar objects. In particular, the He-like line ratios are powerful diagnostic tools for a variety of plasmas such as collisional (e.g., stellar coronae), and/or photo-ionized (e.g., Active Galactic Nuclei) plasmas. I will show how these ratios can be used to infer the density, the electron temperature (and ionization processes) of the plasma. I will review the advantages and limitations of these diagnostics, and the perspectives with Astro-E2, Constellation X and XEUS.

INTRODUCTION

In the X-ray range, the three most intense lines of Helium-like ions (called "triplet", which is in fact a "quadruplet") are: the *resonance* line (r, also called w: $1s^2\,{}^1S_0$ – $1s2p\,{}^1P_1$), the unresolved *intercombination* lines (i, also called $x+y$: $1s^2\,{}^1S_0$ – $1s2p\,{}^3P_{2,1}$) and the *forbidden* line (f, also called z: $1s^2\,{}^1S_0 - 1s2s\,{}^3S_1$). They correspond to transitions between the n=2 shell and the n=1 ground-state shell (see Figure 1).

Gabriel & Jordan [22] were the first to recognize that the ratios of these lines could give important plasma diagnostics on the electron density (n_e) and temperature (T_e).

$$R\,(n_e) \equiv \frac{f}{i} \qquad\qquad G\,(T_e) \equiv \frac{f+i}{r} \qquad (1)$$

These diagnostics have been used for coronal spectra as well as for many laboratory hot plasmas (e.g. tokamak).

Helium-like spectral lines $1s^2$-$1s2l$ have been observed with high spectral resolution for a long time in Solar corona X-ray spectra. During the seventies, several instruments have flown aboard satellites to observe the soft X-ray spectra of Solar flares. Spectra in the 1–25Å range, for example, have been obtained from uncollimated crystal spectrometers on the early *OSO* satellites [18, 50], the *OVI-17* satellite [72], and *OSO 8* [54]. Particular spectral regions (e.g. iron lines at 1.9Å) have been examined with the *intercosmos-4* spectrometer [23] and the *SOLFLEX* instrument on the *P78-1* satellite [19]. Rocket-borne collimated crystal spectrometers have recorded spectra in various wavelength ranges up to 25Å from active regions or weak flares [53, 62, 13], while the collimated *SOLFLEX* instrument aboard *P78-1* has obtained very high resolution spectra from sizeable flares [40]. Thanks to the Flat Crystal Spectrometer (FCS, [55]) and Bent Crystal spectrometers (BCS, [17]) on board the *Solar Maximum Mission (SMM)*,

CP774, *X-Ray Diagnostics of Astrophysical Plasmas,* edited by Randall K. Smith

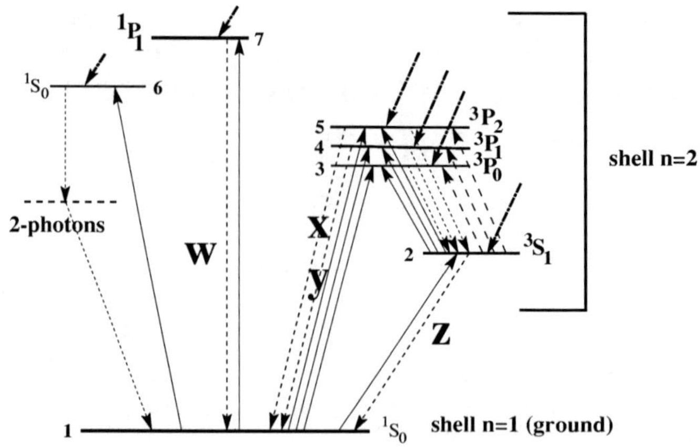

FIGURE 1. Simplified level scheme for Helium-like ions. **w** (or r), **x,y** (or i), and **z** (or f): resonance, intercombination, and forbidden lines, respectively. *Full upward arrows*: collisional excitation transitions, *broken arrows*: radiative transitions (including photo-excitation from $2\,^3S_1$ to $2\,^3P_{0,1,2}$ levels, and 2-photon continuum from $2\,^1S_0$ to the ground level), and *thick skew arrows*: recombination (radiative and dielectronic) plus cascade processes.

many line blends were resolved for the first time allowing new physical information about the emitting plasma to be extracted. In particular the BCS observed He-like iron and calcium. More recently Yohkoh (Solar A) contained also a BCS which pursued similar Solar observations of iron and calcium as *SMM*, extending the observations to lower temperature by using Helium-like sulfur spectra [27, 26]. For active coronae and Solar flares, a temporal analysis of these lines can be carried on during the three consecutive phases: ionization, gradual, recombination (e.g., [42, 43, 44, 20, 60, 61]).

It is now possible, thanks to the spectral resolution of the new generation of X-ray satellites: Chandra and XMM-Newton, to resolve the He-like ion lines and to use these diagnostics in case of extra-solar objects. Indeed, the Helium-like ion line ratios are powerful tools in the analysis of high-resolution spectra of a variety of plasmas such as:

- collisional plasmas: *e.g., stellar coronae* (OB stars, late type stars, active stars, T Tauri, ...)
- photo-ionized or hybrid plasmas (photo-ionization + collisional ionization): *e.g., Active Galactic Nuclei (AGN), X-ray binaries, ...*
- out of equilibrium plasmas: *e.g., SNRs, stellar flares, ...*

The wavelength ranges of the RGS (6-35 Å), the LETGS (2-175 Å), and the HETGS (MEG range: 2.5-31 Å; HEG range: 1.2-15 Å) contain the Helium-like lines from C V (or N VI for the RGS) to Si XIII (Table 1).

These diagnostics have been recently used in numerous observations of extra-solar objects thanks to Chandra and XMM-Newton such as AGN (e.g., NGC 5548: [24]; NGC 4051: [15]; MCG-6-30-15: [33]; NGC 4151: [51], Mrk 3: [63]; NGC 3783: [30];

TABLE 1. Wavelengths in Å of the X-ray He-like ion lines of C V, N VI, O VII, Ne IX, Mg XI and Si XIII (from [69]).

label	line	C V	N VI	O VII	Ne IX	Mg XI	Si XIII
r (*w*)	*resonance*	40.279	28.792	21.603	13.447	9.1681	6.6471
i (*x*)	*inter-*	40.711	29.074	21.796	13.548	9.2267	6.6838
(*y*)	*combination*	40.714	29.076	21.799	13.551	9.2298	6.6869
f (*z*)	*forbidden*	41.464	29.531	22.095	13.697	9.3134	6.7394

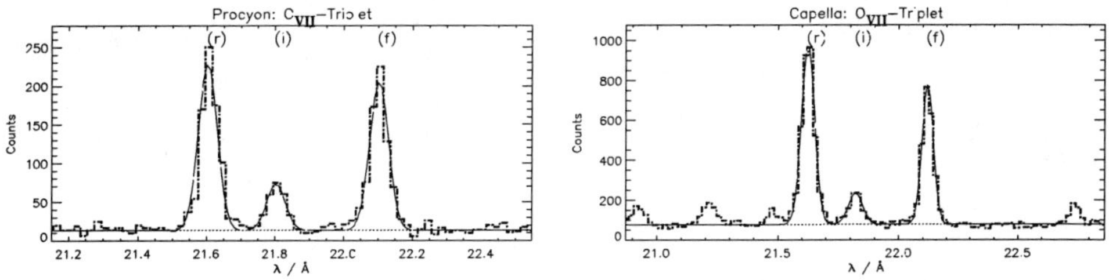

FIGURE 2. Observed spectra (Chandra, LETGS) of the He-like O VII triplet lines (*r*, *i*, and *f*) of the stellar coronae of Capella and Procyon [47].

Mkn 509: [59]); stellar coronae and stellar winds (e.g., Vela X-1: [66]; II Pegasi: [28]; Capella: [47], [2]); and X-ray binaries (SS 433: [36]; 4U 1626-67: [64]; EXO 0748-67: [16]). As an illustration, the spectra of He-like O VII ion of the stellar coronae of Capella and Procyon are shown in Figure 2.

They allow the determination of the ionization process (linked to the temperature) in case of the "Warm Absorber" in AGNs (photo-ionization and/or an additional ionization process such as shock or starburst). However, these diagnostics should be used cautiously in order to avoid any mistaken interpretation of the physical parameters of the observed plasmas. Indeed some neglected atomic processes can lead to wrong determination of the density, the temperature, and the ionization processes: for example, blended dielectronic satellite lines, photo-excitation due to a strong UV radiation field, a resonant scattering (optical depth).

PLASMA DIAGNOSTICS

Density diagnostic

In the low-density limit, all n=2 states are populated directly or via upper-level radiative cascades by electron impact from the He-like ground state, by recombination (radiative and dielectronic) of H-like ions, and for 1s 2s $^1S_0$ and $^3S_1$ levels, by collisional electron ionization from the $1s^2 2s\,^2S_{1/2}$ lithium-like ground level. This later is only important in the out of equilibrium ionization phase such as seen in Solar flares (e.g., [42, 43, 44, 7]).

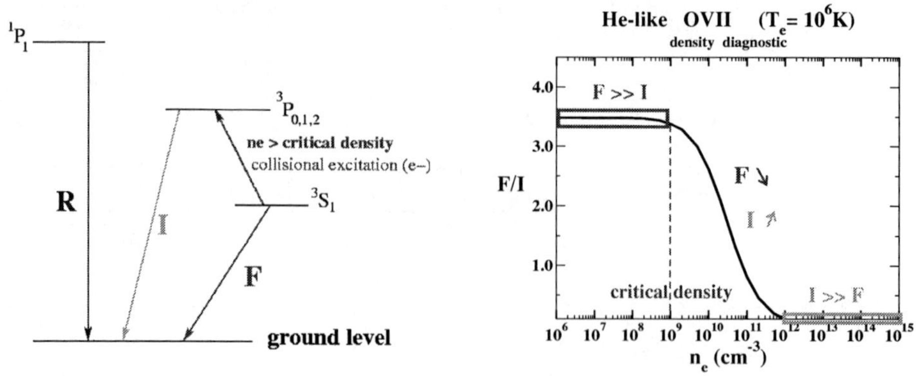

FIGURE 3. R(n_e)=f/i for the He-like ion O VII in case of a collisional plasma, such as stellar coronae. Calculations are taken from [58].

If the suffix λ represents f, i and r, the line emissivity ε_λ can be written as: $\varepsilon_\lambda = n_e \mathscr{C}_\lambda^{He}(T_e)\, N(He-like) \;+\; \mathscr{C}_\lambda^{Li}(T_e)\, N(Li-like) \;+\; \alpha_\lambda^{H}(T_e)\, N(H-like)$, where $\mathscr{C}_\lambda^{He}$, $\mathscr{C}_\lambda^{Li}$, and α_λ^{H} are *effective* rate coefficients, i.e. including radiative cascades inside He-like levels, respectively excitation, ionization and recombination rates; and $N(He-like)$, $N(Li-like)$, and $N(H-like)$ are the population of $1s^2$, $1s^2 2s$ and 1s levels. These n=2 levels decay radiatively directly or by cascades to the ground level. The relative intensities of the three intense lines are then independent of density.
As n_e increases from the low-density limit, some of these states (1s2s $^3S_1$ and $^1S_0$) are depleted by collisions to the nearby states where $n_{crit} C = A$, with C being the sum of the collisional rates depopulating the level, A being the radiative transition probability from n=2 to n=1 (ground state), and n_{crit} being the critical density. Collisional excitation depopulates first the 1s2s $^3S_1$ level (upper level of the *forbidden* line) to the 1s2p $^3P_{0,1,2}$ levels (upper levels of the *intercombination* lines). The intensity of the *forbidden* line decreases while those of the *intercombination* lines increase, hence implying a reduction of the ratio R (according to Eq. 1), over approximately two or three decades of density (see Fig. 3). At high density, the intensity of the *resonance* tends to zero.
For much more higher densities, 1s2s $^1S_0$ is also depopulated to 1s2p$^1P_1$, and the *resonance* line becomes sensitive to the density.

To summarize, for each density domains, one can obtain an information about the density value:

- If the *forbidden* line is much larger than the *intercombination* line, then the density value is less or equal to the value of the critical density, and one can infer an upper limit for the density. For example, for the O VII in a collisional plasma at 10^6 K, from a ratio R of about 3.5, an upper limit for the density of few 10^9 cm^{-3} is inferred (see Fig. 3).
- Above the critical density, where the *intercombination* line increases, one can

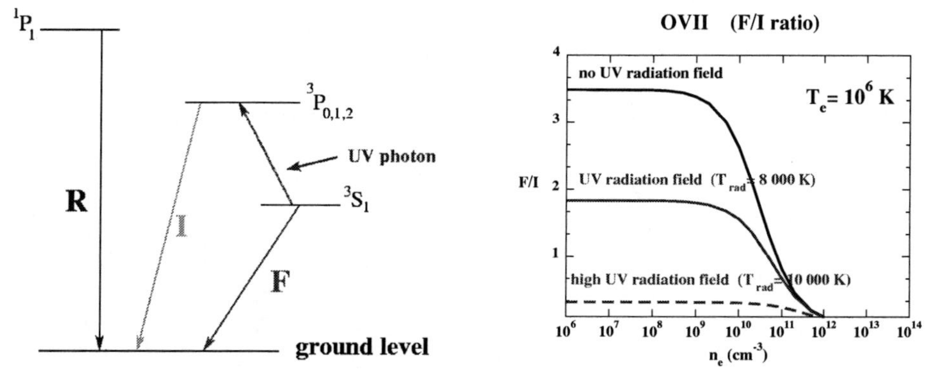

FIGURE 4. The upper and medium/lower curves show the *R* ratio versus density for no UV radiation field and for a significant UV radiation field, respectively.

obtain an accurate determination of the density. For example, for the O VII in a collisional plasma at 10^6 K, a ratio *R* of 2.0 corresponds to a density of $2 \times 10^{10}\,\mathrm{cm}^{-3}$ (see Fig. 3).

- At a density much higher than the critical density, where the *intercombination* line is much intense than the *forbidden* line (which tends to zero). One can obtain a lower limit on the value of the density. For example, for the O VII in a collisional plasma at 10^6 K, for a ratio *R* which tends to zero, a lower limit for the density of about $10^{12}\,\mathrm{cm}^{-3}$ is inferred (see Fig. 3).

In case of an intense **UV radiation field**, for low-Z ions (i.e. C V, N VI, O VII) the photo-excitation between the ^{3}S term and the ^{3}P term is not negligible [58]. Higher-Z ions are much less or not sensitive to this process. This process has the same effect on the *forbidden* line and on the *intercombination* line as the collisional coupling (see Fig. 4 left panel), i.e. lowering of the ratio *R*, and thus could mimic a high-density plasma.
To illustrate this effect, the figure 4 (right panel) displays the $R(n_e)$ ratio for O VII (T_e=10^6 K) in three cases: no UV radiation field, a significant radiation field with radiative temperature (T_{rad}) of 8 000 K, and 10 000 K. For a high value of T_{rad}, $R(n_e)$ is much lower even in the "low density domain", for T_{rad} of 8 000 K, the low density limit is reduced to about 1.8 compared to about 3.5 in case of no UV radiation field. A $R(n_e)$ ratio of about 3.5, gives the indication that the UV radiation field is negligible. A $R(n_e)$ ratio of about 1.8 in case of no radiation field corresponds to the density domain where the ratio is very sensitive to the density value, then an accurate determination of the density can be inferred; i.e. here n_e is about $5 \times 10^{10}\,\mathrm{cm}^{-3}$, while for T_{rad}=8 000 K, this corresponds to the low density domain where only an upper limit for the density can be estimated (here about few $10^9\,\mathrm{cm}^{-3}$). Therefore a possible UV radiation field must be taken into account to avoid any misunderstanding between a high-density plasma and a high radiation field (see e.g., [9, 41, 42, 60, 58, 70, 29, 46, 48]).

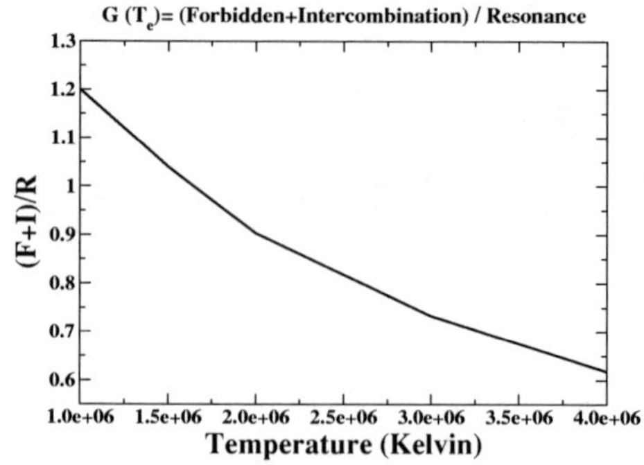

FIGURE 5. G(T_e) for the He-like O VII in case of a collisional plasma, in the low density limit. Calculations are taken from [58].

However, the low-Z He-like ions can be used to determine the location of the X-ray plasma compared to the UV radiation field. This can be used for example to O stars [67, 71], and to chemically peculiar late B stars [5, 6].

Temperature/ionization process diagnostics

The ratio G (see Eq. 1, and Fig. 5) is sensitive to the electron temperature since the collisional excitation rates have not the same dependence with temperature for the *resonance* line as for the *forbidden* and *intercombination* lines To illustrate this effect, figure 5 displays the G ratio versus the electron temperature in case of a collisional plasma. Therefore, as detailed in [57] (see also [45, 35]), the relative intensity of the *resonance* r line, compared to the *forbidden* f and the *intercombination* i lines, indicates which ionization process(es) is dominant: a strong *resonance* line compared to the *forbidden* or the *intercombination* lines corresponds to collision-dominated plasmas. This leads to a ratio of $G = (f+i)/r \sim 1$ (as shown in Fig. 5). On the contrary, a weak *resonance* line corresponds to plasmas dominated by photo-ionization, i.e. without X-ray photo-excitation ($G = (f+i)/r > 4$).

In the case where **X-ray photo-excitation process** is not negligible, the *resonance* line is enhanced, and can then mimic a higher temperature plasma, as for example a hybrid plasma, (i.e. photo-ionization plus collisional ionization; e.g. shock or starburst). As shown by [31], the RGS spectrum of the Seyfert 2 NGC 1068 shows signatures of X-ray photo-excitation (see also [10, 52]), which appears to be also the case for other Seyfert galaxies [63].

Advantages of these diagnostics

These diagnostics are based on the relative intensity of three close lines, and are therefore insensitive to the instrumental calibration. In addition, since they are based on lines from the same ion (of the same element), they are not dependent on the abundance of this element and are produced in the same medium. While low-Z ions (C, N, O) are UV-dependent, higher-Z ions (Ne, Mg, Si, S, Fe) are much less or not sensitive to UV photo-excitation, and so density is the only free parameter. He-like ions are stable over a large domain of T_e in case of collisional plasmas or over a large domain of ionization parameter[1]in case of a photo-ionized plasmas. With the current generation of X-ray satellites, namely Chandra and XMM-Newton, the three lines of low-Z (C, N, O, Ne) are easily resolved. The lines of higher-Z ions (from Mg to Fe) will be completely resolved with the next X-ray satellites (Astro-E2, Constellation-X, XEUS). Each He-like ion is sensitive within different density and electron temperature domains (see figure 6), which is very useful to investigate a possible stratification of the plasma. High-Z ions (above Ne) are little sensitive or not sensitive to UV field and therefore the density value is the only free parameter.

However caution should be taken when using theoretical ratios. For example, He-like lines can be blended with satellites lines (e.g. from Li-like, Be-like ions) in case of high temperature plasma, and/or with line of other elements. Therefore these contributions must be taken into account to infer reliable values for the density and electron temperature [58, 49, 56]. In addition, for high-column density the *forbidden* line is enhanced [8, 14]. For transient plasmas, simulations have been done by [35, 61].

In conclusion, He-like ion line ratios are very powerful plasma diagnostics, but all possible contributions of atomic and physical processes (UV and/or X-ray photo-excitation, blend with satellites lines, high column density, ...) must be taken into account to infer accurate reliable density and temperature values. Several publicly available codes and databases are very useful to interpret He-like ion spectra e.g., APEC/APED [68], CHIANTI [32], Cloudy [21], XSPEC [25], XSTAR [4]. See also [57, 3, 58].

Complementarily, other plasma diagnostics can be used as those based on the line ratios of Fe-L shell ions (e.g., [34, 35, 11, 37, 38, 12, 39]), and the one based on the width of radiative recombination continuum features (e.g., [35, 10, 31, 65])

CONCLUSION AND PERSPECTIVES

Thanks to the current generation of X-ray satellites, Chandra and XMM-Newton, high resolution spectra of extra-solar objects are now possible, allowing applications of X-ray plasma diagnostics such as based on He-like ions as reviewed here, can be used to determine electron density and temperature. These diagnostics are powerful and can be applied for a very large variety of plasmas: photo-ionized plasmas (e.g., AGN, X-ray

[1] The ionization parameter is $\xi \equiv L/(nR^2)$, L is the bolometric luminosity, n is the density and R is the location of the medium relative to the ionizing continuum.

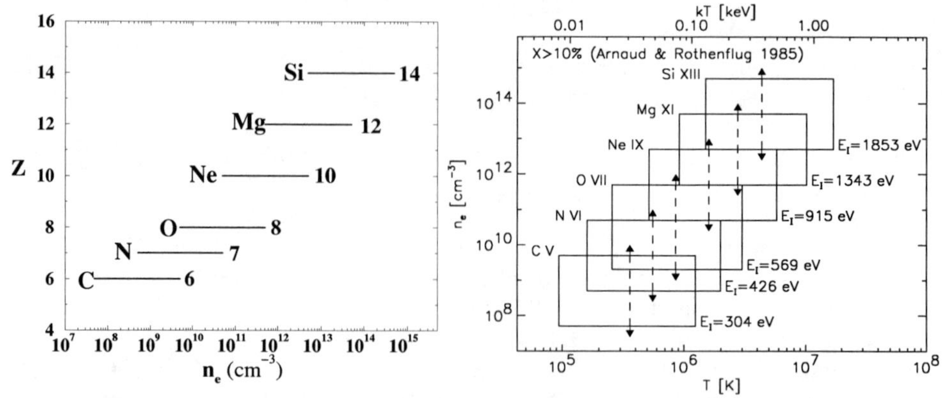

FIGURE 6. *Left panel*: Typical density ranges for which each ion (from C V to Si XIII) are very sensitive to the density. *Right panel*: temperature range where the ionic fraction for each He-like ion is greater than 10% (i.e. in a relatively high abundance to emit significant lines) based on the calculations done by Arnaud & Rothenflug (1985). On the right hand side of each domain, the energy of the intercombination lines of the corresponding He-like ion is reported.

binaries), collisional plasmas (e.g., stellar coronae), out-of equilibrium plasmas (e.g., SNR, Stellar and Solar flares).

The next generation of X-ray satellites (ASTRO-E2, Constellation-X, and XEUS) will benefit from a very high improvement of the spectral resolution with $\Delta E \sim 2$–6 eV up to (or above) 10 keV combined with a higher sensitivity. This will increase the number of objects for which one can apply He-like diagnostics (e.g., T Tauri stars, protostars, weak and/or high-redshift AGN), and extend He-like diagnostics to higher energies (e.g., Fe XXV: probe higher densities and temperatures domains).

We are only at the beginning of the X-ray plasma diagnostic era!

ACKNOWLEDGMENTS

I would like to acknowledge the organizing committee to invite me to give this review about He-like ions. I dedicate this presentation to R. Mewe, who communicated me his passion for X-ray spectroscopy.

REFERENCES

1. M. Arnaud, Rothenflug, *A&AS*, 1985, **60**, 425–457
2. M. Audard, E. Behar, M. Güdel, A. J. J. Raassen, D. Porquet, R. Mewe, C. R. Foley, G. .E. Bromage, *A&A*, 2001, **365**, L329–L335
3. M. .A. Bautista, T. R Kallman, *ApJ*, 2000, **544**, 581–591
4. M. .A. Bautista, T. R Kallman, *ApJS*, 2001, **134**, 139–149

5. E. Behar, M. Leutenegger, R. Doron, M. Güdel, U. Feldman, M. Audard, S. M. Kahn, *ApJ*, 2004, **612**, L65–L68
6. E. Behar, 2005, this volume
7. F. Bely-Dubau, A. .H. Gabriel, S. Volonté, *MNRAS*, 1979, **189**, 801–816
8. S. Bianchi, G. Matt, F. Nicastro, D. Porquet, J. Dubau, *MNRAS*, 2005, in press [astro-ph/0411603]
9. G. R. Blumenthal, G. W. F. Drake, & W. H. Tucker, *ApJ*, 1972, **172**, 205–212
10. A. C. Brinkman, J. S. Kaastra, R. L. J. van der Meer, A. Kinkhabwala, E. Behar, S. M. Kahn, F. Paerels, M. Sako, *A&A*, 2002, **396**, 761–772
11. G. V. Brown, P. Beiesdorfer, H. Chen, M. H. Chen, R. J. Reed, *ApJ*, 2001, **557**, L75-L78
12. G. V. Brown, 2005, this volume
13. A. J. Burek, D. M. Barrus, R. L. Blake, E. E. Fenimore, *ApJ*, 1981, **243**, 660–676
14. L. Chevallier, 2005, this volume
15. M. J. Collinge, W. N. Brandt, S. Kaspi, D. M.Crenshaw, M. Elvis, S. B. Kraemer, C. S. Reynolds, R. M. Sambruna, B. .J. Wills, *ApJ*, 2001, **557**, 2–17
16. J. Cottam, S. M. Kahn, A. C. Brinkman, J. W. den Herder, C. Erd, *A&A*, 2001, **365**, L277–L281
17. J. L. Culhane, C. G. Rapley, R. D. Bentley, A. H. Gabriel, K. J. Phillips, L. W. Acton, C. J. Wolfson, R. C. Catura, C. Jordan, E. Antonucci, *ApJ*, 1981, **244**, L141–L145
18. G. A. Doschek, J. F. Meekins, R. D. Cowan, *ApJ*, 1972, **177**, 261–270
19. G. A. Doschek, R. W. Kreplin, U. Feldman, *ApJ*, 1979, **233**, L157–L160
20. J. G. Doyle, *A&A*, 1980, **87**, 183–187
21. G. F. Ferland, K. T. Korista, D. A. Verner, J. W. Ferguson, J. B. Kingdon, E. M. Verner, *PASP*, 1998, **110**, 761-778
22. A. H. Gabriel & C. Jordan, *MNRAS*, 1969, **145**, 241–248
23. Y. I. Grineva, V. I. Karev, V. V. Krutov, S. L. Mandelstam, L. A. Vainstein, B. N. Vasilyev, I. A. Zhitnik, *Solar Physics*, 1973, **29**, 441–446
24. J. S. Kaastra, R. Mewe, D. A. Liedahl, S. Komossa, A. C. Brinkman, *A&A*, 2000, **354**, L83–L86
25. J.S. Kaastra, R. Mewe, A. .J. J Raassen, *IAU*, Joint discussion, 2003, **17**, 24
26. T. Kato, U. Safronova, A. Shlypteseva, M. Cornille, J. Dubau, J. Nilsen, *Atomic Data and Nuclear Data Tables*, 1997, **67**, 225
27. L. K. Harra-Murnion, K. J. H. Phillips, J. R. Lemen, D. M. Zarro, C. J. Greer, V. J. Foster, R. Barnsley, I. H. Coffey, J. Dubau, F. P. Keenan, A. Fludra, E. Rachlew-Kaellne, T. Watanabe, M. Wilson, *A&A*, 1996, **308**, 670–684
28. D. P. Huenemoerder, C. R. Canizares, N. S. Schulz, *ApJ*, 2001, **559**, 1135–1146
29. S. M. Kahn, M. A. Leutenegger, J. Cottam, G. Rauw, J.-M. Vreux, A. J. F den Boggende, R. Mewe, M. Güdel, *A&A*, 2001, **365**, L312–L317
30. S. Kaspi, W. N. Brandt, H. Netzer, R. Sambruna, G. Chartas, G. P. Garmire, J. A. Nousek, *ApJ*, 2000, **535**, L17–L20
31. A. Kinkhabwala, M. Sako, E. Behar, S. M. Kahn, F. Paerels, A. .C Brinkman, J. S. Kaastra, M. F. Gu, D. A. Liedahl, *ApJ*, 2002, **575**, 732–746
32. E. Landi, U. Feldman, K. P. Dere, *ApJ*, 2002, **574**, 495–503
33. J. C. Lee, P. M. Ogle, C. R. Canizares, H. L. Marshall, N. S. Schulz, R. Morales, A. C. Fabian, K. Iwasawa, *ApJ*, 2001, **554**, L13–L17
34. D. .A. Liedahl, *ApJ*, 1992, **391**, 306–317
35. D. Liedahl, *X-Ray Spectroscopy in Astrophysics*, 189–268
36. H. L. Marshall, C. R. Canizares, N. S. Schulz, *ApJ*, 2002, **564**, 941–952
37. C. W. Mauche, D. A. Liedahl, K. B. Fournier, *ApJ*, 2001, **560**, 992–996
38. C. W. Mauche, D. A. Liedahl, K. B. Fournier, *ApJ*, 2003, **588**, L101-L104
39. C. W. Mauche, D. A. Liedahl, K. B. Fournier, 2005, this volume
40. D. L. McKenzie, P. B. Landecker, R. M. Broussard, H. R. Rugge, R. M. Young, U. Feldman, G. .A. Doschek, *ApJ*, 1980, **241**, 409–416
41. R. Mewe & J. Schrijver, *A&AS*, 1975, **38**, 345–352
42. R. Mewe & J. Schrijver, *A&A*, 1978a, **65**, 99–114
43. R. Mewe & J. Schrijver, *A&A*, 1978b, **65**, 115–121
44. R. Mewe & J. Schrijver, *A&AS*, 1978c, **33**, 311-313
45. R. Mewe, *X-Ray Spectroscopy in Astrophysics*, 109–187
46. N.A. Miller, J. P. Cassinelli, W. L. Waldron, J. .J. MacFarlane, D. H. Cohen, *ApJ*, 2002, **577**, 951

47. J.-U. Ness, R. Mewe, J. H. M. M. Schmitt, A. J. J. Raassen, D. Porquet, J. S. Kaastra, R. L. J. van der Meer, V. Burwitz, P. Predehl, *A&A*, 2001, **367**, 282–296
48. J.-U. Ness, J. H. M. M. Schmitt, V. Burwitz, R. Mewe, A. J. J. Raassen, R. .L. J. van der Meer, P. Predehl, A. C. Brinkman, *A&A*, 2002, **394**, 911–926
49. J.-U. Ness, N. S. Brickhouse, J. J. Drake, D. P. Huenemoerder, *ApJ*, 2003, **598**, 1277–1289
50. W. M. Neupert, M. Swartz, S. .O Kastner, *Solar physics*, 1973, **31**, 171–195
51. P. M. Ogle, H. L. Marshall, J. C. Lee, C. R. Canizares, *ApJ*, 2000, **545**, L81–L84
52. P. M. Ogle, T. Brookings, C. R. Canizares, J. C. Lee, H. L. Marshall, *A&A*, 2003, **402**, 849–864
53. J. H. Parkinson, *Solar physics*, 1975, **42**, 183–207
54. J. H. Parkinson, R. S. Wolff, H. L. Kestenbaum, W. H.-M. Ku, J. R. Lemen, K. S. Long, R. Novick, R. J. Suozzo, M. C. Weisskopf, *Solar physics*, 1978, **60**, 123–136
55. K. J. H. Phillips, B. C. Fawcett, B. J. Kent, A. H. Gabriel, J. W. Leibacher, C. J. Wolfson, L. W. Acton, J. H. Parkinson, J. L. Culhane, H. E. Mason, *ApJ*, 1982, **ApJ**, 256, 774–787
56. K. J. H. Phillips, E. Landi, J. Sylwester, B. Sylwester, B. R. Dennis, J. Dubau, U. Feldman, 2005, this volume
57. D. Porquet & J. Dubau, *A&AS*, 2000, **143**, 495–514
58. D. Porquet, R. Mewe, J. Dubau, A. .J. .J. Raassen, J. .S. Kaastra, *A&A*, 2001, **376**, 1113–1122
59. K. Pounds, J. .N Reeves, P. O'Brien, K. Page, M. Turner, S. Nayakshin, *ApJ*, 2001, **559**, 181–186
60. A. K. Pradhan & J. M. Shull, *ApJ*, 1981, **249**, 821–830
61. A. K. Pradhan, 2005, this volume
62. J. P. Pye, K. D. Evans, R. J. Hutcheon, *MNRAS*, 1977, **178**, 611–618
63. M. Sako, S. M. Kahn, F. Paerels, D. A. Liedahl, *ApJ*, 2000, **543**, L115–L118
64. N. S. Schulz, D. Chakrabarty, H. L. Marshall, C. R. Canizares, J. C. Lee, J. Houck, *ApJ*, 2001, **563**, 941–949
65. N. .J. Schurch, R. S. Warwick, R. E. Griffiths, S. M. Kahn, *MNRAS*, 2004, **350**, 1–9
66. N. S. Schulz, C. R. Canizares, J. C. Lee, M. Sako, *ApJ*, 2002, **564**, L21–L25
67. N. S. Schulz, C. Canizares, D. Huenemoerder, K. Tibbets, *ApJ*, 2003, **595**, 365-383
68. R. K. Smith, N. S. Brickhouse, D. A. Liedahl, J. C. Raymond, *ApJ*, 2001, **556**, L91–L95
69. L. A. Vainshtein & U. I. Safronova, *Atomic Data and Nuclear Data Tables*, 1978, **21**, 49
70. W. L. Waldron & J. P. Cassinelli, *ApJ*, 2001, **548**, L45–L48
71. W. L. Waldron, J. P. Cassinelli, N. A. Miller, J. J. MacFarlane, J. .C. Reiter, *ApJ*, 2004, **616**, 542–561
72. A. B. C. Walker & H. R. Rugge, *A&A*, 1970, **5**, 4–11

The value of density measurements in stellar coronae

Jan-Uwe Ness* and Carole Jordan*

Department of Physics, Rudolf Peierls Centre for Theoretical Physics, University of Oxford, 1 Keble Road, Oxford OX1 3NP

Abstract. The grating instruments on board Chandra and XMM-Newton now allow measurements of electron densities. These rely on the ratios of fluxes in emission lines, where one line depends on both collisional and radiative decay rates. The electron density is required to constrain the physical extent of the emitting region, and large samples of measurements are of interest in the context of trends in coronal activity. Here we discuss the important He I-like ions and the differences in densities that result when different current data bases are used.

Keywords: Density-functional theory, X-ray spectra, X-ray observations, Stellar atmospheres, Stellar activity
PACS: 31.15.Ew, 32.30.Rj, 95.85.Nv, 97.10.Ex, 97.10.Jb

REQUIREMENTS FOR MEASURING PLASMA DENSITIES

The spectrum emitted by an X-ray plasma reflects the physical conditions of the plasma, especially the plasma temperature and the plasma density. While different plasma temperatures affect an X-ray spectrum quite strongly (broad band continuum emission and appearance and disappearance of strong lines originating from ions in different ionization stages), the effects of different densities are rather subtle. This is demonstrated with Fig. 1, where two simulated spectra are shown. For simplicity we chose an isothermal plasma in one case with a high density and in the other with a low density. It can be seen that differences are only detectable in a few emission lines, and therefore density measurements in X-ray plasmas can only be carried out with high-resolution spectroscopy.

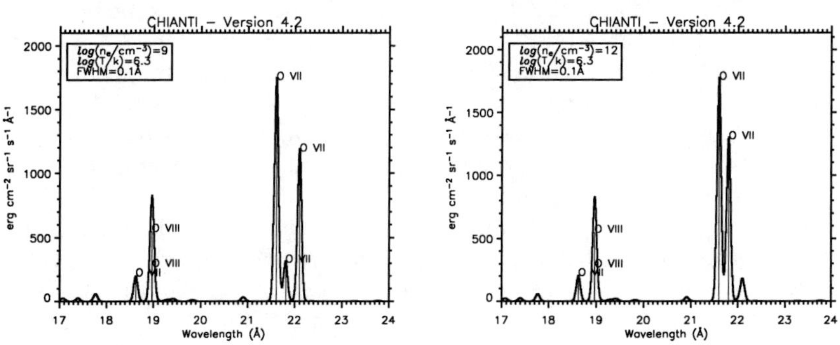

FIGURE 1. Simulations of spectra with low densities (left) and high densities.

CP774, *X-Ray Diagnostics of Astrophysical Plasmas,* edited by Randall K. Smith

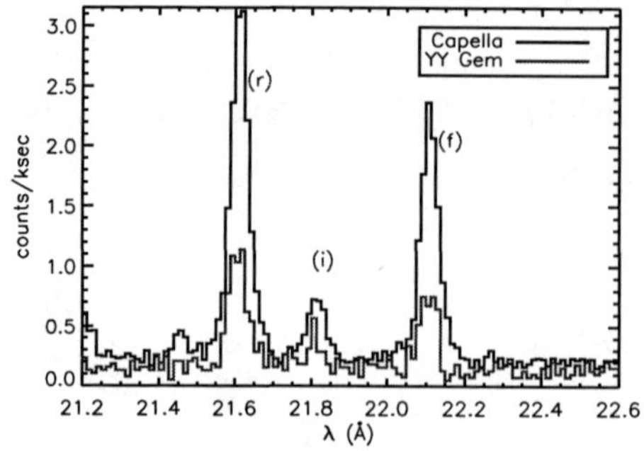

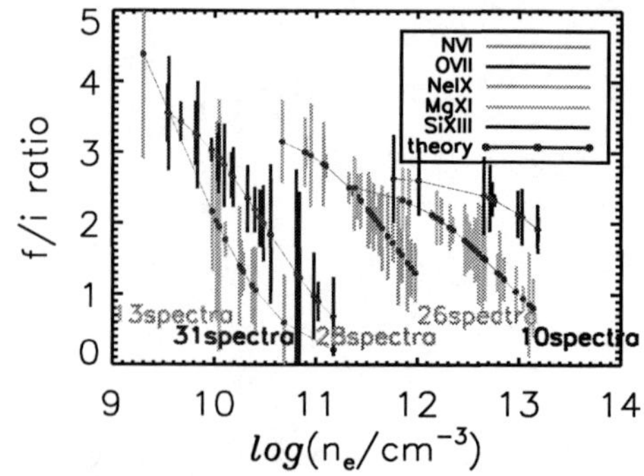

FIGURE 2. Left: Chandra LETGS measurements of the O VII He-like triplets for Capella and for YY Gem (blue), scaled to the distance of Capella. The spectra show different behaviour in the ratio of the lines marked with (i) and (f). **Right:** Density measurement with different He I-like ions for a large sample of stellar coronae. Error bars (1σ statistical errors) are large due to faint intercombination lines in the case of large f/i ratios. Mg XI and Si XIII cannot constrain densities below $10^{12.5}$ cm^{-3}.

DENSITY MEASUREMENTS WITH HE-LIKE TRIPLETS

The He-like ions provide a valuable method of measuring the electron density (n_e), in principle over a wide range of electron temperatures (T_e). The method [see 1] depends on the competition between collisions between the 1s2s ^{3}S and 1s2p ^{3}P levels and the radiative decay in the forbidden line (1s2s ^{3}S – 1s^2 ^{1}S). The ratio of the flux in the forbidden line to that in the intersystem (+magnetic quadrupole) line, then becomes sensitive to n_e, and can be parameterized as $f/i = R_0/(1+n_e/N_c)$, where R_0 is the ratio in the low density limit and N_c is the critical density. Both R_0 and N_c contain atomic data relating to collisional and radiative rates. Fig. 1 shows simulated spectra for O VII at densities of 10^9 cm^{-3} and 10^{12} cm^{-3}, between which the ratio depends on n_e.

The Chandra LETGS measurements of Capella and YY Gem (Fig. 2) show clear differences in f/i ratios, which indicate different densities in the two coronae.

The atomic rates have different dependences on Z - 1, where Z is the nuclear charge, but should depend smoothly on Z - 1, for the ions of interest. Fig. 3 (left) shows the parameters R_0 and N_c as a function of Z - 1. This brings out the differences between the data bases CHIANTI and APEC and shows $\log N_c$ from [2]. Values of R_0 are indicated by different colours; this plot suggests that the data in R_0 for C V and N VI in CHIANTI need to be re-examined, and that those for Si XIII warrant closer examination. Note that $\log N_c$ depends linearly on Z-1.

The right hand panel of Fig. 3 shows the resulting variation of the ratio f/i for O VII, indicating the importance of accurate atomic data when observed ratios are close to the low density limit. Measured densities will always be averages, but the He-like triplets formed at high temperatures probe only high densities (above 10^{12} cm^{-3}), while low-temperature ions measure only lower densities ($\sim 10^{10} - 10^{11}$ cm^{-3}); this leaves two cases unexplored: low densities in hot plasma and high densities in cool plasma.

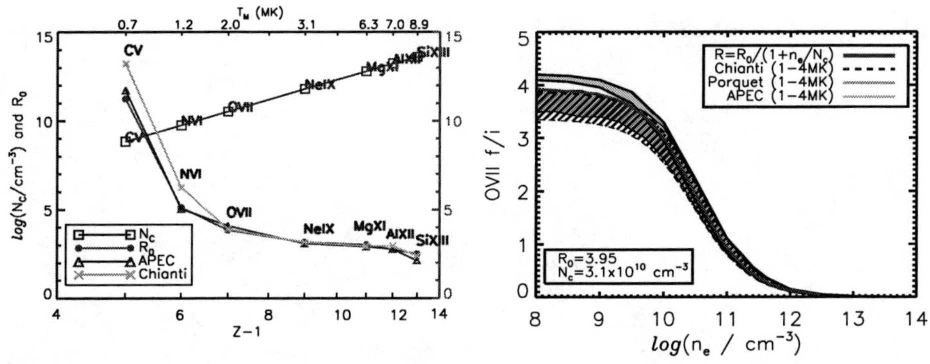

FIGURE 3. Left panel: comparison of R_0 (from different data bases) and N_c [2, in purple]. Right panel: density-dependent f/i line flux ratios for O VII.

WHAT CAN WE LEARN FROM DENSITY MEASUREMENTS?

In stellar coronae the density measurements provide an important link between physical and geometrical properties, through the Emission Measure $\mathrm{EM} = n_e^2 \times V$ (n_e: electron density and V: emitting volume). Densities are also needed to explore the optical depths of the lines: $\tau \propto \int n_e d\ell$, where ℓ is the path length (cm), and hence emitting areas.

RESULTS AND CONCLUSIONS

A survey of density measurements was carried out by [3]. Densities were measured from the He-like triplets (O VII and Ne IX) probing densities at temperatures 2×10^6 K and at 4×10^6 K. Measurements from higher-Z ions (Mg XI and Si XIII) result only in low-density limits $< 10^{12.5}\,\mathrm{cm}^{-3}$ (Fig.2, right). The observed range of f/i ratios shows that real differences in n_e do occur between different coronae.

Accurate atomic data are essential in order to use these and other diagnostics of n_e, thus exploiting the excellent new spectra being obtained. Values of n_e can then be used to constrain physical models of the atmosphere.

ACKNOWLEDGMENTS

JUN acknowledges support from PPARC under grant number PPA/G/S/2003/00091

REFERENCES

1. A. H. Gabriel, and C. Jordan, *MNRAS*, **145**, 241 (1969).
2. G. R. Blumenthal, G. W. F. Drake, and W. H. Tucker, *ApJ*, **172**, 205 (1972).
3. J.-U. Ness, M. Güdel, J. H. M. M. Schmitt, M. Audard, and A. Telleschi, *A&A*, **427**, 667 (2004).

Drs. Jan-Uwe Ness and Klaus Werner watching a presentation.

Diffuse Hot Gas Under X-ray Absorption Line Spectroscopy

Q. Daniel Wang and Yangsen Yao

Department of Astronomy, University of Massachusetts, Amherst, MA 01003

Abstract. We review recent results from X-ray absorption line spectroscopy of diffuse hot gas in and around the Galaxy. Absorption lines of O VII, O VIII, and/or Ne IX have been detected in the spectra of LMC X–3 as well as three AGNs and seven Galactic low-mass X-ray binaries. We have implemented an analysis tool, accounting for line saturation and allowing for joint fits to multiple absorption lines. Our analysis gives no significant detection of the absorption beyond the distance of the LMC. The mean temperature and density of the X-ray-absorbing gas are $\sim 10^{6.4}$ K and 3.6×10^{-3} cm^{-3}, assuming a filling factor of ~ 1 and a typical interstellar metallicity as well as a collisional ionization equilibrium. These assumptions seem to be consistent with constraints obtained from diffuse O VII emission and pulsar dispersion measure observations. The spatial distribution of the hot gas can be characterized by a hot gaseous disk with a scale height of only ~ 1 kpc and an apparent density enhancement in the Galactic central region, consistent with the surface brightness distribution of the Galactic diffuse 3/4-keV background. The X-ray-absorbing species as well as those far-UV tracers such as O VI, N V, and C IV seem to follow a trend from high to low ionization states away from the Galactic plane. These results demonstrate the potential of X-ray absorption spectroscopy as a powerful tool in the study of the hot interstellar medium.

Keywords: X-ray absorption line, hot interstellar medium

INTRODUCTION

One of the major discoveries made with *Chandra* and *XMM-Newton* is the detection of X-ray absorption lines produced by highly ionized species (e.g., O VII and Ne IX) in the spectra of AGNs [6, 7, 13, 14, 16]. These absorption lines are sensitive to hot gas over a broad temperature range, in contrast to the relatively narrow range covered by various far-UV tracers (e.g., O VI; Fig. 1). Existing studies have focused on the search for the so-called warm-hot intergalactic medium (WHIGM), which may account for the "missing" baryons in the present universe [4]. Unfortunately, the sensitivity of the instruments appears to be insufficient for a blind search of weak absorption lines expected for the WHIGM, even in the spectra of the brightest AGNs. The false detection probability in such a search needs to account for the number of statistically independent redshift bins, since the location of a WHIGM feature is not known as a priori. Confusion with outflows from target AGNs can also be a problem [12]. As a result, a firm detection of the WHIGM remains elusive.

We have concentrated on detecting X-ray absorption lines at $z \sim 0$. Our goal is to characterize the physical, chemical, and ionization states as well as the amount and spatial structure of the absorbing hot gas in and around the Galaxy. Achieving this goal is important not only for the understanding of the Galactic eco-system but also for the study of galaxy evolution in general. Structure formation theories predict that a disk

CP774, *X-Ray Diagnostics of Astrophysical Plasmas,* edited by Randall K. Smith

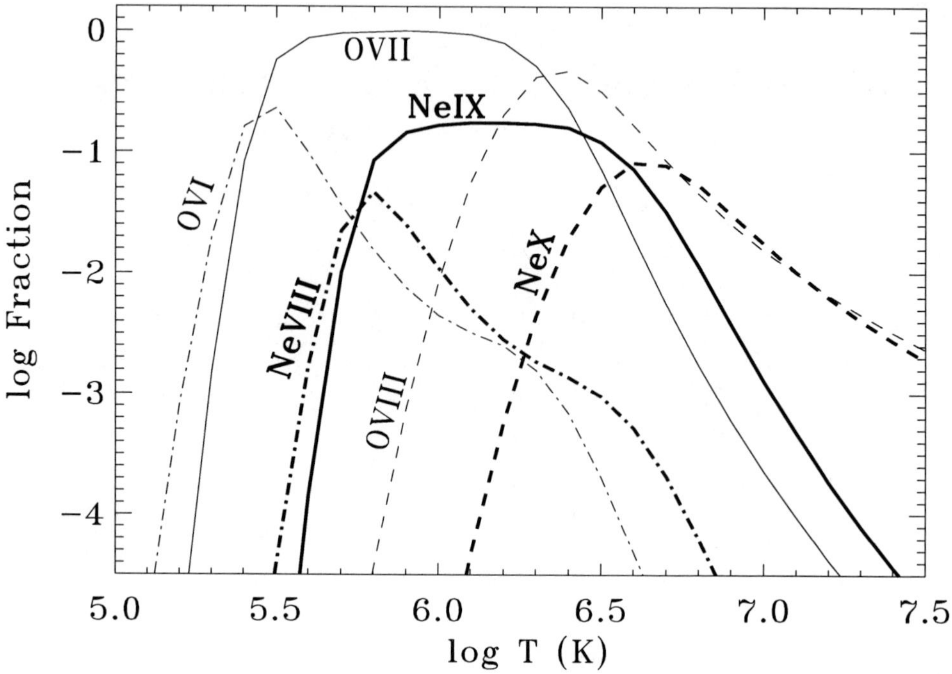

FIGURE 1. The relative ionization fractions of oxygen (*thin lines*) and neon (*thick lines*) vs a CIE plasma temperature. The neon fractions include the neon to oxygen abundance ratio.

galaxy such as our own should still be accreting from the IGM, replenishing the gas consumed by star formation. The accreted gas is expected to be hot. But its properties depend strongly on galactic energy and chemical feedbacks, which are very uncertain. Therefore, we can use the Galaxy as a test of the theories.

X-ray absorption lines can provide key diagnostics of the hot gas. First, we can obtain depth information, if the distances to the background light sources are known. With the detections, or even upper limits, of the lines in various sight-lines to different distances, we can determine the distribution of the gas. Second, the line centroid and width can also be used to constrain the location and nature of the gas. Third, the relative strengths of multiple lines in the same sight-line can be used to probe the physical, chemical, and/or ionization states. Fourth, the strengths of the absorption lines are proportional to column densities, which are related to the gas mass more directly than an emission measure. All these diagnostics are not affected by cool gas absorption, which is often a major uncertainty in the interpretation of soft X-ray emission observations. Clearly, absorption line spectroscopy, complemented by other forms of measurements, is a potentially powerful tool in the study of the hot gas.

2. OBSERVATIONS AND DATA ANALYSIS

Our existing study is based primarily on *Chandra* archival grating observations of Galactic low mass X-ray binaries (LMXBs) as well as those bright AGNs with known X-ray absorption line detections. The inclusion of these Galactic sources is critical to the constraints on the spatial properties of the hot gas. We select LMXBs with $|b| > 2^\circ$ to avoid sources with too heavy cool gas absorption near the Galactic plane. We also exclude sources with identified intrinsic emission/absorption features and use only observations with $S/N > 7$ per bin at ~ 0.9 keV. Our sample consists of ten LMXBs with 17 observations. The LMXB observations, not specifically designed for the absorption line spectroscopy, were mostly taken with the HETG, which is more sensitive to the Ne IX Kα line than the O VII Kα line. In addition, we have recently obtained a 100 ks LETG/HRC observation of the black hole X-ray binary LMC X–3, in collaboration with T. Fang et al. [22]. This observation was designed to separate the Galactic and extragalactic contributions to the X-ray line absorption seen in AGN spectra.

The data analysis procedure has been detailed in [25], including the implementation of an X-ray absorption line profile model in the spectral analysis software XSPEC. This model accounts for line saturation and allows for multiple line joint-fits to simultaneously determine model parameters such as gas temperature, turbulent broadening, and column density. Fig. 2 shows examples of such joint fits. With the spectral resolution of the grating instruments (~ 500 km s^{-1}, FWHM), the detection of an individual line can be used to determine only the upper limit to the line width. But, a strong line such as O VII Kα, if narrow, can be seriously saturated, changing the relative line strengths. A joint fit of multiple lines may thus yield a lower limit to the broadening. With the limited quality and number of the detected lines, our analysis assumes a collisional ionization equilibrium (CIE) of the X-ray-absorbing gas and the Galactic ISM metal abundance given by Wilms et al. [24] (e.g., the oxygen number fraction is 4.90×10^{-4}, compared to the solar abundance 6.76×10^{-4} given by Grevesse & Sauval [9]). We check these assumptions by comparing our results with other related measurements of the hot gas (see the discussion in § 4).

3. SUMMARY OF RESULTS

Table 1 lists the sources that we have studied. Only detections with signal-to-continuum noise ratio greater than 3σ are considered to be significant. We give our fitted temperature T and equivalent hydrogen column density N_H of the X-ray-absorbing hot gas in each of the sight-lines. For the three sources with no significant line detection, the table lists the 90% upper limits to N_H, obtained with T fixed to the mean temperature inferred from the other LMXB sight-lines [25].

Compared to the previous analysis of the AGN data [6, 7, 13, 14, 16], our joint-fits of the multiple lines greatly tighten the constraints on the key parameters. The fitted T values are mostly consistent with each other, although there is a marginal trend for an increasing temperature toward the central region of the Galaxy. On average, N_H appears to increase with the decreasing Galactic latitude. But there are significant variations from one direction to another, even among the high Galactic latitude sight-lines toward

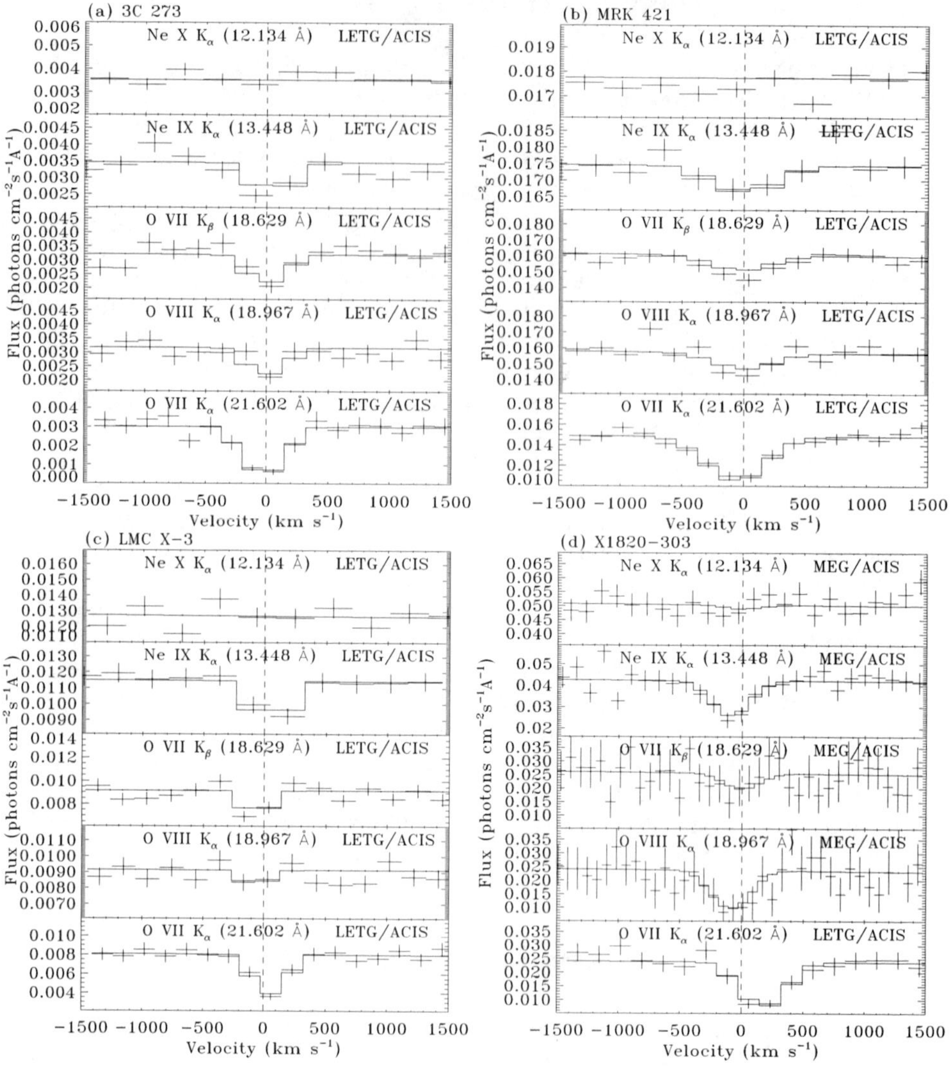

FIGURE 2. The velocity profiles of the joint-fitted absorption lines of *(a)* 3C 273, *(b)* MRK 421, *(c)* LMC X–3, and *(d)* X1820–303 (NGC 6624). The *solid* histograms in each panel represent the best-fit joint-fit model [25]. The vertical *dashed* line marks the zero velocity. A negative velocity indicates a blue shift.

the AGNs. For example, N_H is significantly higher toward 3C273 than toward MRK 421 (Table 1), consistent with the finding of Fang et al. [7].

We detect both O VII and Ne IX Kα lines in the spectra of LMC X–3 (Fig. 2). Because it is located outside the main body of the Large Magellanic Cloud, we do not expect any significant diffuse hot gas contribution from the galaxy. The best-fit O VII line

TABLE 1. Sources and Results.

Source	Galactic l, b (deg)	D (kpc)	Log[$T(K)$]	Log[N_H(cm^{-2})]
MRK 421	179.83, 65.03	1.2E5	6.2(6.1, 6.3)	19.2(19.1, 19.4)
PKS 2125–304	17.73, -52.25	4.8E5	6.33(6.25, 6.41)	19.8(19.5, 20.3)
3C 273	289.95, 64.36	6.3E5	6.23(6.16, 6.31)	19.9(19.7, 20.1)
LMC X–3	273.58, -32.08	50	6.2(5.7, 6.4)	19.7(19.2, 20.6)
X1820–303	2.79, -7.91	7.6	6.4(6.2, 6.5)	20.0(19.8, 20.2)
X1728–169	8.51, 9.04	5.0	6.3(6.0, 6.5)	19.9(19.7, 22.2)
X1837+049	36.12, 4.84	8.4	6.45(6.37, 6.54)	20.1(19.9, 20.2)
X2142+380	87.33, -11.22	7.2	6.3(6.2, 6.4)	19.7(19.5, 19.9)
X1118+480	157.66, 62.32	1.8	6.4(f)	$<$19.6
X0614+091	200.88, -3.36	≤ 3	6.4(f)	$<$19.5
X0929–314	260.11, 14.22	≥ 5	6.4(f)	$<$19.7
X1636–536	332.92, -4.82	6.5	6.3(5.7, 6.8)	20.2(19.5, 22.3)
X1735–444	346.05, -6.99	7.1	6.5(6.3, 6.6)	20.2(20.0, 20.4)
X1702–363	349.10, 2.75	5.0	6.0(5.6, 7.1)	19.9(19.4, 20.3)

centroid, -6(-91, 214) $\mathrm{km\,s^{-1}}$, indeed differs significantly from the systemic velocity 310 $\mathrm{km\,s^{-1}}$ of the LMC. Furthermore, the *XMM-Newton* RGS spectroscopy has ruled out a line-driven stellar wind as the power source for the X-ray emission of LMC X–3 [15]. Therefore, the absorption lines seen toward LMC X–3 is most likely Galactic in origin. The N_H value of LMC X–3 is consistent with those of the AGNs within 90% error bars (Fig. 3). We may further improve the comparison between the LMC X–3 and MRK 421 sight-lines, which appear comparable in the *ROSAT* diffuse 3/4-keV intensity map [20]. We jointly fit the LMC X–3 and MRK 421 spectra, assuming that the thermal and turbulent properties of the absorbing gas to be the same in the two sight-lines. Whereas the result on MRK 421 remains essentially the same as in Table 1, the constraint on N_H $(= 10^{19.3\pm0.2})$ in the sight-line to LMC X–3 is greatly tightened. The comparison between these two sight-lines shows no evidence for significant X-ray line absorption by hot gas beyond the LMC!

Combining both the Galactic and extragalactic N_H measurements in Table 1, we characterize the spatial distribution of the hot gas. Fig. 3 shows two simple model fits to the measurements: $n_H = 5.6^{+2.8}_{-1.9} \times 10^{-3}\ \mathrm{cm^{-3}} \exp(-|z|/1.0^{+0.6}_{-0.4}\ \mathrm{kpc})$ (assuming a disk morphology) or $n_H = 6.9^{+2.8}_{-2.5} \times 10^{-2}\ \mathrm{cm^{-3}} \exp(-R/2.6^{+0.6}_{-0.4}\ \mathrm{kpc})$ (a sphere morphology), where z and R are the distances from the Galactic plane and center, respectively. The reality should be between these extremes. The total mass of the hot gas, estimated from the sphere model, is $\sim 8 \times 10^8 M_\odot$.

4. COMPARISONS WITH OTHER MEASUREMENTS

We use complementary information from other relevant measurements to further constrain the properties of the hot gas:

• *X-ray emission lines*

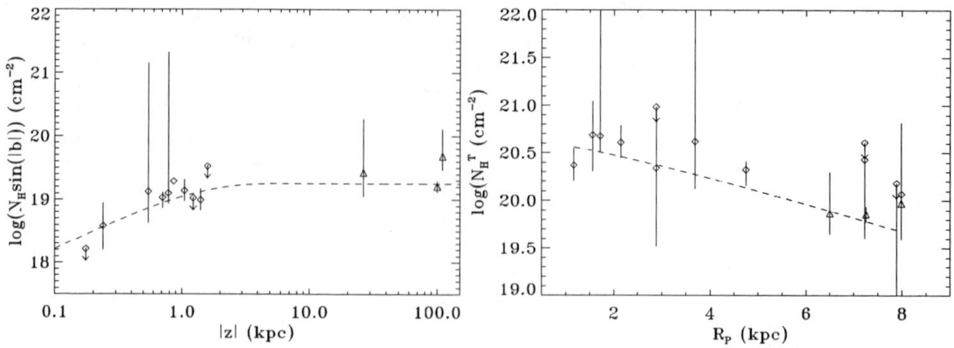

FIGURE 3. *Left panel*: The vertical hot gas column density as a function of the distance from the Galactic plane; *right panel*: the total hot gas column density N_H^T as a function of the off-GC impact radius R_p in the spherical distribution model. *Diamonds* mark the measurements toward the LMXBs and *triangles* represent the line-of-sights to the extragalactic sources.

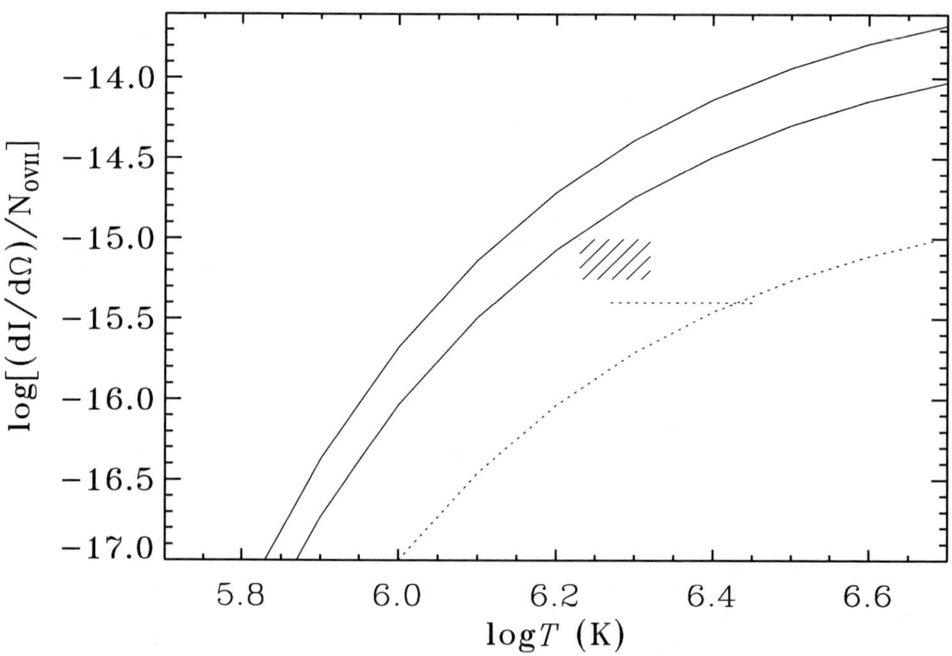

FIGURE 4. The ratio of O VII emission to absorption as a function of the CIE gas temperature. *Solid lines* outline the prediction based on the hot gaseous disk model with its parameter uncertainties. The shaded region shows the 90% measurement uncertainties inferred from the X-ray absorption line fit to MRK 421 (Table 1). In comparison, the dotted horizontal line and curve approximately mark the temperature range and the prediction of a uniform hot gas distribution to a distance of 140 kpc, as used by Rasmussen et al. [16].

Fig. 4 illustrates the theoretical dependence of the O VII emission to absorption ratio on both the temperature and the spatial distribution of the assumed CIE gas. To estimate this ratio observationally, we use a surface brightness $dI/d\Omega = 4.8(\pm 0.8)$ ph $s^{-1}cm^{-2}sr^{-1}$ of the O VII emission line triplet, as reported by McCammon et al. [11]. This emission measurement with an energy resolution of ~ 9 eV is from a sounding rocket experiment that covered a field of ~ 1 sr (centered at $l = 90°, b = +60°$). The sight-line to MRK 421 is adjacent to this field and can best be used to estimate both the O VII column density and the temperature of the hot gas. Fig. 4 shows that our estimated ratio and temperature are marginally inconsistent with the theoretical prediction based on the hot gaseous disk model, which is constrained with the set of the X-ray absorption line measurements (§ 3 and Table 1). This inconsistency, if confirmed, would indicate that the hot gas is in a non-CIE state with over-ionized oxygens, which may be expected from a fast adiabatically cooled outflow [3]. We can probably rule out the model of a uniform hot gas distribution over a distance $\gtrsim 140$ kpc as preferred by Rasmussen et al. [16]. Their results were obtained from the *XMM-Newton* RGS observations of MRK 421, PKS 2125-304, and 3C 273. These later two sight-lines include local X-ray enhancements toward inner regions of the Galaxy, as will be discussed next. Therefore, their estimated mean temperature and column density are higher than those from the MRK 421 sight-line alone, as adopted here. Furthermore, our hot gas temperature constraint is much tighter because of the multiple line joint fit. More detailed discussion on this comparison will be presented in Wang et al. [22].

- *Broad-band diffuse X-ray intensity*

We further find an apparent correlation between the X-ray-absorbing hot gas column density and the diffuse 3/4-keV band intensity in the AGN fields. In particular, the excess column of the X-ray-absorbing gas toward 3C 273 is most likely due to the passage of the sight-line through a large-scale soft X-ray emission enhancement that is probably associated with Radio Loop I [20]. The nature of this enhancement is controversial. One proposal is that it is generated by a bipolar wind or a giant explosion from the Galactic nuclear region [21, 2]. In this case, the physical size of the enhancement should be comparable to our distance to the Galactic center (~ 8 kpc). But, part of the enhancement is apparently due to the North Polar Spur, which is the Galactic northeastern ridge of the Loop and has been shown to be a local structure at $D \sim 100 \pm 20$ pc [1]. We can use the X-ray emission and absorption combination to directly constrain the line-of-sight size (d) of the enhancement. We estimate its column density from the N_H difference between the 3C 273 and MRK 421 sight-lines (Table 1), which are at nearly identical Galactic latitudes. The N_H difference gives a 90% confidence column density $n_e d = (3 - 11) \times 10^{19}$ cm^{-2}. The X-ray-emitting gas in Loop I has an average emission measure $n_e^2 d \sim (1-2) \times 10^{-2}$ cm^{-6} pc [5, 23]. We thus obtain $d \gtrsim 5$ kpc (the upper limit 127 kpc probably does not mean much). This lower limit could be smaller (or larger), if the hot gas is in an over-ionized (or under-ionized, relative to the electron temperature) non-CIE state. But it may be hard to reduce the d estimate by a factor of ~ 50 to make it consistent with the local scenario. Therefore, we tentatively suggest that the enhancement along the sight-line to 3C 273 is most likely associated with outflows from the Galactic central region. The sight-line to PKS 2125-304, also in an inner Galactic direction with a general 3/4-keV background enhancement, could similarly be affected by the outflows.

• *Far-UV absorption lines*

Absorption lines of highly ionized species have been studied extensively in far-UV. The most relevant species are O VI, N V, C IV, and S IV, ordered with decreasing ionization energy. While a significant fraction of S IV may be produced by photo-ionization, most of the other species are likely due to collisional ionization. The O VI population, for example, peaks at $T \sim 3 \times 10^5$ K (Fig. 1), near the peak of the cooling curve of a CIE plasma. Therefore, the species may reside primarily at the interfaces of cool clouds with hot gas, consistent with the finding that about half of distinct high-velocity O VI absorbers have H I counterparts [19]. In the solar neighborhood, O VI, N V, C IV, and S IV have the vertical scale heights of $2.7 \pm 0.4, 3.9 \pm 1.4, 4.4 \pm 0.6$, and 5.1 ± 0.7 kpc, respectively (e.g., Savage et al. [17, 18]). The increasing scale height with the decreasing ionization energy of the species is striking. Interestingly, our estimated scale height ($\sim$ 1 kpc) of the X-ray-absorbing species (primarily O VII and Ne IX) follows the same trend. The gas could be responsible for $\sim$ 10% of the O VI density $\sim 1.7 \times 10^{-8}$ cm^{-3} at the Galactic midplane [10] and less in regions away from it. But, the tightly constrained mean velocity dispersion of the O VI-absorbing gas agrees well with the thermal velocity $\sim$ 50 km s^{-1} of a gas at $T \sim 10^{6.4}$ K, which may indicate a link between the two species.

• *Pulsar Emission Measure*

We can obtain a fundamental constraint on the total amount of hot gas from pulsar dispersion measures (e.g., [8]). Interestingly, the vertical scale height of free electrons, inferred from the dispersion measures, is also about 1 kpc, essentially identical to that of the X-ray-absorbing gas. The electron column density is $N_e = 0.6 \times 10^{20}$ cm^{-2} (Galactic pulsars only) and $=2 \times 10^{20}$ cm^{-2} (including MC pulsars). Assuming that a fraction ξ of the electrons is in the hot gas, as traced by the X-ray-absorbing species, we constrain its metal abundance to be $\sim (6-30)\% \xi^{-1}$ of the assumed ISM value. If the bulk of the electrons is in photo-ionized warm gas, as is normally expected, the abundance can then be close to the typical ISM value or even higher. Therefore, the hot gas is unlikely to be accreted recently from the IGM.

5. CONCLUSIONS

Our measurements of X-ray absorption lines in the spectra of both Galactic and extra-galactic sources, together with other relevant observations, have yield useful constraints on the hot gas in and around the Galaxy:

- The mean absorbing gas temperature and density are $\sim 10^{6.4}$ K and $\sim 3.6 \times 10^{-3}$ cm^{-3} (assuming a filling factor of $\sim$ 1);
- The gas has a vertical exponential scale height of $\sim$ 1 kpc, similar to that of free electrons inferred from pulsar dispersion measures, but significantly smaller than those of far-UV tracers (O VI, N V, C IV, and S IV);
- The metal abundance of the hot gas is probable comparable to the Galactic ISM value;

- The distribution of the hot gas is quite patchy, especially in inner Galactic regions with strong enhancements that may be associated with outflows from the Galactic center;
- The total absorbing gas mass is $\sim 1 \times 10^9 M_\odot$;
- No evidence is found for significant X-ray absorption beyond the LMC (probably $\lesssim 10^{19}$ cm^{-2}, assuming the Galactic ISM abundance).

ACKNOWLEDGMENTS

The authors thank Frits Paerels, Todd Tripp, and Jelle Kaastra for useful discussions. The project is supported by NASA.

REFERENCES

1. Bingham, R. G. 1967, MNRAS, 137, 157
2. Bland-Hawthorn, J., & Cohen, M. 2003, ApJ, 582, 246
3. Breitschwerdt, D., & Schmutzler, T. 1999, A&A, 347, 650
4. Cen, R. & Ostriker, J. P. 1999, ApJ, 514, 1
5. Egger, R. J., & Aschenbach, B. 1995, A&A, 294, L25
6. Fang, T., et al. 2002, ApJ, 572, L127
7. Fang, T., Sembach, K. R., & Canizares, C. R. 2003, ApJ, 586, L49
8. Gómez, G. C., et al. 2001, AJ, 122, 908
9. Grevesse, N., & Sauval, A.J. 1998, Space Science Reviews 85, 161
10. Jenkins, E. B., Bowen, D. V., & Sembach, K. R. 2001, in Proc. XVIIth IAP Colloq. Gaseous Matter in Galactic and Intergalactic Space, ed. R. Ferlet, M. Lemoine, J. M. Desert, & B. Raban (Paris: Frontier Group), 99
11. McCammon, D., et al. 2002, ApJ, 576, 188
12. McKernan, B., Yaqoob, T., & Reynolds, C. S. 2004, astro-ph/0408506
13. Nicastro, F., et al. 2002, ApJ, 573, 157
14. Nicastro, F. 2003, (astro-ph/0311162)
15. Page, M. J., et al. 2003, MNRAS, 345, 639
16. Rasmussen, A., Kahn, S., & F. Paerels 2003, (astro-ph/0301183)
17. Savage, B. D., et al. , 1997, AJ, 113, 2158
18. Savage, B. D., et al. , 2000, ApJ, 538, 27
19. Sembach, K. R., et al. 2003, ApJS, 146, 165
20. Snowden, S. L., et al. 1997, ApJ, 485. 125
21. Sofue, Y. 1996, ApJ, 458, 120
22. Wang, Q. D., et al. 2004, in preparation
23. Willingale, R., et al. 2003, MNRAS, 343, 995
24. Wilms, J., Annen, A., & McCray, R. 2000, ApJ, 542, 914
25. Yao, Y., & Wang, Q. D. 2004, ApJ, submitted

Notetaking at the XDAP workshop.

IV: NON-EQUILIBRIUM PLASMA DIAGNOSTICS

X-ray Diagnostics of Ionizing Plasmas in Collisionless Shocks

Cara E. Rakowski

Harvard-Smithsonian, Center for Astrophysics, 60 Garden St. MS-70, Cambridge MA 02138

Abstract.
Although collisional ionization equilibrium conditions have been much more thoroughly studied, many astrophysical situations involve timescales that are shorter than the time it would take for ionization equilibrium to have been achieved. In plasmas out of equilibrium ionization the ion fractions cannot be assumed to be a simple function of the electron temperature, but rather must be independently assessed before comparisons can be made of the relative abundances of different elements. The emission spectrum is a function of both the temperature of the surrounding electrons, kT_e as well as the ionization timescale, $n_e t$ which, together with kT_e determines the current charge state distribution. The accuracy of measuring an ionization timescale from the X-ray spectrum depends strongly on our knowledge of the ionization and recombination rates as well the emission processes. However, lines that are only strong in a rapidly ionizing plasma could provide a separate proof of non-equilibrium conditions distinct from diagnostics that are indicative of the temperature or ion fractions. In this talk I will present examples of both traditional temperature diagnostics as well as lines that would be undetectable in anything other than a rapidly ionizing plasma.

Keywords: Collisionless Shocks; Non-equilibrium Ionization; Supernova Remnants; Li-like ions
PACS: 52.70.La, 96.50.Fm, 98.38.Mz, 32.30.Rj

OUTLINE

This talk is given from the perspective of an X-ray astronomer studying supernova remnants. The continuing need for accurate atomic calculations of processes that are important for non-equilibrium ionization (NEI) will be motivated by a brief outline of some of the physical questions surrounding supernova remnants (SNRs). NEI plasma models are essential for the proper interpretation of SNR observations. Here I will outline some of the pitfalls of inappropriately assuming collisional ionization equilibrium (CIE). Following this introduction, a sampling of the basic concepts in atomic physics that can diagnose non-equilibrium conditions will be presented. In closing, some examples of common complications in the application of NEI models to physical situations will be addressed.

OBSERVATIONS OF SUPERNOVA REMNANTS

The young age of most remnants and the extremely low density of the ambient medium into which the hot supernova blast-wave is expanding, conspire to keep the ion fractions from reaching an equilibrium. Hence accurate interpretation of SNR spectra require modeling the emission from a non-equilibrium ionization plasma. The following are

CP774, *X-Ray Diagnostics of Astrophysical Plasmas,* edited by Randall K. Smith

two broad classes of fundamental physics questions that studies of supernova remnants are attempting to answer.

Supernova Explosion Mechanisms

Supernova (SN) explosions are complex systems to model. The physics behind both the core-collapse of massive stars (type II SN) and thermonuclear explosions of white dwarfs (type Ia SN) still have un-answered questions. Can bipolar jets themselves revitalize the supernova shock front in core-collapse SN, or are they secondary? In either case, what drives the jet? For type Ia explosions, models need to match both the range of elements produced and their apparently stratified distribution. Current 1D models invoke an ad-hoc transition from a sub-sonic to a super-sonic burning front to reproduce the elemental composition. Three-dimensional models, on the other-hand, show strong turbulent instabilities such that they have difficulty retaining a stratified ejecta distribution. Furthermore, as primary sources of higher element production in the universe, the nucleosynthesis that occurs in core-collapse and type Ia SN is of interest both for it's role in cosmic chemical evolution and as a possible means to test fundamental questions in nuclear physics.

In order to constrain models of supernova explosions, studies of SNRs need to yield accurate measurements of the relative production of different metals. Individual knots of ejecta can provide a snapshot of the nucleosynthetic yield from a single density and time within the explosion. The spatial distribution and velocities of such knots can be compared to the three-dimensional simulations (see Laming et al., this proceedings, for on-going investigations in Cassiopeia A). The energetics, age, and total mass of the remnants are all important potential inputs to these comparisons.

Cosmic-ray Production and Collisionless Shocks

The shock waves in supernova remnants are also widely thought to be the primary galactic source for high energy cosmic-rays at least up to the "knee" energy. The ambient density around SNRs is extremely low. In fact, SNR shocks are classified as "collisionless" because the width of the shock is smaller than the mean free path for collisions. The shock is not mediated by particle collisions but rather by the collective magnetic field of all of the charged particles. The heating of particles at the shock front is the result of magneto-hydrodynamic waves excited by different plasma instabilities. Electrons and protons are not effected by the same instabilities because of their huge difference in mass. Hence, collisionless shock heating may or may not result in equilibrated electron and proton temperatures.[1] The particle acceleration process is highly non-linear as various plasma instabilities feed off one another or are present only under specific

[1] The plasma immediately behind the shock will always be out of ionization equilibrium, since ionizations require collisions (except for auto-ionization). However if the shock heating of electrons has not been effective, the plasma may also be out of temperature equilibrium

conditions and the existence of highly accelerated particles modifies the profile of the shock itself. Theoretical calculations require numerical simulations, which must then be both interpreted and tested against observations. The proximity of the solar wind has enabled extensive studies that have provided most of the basis of comparison for theory (for a recent review of the MHD theory, simulations and solar wind observations see Lembege et al. [1]). However, the shocks in SNR that are energetic enough to accelerate cosmic-rays are also orders of magnitude faster than solar-wind shocks. At such high Alfvén Mach numbers the instabilities that dominate the shock heating may be different than those in the solar wind shocks. Hence the partition of energy by plasma instabilities at the shock into the thermal and high-energy populations of electrons, protons and ions remains an active area of investigation. Evidence from multiple wave-bands is crucial to constrain the particle dynamics of the shock. The width of optical and ultraviolet lines from thermal ions can be measured to determine their temperatures (modulo geometric effects). Evidence of high energy cosmic-ray electrons can be seen through synchrotron emission in the radio and sometimes X-ray bands. The temperature of the thermal electrons can be deduced from X-ray observations, using both the shape of the bremsstrahlung continuum and temperature diagnostics of the emission lines. Combined with a proper motion measurement at a known distance these observations can be combined to create a complete picture of the shock dynamics and collisionless heating.

The partition of energy in collisionless shocks is important not only for its influence on galactic cosmic-ray production, but also to correctly interpret the dynamics of the object being studied. While the best science is done with multi-wavelength analysis, sometimes we are limited to a single wave-band. One would like to be able to interpret the electron temperature and derive, for instance, a shock velocity to compare to proper motions and determine a distance. However orders of magnitude may separate the velocities predicted under the two extreme assumptions of purely thermal, fully equilibrated electrons and protons versus inefficient electron-heating or extremely efficient cosmic-ray production. Hence, in depth study of the particle energy distributions in those remnants where information from multiple wave-bands is available is crucial for the interpretation of the dynamics of all remnants.

Perils of Assuming CIE

In an ionizing plasma the charge state distribution of the ions is not as fully stripped as the equilibrium distribution at that electron temperature, kT_e. Hence, as Masai [2], Liedahl [3] and others warn, one can often incorrectly model a single NEI plasma with two CIE components, one for the continuum radiation (at kT_e) and one for the line emission at a lower temperature. Obviously this can result in the false identification of two shocks along the line of sight. Furthermore, the emissivities of the lines and the continuum generally increase as a function of temperature. Hence, neither of these components is likely to faithfully predict the emission measure ($\int n_e dV$) needed for the flux seen. The true composition and density of the plasma will not be accurately assessed. In particular, the low-temperature component that was needed to provide the less highly ionized species will require higher absolute elemental abundances to produce strong lines. Even modeling the ion fractions from line ratios will not be accurate because generically the relative emissivities of the lines depend on the electron temperature, not just the charge state distribution.

In general, X-ray spectra provide two ways of determining the electron temperature: the bremsstrahlung continuum and diagnostic line ratios. If lines from lower ionization stages exist in what appears to be a much hotter continuum, this signals NEI. However, two CIE components can offer an alternative explanation, especially for low-resolution spectra. To complicate matters further, there are instances where we do expect more than one type of shock within the spatial region examined (e.g. forward and reverse, density variations within a shocked cloud...). Hence, ideally we would like to be able to identify lines that are only strong while a plasma is rapidly ionizing, in order to distinguish between possible explanations. The basic physical mechanism behind some such diagnostics are discussed shortly, but first a review of the important He-like "triplet" is needed for context.

He-like K-shell Temperature Diagnostics

For any given element, He-like ions are present across a wide range of temperatures. Prominent K-shell (n=2) lines from He-like species provide an opportunity to determine the temperature of individual elements independently from each other and the X-ray bremsstrahlung continuum (see e.g. Porquet et al. [4] concerning He-like lines in collisionally excited plasmas and Oelgoetz and Pradhan [5] for a recent in-depth discussion of the Kα complex in He-like iron in transient plasmas). Transitions from the n=2 shell back to the ground state form the He-like "triplet"[2] which is actually four lines: w (resonance, r), x and y (intercombination, i), z (forbidden, f). Decay from the singlet P state ($1s2p\ ^1P_1 \rightarrow 1s^2\ ^1S_0$) is called the resonance line. The decays from the triplet P states (or intercombination lines) are split into two lines by spin: x ($1s2p\ ^3P_2 \rightarrow 1s^2\ ^1S_0$) and y ($1s2p\ ^3P_1 \rightarrow 1s^2\ ^1S_0$). Decays from the 1s2s states to the ground either require the emission of 2 photons for $1s2s\ ^1S_0$ or a magnetic transition for the forbidden line $1s2s\ ^3S_1$.

A particularly succinct explanation of why certain ratios between the w, x, y, z lines are diagnostic to temperature or density is given by Bautista and Kallman [6] The ratio of the forbidden to intercombination lines is sensitive to density because above a certain density the meta-stable ^3S state can be excited to the ^3P state, increasing the intercombination lines at the expense of forbidden transitions. The "G" ratio of r/(i+f) is sensitive to temperature because at high temperatures all the states are collisionally excited, but at low temperatures recombination becomes more important. The triplet states will be preferentially populated through recombinations simply because of their greater statistical weight, thus increasing the intercombination and forbidden line strengths. Since all the lines considered are from one species, He-like line diagnostics are independent of the ionization state if collisional ionization equilibrium holds. Thus one expects that in CIE the inferred temperature should match the value measured from the bremsstrahlung

---

[2] This nomenclature, while conventional, is misleading since it only refers to their appearance at moderate spectral resolution and has nothing to do with triplet levels of the excited states of the ion.

continuum. In high resolution spectra these ratios can be used to determine the local electron temperature near each element, thus potentially spectrally identifying stratification of the ejecta. Out of equilibrium, contributions from ionizations and recombinations into excited states of the He-like ion must be taken into account, such that the electron temperature determined from the He-like triplet under the incorrect assumption of CIE may not match that from the continuum. The non-equilibrium ionization values of the measured He-like ratios can be modeled, but will be a function of both $n_e t$ and kT_e.

SIGNATURES UNIQUE TO PLASMAS OUT OF CIE

Unusual Ion Fractions

The conceptually simplest sign of a non-equilibrium plasma is an unusual charge state distribution. At all but the highest temperatures where the atoms are mostly fully stripped, the equilibrium ion fractions will generally be weighted towards filled valence shells, such as He-like or Ne-like ions. This follows from the fact that it is particularly easy to remove the last remaining electron in a shell, whereas the ionization potential from a filled shell is considerable. For any given element there is a large range of temperatures over which the He-like species dominates (see for instance [4] for the relevant temperature ranges as a function of atomic number). Conversely, Li-like (or Na-like) will always be rare in the steady-state.

The same pattern cannot be true at all times for ionizing plasmas, simply because any given atom must pass through the Li-like or Na-like state in order to reach the more stable He-like (or Ne-like) state. In his review, Liedahl [3] presents the example of the ionization of oxygen behind a step-jump change in temperature (or shock) where the electrons are initially still cold but equilibrate on a Coulomb timescale with the protons. At any temperature in CIE the Li-like O VI fraction never exceeds much more than 0.2. By contrast, the O VI fraction reaches a peak of near 0.6 while the plasma is ionizing. Similarly a high Li-like fraction, greater than 0.4, will also be obtained if the electron and proton temperatures were equilibrated at the shock front (R. Smith, personal communication).

Strong Fluorescent and Satellite Lines from Inner-shell Excitation

What lines originate from Li-like species that one can use to identify an unusually high Li-like fraction? Consider a Li-like ion in it's ground state $1s^2 2s$. As already mentioned in the discussion of the equilibrium ion fractions, the ionization potential needed to strip the 2s electron is small. This has two consequences. First that transitions from $1s^2 nl$ to $1s^2 2s$ will have significantly smaller energies and longer wavelengths than the He-like resonance line. This large difference in wavelength complicates any diagnostic comparison of them because of differences in absorption and detection efficiency. The more interesting consequence is that for Li-like ions, exciting an electron from the 1s level requires more energy than freeing the 2s electron. Cases like this, of transitions from excited states whose energies are higher than the "first ionization potential," are called "fluorescent" lines. One might not think that such bound states could even exist, no matter how short-lived, but they will occur if an inner shell electron is excited.

Inner-shell excitation (or ionization) will only be important in non-collisional equilibrium conditions. In any CIE plasma where Li-like species exist, it is necessarily the case that the average energy of the surrounding electrons is similar to or smaller than the first ionization potential. Hence only the tail of the electron distribution can excite an inner shell electron, and the bulk of the collisions will involve ionization, recombination or excitation of the outermost electron. However, if this low temperature plasma is shocked such that the free electrons are suddenly much more energetic, then things change. Above the threshold energy for ionization, electron collision rates drop steeply. These high energy electrons are actually more likely to excite or ionize the inner shell electrons than they are to even notice the outermost electron. Hence far from equilibrium, fluorescent lines from inner-shell excitation may be important. Similar logic holds for an ambient UV field exciting these transitions in photo-ionization equilibrium (PIE) which is the case most commonly considered for fluorescence.

The fluorescent line resulting from the excitation of a 1s electron in a Li-like atom to the 2p state is referred to as the "q" line. This transition, $1s2s2p \rightarrow 1s^2 2s$, is identical to the resonance transition in the He-like ion, $1s2p \rightarrow 1s^2$ except for the addition of a "spectator electron" in the 1s level. The spectator electron reduces the effective potential of the nucleus from the perspective of the outermost electron thus decreasing the energy of the transition and shifting the wavelength long-wards. This particular fluorescent line is just one example of the class of "satellite lines" formed from the existence of perturbing spectator electrons to a parent transition. As one might imagine, additional spectator electrons in lower ionization states, do not have as large of an effect on the effective potential as the first spectator does. Hence in practice, the Li-like satellite "q" line will be contaminated by unresolved contributions from lower ionization states.

Forbidden Line Enhancement from Inner-shell Ionization

The surrounding sea of hot electrons under non-ionization equilibrium conditions (or the ambient UV field in PIE) will not only excite the inner-shell electrons but also release them. Because the population of excited states is negligible at most astrophysical densities, any inner-shell ionization will be out of the ground state. Thus inner-shell ionization of a Li-like ion will leave behind a 1s and a 2s electron. The resulting He-like atom must then be in either the singlet or triplet S states. The $^1S_0$ state decays by two photons, but the $^3S_1$ state is the upper level of the forbidden (z) line. Hence a large population of Li-like ions that are ionizing to the more stable He-like species will contribute significantly to the strength of the forbidden line, especially under high temperature conditions such that inner-shell ionization is actually more probable than outer-shell ionization.

INTERPRETING NEI CONDITIONS

The atomic processes involved in NEI are not conceptually difficult. However, there are a number of issues which need to be addressed when developing models to interpret NEI conditions in real astrophysical situations. Among these are:

- Ionization fractions of different species rise and fall on the path to collisional ionization equilibrium. Therefore there is no one species which uniquely determines the ionization timescale [3].
- Observationally, even at high spectral resolution, it can be difficult to identify all the weak lines in a spectrum. As Brickhouse et al. [7] demonstrate, neglecting high n lines leads to a misinterpretation of the low level flux as continuum emission, biasing the abundance measurements to lower values.
- High order corrections are required to calculate accurate transition wavelengths. Therefore it is vital that atomic theory be verified experimentally for secure identification of rare lines (e.g. [8, 9], and many contributions to this proceedings).
- $n>3$ satellite lines differ very little in wavelength from their parent transitions. Their contributions to the standard line ratios cannot be neglected. Many observational examples of this ambiguity exist in the literature [10, 11, 12] A clear illustration of the effect of dielectronic recombination satellite lines on the inferred temperatures from diagnostics of Ne-like Fe XVII are shown in figures 1-4 of Raymond and Smith [12].
- Raymond and Smith [12] also illustrate the difficulty in assessing the average temperature in a multi-temperature plasma. As they explain, if there exist a range of temperatures then the emission in any given line is primarily from the range of temperatures where that line is intrinsically strongest. Hence a measurement of the temperature using a line ratio who's numerator is strongest at low temperatures will necessarily be weighted towards lower temperatures. Examples of the differences between the temperatures derived from different diagnostic ratios are shown in table 2 of Raymond and Smith [12].
- Solar flares provide a wealth of high signal to noise observations against which to test the atomic calculations of NEI processes (recommended reading: [13, 14]). However, solar flares (and solar coronae) are also inherently multi-temperature which confuses their interpretation.
- Similarly the time evolution of the SNR shocks effects their interpretation. While Itoh [15] and Cox and Anderson [16] are generally considered to be the classic references for calculating spectra from evolving SNRs, Masai [2] gives a particularly lucid examination of the effect of non-equilibrium ionization on the X-ray emission from SNRs in conjunction with the evolution of the forward and reverse shocks.
- In studies of collisionless heating by plasma instabilities, one would like to know the temperatures of the particles immediately behind the shock front. However the X-ray emissivity increases downstream of the shock, after the ions have been stripped down to highly ionized states that emit brightly in the X-ray band. Hence the temperature measured will be representative of a time after coulomb collisions have ionized some of the gas and begun equilibrating the temperatures of the

electrons and protons. The current ionization state must be used to calibrate $n_e t$ from which the range of possible initial electron temperatures can then be deduced. In any expanding shock-front, adiabatic expansion will decrease the overall average particle temperature as well as the density (which effects $n_e t$). Furthermore as more material is swept up by the shock, it decelerates, so that the temperature interior is not directly comparable to the current shock velocity. To deduce the range of possible original shock temperatures for a given parcel of gas, one must use the ionization itself to calibrate the timescale, $n_e t$, along which both the ionization and any coulomb equilibration of the temperatures evolve.

Over the years a number of workshops and conferences have been held on the topic of X-ray spectroscopy. In my opinion, the single most useful introduction to the atomic processes, instrumentation and observational methods is the proceedings of the 1997 European Astrophysics Doctoral Network School "X-ray Spectroscopy in Astrophysics" edited by Jan van Paradijs and Johan A. M. Bleeker. All the talks within are both thorough and clear. For the purposes of this talk, the papers by Mewe [17] and Liedahl [3] were especially educational and I highly recommend reading them for further detail and explication of the atomic processes covered here as well as X-ray diagnostics in general. I would also like to refer the reader to a recent Annual Reviews in Astronomy and Astrophysics article by Paerels and Kahn [18] who have compiled a brief review of the X-ray spectroscopy conducted with the grating instruments aboard *Chandra* and *XMM-Newton* in the first few years after their launches.

ACKNOWLEDGMENTS

In the course of researching the subject of non-equilibrium ionization, I had the opportunity to read work by the majority of the speakers at the conference. I would like to thank all of them for their contributions to the field, and the conference organizers for giving me the excuse to immerse myself in the subject. Hopefully I have managed to impart to the reader some of the understanding I gained in the course of preparing this talk. Many thanks to K.K. Dyer for reviewing this manuscript. This work was supported by NASA grant GO2-3081X.

REFERENCES

1. B. Lembege, J. Giacalone, M. Scholer, T. Hada, M. Hoshino, V. Krasnoselskikh, H. Kucharek, P. Savoini, and T. Terasawa, *Space Science Reviews*, **110**, 161–226 (2004).
2. K. Masai, *Astrophys. J.*, **437**, 770–780 (1994).
3. D. A. Liedahl, "The X-Ray Spectral Properties of Photoionized Plasma and Transient Plasmas," in *X-Ray Spectroscopy in Astrophysics, Lectures held at the Astrophysics School X*, edited by J. van Paradijs, and J. A. M. Bleeker, Springer Verlag, Berlin, 1999, vol. 520 of *Lecture Notes in Physics*, pp. 189–268.
4. D. Porquet, R. Mewe, J. Dubau, A. J. J. Raassen, and J. S. Kaastra, *Astron. Astrophys.*, **376**, 1113–1122 (2001).
5. J. Oelgoetz, and A. K. Pradhan, *Mon. Not. R. Astron. Soc.*, **354**, 1093–1102 (2004).
6. M. A. Bautista, and T. R. Kallman, *Astrophys. J.*, **544**, 581–591 (2000).

7. N. S. Brickhouse, A. K. Dupree, R. J. Edgar, D. A. Liedahl, S. A. Drake, N. E. White, and K. P. Singh, *Astrophys. J.*, **530**, 387–402 (2000).
8. J. K. Lepson, P. Beiersdorfer, E. Behar, and S. M. Kahn, *Astrophys. J.*, **590**, 604–617 (2003).
9. V. Decaux, P. Beiersdorfer, S. M. Kahn, and V. L. Jacobs, *Astrophys. J.*, **482**, 1076–+ (1997).
10. K. J. H. Phillips, J. R. Lemen, R. D. Cowan, G. A. Doschek, and J. W. Leibacher, *Astrophys. J.*, **265**, 1120–1134 (1983).
11. J. R. Lemen, K. J. H. Phillips, R. D. Cowan, J. Hata, and I. P. Grant, *Astron. Astrophys.*, **135**, 313–324 (1984).
12. J. C. Raymond, and B. W. Smith, *Astrophys. J.*, **306**, 762–766 (1986).
13. G. A. Doschek, *Astrophys. J. Suppl. Ser.*, **73**, 117–130 (1990).
14. G. A. Doschek, A. Fludra, R. D. Bentley, J. Lang, K. J. H. Phillips, and T. Watanabe, *Astrophys. J.*, **358**, 665–673 (1990).
15. H. Itoh, *Publ. Astron. Soc. Japan*, **30**, 489–498 (1978).
16. D. P. Cox, and P. R. Anderson, *Astrophys. J.*, **253**, 268–289 (1982).
17. R. Mewe, "Atomic Physics of Hot Plasmas," in *X-Ray Spectroscopy in Astrophysics, Lectures held at the Astrophysics School X*, edited by J. van Paradijs, and J. A. M. Bleeker, Springer Verlag, Berlin, 1999, vol. 520 of *Lecture Notes in Physics*, pp. 109–187.
18. F. B. S. Paerels, and S. M. Kahn, *Annu. Rev. Astron. Astrophys.*, **41**, 291–342 (2003).

Drs. Ann Hornschemeier, Keith Arnaud, and Kazimierz Borkowski during a lunch break.

K-Alpha Emission Spectra From Non-Equilibrium Ionizing Plasmas

V. L. Jacobs

Center for Computational Materials Science, Code 6390.2
Materials Science and Technology Division,
Naval Research Laboratory Washington, DC 20375-5000

V. Decaux and P. Beiersdorfer

Lawrence Livermore National Laboratory,
P. O. Box 808, Livermore, CA 94551

Abstract. Kα X-ray emission spectra from highly charged Fe ions have been theoretically predicted using a detailed and systematic spectral model. Account has been taken of the fundamental atomic radiative-emission processes associated with inner-shell electron collisional excitation and ionization, as well as dielectronic recombination. Particular emphasis has been directed at extreme non-equilibrium or transient-ionization conditions, which can occur in astrophysical and tokamak plasmas. Good agreement has been found in comparisons with spectral observations on the EBIT-II electron beam ion trap at the Lawrence Livermore National Laboratory. We have identified spectral features that can serve as diagnostics of the electron density, the line-formation mechanism, and the charge-state distribution.

Keywords: Autoionization, Radiative transitions, Inner-shell-electron collisional excitation and ionization, X-ray spectra, Equilibrium and non-equilibrium plasmas, Electron-ion interactions.
PACS: 52.72.+v, 32.80.Dz, 32.70.Fw, 32.30.Rj

I. INTRODUCTION

Kα emissions associated with 2p → 1s inner-shell-electron radiative transitions in highly ionized Fe ions occur as prominent satellite-line features in the X-ray spectra of laboratory and astrophysical plasmas. We have developed a detailed atomic spectral model [1] for the Kα radiative emissions produced by the iron ions from Fe XVIII to Fe XXIV, which occur in the spectral range from 1.84 Å to 1.94 Å. From an analysis of the Kα emission spectra, basic physical properties within the emitting region can be deduced, including temperatures, densities, charge-state populations, and electric- and magnetic-field distributions. In the development of spectral models for atomic radiative emissions in plasmas, the plasma electrons have been traditionally represented by a Maxwellian electron velocity (or energy) distribution function. In previous applications [1, 2] of our detailed atomic-physics model, computer simulations have been carried out for low-density (solar-corona), intermediate-density (tokamak), and high-density (laser-produced) plasmas using Maxwellian electron

CP774, *X-Ray Diagnostics of Astrophysical Plasmas,* edited by Randall K. Smith
2005 American Institute of Physics 0-7354-0259-0/05/$22.50

energy distributions. In this investigation, we present new results obtained from an extension of our previously reported investigations, in which an arbitrary electron energy distribution can be accommodated. This extension is necessary for the computer simulation of the Kα radiative emission from extreme non-equilibrium or transient-ionization environments.

II. ELEMENTARY RADIATIVE TRANSITIONS

Elementary transitions involving autoionizing resonances of multiply-charged ions are known to play important roles in the determination of the atomic-level population densities and charged-state distributions [3]. Radiative transitions from autoionizing states appear in the emission spectra as distinctive satellite features [4], in close proximity to a resonance line of the ion with one additional electron removed, e. g., the recombining ion in the two-step dielectronic recombination processes described by Burgess [5]. The spectral features produced by radiative transitions from autoionizing states may be observable as distinct spectroscopically resolvable satellite lines, as unresolvable enhancements of the associated resonance line, or as blends consisting of many overlapping satellite lines. For the precise computer simulation of the high-resolution Kα emission spectra, it is necessary to take into account the separate contributions from the individual fine-structure components of the 2p → 1s inner-shell-electron radiative transitions. Accordingly, the states involved in these radiative transitions must be specified in complete detail, as individual fine-structure states (hyperfine structure will be ignored). The autoionizing states of interest may be described as being approximately associated with the electronic configurations $1s^1 2s^r 2p^s$ (with $0 \leq r \leq 2$ and $0 \leq s \leq 6$) in the irons ions from Fe XVIII to Fe XXIV, with the understanding that these identifications cannot have a rigorous meaning within the framework of a fully-relativistic multi-configuration representation for the many-electron atomic states.

III. EXCITATION OF AUTOIONIZING STATES IN PLASMAS

In high-temperature plasmas, the autoionizing states can be populated by the radiationless electron-capture process, which is the first step in the two-step dielectronic recombination process. This indirect process is usually the dominant recombination mechanism for multiply-charged ions at low densities. In low-density plasmas, the initial ions undergoing radiationless electron capture can be assumed to be predominantly in their ground states (or perhaps in a metastable state). However, the dominant autoionization process may correspond to a transition leading to the formation of an excited state of the residual ion [6]. In addition, autoionization into excited states following radiationless electron capture can provide an important indirect (resonant) contribution to the total rate of electron impact excitation [7].

While radiationless electron capture is often the dominant excitation mechanism for the Kα emission in a high-temperature plasma, which is then commonly referred to as dielectronic-recombination satellite emission, it is also necessary to take into account the population of the autoionizing states as a result of the inner-shell-electron electron-impact excitation process. In addition, autoionization (instead of radiative decay) following inner-shell-electron excitation can provide an appreciable indirect (resonant) contribution to the total rate for electron impact ionization [8].

In the development of our earlier atomic spectral model [1] for the Kα emission of highly-charged Fe ions, we considered only radiationless electron capture and inner-shell-electron collisional excitation. In the present investigation, we have taken into account the additional population of the autoionizing states due to inner-shell-electron electron-impact ionization. Radiationless electron capture is expected to play the dominant role in plasmas under conditions near a steady-state ionization-recombination balance, e. g., as described by the low-density corona-model approximation discussed by Griem [9]. However, inner-shell-electron collisional excitation and ionization can provide the dominant Kα line-formation mechanism for extreme non-equilibrium, transient-ionization environments.

IV. SPECTRAL MODEL FOR NON-EQUILIBRIUM PLASMAS

In spectral models for atomic radiative emissions following electron-ion collisions in plasmas, the electron velocity distribution has been traditionally represented by a local-thermodynamic-equilibrium (Maxwellian) distribution function. The radiative emission spectra can then be predicted as functions of the local electron temperature. We have employed a natural extension of the traditional description, in which account can be taken of an arbitrary single-electron velocity (or energy) distribution function. Our investigation of the emission spectra produced by extreme non-equilibrium electron energy distributions has been carried out by means of computer simulations of the Fe Kα emission spectra observed from the EBIT-II electron beam ion trap facility at the Lawrence Livermore National Laboratory [10, 11]. Since the electron beam in EBIT-II is nearly mono-energetic, a precise investigation of the various fundamental Kα radiative emission processes can be carried out. In order to observe the contributions from the dielectronic recombination satellites, which are produced by radiationless electron capture, it had been necessary either to select the electron-beam energy at a particular resonance energy [12] or to vary the electron-beam energy in a continuous manner through the resonance region [13]. In the present investigation, Fe Kα radiative emission spectra have been theoretically predicted for two different non-resonant electron-beam energies, for which high-resolution spectra have been observed from EBIT-II [14]. These non-resonant energies, for which the dielectronic recombination process does not contribute, occur above and below the K-shell ionization threshold. From the comparison between the observed high-resolution Fe Kα radiative emission spectra [14] and the corresponding theoretically predicted spectra, the contributions from inner-shell-electron collisional excitation and ionization have been separately investigated, and the sensitivities of various Kα radiative emission lines to density variations and transient ionization conditions have been assessed.

V. DENSITY DEPENDENCE OF THE EMISSION SPECTRA

The decomposition of the Kα emission spectra into separate contributions associated with the three fundamental electron-ion collision processes cannot be made in an unambiguous manner and is strictly applicable only for low densities, for which the simple corona-model approximation is valid. With increasing electron density, the population densities of the radiating autoionizing states can be altered as a result of electron collisional transitions among the low-lying bound fine-structure states. A complete density-dependent description of the radiative emission spectra can be provided only on the basis of a detailed (possibly time-dependent) collisional-radiative model. In a systematic collisional-radiative model, the population densities of both the bound and autoionizing states would be self-consistently determined, taking into account the multitude of elementary collisional and radiative transitions together with the relevant autoionization processes.

In order to provide an approximate description of the density dependence of the Kα emission spectra, which would circumvent the need to consider the enormous collisional-radiative matrices corresponding to the multitude of the relevant fine-structure states, we have introduced a hierarchy of simple statistical models [1] for the distribution of the population of the initial ions among the various low-lying (bound) fine-structure states. At low densities, the initial-ion population can be assumed to be confined entirely to the lowest-lying fine-structure states of the ground-state electronic configurations. An approximate description of the population distribution for the moderate densities characteristic of tokamak plasmas, which is believed to be appropriate for the intermediate-density regime $10^{12} \leq N_e \leq 10^{14}$ cm^{-3}, can be provided by assuming that the collisional transitions are capable of establishing a statistical-equilibrium distribution of the initial-ion population among all fine-structure states belonging to the ground-state electronic configuration. At higher densities, which are encountered in laser-produced plasmas, we have assumed that a statistical-equilibrium distribution can be established between all fine-structure states of the ground-state and certain low-lying excited-state electronic configurations [1]. The collisional transitions among the autoionizing states, which can lead to strong density variations in dense laser-produced plasmas [15-17], have not been taken into account in our Kα spectral simulations.

VI. SPECTRAL SIMULATIONS

Employing the decomposition of the total Kα spectral intensity into the separate contributions associated with the three fundamental electron-ion collisional interactions, we have evaluated the individual radiative emission coefficients for the nearly mono-energetic electron beam of the EBIT-II experiment. In the low-density regime, the spectral line-shape functions are determined by Doppler broadening and by the broadening due to the permissible autoionization and spontaneous radiative emission processes. The initial Fe ions are assumed to be either in the ground fine-structure state or statistically distributed among all fine-structure states belonging to the ground-state electronic configuration. We have determined the charge-state distributions of the radiating atomic ions for both equilibrium and non-equilibrium environments. While the charge-state distributions in low-density high-temperature equilibrium plasmas are determined predominately by the (possibly dynamical) balance between electron-impact ionization (including autoionization following inner-shell-electron excitation) and direct radiative together with dielectronic recombination, a different physical situation is usually encountered in the description of the charge-state balance in extreme non-equilibrium or transient-ionization plasmas. An even more substantial departure from the familiar corona-equilibrium charge-state distributions can be produced in photoionized plasmas, for which radiative excitation, photoionization, and charge-transfer processes can play a significant role in the population of the autoionizing states. In the creation of sufficiently deep inner-shell-electron vacancies in heavy atomic ions, spontaneous decay can occur through a complex cascade process involving a sequence of radiative, Auger, and Coster Kronig transitions, which can result in a state of multiple ionization [18].

$K\alpha$ spectral simulations have been carried out for both equilibrium and non-equilibrium conditions [19]. To obtain spectral predictions that can be compared with the observed EBIT-II spectra, it has been necessary to include corrections to the theoretically predicted wavelengths. The wavelength corrections are in part due to an inadequate description in the atomic structure of the electron-electron correlations but are primarily attributable to the radiative corrections of quantum electrodynamics. In order to obtain agreement with the experimental spectra obtained from EBIT-II, the effects of electron density cannot be neglected. We have investigated the contributions of the various fundamental excitation processes under equilibrium and non-equilibrium conditions, and we have identified spectral features that are sensitive to density variations and to extreme non-equilibrium, transient-ionization conditions.

VII. FUTURE EXTENSIONS

In future extensions of this investigation, it would be desirable to include spectral contributions from more highly excited autoionizing states, which may blend with the features already included or produce new features. To provide a detailed determination of the populations of the fine-structure states, as functions of density, a systematic and self-consistent collisional-radiative model [3] should be developed. The ionization and recombination rate coefficients evaluated in this investigation are the low-density (corona-model) limits of effective transition rates, which are most rigorously introduced as a result of a reduced kinetic-theory description [3, 20] of the multitude of atomic collisional and radiative interactions. It will also be desirable to implement a more detailed description of the radiative emission processes [21], taking into account the angular distribution and polarization of the emitted photons that can be produced by directed-electron excitation. To obtain theoretical predictions beyond the isolated-resonance approximate, it is necessary to include quantum-mechanical interference effects, which have been investigated for radiative and dielectronic recombination [22, 23]. With increasing spectral resolution of the satellite observations, it may also be necessary to employ more sophisticated descriptions of the spectral line shapes, e. g., beyond the lowest-order approximation of QED perturbation theory. Using a density-matrix approach [24], a self-consistent treatment can be provided for the atomic-population kinetics and the spectral-line shapes.

ACKNOWLEDGMENTS

This work has been supported by a NASA Astronomy and Physics Research and Analysis grant to LLNL, and performed under the auspices of the U. S. Department of Energy by the University of California Lawrence Livermore National Laboratory, under contract number No. W-7405-ENG-48. The work of V. L. Jacobs has been supported by the U. S. Department of Energy, through Interagency Agreement No. DE-AI02-93-ER-54198 with the Naval Research Laboratory, and by the Office of Naval Research.

REFERENCES

1. Jacobs, V. L., Doschek, G. A., Seely, J. F., and Cowan, R. D., *Phys. Rev. A* **39**, 2411 (1989).
2. Beiersdorfer, P., Phillips, T., Jacobs, V. L., Hill, K. W., Bitter, M., von Goeler, S., and Kahn, S. M., *Astrophys. J.* **409**, 846 (1993).
3. Jacobs, V. L., *J. Quant. Spectrosc. Radiat. Transfer* **54**, 195 (1995).
4. Bhalla, C. P., Gabriel, A. H., and Presnyakov, L. P., *Mon. Not. R. Astron. Soc.* **172**, 359 (1975).
5. Burgess, A., *Astrophys. J.* **139**, 776 (1964).
6. Jacobs, V. L., *Astrophys. J.* **296**, 121 (1985).
7. Jacobs, V. L., Hagelstein, P. L., Chen, M., Jung, R. K., and Seely, J. F., *Phys. Rev. A* **41**, 1041 (1990).
8. Goldberg, L., Dupree, A. K., and Allen, J. W., *Ann. Astrophys.* **28**, 589 (1965).
9. Griem, H. R., *Principles of Plasma Spectroscopy*. Cambridge University Press (1997).
10. Levine, M. A., Marrs, R. E., Henderson, J. R., Knapp, D. A., and Schneider, M. B., *Phys. Scr.* **T22**, 157 (1988).
11. Levine, M. A., Marrs, R. E., Bardsley, J. N., Beiersdorfer, P., Bennett, C. L., Chen, M. H., Cowan, T., Dietrich, D., Henderson, J. R., Knapp, D. A., Osterheld, A., Penetrante, B. M., Schneider, M. B., and Scofield, J. *H., Nucl. Instrum. Methods Phys. Res. Sect. B* **43**, 431 (1989).
12. Beiersdorfer, P., Phillips, T. W., Wong, K. L., Marrs, R. E., and Vogel, D. A., *Phys. Rev. A* **46**, 3812 (1992).
13. Knapp, D. A., Beiersdorfer, P., Chen, M. H., Scofiled, J. H., and Schneider, D., *Phys. Rev. Letters* **74**, 54 (1995).
14. Decaux, V., Beiersdorfer, P., Kahn, S. M., and Jacobs, V. L., *Astrophys. J.* **482**, 1076 (1997).
15. Vinogradov, A. V., Skobelev, I. Yu., and Yukov, E. A., *Sov. Phys. J. E. T. P.* **45**, 925 (1977).
16. Jacobs, V. L., and Blaha, M., *Phys. Rev. A* **21**, 525 (1980).
17. Zigler, A. et al., *Phys. Rev. A* **45**, 1569 (1992).
18. Jacobs, V. L., and Rozsnyai, *B. F., Phys. Rev. A* **34**, 216 (1986).
19. Decaux, V., Jacobs, V. L., Beiersdorfer, P., Liedahl, D. A., and Kahn, S. M., *Phys. Rev. A* **68**, 012509 (2003).
20. Jacobs, V. L., and Davis, J., *Phys. Rev. A* **18**, 697 (1978).
21. Jacobs V. L., and Filuk, A. B., *Phys. Rev. A* **60**, 1975 (1999).
22. Haan, S. L., and Jacobs, V. L., *Phys. Rev. A* **40**, 80 (1989).
23. Jacobs, V. L., Cooper, J., and Haan, S. L., *Phys. Rev. A* **36**, 1093 (1987).
24. Jacobs, V. L., Cooper, J., and Haan, S. L., *Phys. Rev. A* **50**, 3005 (1994)

"Using' what we got": Calculating X-ray Spectra for Non-Equilibrium Plasmas

Daniel Dewey

M.I.T., Cambridge, MA 02139 USA

Abstract. It is desirable to calculate the X-ray emission spectrum from a (region of a) plasma that is not in ionization equilibrium, e.g., with application to supernova remnants, SNRs. Such a plasma will generally have ionization fractions that are not directly related to the local electron temperature and, in the case of SNR ejecta, may contain little or no H and He. This work describes the use of a currently available database, ATOMDB, and software, ISIS, to generate spectra for a non-equilibrium ionization, NEI, plasma which is specified by explicit densities and ionization fractions and an independent electron temperature. An example spectrum is shown and discussed. The approach here does have some deficiencies currently: not all relevant inner-shell lines are included, there are no Cr lines, and continua spectra are not given on an ion-by-ion basis. Finally, in the low density, NEI case, it may be useful to consider the emission components that arise from each "quiescent" ion.

Keywords: atomic processes, hydrodynamics, plasmas, radiation mechanisms: thermal, supernova remnants, X-rays: ISM
PACS: 32.30.Rj, 98.38.Mz

There are many cases where it is desirable to calculate the X-ray emission from a plasma that is not in collisional-equilibrium ionization, CEI. We have for example: the outputs of hydrodynamical models of Type Ia supernova remnants, SNRs[1], the emission from specific knots in a Type II SNR[2], the idealized models such as a plane parallel shock or a Sedov blast wave, and even the case of *ab initio* models where the plasma is simply built to agree with observational data. In all these cases the specifics of the plasma are known and the resulting X-ray spectrum is desired.

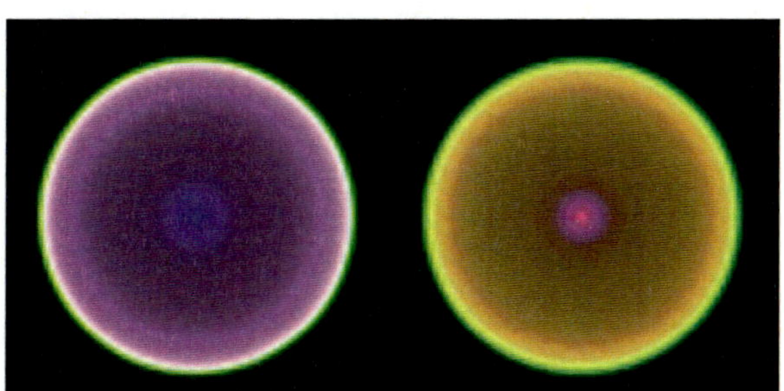

FIGURE 1. 2D projections of Type Ia ejecta models. Badenes et al.[1] models DDTa (left) and DDTe (right) are shown at $t = 1000$ yr. The image color values, R:G:B, are determined by the element column densities of Fe:Si:Ni; e.g., the colors indicate: blue-Ni, purple-Fe+Ni, red-Fe, orange-Fe(+Si), yellow-Si+Fe, green-Si. The image brightness is weighted by n_e, emphasizing ionized regions.

CP774, *X-Ray Diagnostics of Astrophysical Plasmas,* edited by Randall K. Smith

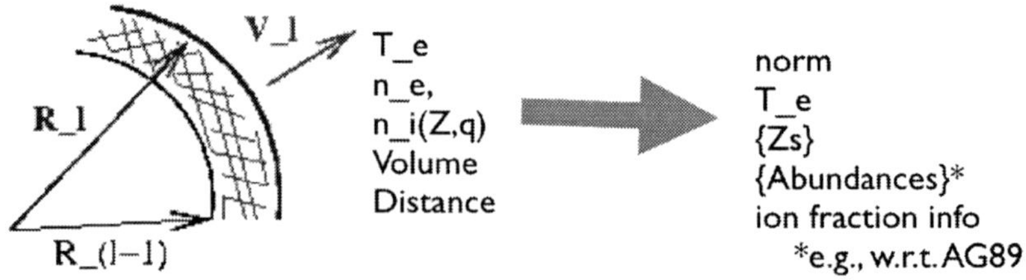

FIGURE 2. Transforming the parameters of the plasma model. Hydrodynamical simulations, e.g., for Type Ia SNR, generally give explicit, local plasma values (left.) The input to our ISIS code is a related set of parameters (right) with the NEI information contained in the explicit ion fractions.

Here we apply current CEI atomic data and software to approximate emission from a non-equilibrium ionization, NEI, plasma. As a specific example we use the models of Badenes et al.[1] which describe Type Ia SNR ejecta. These are 1D hydrodynamical models of 3D spherically symmetric explosions and give the plasma composition and properties in discrete shells or layers as a function of time. The images in Figure 1 show visualizations of the composition variation for two different models. The goal inspiring this work is to calculate and visualize the X-ray emission from such a model.

Given model parameters, in this case for each layer or spherical shell of the Type Ia model, Figure 2, we can convert them into parameters used by standard codes; here we use the Interactive Spectral Interpretation System, ISIS[3] and the ATOMDB spectroscopy database[4]. Using as an example layer 90 of a DDTe model at 1000 yr. for which $T_e \approx 14$ MK, the norm and the abundances are determined. This layer's elemental abundances expressed with respect to "solar" are found to be in the range of 500–3500 for Si, S, Ar, Ca, Cr, and Fe. Note that these equivalent abundances are based on the electron density: $A_z = (n_z/n_e)_{\rm model}/(n_z/n_e)_{\rm solar}$ so that the presence of H and He is not assumed. Likewise the value of n_e is self-consistant with the densities and ionization states of the plasma's ions, i.e., $n_e = \sum_z n_z \bar{q}_Z$.

It is straight forward to calculate the modified-abundance CEI spectrum for this layer, shown by the light curve in Figure 3. The final step in generating the spectrum is to include the effect of the actual ion-fractions by scaling the line emission from each ion by the ratio of the assumed CEI ion fraction to the desired model ion fraction. The difference between CEI and NEI ion populations at the same temperature can be large – for this layer over 80% of the NEI model Si atoms are in the ion states Si IX, Si X, and Si XI; whereas the CEI case has >99% of the Si atoms in Si XIII and higher ionization states. As might be expected, the final NEI spectrum, the dark curve in Figure 3, has very little K-shell emission from the elements present, i.e., shortward of 7 Å.

Clearly this approach currently has some deficiencies, most importantly the lack of inner-shell lines from low-ionization states, e.g., Si-K-shell lines from Si X in the presence of high-T electrons. Relatedly, the continua would ideally be given on an ion-by-ion basis as well. Of relevance to the Type Ia models, Cr lines are not currently in ATOMDB. Finally, in the low density, non-equilibrium case considered here, it may be useful to consider, tabulate, and label the emission components that arise from each

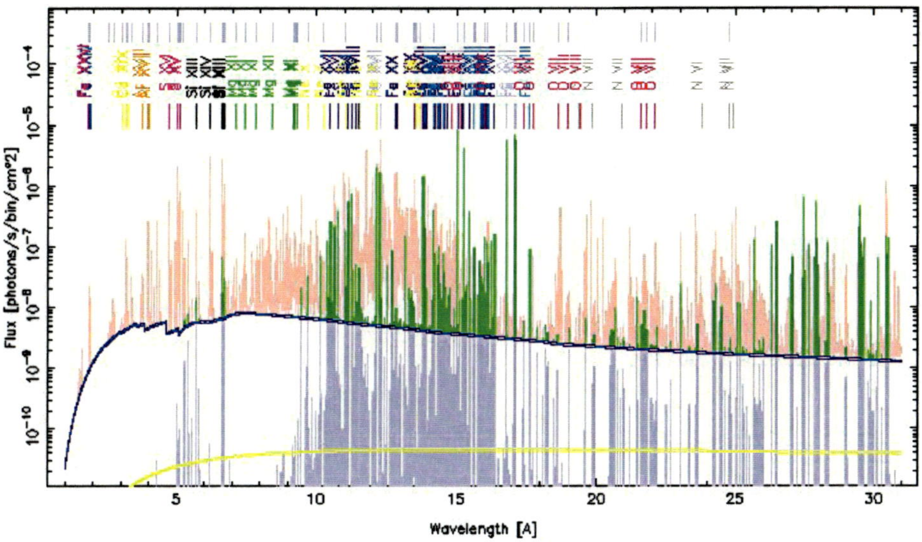

FIGURE 3. Example of an NEI spectrum. The spectrum for layer 90 of the DDTe 1000 yr. model is shown (light,pink) assuming CEI conditions for $T_e \approx 14$ MK. Scaling by the actual model ion fractions gives the spectrum (dark,green) dominated by lines in the 10–17 Å range.

"quiescent" ion population as opposed to the actual emitting ion which may only form as a result of an interaction, e.g., one of the "three fundamental binary electron-ion collision processes" [5]. In this way Si X would be seen as the parent of "Si IX lines" via dielectronic recombination, and O IX would be viewed as the source of some "O VIII" emission; in each case, the emission scales with the quiescent parent ion population.

ACKNOWLEDGMENTS

Thanks to John Houck for helpful comments, to Carles Badenes for model data and stimulating discussions, and to Kazik Borkowski for expert insight. This work was supported by NASA through contract NAS8-01129.

REFERENCES

1. C. Badenes, E. Bravo, K. J. Borkowski, and I. Dominguez, "Thermal X-ray Emission from Shocked ejecta in Type Ia Supernova Remnants: ...", ApJ, 593, 358 (2003).
2. J. M. Laming and U. Hwang, "On the Determination of Ejecta Structure and Explosion Asymmetry from the X-Ray Knots of Cassiopeia A", ApJ, 597, 347 (2003).
3. ISIS software: `http://space.mit.edu/CXC/ISIS/`
4. ATOMDB atomic database: `http://cxc.harvard.edu/atomdb/`
5. V. Jacobs, V. Decaux, and P. Beiersdorfer, "K-alpha Emission Spectra from Non-Equilibrium Ionizing Plasmas", these proceedings.

Production and Decay of Atomic Inner-Shell Vacancy States

T. W. Gorczyca

Department of Physics, Western Michigan University, Kalamazoo, MI 49008, USA

Abstract. Our recent investigations on the accuracy of the existing 1s-vacancy fluorescence and Auger data base, which is crucial for the interpretation of spectra from X-ray photoionized plasmas and for supernova remnants under conditions of non-equilibrium ionization, is reviewed. In particular, effects such as non-hydrogenic scaling, correct initial state averaging, configuration interaction (CI), and spin-orbit mixing, which were not included in the earlier calculations but sometimes affect profoundly our computed fluorescence yields, is discussed. Furthermore, the production of 1s-vacancy states through photoabsorption is reviewed, with emphasis on orbital relaxation and spectator Auger broadening of the resonance edge. How our recent oxygen photoabsorption data contribute to elemental abundance studies of the interstellar medium will also be discussed.

Keywords: atomic data — atomic processes — line:formation — X-rays:general
PACS: 31.25.Jf,32.30.Rj,32.50.+d,32.80.Dz,32.80.Fb,33.60.Fy,52.72.+v,95.30.Dr,95.30.Ky,95.85.Nv

INTRODUCTION

The absorption and emission of X-rays from atomic ions are important processes which take place in astrophysical plasmas throughout the cosmos. The talk given at this workshop reviewed two of the more basic absorption and emission processes: K-shell photoabsorption, focusing on the ubiquitous neutral oxygen spectrum, and K-shell vacancy radiative decay, focusing on the fluorescence yields of second-row ions. Particular emphasis is placed on the need for higher-order atomic descriptions for obtaining reliable atomic data required for plasma modeling.

K-SHELL PHOTOABSORPTION OF OXYGEN

An accurate description of the photoionization of oxygen is important for a number of atmospheric and astrophysical applications. This has been discussed in detail by McLaughlin and Kirby [1]. In particular, with the recent launch of the *x-ray* satellites *Chandra* and *XMM Newton* and in the coming decade the launch of *Constellation - X*, the need for accurate theoretical photoabsorption cross section data is paramount. For the calculations by McLAughlin and Kirby [1], the earlier valence shell work of Bell and co-workers [2] was extended to treat the photoionization processes near the $1s^{-1}$ threshold. For valence shell photoionization of atomic oxygen, these extended wavefunctions produced results in suitable harmony with recent measurements performed by Wilhelmi *et al.* [3]. For the inner-shell photoionization processes, while there was good overall qualitative agreement between the theoretical results of McLaughlin and Kirby [1] and the experimental results of Stolte *et al.* [4] and that of Menzel *et al.* [5], significant dis-

CP774, *X-Ray Diagnostics of Astrophysical Plasmas,* edited by Randall K. Smith

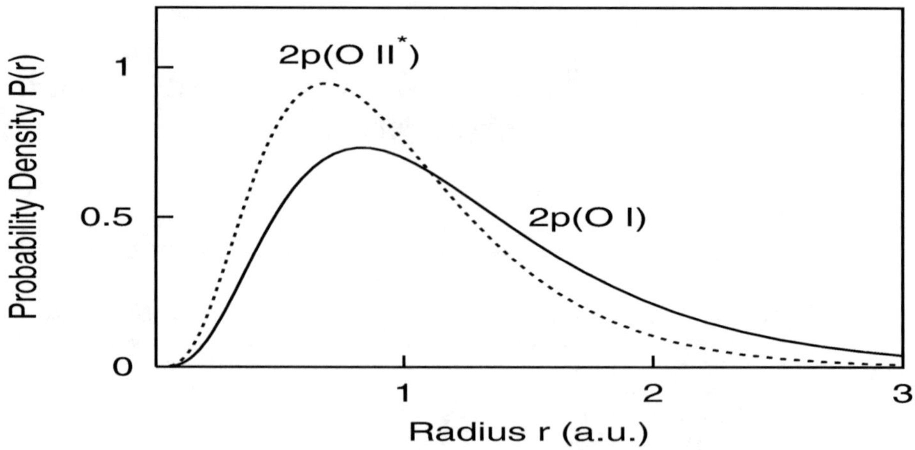

FIGURE 1. Spatial distribution of the Hartree-Fock [7] $2p$ orbitals in both neutral oxygen, with 8 electrons in the O I($1s^22s^22p^4$) ground state, and K-vacancy singly-ionized oxygen, with 7 electrons in the O II*($1s2s^22p^4$) excited state, showing massive relaxation effects.

crepancies were apparent, mostly for the positions of the inner-shell excited resonances and the near-threshold resonance profiles. We have readdressed [6] the main features of photoionization processes in the vicinity of inner-shell thresholds by focusing on the description of K-shell excited states, and on their subsequent Auger decay.

K-Shell Vacancy: Orbital Relaxation

In order to model inner shell photoexcitation accurately, it is extremely important to represent states with a $1s^{-1}$ vacancy very well. One difficulty in this task is choosing a suitable set of atomic orbitals to be used in the atomic structure description, the reason being that massive *orbital relaxation* takes place when a K-shell electron is removed. This phenomenon is illustrated in Fig. 1, where the $2p$ orbitals are shown when one or two $1s$ orbitals are occupied. It can be seen that the $2p$ orbital in neutral oxygen, which is screened by two $1s$ electrons, is much more diffuse than the $2p$ orbital in inner-shell-excited, singly-ionized oxygen, which is only screened by one $1s$ electron. Thus, it is necessary to somehow describe both initial and final states with a single set of orbitals.

To do this we created our basis of orbitals and configurations as follows. First, a Hartree-Fock (HF) calculation [7] was performed for the configuration-averaged $1s^22s^22p^3$ state to obtain the first three orbitals ($1s$,$2s$ and $2p$), giving a first-order orbital description for the lowest-energy participator decay route in Eq. 2. Next, a multi-configuration Hartree-Fock (MCHF) calculation [7] was carried out in order to optimize the energy of the $1s^22s2p^4(^4P)$ term, including the next most important configuration $1s^22s^22p^2\overline{3d}$, which determined an additional $\overline{3d}$ pseudo-orbital. Following this an MCHF calculation for the $1s2s^22p^4(^4P)$ term was then performed by including all configurations consistent with single promotions, most importantly the $1s2s^22p^3\overline{3p}$ ($\sim 5.5\%$) and $1s2s2p^4\overline{3s}$ ($\sim 1\%$) configurations that allow for relaxation of the $2p$ and $2s$ subshells due to the $1s^{-1}$ vacancy; all configurations consistent with single, double, and triple promotions Having determined the orbital basis $\{1s, 2s, 2p, \overline{3s}, \overline{3p}, \overline{3d}\}$, a configuration basis consistent with single and double promotions out of the $1s^22s^22p^3$, $1s^22s2p^4$, $1s^22p^5$, and $1s2s^22p^4$ configurations was used to describe the O^+ target wavefunctions listed in Table 1. An R-matrix boundary radius of 7.3 a.u. was required to contain the electron density from the states made up of these orbitals that are included in our work. For the corresponding $(N+1)$-electron configuration description of the atomic oxygen states, such as the $1s^22s^22p^4$ ground state and the lowest-lying inner shell resonance states $1s2s^22p^5$; we included all configurations consistent with single, double, and triple promotions out of the $2s^22p^4$ ground state for the corresponding neutral states of oxygen. All R-matrix results were obtained using the Belfast suite of codes[8, 9].

K-Shell Vacancy: Spectator Auger Decay Broadening

Another difficulty in describing inner-shell processes is the inclusion of spectator Auger decay. Near the $1s^{-1}$ thresholds of atomic oxygen, single-configuration considerations help to reveal this process. Following inner-shell photoexcitation of the atomic oxygen ground state,

$$h\nu + 1s^22s^22p^4 \longrightarrow 1s2s^22p^4np \, , \tag{1}$$

there are two competing decay routes. First, there is *participator* Auger decay

$$\begin{aligned} 1s2s^22p^4np &\longrightarrow 1s^22s^22p^3 + e^- \\ &\longrightarrow 1s^22s2p^4 + e^- \, , \end{aligned} \tag{2}$$

where the valence electron np takes part in the autoionization process; the decay rate therefore scales as $1/n^3$. More importantly, there is *spectator* Auger decay

$$\begin{aligned} 1s2s^22p^4np &\longrightarrow 1s^22s^22p^2np + e^- \\ &\longrightarrow 1s^22s2p^3np + e^- \\ &\longrightarrow 1s^22p^4np + e^- \, , \end{aligned} \tag{3}$$

where the valence electron np (εp for the case of above-threshold continuum electrons) does not take part in the autoionization process; this decay rate is independent of n. Spectator Auger decay is therefore the dominant decay route as $n \longrightarrow \infty$. A constant spectator width also leads to a broadening of the higher-n resonances with a smooth transition from the below-threshold to the above-threshold photoabsorption cross section.

To account for the spectator Auger decay in Eq. 3, an optical potential method was employed, based on similar radiative optical potential ideas [10, 11]. This approach has been rather successful in describing inner shell resonance spectra in Ar [12] and Ne [13]; further details are given in the last two references. Briefly, the core energies of the closed

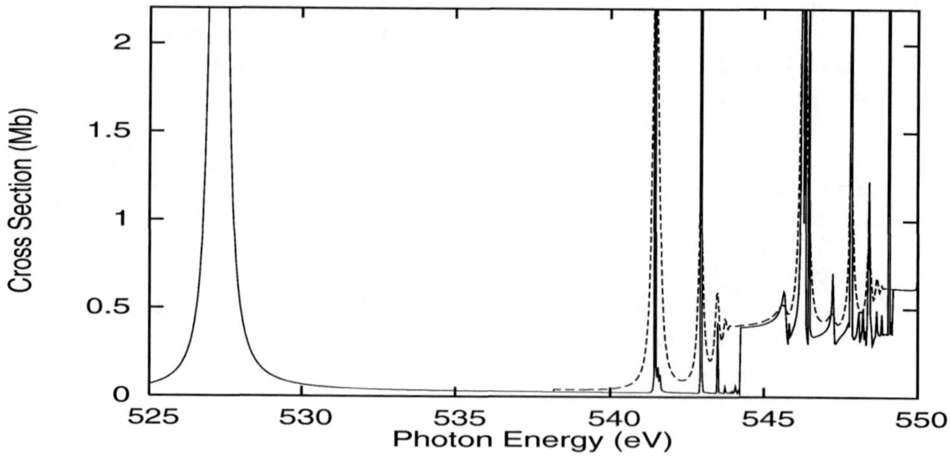

FIGURE 2. Standard R-matrix (solid curve) and optical potential (dashed curve) results for the total photoionization cross section of atomic oxygen in the vicinity of the K-edge.

channels $O^*(1s2s^22p^4(^{2S_c+1}L_c)np)$ are adjusted to allow for Auger decay to spectator channels not included in the close coupling expansion:

$$E_{1s^{-1}} \rightarrow E_{1s^{-1}} - i\Gamma_{1s^{-1}}/2\ , \tag{4}$$

where $\Gamma_{1s^{-1}}$ is the spectator Auger decay rate to higher n in Eq. 3, computed in lowest-order perturbation theory using the program AUTOSTRUCTURE [14].

Results from a standard R-matrix calculation, *including* core relaxation effects in the atomic description, are shown in Fig. 2, and are compared to results with identical atomic structure but also including an optical potential. Resonances in the former are seen to scale as n^{-3}, rendering their full resolution impossible for any finite energy mesh, and therefore leading to an underestimation of the resonance contribution below threshold. The optical potential results, on the other hand, show broader Lorentzian features, with characteristic widths that are independent of n, giving a resonant cross section below each threshold that matches smoothly to the above-threshold value.

Results

To produce a reliable theoretical photoabsorption cross section for oxygen near the K-edge, we first obtained a suitably converged representation of the $1s2s^22p^4$ states of O^+, and then included implicitly the spectator Auger channels in Eq. 3 by using an optical potential. Results from our R-matrix optical potential calculation, as shown in Fig. 3, illustrate remarkable overall agreement in the resonance profiles as compared to the total experimental photoionization cross section results for atomic oxygen [4]. Also shown are the earlier results of McLaughlin and Kirby [20], and it is seen that those results do not reproduce the observed features, predicting resonance positions too high

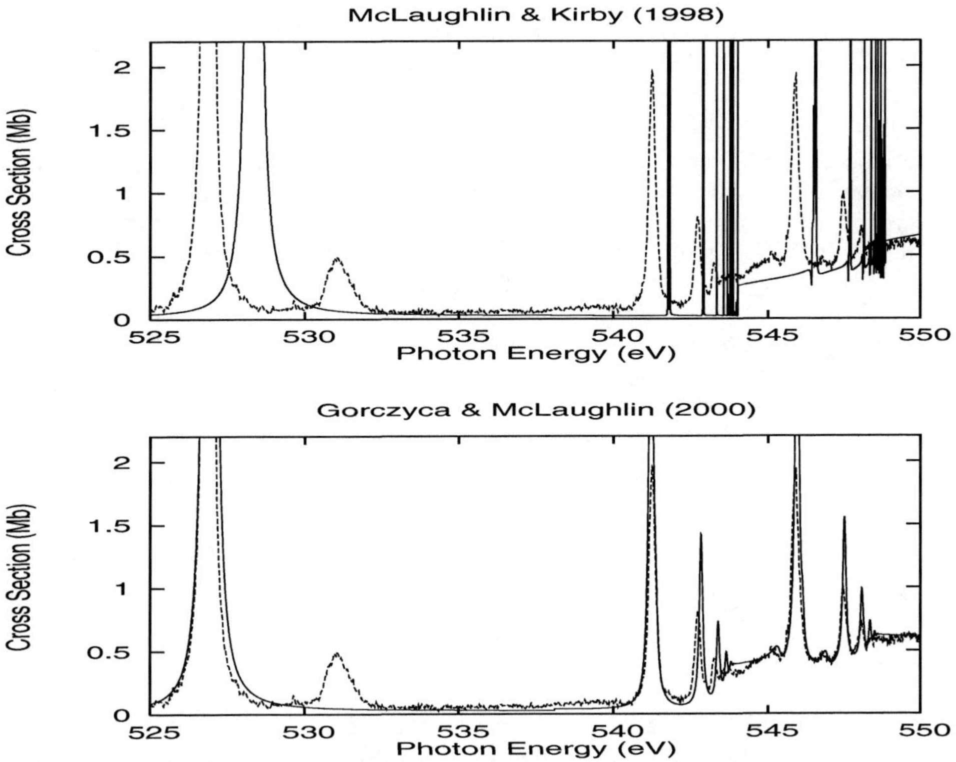

FIGURE 3. Comparison of theoretical (solid curves) and experimental (Ref. [4], dashed curve) results for the total photoionization cross section of atomic oxygen. The upper panel shows the results of McLauglin and Kirby [1], which lacked relaxation and spectator Auger broadening effects; these were made available in the Chandra Database. The lower panel shows the results of Gorczyca and McLaughlin [6], which included relaxation effects via the use of additional pseudoorbitals and spectator Auger broadening via an optical potential. Note the near perfect agreement between theory and experiment in the latter. Both theoretical results were convoluted with a 182 meV FWHM Gaussian to simulate the experimental energy resolution. The broad feature seen in the experimental cross section at $h\nu = 531$ eV is due to the $1s \rightarrow \pi^*$ resonance in molecular oxygen and is not considered in the calculations.

in energy due to their omission of relaxation effects, and showing unphysically narrow (and unresolvable) resonance widths near threshold due to their omission of spectator Auger decay.

K-Shell Photoabsorption of Oxygen as an Elemental Abundance Diagnostic of the ISM

Having completed the absorption spectrum for neutral oxygen, we then used the same techniques to calculate the photoabsorption cross sections. These are shown in Fig. 4 compared to the independent-particle results of Reilman and Manson [16], which do not have any resonances included, and the R-matrix results of Pradhan et al., which only treated the $1s \rightarrow 2p$ resonance. Clearly it is important to include all processes for an accurate description of the photoabsorption process.

It is important to obtain photoabsorption cross section for all ionization stages of oxygen since all of these are present in the interstellar medium (ISM) through which photons are absorbed. A recent study [18] used all these results to infer the elemental abundances (see Fig. 5), where the observed intensity as a function of wavelength can be determined from the absorption cross sections as a function of photon energy as

$$I(\lambda) \;=\; I_0 e^{-\sigma(\frac{hc}{\lambda})N}\,, \tag{5}$$

where N is the column density $N = \int dx n(x)$ and $n(x)$ is the density of oxygen ions. By choosing the optimum parameters N to align theoretical and observed results, it was inferred [18] that the column densities for neutral oxygen were in the range $N_O = 10^{18} - 10^{19}$ cm^{-2} for the X-Ray binaries whose observed spectra are shown in Fig. 5.

ASSESSMENT OF THE FLUORESCENCE AND AUGER DATABASE USED IN PLASMA MODELING

In collisionally ionized or X-ray photoionized plasmas, high-energy electrons or photons lead to the production of $1s$-vacancy ionic states which then decay via sequential emission of single or multiple electrons and/or photons. The exact strengths of these competing processes determine fundamentally important quantities of the plasma such as the ionization balance and the observed spectra of emitted and/or absorbed photons. Hence, interpreting the properties of these plasmas requires accurate atomic physics calculations for the various autoionization and radiative rates. For this reason, we have recently assessed [19] the accuracy of the available data base [20] that provides such computed (or inferred) Auger rates and fluorescence yields to the astrophysics community. The accuracy of these atomic data are crucial to the interpretation of the spectra of photoionized plasmas such as are found in X-ray binaries and active galactic nuclei. These data are also important for supernova remnants (SNRs) under conditions of non-equilibrium ionization (NEI).

Two of the more widely used spectral codes for modeling photoionized plasmas are CLOUDY [21] and XSTAR [22]. A commonly used code for modeling NEI in SNRs is that of Borkowski *et al.* [23]. These all in turn rely on the table of electron and photon emission probabilities compiled by Kaastra and Mewe [20]. This comprehensive data base considers the sequential multiple electron and/or photon ejections for all stages of all $1s$-vacancy ions in the periodic table up through zinc. In order to produce such a massive array of numbers, however, certain approximations, questionable from a purely

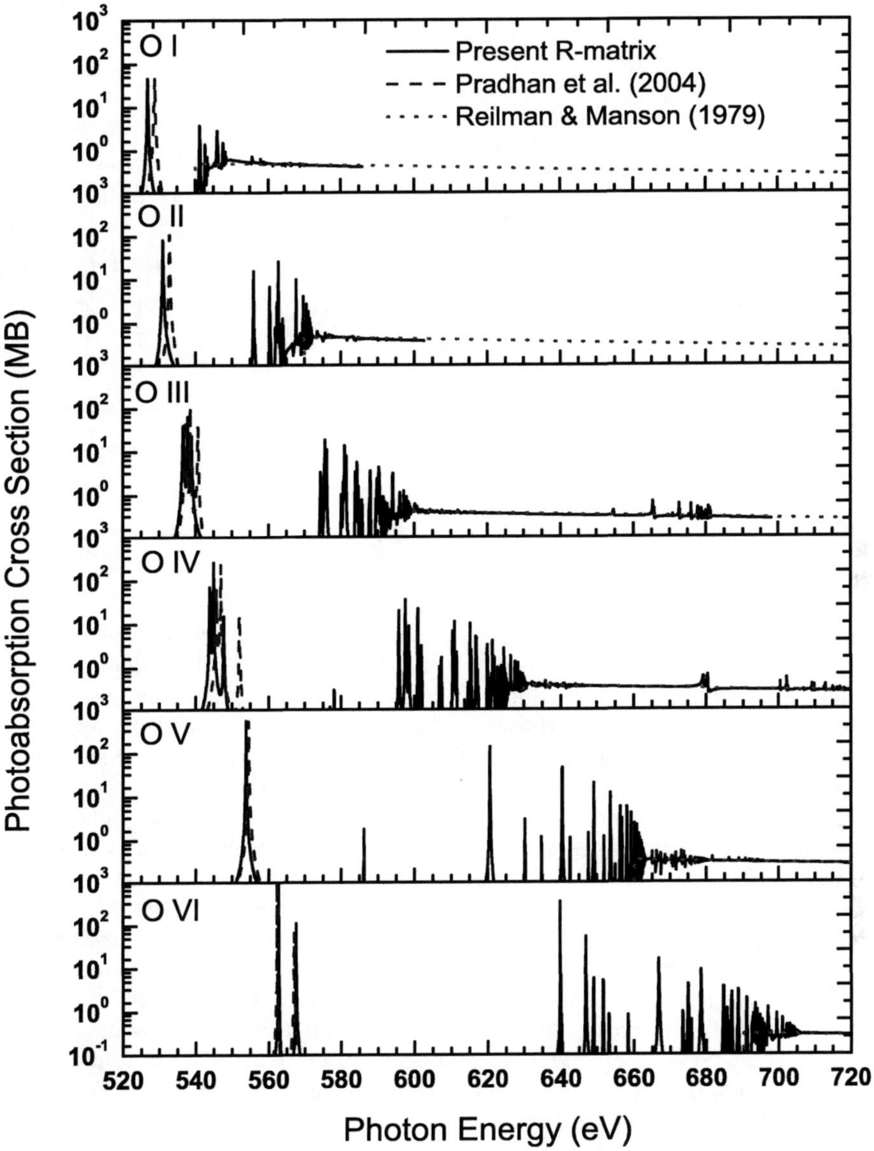

FIGURE 4. K-shell photoabsorption of all oxygen ions. Our present R-matrix results [15] are compared to the independent-particle results of Reilman and Manson [16], which do not have any resonances included, and the R-matrix results of Pradhan et al., which only treated the $1s \rightarrow 2p$ resonance.

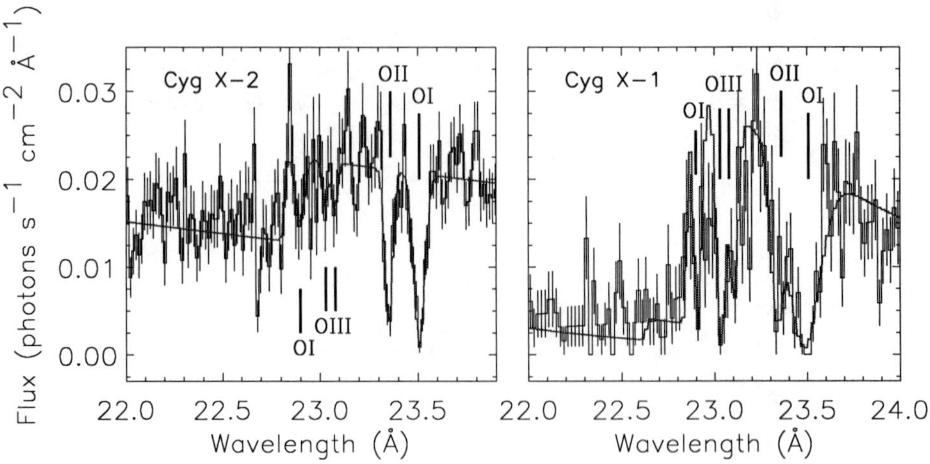

FIGURE 5. Flux spectra of the oxygen edge region for Cyg X-2 and Cyg X-1 ObsID 107. Line identifications are also shown.

theoretical atomic physics standpoint, were invoked. First, the only rigorously computed atomic rates were taken from the early works of [24, 25, 26, 27] for *singly-charged ions*, which furthermore neglected configuration interaction (CI) and spin-orbit effects. Due to the limited computational resources available at the time, and the approximations thus needed to perform such calculations, even these cannot be considered as reliable as those that can be carried out with today's state-of-the-art capabilities. Second, these singly-ionized results were then scaled along entire isoelectronic sequences, assuming constant autoionization rates and oscillator strengths; this approximation is least valid for near-neutrals.

A third approximation used in Ref. [20] is that the electron and photon emission yields were computed using radiative and autoionization rates that were configuration averaged over possible terms, and the fluorescence yield was then given as a ratio of the averaged radiative rate to the sum of the averaged radiative and averaged Auger rate. For modeling purposes, however, this is incorrect; the actual required value is the average of the term-specific yields - an average of ratios rather than a ratio of averages. In other words, the relative probability of producing each specific inner-shell-vacancy term, and its subsequent term-specific decay, needs to be considered, and this was not done correctly for the data compiled by Kaastra and Mewe [20] (see also Ref. [28] for a further discussion of this importance).

In a recent study [19], we investigated the validity of the above three approximations in order to assess the accuracy of the resultant data base of Kaastra and Mewe [20]. To this end, we first studied the simplest $1s$-vacancy system that can radiate (in a single-configuration approximation, as we will discuss further below) via a $2p \rightarrow 1s$ dipole-allowed transition. This is the removal of a $1s$ electron from the ground-state B-like sequence, or rather the $1s2s^22p$ Be-like inner-shell excited sequence. This sequence is

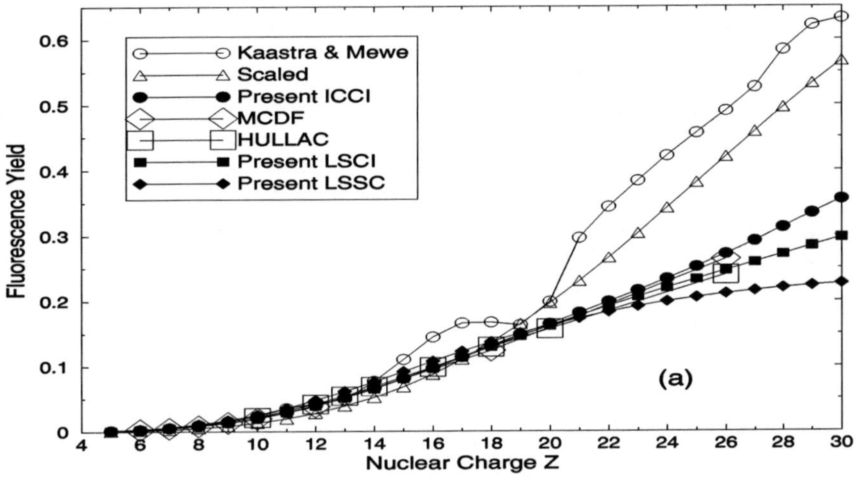

FIGURE 6. (a) Comparison of various computed and inferred fluorescence yields ξ for Be-like $1s2s\,^22p$ ions: present LS results in the single-configuration (SC) approximation – solid diamonds; present LS results with configuration interaction (CI) included – solid squares; present IC results with CI included – solid circles; HULLAC results from [29] – open squares; MCDF results from [30] – open diamonds; Kaastra and Mewe [20] – open circles; results when we try to reproduce the method of Kaastra and Mewe [20] – open triangles.

further simplified by the fact that only one electron, or one photon, can be emitted. We then study of the simplest closed-(outer)shell case of F-like ions, corresponding to $1s$ vacancies from the Ne-like sequence. The intervening B-, C-, N-, and O-like sequences are next investigated, and we finally discuss the Li-like sequence, a case where no fluorescence occurs in a low-order atomic physics description.

Be-like fluorescence yields

In all of our studies, we employed the AUTOSTRUCTURE code [14] to compute accurate atomic transition rates (radiative and Auger), and considered successively the effects of configuration-interaction (CI), the spin-orbit interaction, and averaging over inner-shell vacancy states. Our results for the Be-like isoelectronic sequence are shown in Fig. 6 at successive single-configuration, nonrelativistic (LSSC), configuration-interaction, non-relativistic (LSCI), and configuration-interaction, semi-relativistic intermediate-coupling (ICCI) levels of approximations. Also shown are the results of Kaastra and Mewe [20], the results we obtained when trying to reproduce their method, independent multiconfiguration Dirac-Fock (MCDF) results [30], and HULLAC [31] results [29]; the former two show large discrepancies with our recommended ICCI results while the latter two are in quite good agreement.

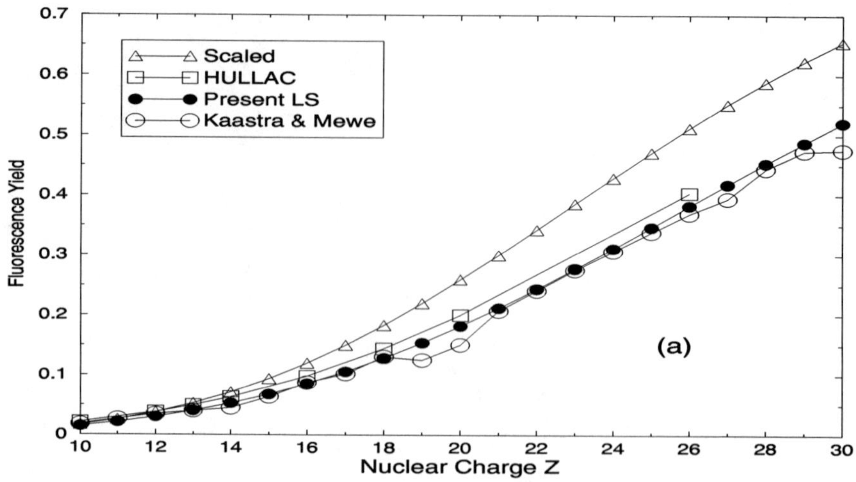

FIGURE 7. (a) Comparison of various computed and inferred fluorescence yields ξ for F-like $1s2s^2 2p^6(^2S)$ ions: present LS results – solid circles; Kaastra and Mewe [20] – open circles; results when we try to reproduce the method of Kaastra and Mewe [20] – open triangles; HULLAC results of [29] – open squares.

The inaccuracies we have discovered in the reported results of Kaastra and Mewe [20] for Be-like ions can be summarized as follows:

1. The computed atomic data for B II are such that the radiative and autoionization rates have been averaged over the ^1P and ^3P configurations, whereas the desired quantity for plasma modeling applications is not the same thing, differing qualitatively and quantitatively, especially in the asymptotic high-Z limit.
2. The hydrogenic scaling assumed is invalid. The autoionization rates, the emission oscillator strengths, and even their ratio are not independent of nuclear charge Z.
3. The tabulated data of Kaastra and Mewe [20] do not seem to follow the results we obtain when we try to reproduce their stated method.
4. The calculations of Kaastra and Mewe [20] neglected CI and spin-orbit effects as they scaled to higher Z.

F-like fluorescence yields

Given the differences uncovered for the $1s2s^2 2p$ Be-like sequence, where only one $2p$ orbital is filled, we next looked at the $1s2s^2 2p^6$ F-like case with a filled $2p$ shell. As such, the above items 1 and 4 are not issues since there is only one inner-shell vacancy term (^2S). However, points 2 and 3 still apply for the F-like sequence. One can see in Fig. 7 that the discrepancy between our results and those of Kaastra and Mewe [20] are now much less profound, although there still exists unphysical dips and we cannot reproduce their data through their stated method.

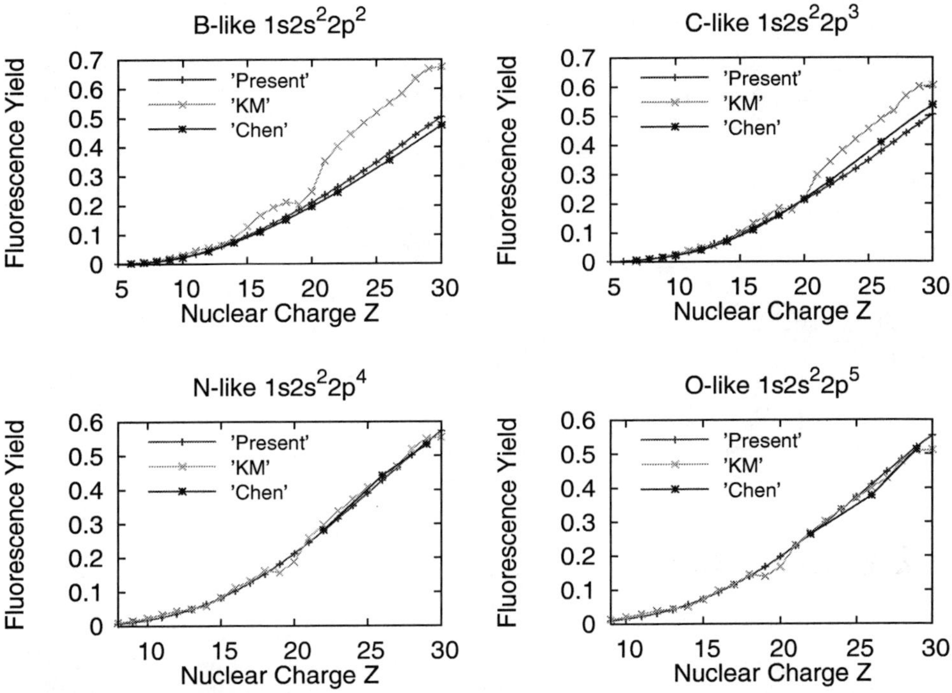

FIGURE 8. Comparison of various computed and inferred fluorescence yields ξ for B-like $1s2s^22p^2$, C-like $1s2s^22p^3$, N-like $1s2s^22p^4$, and O-like $1s2s^22p^5$ ions: present IC results, those of Kaastra and Mewe [20], and Chen's MCDF results [33, 34, 35].

B-like, C-like, N-like, and O-like fluorescence yields

It is seen that the differences between our results and those of Kaastra and Mewe [20] were larger at the left side of the second row (Be-like) than at the right side (F-like). Indeed, more recent calculations [32] for the intervening B-, C-, N-, and O-like sequences confirms this trend (see Fig. 8). This can be understood by considering the structure of all second-row systems, as shown in Table 1, and the percentage of K-vacancy states which cannot radiate in the LSSC approximation. The incorrect averaging used by Kaastra and Mewe [20] erroneously leads to an asymptotic high-Z behavior of 100% whereas they correct value is given in the right column of Table 1; CI and spin-orbit effects increase this value (especially for the Li-like sequence, as we discuss next), but not as much as the incorrect averaging method does. Hence, the results of Kaastra and Mewe overestimate the correct fluorescence yield more at the left end of the second row, as we have seen in Figs. 6-8.

TABLE 1. Breakdown of allowed (within a single-configuration, non-relativistic approximation) fluorescing K-vacancy states of 2nd-row elements

Series	Final Decay States	Initial K-Vacancy States*	% Allowed Fluorescence
Li	$1s^22s$	$1s2s^2([\mathbf{^2S}])$	0.0†
Be	$1s^22s^2$	$1s2s^22p(^1P^o,[\mathbf{^3P^o}])$	25.0
B	$1s^22s^22p$	$1s2s^22p^2([\mathbf{^4P}],^2P,^2D,^2S)$	60.0
C	$1s^22s^22p^2$	$1s2s^22p^3([\mathbf{^5S^o}],^3S^o,^3D^o,^1D^o,^3P^o,^1P^o)$	87.5
N	$1s^22s^22p^3$	$1s2s^22p^4(^4P,^2P,^2D,^2S)$	100.0
O	$1s^22s^22p^4$	$1s2s^22p^5(^3P^o,^1P^o)$	100.0
F	$1s^22s^22p^5$	$1s2s^22p^6(^2S)$	100.0

* States which *cannot* radiate in a single-configuration, non-relativistic approximation are bold-faced and in square brackets

† The fluorescent yield for *all* Li-like ions is listed as 0.0% by Kaastra and Mewe [20]; we find that configuration-interaction (CI) effects increases this to about 20% for highly-charged ions.

Non-zero Li-like fluorescence yields: beyond the independent particle approximation

It is also seen in Table 1 that, in a single-configuration, non-relativistic (LSSC) approximation, K-shell vacancy states of Li-like ions cannot radiate, that is, the fluorescence yields of the $1s2s^2(^2S)$ state, with wavefunction

$$\Psi^{LSSC} = 1s2s^2(^2S) \quad \Rightarrow \quad FY = 0\,, \tag{6}$$

is identically zero. Indeed, the tabulated results of Kaastra and Mewe [20] list zero for all these ions. This changes radically when configuration interaction (CI) is considered. The predominant CI for second-row elements is the intrashell $2s^2 \rightarrow 2p^2$ promotion, leading to a CI wavefunction

$$\Psi^{LSCI} = c_1\,1s2s^2(^2S) + c_2\,1s2p^2(^2S) \quad \Rightarrow \quad FY > 0\,, \tag{7}$$

where $c_1^2 \approx 0.1$ throughout the Li-like isoelectronic sequence. The $1s2p^2 \rightarrow 1s^22p$ radiative rate eventually dominates over the Auger rate, giving an asymptotic $Z \rightarrow \infty$ fluorescence yield of 1.0; we find the value for iron ($Z = 26$) to be about 0.2, a substantial fraction compared to the LSSC (and Kaastra and Mewe) result of 0.0!

ACKNOWLEDGMENTS

The work presented here was the result of several collaborations. Special thanks are due to N. R. Badnell, M. A. Bautista, E. Behar, M. H. Chen, I. Dumitriu, J. Garcia, M. F. Hasoglu, T. R. Kallman, C. N. Kodituwakku, K. T. Korista, B. M. McLaughlin, C. Mendoza, P. Palmieri, F. Robicheaux, D. W. Savin, and O. Zatsarinny for their input to these studies. T.W.G. and coworkers at Western Michigan University were supported by the National Science Foundation through a grant for the Institute for Theoretical Atomic and Molecular Physics at Harvard University and Smithsonian Astrophysical Observatory, by the NASA Space Astrophysics Research and Analysis Program, by the NASA Astronomy and Physics Research and Analysis Program, and by the NASA Solar and Heliospheric Supporting Research and Technology program.

REFERENCES

1. B. M. McLaughlin and K. P. Kirby, J. Phys. B **31**, 4991 (1998).
2. K. L. Bell, P. G. Burke, A. Hibbert, and A. E. Kingston, J. Phys. B **21**, 2319 (1989).
3. O. Wilhelmi, G. Mentzel, B. Zimmermann, K.-H. Schartner, H. Liebel, H. Schmoranzer, and B. M. McLaughlin, Phys. Rev. A **60**, 3702 (1999).
4. W. C. Stolte, J. A. R. Samson, O. Hemmers, D. Hansen, S. B. Whitfield, and D. W. Lindle, J. Phys. B **30**, 4489 (1997).
5. A. Menzel, S. Benzaid, M. Krause, C. D. Caldwell, U. Hergenhahn, and M. Bissen, Phys. Rev. A **54**, R991 (1996).
6. T. W. Gorczyca and B. M. McLaughlin, J. Phys. B **33**, L859 (2000).
7. C. Froese Fischer, Comput. Phys. Commun. **64**, 369 (1991).
8. P. G. Burke and K. A. Berrington, *Atomic and Molecular Processes: An R-matrix Approach* (IOP Publishing, Bristol, 1993).
9. K. A. Berrington, W. B. Eissner, and P. H. Norrington, Comput. Phys. Commun. **92**, 290 (1995).
10. A. P. Hickman, J. Phys. B **17**, L101 (1984).
11. F. Robicheaux, T. W. Gorczyca, M. S. Pindzola, and N. R. Badnell, Phys. Rev. A **52**, 1319 (1995).
12. T. W. Gorczyca and F. Robicheaux, Phys. Rev. A **60**, 1216 (1999).
13. T. W. Gorczyca, Phys. Rev. A **61**, 024702 (2000).
14. N. R. Badnell, J. Phys. B **19**, 3827 (1986).
15. J. Garcia, C. Mendoza, M. A. Bautista, T. W. Gorczyca, T. Kallman, and P. Palmeri, Astronomy and Astrophysics, submitted.
16. R. F. Reilman and S. T. Manson, Astrophys. J. Suppl. Ser. **40**, 815 (1979).
17. A. K. Pradhan, G. X. Chen, F. Delahaye, S. N. Nahar, and J. Oelgoetz, Mon. Not. Roy. Acad. Suppl. **341**, 1268 (2003).
18. A. M. Juett, N. S. Schulz, and D. Chakrabarty, Astrophysics Journal **612**, 308 (2004).
19. T. W. Gorczyca, C. N. Kodituwakku, K. T. Korista, O. Zatsarinny, N. R. Badnell, E. Behar, M. H. Chen, and D. W. Savin, Astrophysical Journal **592**, 636 (2003).
20. J. S. Kaastra R. and Mewe, Astrophysical Journal Suppl. **97**, 443 (1993).
21. G. J. Ferland, K. T. Korista, D. A. Verner, J. W. Ferguson, J. B. Kingdon, and E. M. Verner, PASP **110**, 761 (1998).
22. T. R. Kallman and J. H. Krolik, *XSTAR, a Spectral Analysis Tool*, unpublished (1997).
23. K. T. Borkowski, W. J. Lyerly, and S. P. Reynolds, Astrophysical Journal **548**, 820 (2001).
24. E. J. McGuire, Phys. Rev. **185**, 1 (1969).
25. E. J. McGuire, Phys. Rev. A **2**, 273 (1970).
26. E. J. McGuire, Phys. Rev. A **3**, 587 (1971).
27. E. J. McGuire, Phys. Rev. A **5**, 1052 (1972).
28. M. H. Chen, B. Craseman, and D. L. Matthews, Phys. Rev. Lett. **34**, 1309 (1985).
29. E. Behar and H. Netzer, Astrophysical Journal **570**, 165 (2002).
30. M. H. Chen, Phys. Rev. A **31**, 1449 (1985).
31. A. Bar-Shalom, M. Klapisch, W. H. Goldstein, and J. Oreg, *The HULLAC Code for atomic physics* (unpublished) (1998).
32. I. Dumitriu, M. F. Hasoglu, K. T. Korista, T. W. Gorczyca, N. R. Badnell, and D. W. Savin, unpublished.
33. M. H. Chen and B. Crasemann, Phys. Rev. A **35**, 4579 (1987).
34. M. H. Chen et al., At. Data Nucl. Data Tables **65**, 289 (1997).
35. M. H. Chen, private communication.

Dr. Mario Jimenez-Garate presenting his results on accretion disk models.

Stellar Cluster Winds: a Non-equilibrium Ionization Model

L. Ji*, Q. Daniel Wang,* and John Kwan*

Department of Astronomy, University of Massachusetts, Amherst, MA, 01002

Abstract. We have constructed a 1-D non-equilibrium ionization model for steady stellar cluster winds. This model is based on a self-consistent treatment of the dynamical and thermal wind structures and is implemented in XSPEC. The physical quantities (mass-loss rate, terminal velocity, and average metallicity) of interest can be obtained directly by fitting the X-ray spectral data of cluster winds. The model predictions for the NGC3603 cluster are presented and compared with other independent estimates. We also present preliminary results on the modeling of the X-ray emission from the Arches cluster, which shows a strong 6.4 keV FeKα line.

Keywords: wind: stellar cluster, non-equilibrium ionization:spectrum
PACS: 95.30.Qd

INTRODUCTION

Mass-loss from a massive star remains one of the largest uncertainties in stellar evolution. X-ray observations of massive stellar clusters can potentially be a powerful tool to make significant advance in this field. In a compact stellar cluster, winds from individual stars are expected to collide, shock-heat to high temperatures and merge into a so-called stellar cluster wind.

Observationally, such winds have been detected as extended X-ray emission features around stellar clusters (e.g. NGC3603, [3]; Arches cluster at the Galactic center and R136 cluster in 30 Dor, [4], [5]). In the past, models based on the assumption of collisional ionization equilibrium (CIE) were developed, including hydrodynamic simulations. But such models have not been tested quantitatively. One also suspects that the gas in a cluster wind may be in a highly non-equilibrium state, because of the rapid shock-heating in the wind-wind collision and fast adiabatic cooling in the expansion.

We are conducting a comprehensive modeling of the X-ray emission from cluster winds, aiming to characterize their thermal, chemical, and ionization states.

PHYSICAL MODELS AND ASSUMPTIONS

For the wind dynamics, we assume: (1) either an exponential or uniform distribution of stars within a radius; (2) a steady and spherically symmetric cluster wind; (3) mass and energy injections are proportional to the stellar mass distribution; (4) radiative cooling is negligible.

Our spectrum calculation is based on the most up-to-date *CHIANTI* atomic database [7]. To save computational time, we have combined those fine structure levels together

CP774, *X-Ray Diagnostics of Astrophysical Plasmas,* edited by Randall K. Smith

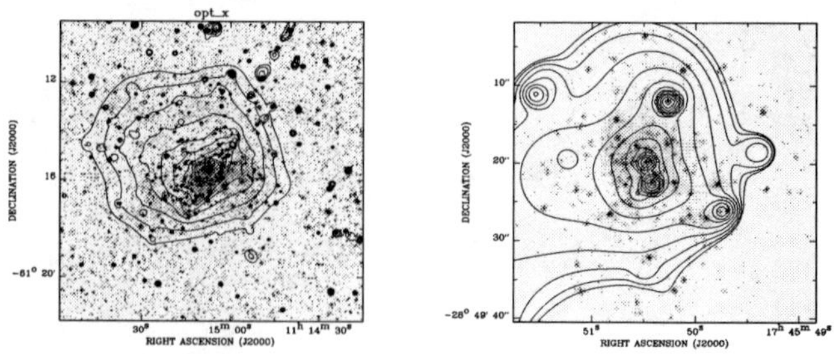

FIGURE 1. **NGC3603**(left): *Chandra* ACIS-I 1.0-9.0 keV intensity contours overlaid on an optical image (gray-scale). **Arches(right)**: *Chandra* ACIS-I 1.0-8.0 keV intensity contours on a grey image in near-IR [1]

according to their degeneracies and have re-calculated the collisional excitation rates and A-rates for each ion. The ionization structure is determined by the processes of collisional ionization, autoionization, radiative [6] and di-electronic recombinations ([2] and references therein). Stellar winds before being shocked are assumed to be cold ($T \sim 10^4$ K).

The model-predicted spectra from a given projected annular region have been calculated and implemented into XSPEC as a table model. There are four model parameters: mass loss rate $\dot{M}_0$, stellar wind terminal velocity V_t, abundance Z, and the scale radius R_s (on cluster radius for a uniformly distributed model).

OBSERVATIONS AND APPLICATIONS

NGC3603 (Figure1, left): We assume that at the center of the star cluster, the gas was heated to the high temperature and almost completely ionized. Due to the fast adiabatic expansion, recombination is delayed. An observed spectrum has been extracted from the annular region with the projected radius between 19″ to 82.5″.

The single temperature mekal model favors a lower metallicity ($Z \sim 0.1 Z_\odot$, $\chi^2/dof = 148.4/159$; $Z = Z_\odot$, $\chi^2/dof = 227.1/160$). However, for the 1-D non-equilibrium ionization model with either exponential or uniform stellar distribution, we found that the fitting is not sensitive to the metallicity (illustrated in Figure2 for exponential stellar distribution case). Between the two stellar distribution cases, we found that the exponential stellar distribution case gives more reasonable constrains on the physical parameters. For example, assuming $R_s = 0.68$pc, (2.8pc for uniformly distributed wind model [5]). For this case, the best fit parameters with 90% confidence ranges are: $\dot{M}_0 = 1.44^{+0.052}_{-0.050} \times 10^{-4} \mathrm{M}_\odot\ \mathrm{yr}^{-1}$; $V_t = 1.86^{+0.077}_{-0.084} \times 10^3 \mathrm{km\ s}^{-1}$; $Z = 0.17^{+0.19}_{-0.16} Z_\odot$; $N_H = 0.68^{+0.060}_{-0.057} \times 10^{22} \mathrm{cm}^{-2}$. These best values for $\dot{M}_0$ and V_t are smaller than those ($2.3 \times 10^{-4} \mathrm{M}_\odot\ \mathrm{yr}^{-1}$, $2.844 \times 10^3 \mathrm{km\ s}^{-1}$) estimated by Stevens et.al [5]. More detailed work is in preparation.

ARCHES CLUSTER (Figure1, right): The diffuse X-ray spectrum of the Arches cluster shows a strong 6.4 keV FeKα line (Figure2, right panel), which indicates that the hot gas must contain large amounts of weakly ionized iron species. With an exponential

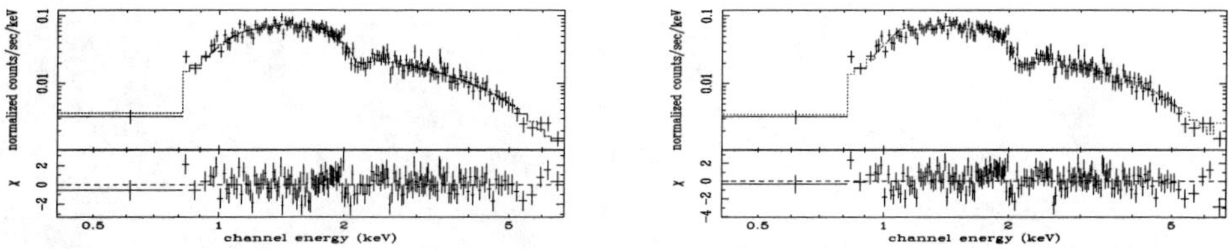

FIGURE 2. Fits to the ACIS-I spectrum of NGC3603 with the exponentially distributed wind model. On the left, the abundance is a free parameter (best value is 0.17 with $\chi^2/dof = 148.6/159$); on the right, the abundance is set to solar and $\chi^2/dof = 180.0/160$.

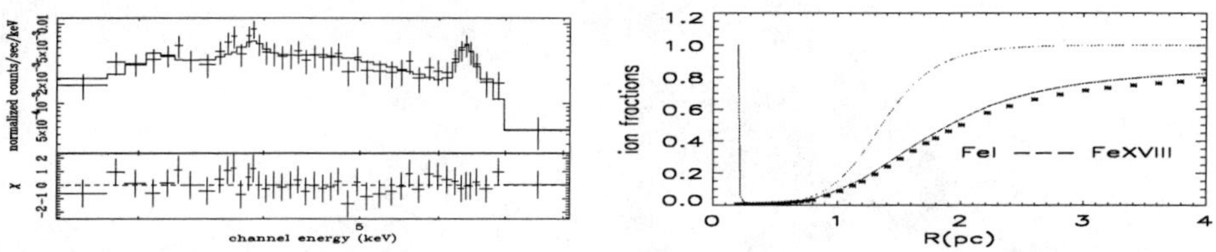

FIGURE 3. **Left**: Diffuse X-ray spectrum of the Arches cluster fitted with a non-equilibrium model . **Right**: The total ion factions from FeI to FeXVIII as a function of the radius at different ionization cases: red solid line assumes metal is neutral at the beginning, blue dot-dash line assumes thermal equilibrium for all the radius, and black asterisk lines assumes thermal equilibrium only at the beginning.

star cluster wind model, we assume a mean location of massive stars at the core radius (~ 0.2pc) and start our ionization calculation there. And we assume electron temperature evolves according to the Coloumb interaction with ions. Due to the lack of innershell ionization lines in our atomic database, here we only show the model predicted ion fractions.

Assuming wind parameters, $\dot{M}_0 = 1 \times 10^{-3} M_\odot$ yr^{-1}, and V_t =1000 km s^{-1}, which are consistent with the observational data, we show the ion fraction under the different ionization cases in Figure 3. The weakly ionized irons are produced in our test model. More extensive investigations are ongoing.

ACKNOWLEDGMENTS

We thank Z.Y. Li for the use of his galactic wind model and Y.S.,Yao, S.K. Tang for the helpful discussions.

REFERENCES

1. Figer, D.F., Morris, M., Geballe, T.R., et. al, 1999, *ApJ*, 525, 759
2. Mazzotta, P., Mazzitelli, G., Colafrancesco, S., Vittorio, N., 1998, *A&AS*, 117, 920
3. Moffat, A.F.J., Corcoran, M.F., Stevens, I.R., et. al 2002, *ApJ*, 573, 191
4. Rockefeller, G., Fryer, C.L., Melia, F., Wang, Q.D., 2004, *astro-ph/0409704*
5. Stevens, I.R., & Hartwell, J.M. 2003, *MNRAS*, 339, 280
6. Verner, D.A., & Ferland, G.J., 1996, *ApJS*, 103, 467
7. Young, P.R., Del Zanna, G., Landi, E. et al. 2003, *ApJS*, 144, 135

Mr. Hosung Kang showing his poster to Dr. Kristy Dyer.

High Resolution X-ray Spectroscopy of SN 1006

Jacco Vink

SRON National Institute for Space Research, Sorbonnelaan 2, 3584CA, Utrecht, The Netherlands

Abstract. I discuss the high resolution *XMM-Newton* Reflection Grating Spectrometer (RGS) spectrum of SN 1006. SN 1006 is one of the best examples of a supernova remnant that is far out of ionization equilibrium. Moreover, optical, UV and X-ray data indicate that it is also out of temperature equilibrium. I discuss the X-ray evidence for this.

In addition I discuss the lower resolution RGS spectrum of the eastern rim of SN 1006. Despite the lower resolution, the spectrum contains significant evidence for an asymmetric expansion velocity. Two likely solutions fit the O VII triplet. One with no significant thermal broadening and a shell velocity of ~ 6500 km s^{-1}, and one with significant broadening and a shell velocity of 9500 km s^{-1}. The first solution seems the most plausible, as it is consistent with radio expansion measurements, which suggests a decelerated shell.

INTRODUCTION

The study of supernova remnants lies at the crossroads of astrophysics, combining the interests in the end products of stellar evolution with the investigation of the interstellar medium and cosmic ray acceleration. This is all the more interesting now that Type Ia supernovae [e.g. 1] have become central to the study of the geometry of the universe and gamma ray bursts are known to originate from a subset of core collapse supernovae [2].

However, of more interest for these proceedings is that supernova remnants are objects in which interesting plasma processes take place. Supernova remnant shocks move through a medium with a typical density of 1 cm^{-3}. At those low densities Coulomb (particle-particle) interactions are rare, leading to typical collision time scales of several thousand years. Shock heating does, however, occur in supernova remnants, otherwise we would not observe X-ray emission from hot plasmas. This implies that the heating process takes place through collective effects, involving plasma waves. Apart from being collisionless, supernova remnant shocks are also the most probable source of cosmic rays, at least for energies up to $\sim 10^{15}$ eV, but possibly even up to $\sim 10^{18}$ eV [6, 7].

The fact that supernova remnant shocks are collisionless and are the locations of particle acceleration has two important consequences. First of all, we can no longer assume that different particles species (protons and other ions and electrons) are in temperature equilibration. In an extreme case, and ignoring particle acceleration, the temperature, kT_i, of each plasma component, i, is:

$$kT_i = \frac{2(\gamma-1)}{(\gamma+1)^2} m_i v_s^2 = \frac{3}{16} m_i v_s^2, \qquad (1)$$

where γ is the adiabatic index, m_i, is the particle mass, and v_s is the shock velocity. For full equilibration this is $kT = 3/16 < m > v_s^2$. In addition, we have to consider that

CP774, *X-Ray Diagnostics of Astrophysical Plasmas,* edited by Randall K. Smith

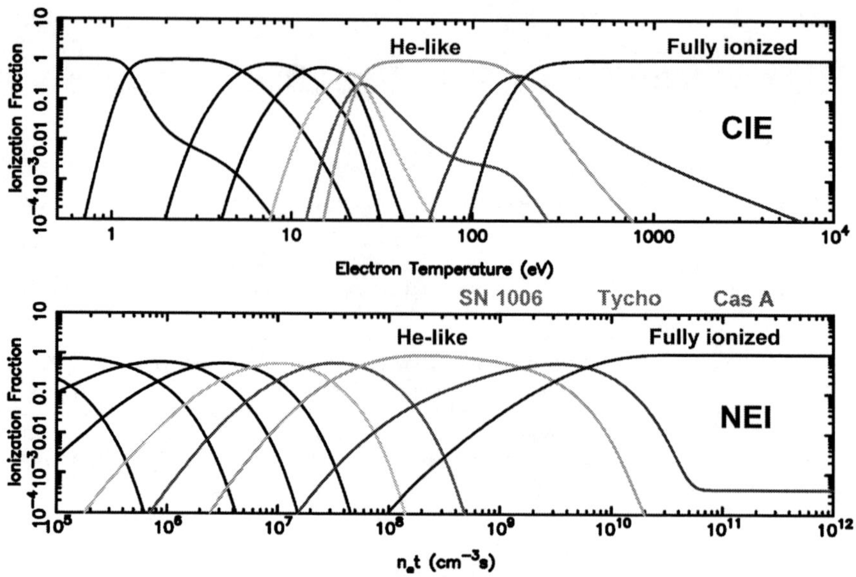

FIGURE 1. An illustration of the effects non-equilibration ionization. Although the top and bottom panels look very similar, the top panel shows the oxygen ionization fraction as a function of electron temperature for the case of collisional ionization equilibrium (CIE), whereas the bottom panel shows how in the case of recently heated plasmas (NEI) the ionization fraction at a fixed temperature of $kT_e = 1.5$ keV changes as a function of ionization age, which is the product of electron density and time since the plasma was heated ($n_e t$). Note the sturdiness of the He-like ion, which is present over a wide temperature or $n_e t$ range. The calculations are based on the simple to use, but somewhat outdated ionization and recombination rates in [3].

efficient cosmic ray acceleration, and possible magnetic field generation, will take away energy, lowering the plasma temperature(s) of the thermal plasma components [8].

As a result measuring the plasma temperature through X-ray spectroscopy, which usually will give us the *electron* temperature, will no longer inform us directly about the shock velocity. This has been known for quite some time [e.g. 9, 10], but until recently it was ignored consciously [e.g. 11] or subconsciously, as it was difficult to assess the amount of temperature equilibration from the observational data.

However, another form of non-equilibration, namely non-equilibration of ionization (NEI), has received more attention over the last three decades, because its consequences could be more easily deduced from X-ray spectra of sufficient quality [e.g 12, 13, 14, 11].

The concept of NEI is relatively simple [9, 15, 16]. NEI is important in supernova remnants as in the relatively recently heated plasmas the ions have not had sufficient time to ionize up to the ionization level at which the number of ionizations is compensated by the number of recombinations (i.e. collisional ionization equilibration or CIE). The effect of NEI is illustrated in Fig. 1. Observationally NEI gives rise to a mismatch between the electron temperature derived from line emission ratios, and the electron temperature

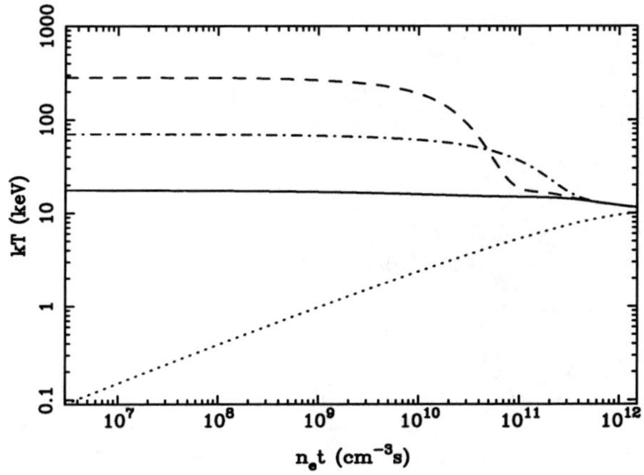

FIGURE 2. An illustration of the effect of non-equilibration of temperatures, taking into account ionization effects, but not adiabatic expansion. The temperature of the electrons (dotted), protons (solid), helium (dashed-dotted) and oxygen ions (dashed) is proportional to their mass ratio behind the shock. Due to Coulomb interactions the temperatures slowly equilibrate. It is interesting to note that the oxygen-proton equilibration will eventually be faster than the helium-proton equilibration, as the oxygen ions acquires a higher charge and the interactions scale with Z^2 [4, 5].

derived from the continuum shape, which reflects the actual electron temperature. In addition, spectra of NEI plasmas will display lines that are unique for NEI, and are the result of inner shell excitations and ionizations. The Fe-K line emission is most widely used as a diagnostic tool for NEI conditions, as for a plasma with $kT_e \gtrsim 2$ keV the centroid of the $n = 2 \rightarrow 1$ line emission shifts gradually from ~ 6.4 keV to ~ 6.7 keV for Fe XXV [17, 18].

Here I discuss high resolution spectra of a supernova remnant that is one of the most extreme cases of non-equilibration of ionization and temperatures, SN 1006.

THE *XMM-NEWTON*-RGS SPECTRA OF SN 1006

The young historical remnant SN 1006 is thought to be a Type Ia supernova remnant on account of its high Galactic latitude (+14.6) and the long duration of the historical light curve [20]. The high latitude also explains the relatively low density of the local medium of ~ 0.1 cm^{-3} [21, 22, 23, 24]. This makes SN 1006 the ideal candidate for studying the two non-equilibration effects, as it has an ionization parameter of $n_e t \sim 2 \times 10^9$ cm^{-3}s. The distance to SN 1006 is 2.1 kpc [25].

XMM-Newton observed SN 1006 several times, both as a guaranteed time and guest observer target. I will discuss here the guest observer observations of August 2002, which consisted of two pointings. One pointing was on the bright knot at the northwestern edge of the remnant, the other was on a narrow feature on the eastern rim (Fig. 3). SN 1006 is a large remnant (30′), but the pointings were optimized in order to obtain high resolution spectra of bright, small structures. Moreover, the pointings were on the

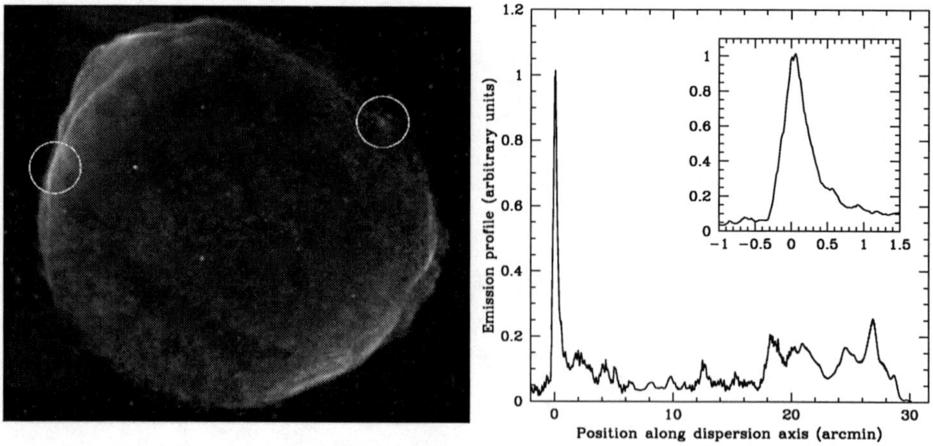

FIGURE 3. *Chandra* mosaic of SN 1006. The circles indicate the *XMM-Newton* pointings discussed here. The emissivity profile is taking in the dispersion direction of the RGS [from 19].

rest of the remnant where it is unlikely that there are bulk plasma motions in our line of sight. This is necessary in order to allow estimates of the thermal line broadening.

A sufficiently high resolution spectrum was obtained for the northwestern pointing, but for the eastern rim, only the non-thermal X-ray emission is very narrow, whereas the thermal emission is much more extended, degrading the spectral resolution. This is somewhat unfortunate, as the electron and ion temperatures of the plasma may be influenced by efficient cosmic ray acceleration, and non-thermal emission from the rims of SN 1006 indicate a very efficient electron cosmic ray acceleration at the west and eastern rims, but not at the northwestern edge [26, 27].

The X-ray spectrum of the northwestern knot

The *XMM-Newton* Reflection Grating Spectrometer [RGS, 28] spectrum of the northwestern knot shows many spectral details that cannot be seen in CCD spectra. The most dominant emission is coming from O VII, in particular from the O VII triplet, but also O VII line emission from higher quantum levels up to the series limits. The line emission around 19 Å is a mix of O VII Heβ emission and O VIII Lyα emission. Lines from Fe-L transitions, in particular Fe XVII which has lines around 15 Å and 17 Å are conspicuously absent. This may be surprising given the fact that SN 1006 is a likely Type Ia remnant, which produce $\sim 0.7\ M_{\odot}$ of Fe. However, for the low ionization age of SN 1006 Fe is expected to have a peak ionization state of Fe XIII, so no Fe XVII emission is to be expected.

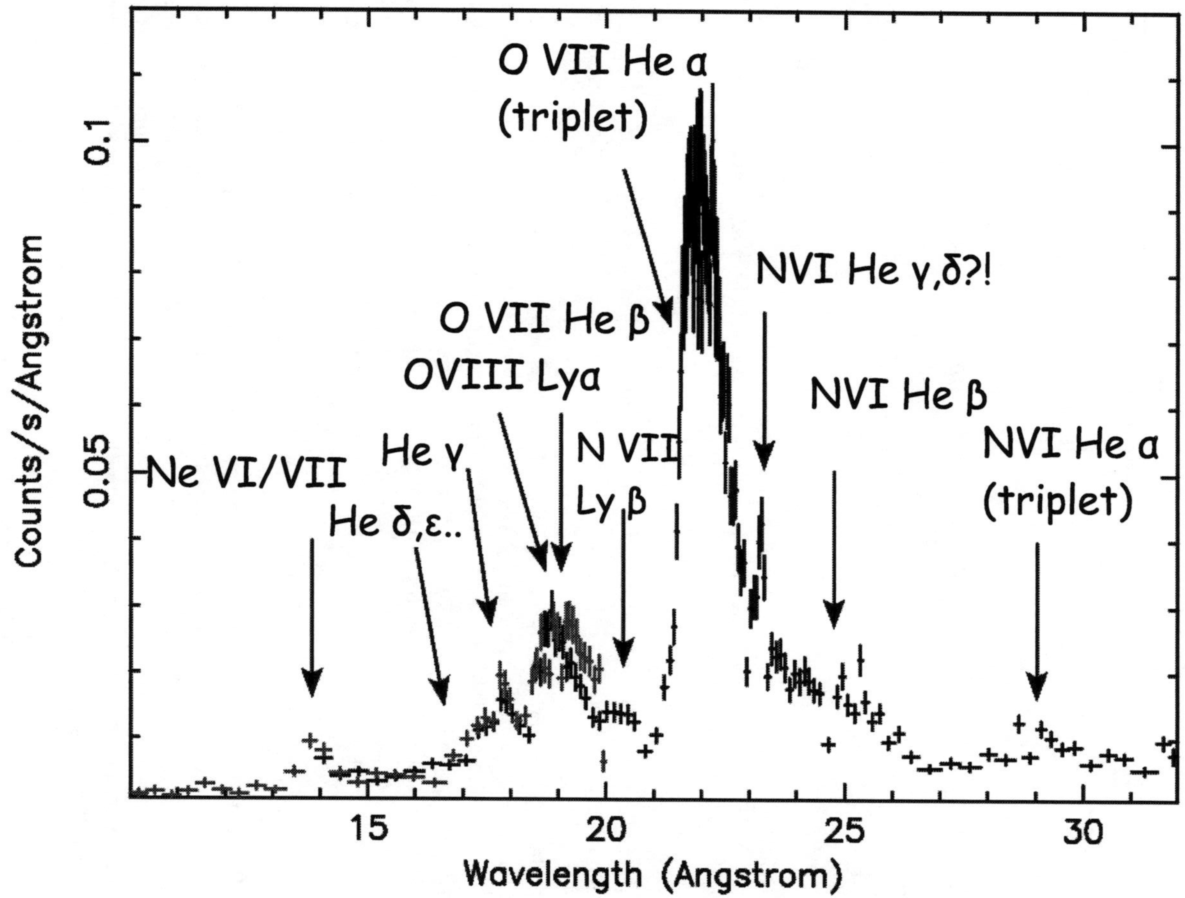

4: The *XMM-Newton*-RGS spectrum of the northwestern knot.

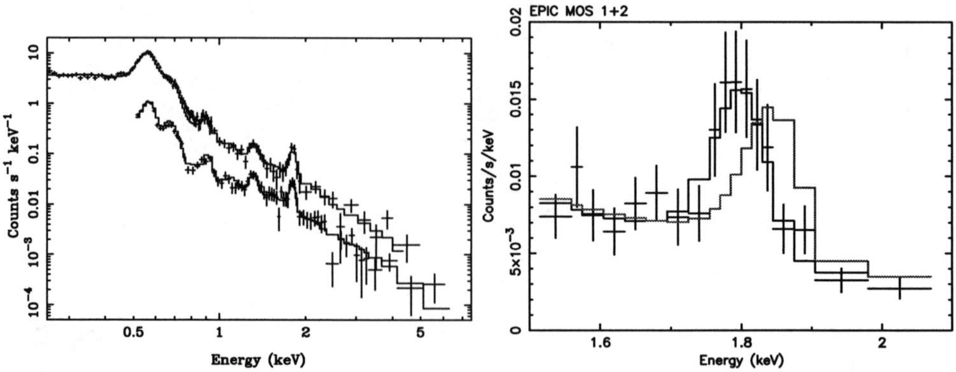

FIGURE 5. The *XMM-Newton* PN and MOS spectra (left) and a detail of the Si emission as observed by the MOS detectors (right). The shifted line shows the position of the Si emission if most of the Si would have been in the He-like state.

The CCD spectra were modeled using the SPEX[1] [29] spectral code. The detailed results can be found in [19] and confirm the low ionization age of the plasma, $n_e t = 2.4 \times 10^9$ cm^{-3}s, whereas $kT_e = 1.5 \pm 0.2$keV. There is a relative overabundance of Si, which suggests that the knot is an ejecta knot, and not a shocked pre-existing density enhancement in the interstellar medium. Interestingly, there is a weak enhancement of X-ray emission in front of the knot as well, that seems to emit slightly harder X-ray emission. It is not clear how these relate to each other.

The low ionization age of the plasma can be deduced not only by globally fitting the spectrum, but also by measuring the centroid of each individual line, or by modeling the O VII triplet. For instance the EPIC PN and MOS spectra clearly indicate that the centroid of the Si emission is 1.80 ± 0.01 keV instead of at 1.85 keV, which is the centroid expected if Si is in the He-like state (Fig. 5). Also the Ne and Mg line centroids are at lower energies than the He-like. Purely combining the information of the centroids, and assuming $kT_e = 1.5$ keV one also finds an ionization age of $n_e t \sim 2 \times 10^9$ cm^{-3}s (Fig. 6). The most important inner shell lines are present in the SPEX spectral code, but in general one should be aware that spectral codes are not always complete in this sense, as most atomic databases were initially compiled with CIE plasmas in mind. Moreover, the line parameters may also be less accurate, as less experimental data exists to measure the line properties.

Non-equilibration of temperatures in the northwestern knot

One of the main goals of the *XMM-Newton* observations was to measure the amount of non-equilibration of electron and ion temperatures, by measuring the thermal Doppler

[1] `http://www.sron.nl/divisions/hea/spex/version2.0/index.html`

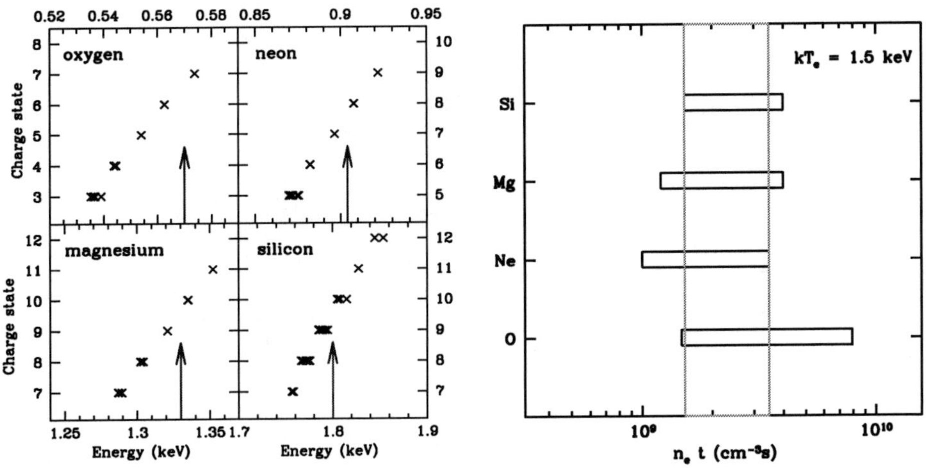

FIGURE 6. Line energy versus charge for O, Ne, Mg, and Si. The crosses give the expected line energies for different ionization states (from Dr. M.F. Gu's FAC code) and the arrows approximately indicate the observed centroid energies.

broadening of O VII and O VIII line emission. So far evidence for non-equilibration in supernova remnants comes mainly from measuring the Hα line width and broad to narrow line ratios. This has been done for several remnants including SN 1006 [30, 31]. The amount of non-equilibration has also been measured in the UV from C IV, N V and O VI lines by *HUT* [32] and very recently by *FUSE* [33]. The optical and UV results indicate that there is no equilibration at the shock front, although the *FUSE* results suggests that the temperature of different species is not proportional to the species mass, as implied by eq. 1 and the *HUT* spectrum.

The O VII Heα emission consists of the He triplet, plus a potential contribution from an O VI line very close to the O VII forbidden line emission. In order to measure the Doppler broadening models for the line ratios for a grid of $n_e t$ values were fitted, using parameters obtained from the flexible atomic code (FAC) of Dr. M. F. Gu [34]. The spectrum of the RGS1 O VII spectrum and the best fit models are shown in Fig. 7. The best fit suggests a very significant broadening at a confidence level of 9.8σ. Allowing full freedom of the line normalizations still indicates thermal broadening, at a still high confidence level of 6.5σ. The broadening itself is $\sigma_E = 3.4 \pm 0.4$ eV. The published results are based on a response matrix that was corrected for the emissivity profile shown in Fig. 3. However, the fitting program XSPEC also contains the convolution model *rgsxsrc* made by Dr. A. Rasmussen, which uses an input image to determine the emissivity profile. Using *rgsxsrc* gives very similar results.

The derived value for the broadening implies an oxygen temperature of $kT = 530 \pm 150$ keV, which looks incredibly high, but is in effect consistent with eq. 1 if the gas was shocked with a velocity of ~ 5000 km s^{-1}. Of course the oxygen temperature is more than a factor 300 higher than the electron temperature, clearly indicating non-equilibration of temperatures. The measured O VII line broadening is more than reported

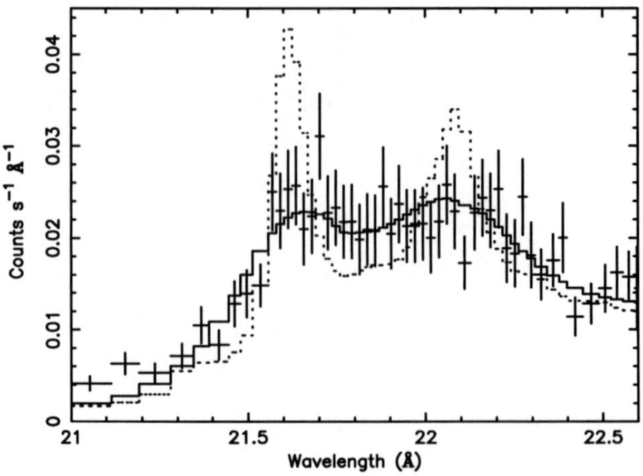

FIGURE 7. The O VII triplet from the northwestern not as observed by the RGS1. The dotted line shows the best fit spectrum with (solid) and without including thermal Doppler broadening.

for O VI UV emission from other regions of the northwestern rim [2] [33]. There are several reasons why there could be a discrepancy. The knot may be an ejecta knot, as implied by the high Si abundance, and the heating may have been caused by the reverse shock, which has a different speed than the shock. Another possibility is that the plasma of the knot was shocked earlier than the regions observed in O VII, after all the X-ray measurements concern more highly ionized material, which is situated further downstream from the shock.

What about the eastern rim?

The eastern and western rims of SN 1006 are likely sites of efficient cosmic ray shock acceleration [26]. If the cosmic ray pressure dominates over the thermal pressure, eq. 1 is no longer valid. It will result in an average plasma temperature that is lower. Moreover, if the cosmic ray is very efficient and the cosmic ray power law index is equal to or flatter than -2, the energetic cosmic rays that escape the shock will drain the shock of its energy and the shock will become (cosmic) radiative, resulting in much higher shock compression ratios [35].

Given the interesting physics at the eastern and western rim, it is unfortunate that the RGS spectrum from the eastern rim does not have a high spectral resolution, especially since no UV emission lines from the eastern rim have been detected by *FUSE* [33]. It is therefore still worthwhile to obtain as much information from the RGS spectrum as possible. I will therefore discuss here a preliminary analysis of RGS spectrum.

The effective spectral resolution is mainly determined by the width of the shell, about

---

[2] Unfortunately in [33] the measured broadening in σ is mistakenly taken to be a FWHM.

$5'$, resulting in an effective resolution for the first order spectrum of ~ 0.6 Å. This is not sufficient to resolve the O VII triplet, which is also for the eastern rim the dominant line emission. However, there is still high spectral resolution information in the spectrum, which results from the sharp increase of the emission at the edge of the remnant.

The O VII triplet emission is displayed in Fig. 8. It shows that a model that does only contain the triplet lines and an instrumental response that only takes into account the spatial emission profile does not give an adequate fit. The long wavelength side seems to fit reasonably well, but there is a dip in the data around 21.5 Å, and the model underpredicts the emission around 21.3 Å. Note that here the interior of remnant is dispersed in an opposite direction compared to the spectrum of the northwestern knot. So the edge of the remnant determines the shape of the long wavelength side. This suggests that discrepancy is caused not so much by line broadening, but by line shifts. The plasma motions at the edge of the remnant, will be mostly in the plane of the sky, whereas going to the interior of the remnant radial motions will give rise to Doppler shifts. As we are observing both the front and backside of the shell, both blueshifts and redshifts should be taken into account. The redshifts are not directly noticeable (they blend in with the rest of the triplet emission), but the blueshifts can be seen as the mismatch between model and spectrum around 22.3 Å.

I have attempted to fit a simple model for the O VII triplet, taking into account the increase in blueshifts and redshifts moving interior to the remnant (corresponding to shorter wavelength). As it is a mixing of spectral and spatial information, the method consisted in making a grid of response matrices in which the spatial and redshift and blueshift were incorporated. The input to the model consisted of the projected plasma velocity, which increased toward the interior, and a parameter that determined how much of the emission was coming from the near and far side of the remnant. This is a simplification, as velocity gradients are presents and the morphology of the shell may be complicated. It was nevertheless possible to obtain acceptable fits to the O VII triplet emission (Fig. 8).

It turns out there are two good fitting models. One with no, or only a small additional (thermal) broadening ($\sigma_E = 0.4 \pm 0.3$ eV), and a shell with velocity 6500 ± 830 km s^{-1} (68% confidence range, Fig. 8). The other solution is for a much higher shell velocity of > 9000 km s^{-1}, and quite a substantial additional broadening of $\sigma_E = 3 \pm 1$ eV. In both cases the best solution suggests that about 70% emission comes from material moving toward us ($> 55\%$ at a 3σ confidence). The solution with a large broadening gives a better fit (Fig. 8), but is less plausible. The reason is that a bulk velocity of > 9000 km s^{-1}would imply that the shock heated plasma is still in free expansion, whereas a solution with $v \sim 6500$ km s^{-1} fits in much better with radio proper motions [36]. The average shock velocity of SN 1006, given its age and distance is $< v >=$ 9375 km s^{-1}. The radio expansion measurements suggest that $v_s = (0.48 \pm 0.13) < v >=$ (4500 ± 1220) km s^{-1}. Both solutions exclude the possibility that the plasma velocity is similar to what has been measured for the northeastern Hα filaments $v_s = 2800$ km s^{-1} [30]. This is consistent with the smaller radius in the northwest and the finding of an increased density as indicated by 21cm line measurements [37].

The shape of the O VII triplet suggests that more material is blueshifted than redshifted. This asymmetry is reminiscent of the difference in forward shock and reverse shock velocity found in absorption toward a UV star projected on the center of SN 1006

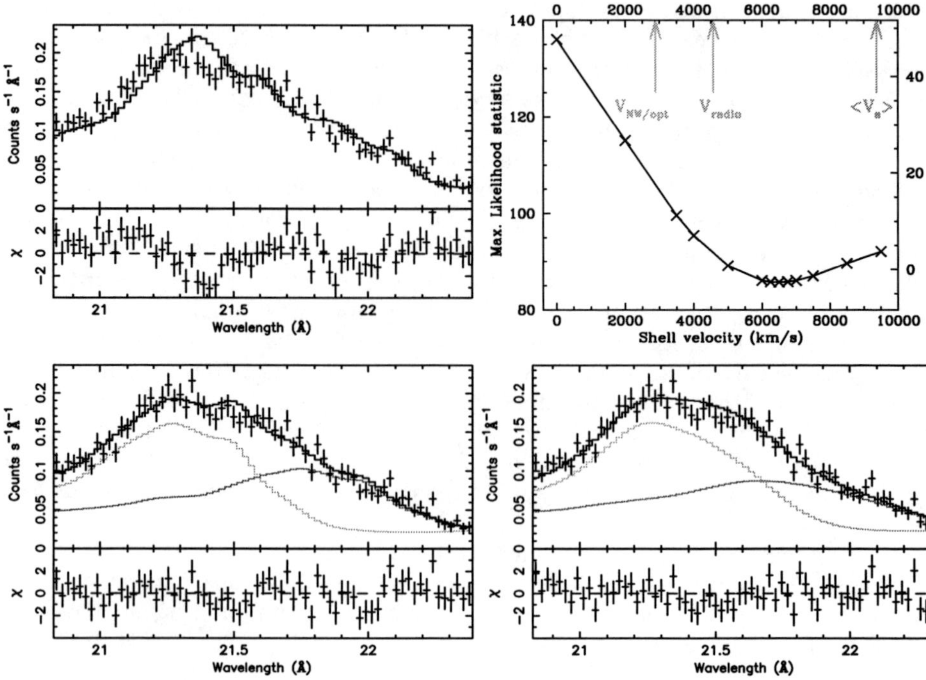

FIGURE 8. The O VII triplet from the eastern rim as observed by RGS1. Top left: The best fit model with only taking into account the spatial profile. Top right: the likelihood statistic as a function of shell velocity, for a front back asymmetry of 70%. The arrows indicate other velocity measurements from respectively the northwest, radio and the average velocity. Bottom: The left panel shows the best fit solution for a shell velocity of 6500 km s^{-1}with no broadening. The right panel shows a solution with a velocity of 9500 km s^{-1} including substantial thermal broadening.

[38]. The reverse shock appears to have penetrated more into the inner ejecta layers of the front side than on the backside. Such an asymmetry is likely to be the result of a higher density at the front side, which will also result in more shocked heated plasma at the front side.

CONCLUDING REMARKS

SN 1006 is from many points of view an interesting supernova remnant. It is one of the few historical remnants, and it was the first remnant for which it was established that the particle acceleration by the shock was efficient enough to produce X-ray synchrotron emission. The fact that SN 1006 expands into a low density medium makes it also an ideal laboratory to study the effects of non-equilibrium ionization and non-equilibration of temperatures, both of which depend on the value of $n_e t$, which is around 2×10^9 cm^{-3}s for SN 1006, one of the lowest values found in a supernova remnant.

The measurements presented here show the diagnostic power of high resolution spectroscopy when it comes to characterizing the ionization state, temperature and velocities of hot plasma. The results on a remnant like SN 1006 raises the expectation for the results from micro-calorimeters carried by *Astro-E2*, and *Constellation-X/Xeus*, which will make it much more straightforward to measure Doppler motions from extended sources.

I thank the organizers of the XDAP 2004 workshop for their invitation and financial support.

REFERENCES

1. Tonry, J. L., et al., *ApJ*, **594**, 1–24 (2003).
2. Stanek, K. Z., et al., *ApJL*, **591**, L17–L20 (2003).
3. Shull, J. M., and van Steenberg, M., *ApJS*, **48**, 95–107 (1982).
4. Zeldovich, Y., and Raizer, Y. P., *Elements of gasdynamics and the classical theory of shock waves*, New York: Academic Press, 1966, edited by Hayes, W.D.; Probstein, Ronald F., 1966.
5. Huba, J. D., NRL Plasma Formulary (2002).
6. Bykov, A. M., *Advances in Space Research*, **33**, 366–375 (2004).
7. Vink, J., *Adv. of Space Research*, **33**, 356–365 (2004).
8. Hughes, J. P., Rakowski, C. E., and Decourchelle, A., *ApJL*, **543**, L61–L65 (2000).
9. Itoh, H., *PASJ*, **29**, 813–830 (1977).
10. Itoh, H., *ApJ*, **285**, 601–606 (1984).
11. Jansen, F., et al., *ApJ*, **331**, 949–957 (1988).
12. Winkler, P. F., et al., *ApJ*, **245**, 574–580 (1981).
13. Gronenschild, E. H. B. M., and Mewe, R., *A&AS*, **48**, 305–331 (1982).
14. Hughes, J. P., and Helfand, D. J., *ApJ*, **291**, 544–560 (1985).
15. Mewe, R., and Schrijver, J., *A&A*, **87**, 261–268 (1980).
16. Liedahl, D. A., *Lecture Notes in Physics, Berlin Springer Verlag*, **520**, 189–+ (1999).
17. Palmeri, P., et al., *A&A*, **410**, 359–364 (2003).
18. Mendoza, C., Kallman, T. R., Bautista, M. A., and Palmeri, P., *A&A*, **414**, 377–388 (2004).
19. Vink, J., Laming, J. M., Gu, M. F., Rasmussen, A., and Kaastra, J., *ApJL*, **587**, 31–34 (2003).
20. Stephenson, F. R., and Green, D. A., *Historical supernovae and their remnants*, Oxford: Clarendon Press, 2002.
21. Vink, J., Kaastra, J. S., Bleeker, J. A. M., and Preite-Martinez, A., *A&A*, **354**, 931–937 (2000).
22. Allen, G. E., Petre, R., and Gotthelf, E. V., *ApJ*, **558**, 739–752 (2001).
23. Dyer, K. K., Reynolds, S. P., and Borkowski, K. J., *ApJ*, **600**, 752–768 (2004).
24. Long, K. S., et al., *ApJ*, **586**, 1162–1178 (2003).
25. Winkler, P. F., Gupta, G., and Long, K. S., *ApJ*, **585**, 324–335 (2003).
26. Koyama, K., et al., *Nat*, **378**, 255–+ (1995).
27. Rothenflug, R., et al., *A&A*, **425**, 121–131 (2004).
28. den Herder, J. W., et al., *A&A*, **365**, L7–L17 (2001).
29. Kaastra, J. S., Mewe, R., and Nieuwenhuijzen, H., "A new spectral code," in *Proc. of the 11th Coll. on UV and X-ray, UV and X-ray Spectroscopy of Astrophysical and Laboratory Plasmas, ed. K. Yamashita & T. Watanabe (Tokyo: Universal Academy Press)*, 1996, p. 411.
30. Ghavamian, P., Winkler, P. F., Raymond, J. C., and Long, K. S., *ApJ*, **572**, 888–896 (2002).
31. Rakowski, C. E., Ghavamian, P., and Hughes, J. P., *ApJ*, **590**, 846–857 (2003).
32. Laming, J. M., Raymond, J. C., McLaughlin, B. M., and Blair, W. P., *ApJ*, **472**, 267–+ (1996).
33. Korreck, K. E., Raymond, J. C., Zurbuchen, T. H., and Ghavamian, P., *ApJ*, **615**, 280–285 (2004).
34. Gu, M. F., *ApJ*, **593**, 1249–1254 (2003).
35. Ellison, D. C., Decourchelle, A., and Ballet, J., *A&A*, **413**, 189–201 (2004).
36. Moffett, D. A., Goss, W. M., and Reynolds, S. P., *AJ*, **106**, 1566–1572 (1993).
37. Dubner, G. M., et al, *A&A*, **387**, 1047–1056 (2002).
38. Hamilton, A. J. S., Fesen, R. A., Wu, C.-C., Crenshaw, D. M., and Sarazin, C. L., *ApJ*, **481**, 838–+ (1997).

V: XAFS AND CHARGE EXCHANGE DIAGNOSTICS

Prospects for determining the grain composition of the interstellar medium with Chandra and Astro E2

Julia C. Lee* and Bruce Ravel†

*Chandra Fellow; Harvard University, Center for Astrophysics, 60 Garden Street MS-4, Cambridge MA 02138
†Center for Corrosion Chemistry and Engineering, Code 6134, Naval Research Laboratory, Washington DC 20375

Abstract. The ability of high resolution X-ray spectroscopy to directly probe the grain composition of the ISM is discussed. Using experimental data taken at the National Synchrotron Light Source at Brookhaven National Laboratory and the Advanced Photon Source at Argonne National Laboratory, current and near-future mission prospects for (1) distinguishing gas from dust phase absorption, and (2) determining the chemical composition of interstellar grains, through the details of X-ray absorption fine structure is assessed. Details, in-depth discussions and complete references can be found in Lee & Ravel 2005 [8].

Keywords: dust: XAFS, ISM, composition – ISM: dust composition – technique: solid state – X-rays: ISM, dust
PACS: 32.30

INTRODUCTION

The spectral resolution of Chandra and XMM-Newton is such that we are detecting many more spectral features from X-ray hot gas than the expected Hydrogen- and Helium-like transitions of astrophysically abundant elements. The most notable of these include (1) the detection of inner shell resonance lines of oxygen, magnesium, silicon and iron from ionization states typically probed in the UV, IR or optical [5, 7, 10, 13], and (2) spectral features attributed to solid state effects from dust. While dust is the primary focus of this proceeding, we point out that the detection of inner shell resonance lines from lower ionized states are significant since, in the context of AGN studies, the ability to measure these lines in the X-rays have strong implications for allowing us to tie the X-ray data to the vast body of work that has been done in other wavebands (e.g. UV, IR, optical). As such, these lines provide a powerful alternative for (1) assessing the UV properties of dust enshrouded sources where UV data is unobtainable, and (2) where UV data is available, we can assess the kinematic associations, if they exist, of the absorbers across many wavebands. Such studies have benefitted from atomic calculations [1, 3, 12, e.g.]. For astrophysical studies of dust, X-rays provide a unique probe of the solid state nature of grains.

CP774, *X-Ray Diagnostics of Astrophysical Plasmas,* edited by Randall K. Smith

Why X-ray Studies of Dust

With the exception of X-ray and infrared studies, the presence of dust in astrophysical environments is largely inferred in probes by other wavebands. Dust particles smaller than 10 μm are partially transparent to X-rays so that the measured absorption in this energy band should be sensitive to *all atoms in both gas and solid phase*. Therefore, high resolution X-ray spectroscopy provides a powerful and unique tool for determining the state and composition of the material in the ISM. For interstellar grains, we would detect wave-like modulations near the photoelectric absorption edge, or X-ray absorption fine structure (XAFS), which are the result of interference effects produced by interactions between neighboring atoms within the molecule, or within the atom itself as out-going electrons interact with higher (e.g. $n \gtrsim 100$) discrete energy levels. (The strength of the edge discontinuity alone can only give information about the absorbing column and cannot distinguish gas from dust-phase absorption; e.g. Wilms et al. 2000 [19].) By probing the details of the atomic and molecular structure of solids using X-rays, we, in theory have the ability to distinguish gas from dust, and determine the chemical composition of grains. Therefore, X-rays studies of dust will be a great complement to the many studies in other wavebands (see e.g. Draine 2003 [2] and Whittet 2004 [18] and references therein for reviews).

X-ray ISM studies

The first attempt to quantify absorption in the ISM using X-rays was by Schattenburg & Canizares (1986) [14] using the Einstein Focal Plane Crystal Spectrometer. However, it wasn't until Chandra that significant progress in this area has been made [4, 11, 15, 16, 17]. And yet, due partly to insufficient throughput, we can only study this in detail with a handful of sources. To date, several tentative reports of ISM grain types have been suggested through the detection of XAFS in X-ray spectra of highly absorbed bright X-ray binaries [9, 17].

THE THEORY AND PRACTICE BEHIND MEASURING XAFS TO DETERMINE MOLECULAR COMPOSITION.

Theory

The basic idea behind XAFS is largely based on the principles of the photoelectric effect. When an incident X-ray photon has energy equal to that of the binding energy of a core electron (e.g. K, L, M shell), the probability for absorption increases so that the observed spectral effect is a large photoelectric edge step. For isolated atoms, the promotion of this electron to the continuum results in an increase in the probability for absorption, giving rise to an absorption edge in the bound-free process for single atoms. Inner shell resonance lines are the result of the bound-bound case in isolated atoms where there is only enough energy to excite a deep core electron to the next

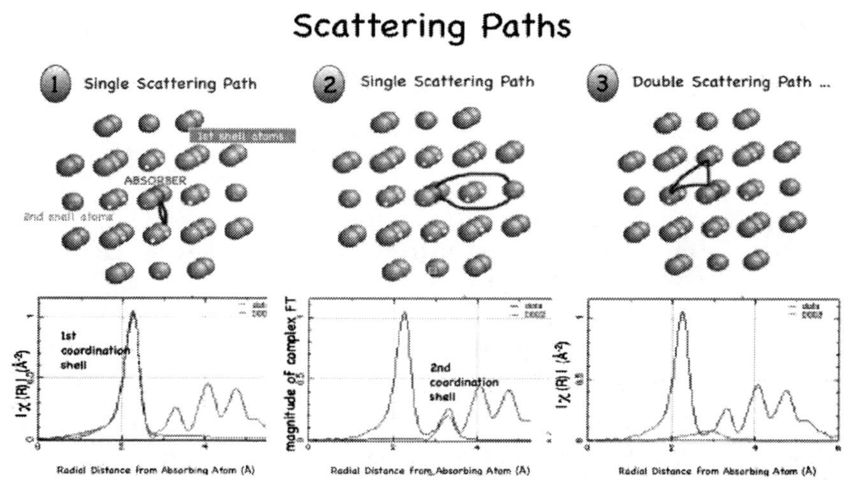

FIGURE 1. A simple illustration of the relationship between scattering paths and the magnitude of the Fourier transform ($|\chi(R)|$ (Å^{-2})) in an XAFS measurement (© 2002 Bruce Ravel). (1) The first peak corresponds to the nearest neighbor shell from the absorbing atom. (2) Similarly, the second peak corresponds to the 2nd shell. (3) More complex scattering paths contribute to the overall spectrum. Our analysis accounts for all the important scattering paths; the scattering approximation is summed over *all* atoms surrounding the absorber.

quantum level. A similar bound-bound process in solids, whereby the ejected electron gets excited into a higher unoccupied quantum level of the atom, or the conduction band of the molecule itself gives rise to XAFS, which are seen as spectral modulations at energies above the edge.

Mathematically, this can be simply stated using Fermi's Golden Rule to describe the probability for absorption, since X-ray absorption is a transition between two quantum states :

$$\mu(E) = \mu_0(E)\,[1+\chi(E)] \propto |< i|H|f >|^2 \tag{1}$$

Here, the *initial* state i describes an X-ray of certain energy E_0 and a core electron in a stable atom and the *final* state f describes the resultant core hole and a photoelectron. The operator H acting on this is known as the interaction term and represents the change between two energy-momentum states. For isolated atoms, the fine structure $\chi(E)$ does not play a role. In contrast, when there are other atoms involved (as e.g. in molecules), the final state is modified by the back-scattering effect from the neighboring atoms such that

$$\chi(E) = \frac{\mu(E) - \mu_0(E)}{\Delta\mu_0(E)} \propto |< i|H|\Delta f >|^2 \tag{2}$$

where Δf describes the change in the photo-electron final state due to backscattering from neighboring atoms.

Practice

In practice, the way to assess XAFS in spectra is to isolate the oscillatory structure $\chi(E)$ from the absorption cross-section $\mu(E)$ by removing the edge step. Because XAFS are an interference effect, and depends on the wave nature of the photo-electron, we can re-express the scattering term in units of photoelectron wave number, i.e. $\chi(E) \rightarrow \chi(k)$ where $k = \sqrt{\frac{2m}{\hbar^2}(E - E_0)}$. The Fourier transform of this is related to the radial distribution function of atoms around the absorber, and thus can be analyzed to identify the distance to, and species of the neighboring atoms (see e.g. Fig. 1; [6]). As such, analysis of photoelectric edge spectra should be able to elucidate the content of solid-state material (e.g. dust in astrophysical environments).

THE LIMITING EFFECTS OF NOISE AND SPECTRAL RESOLUTION

In order to assess the limiting effects of noise and spectral resolution for detecting significant XAFS signal from the Astro E2 XRS, we made laboratory synchrotron XAFS measurements of likely candidates for ISM dust that are iron-based. After convolving the synchrotron data to the approximate 10 eV resolution of the XRS, and adding random fluctuation to simulate noise at the 5% and 10% level, we find that at the spectral resolution of the XRS, noise is by far the most serious impediment to our ability to speciate interstellar grains. Given sufficient signal, we can definitively distinguish dust from gas, and have good prospects for discerning some chemical compositions. See Lee & Ravel 2005 (Figs 1 & 2) for details.

CONCLUSION : CAN WE DETERMINE THE GRAIN COMPOSITION OF THE ISM WITH CHANDRA AND ASTRO E2 ?

The most important conclusions are itemized below.

- While our prospects are good, noise can be a serious impediment so that adequate integration times should be considered. Given sufficient signal, we will have the ability to definitively separate gas from dust, and have good prospects for determining the chemical composition of certain grain types.
- Because X-ray astrophysical studies of dust rely on pointings towards bright X-ray binaries with large amounts of absorbing column, soft X-ray XAFS studies with the Chandra and/or XMM gratings, while not impossible, are made more difficult by

the presence of ionized absorption / emission lines originating from the binaries' own environment.

- The sparsity of atomic lines at energies > 6 keV compared with the low energies, will make the problem better handled with Astro E2, for iron-based dust.
- Because the study of XAFS is largely empirical, space-based measurements should be complemented with (1) atomic calculations and (2) XAFS data taken at synchrotron beamlines to determine the exact chemical state of the astrophysical dust.
- Both **area** and **spectral resolution** warrant serious consideration for the planning of future X-ray missions.

ACKNOWLEDGMENTS

JCL thanks and acknowledges support from the Chandra fellowship grant PF2–30023 – this is issued by the Chandra X-ray Observatory Center, which is operated by SAO for and on behalf of NASA under contract NAS8–39073.

REFERENCES

1. Behar, E., Sako, M., & Kahn, S. M. 2001, ApJ, 563, 497
2. Draine, B. T. 2003, ARAA, 41, 241
3. Gu, M. F. 2003, ApJ, 582, 1241
4. Juett, A. M., Schulz, N. S., & Chakrabarty, D. 2004, ApJ, 612, 308
5. Kaspi, S., et al. 2002, ApJ, 574, 643
6. Koningsberger, D. C. & Prins, R. 1988, X-ray Absorption: Principles, Applications, Techniques of EXAFS, SEXAFS, and XANES; Editors Koningsberger & Prins (John Wiley & Sons)
7. Lee, J. C., Ogle, P. M., Canizares, C. R., Marshall, H. L., Schulz, N. S., Morales, R., Fabian, A. C., & Iwasawa, K. 2001, ApJ, 554, L13
8. Lee, J. C., & Ravel, B. 2005, ApJ, in press (astro-ph/0412393)
9. Lee, J. C., Reynolds, C. S., Remillard, R., Schulz, N. S., Blackman, E. G., & Fabian, A. C. 2002, ApJ, 567, 1102
10. Netzer, H., et al. 2003, ApJ, 599, 933
11. Paerels, F., et al. 2001, ApJ, 546, 338
12. Pradhan, A. K. 2000, ApJ, 545, L165
13. Sako, M., et al. 2003, ApJ, 596, 114
14. Schattenburg, M. L. & Canizares, C. R. 1986, ApJ, 301, 759
15. Schulz, N. S., Cui, W., Canizares, C. R., Marshall, H. L., Lee, J. C., Miller, J. M., & Lewin, W. H. G. 2002, ApJ, 565, 1141
16. Takei, Y., Fujimoto, R., Mitsuda, K., & Onaka, T. 2002, ApJ, 581, 307
17. Ueda, Y., Mitsuda, K., Murakami, H., & Matsushita, K. 2004, in press (astro-ph/0410655)
18. Whittet, D. C. B. 2003, Dust in the Galactic Environment (first edition 1992) (IOP Publishing Ltd)
19. Wilms, J., Allen, A., & McCray, R. 2000, ApJ, 542, 914

Dr. Svetlana Kotochigova presenting her results on high-accuracy models of iron emission line wavelengths.

Probing the interaction between comets and the solar wind

D. Bodewits[1], R. Hoekstra

KVI atomic physics, Rijksuniversiteit Groningen, Zernikelaan 25, 9747 AA Groningen, The Netherlands

A.G.G.M. Tielens

Kapteyn Institute/SRON, Rijksuniversiteit Groningen, PO Box 800, 9700 AV Groningen, The Netherlands

Abstract. The fast ions in the solar wind interact with neutral molecules and atoms in the coma of comets through charge exchange processes. The resulting excited species decay through ultraviolet or X-ray emission. Conversely, observations of cometary far ultraviolet and X-ray emission can provide detailed insight in the interaction between comets and the solar wind. Based on an extensive interaction model, we identify the principal experimental and observational needs for the interpretation of cometary XUV emission.

Keywords: Electron capture, charge transfer, comets, solar wind, X-ray, Far Ultraviolet

PACS: 96.50.Ek, 52.70.La, 52.70.Kz, 34.70.+e

INTRODUCTION

Charge exchange occurs whenever ions collide with neutrals. Electrons are captured into excited states that subsequently decay to the ground state through the emission of one or more photons. With the launch of a new generation of satellite observatories, XMM and Chandra, X-ray and FUV (XUV) emission driven by electron capture by solar wind ions has been found on a large variety of solar system objects, such as comets, Mars, the interstellar medium and the heliosphere [1-4].

In 1996, ROSAT observed strong X-ray emission from comet Hyakutake [1]. Thereafter, many more comets were discovered in the (soft)-X-ray region [5,6] and with the detection of line emission from comet Linear S4 [7], charge exchange became the generally accepted scenario [8-10]. Cometary XUV emission results from electron transfer between highly-charged solar wind minor ions and neutrals in the outflowing cometary gas, see Figure 1. At solar wind velocities, charge transfer dominates the interaction of multiply charged ions, A^{q+}, with neutral gas, B. These charge-transfer

[1] Email: bodewits@kvi.nl

CP774, *X-Ray Diagnostics of Astrophysical Plasmas,* edited by Randall K. Smith

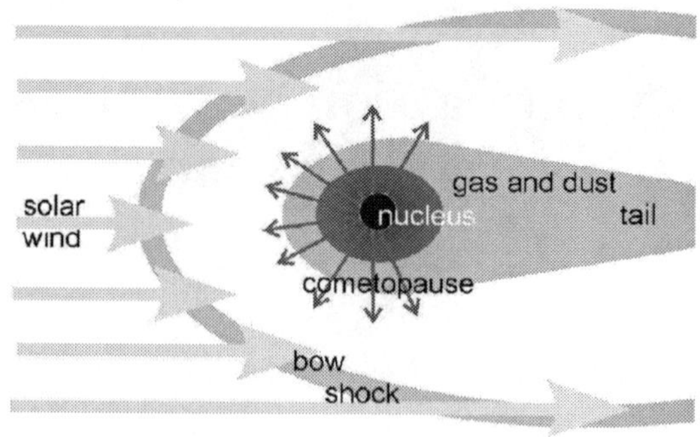

FIGURE 1. The interaction between comets and the solar wind. Gas is sublimated off the comet nucleus, spherically expanding the coma until it is destroyed by sun light. Solar wind ions penetrate from the left, gradually losing their charge state by electron capture. Figure taken from [24].

processes are a very strong source of energetic photons because mainly excited states $A^{(q-1)+}(nl)$, are populated, which subsequently decay through photon emission:

$$A^{q+} + B \rightarrow A^{(q-1)+}(nl) + B^{+}$$

and

$$A^{(q-1)+}(nl) \rightarrow A^{(q-1)+}(n'l') + h\nu.$$

This results in an emission line spectrum. Since the population of specific *nl* states depends strongly on both the collision energy and on the electron- donor species (e.g. [11-13]), the line emission provides a "fingerprint" of the underlying electron capture reactions. Cometary FUV and X-ray emission can therefore be used to probe the interaction between comets and solar wind throughout the inner solar system [14-17].

To make full use of the diagnostic qualities of comets as solar wind probes, experimental data on typical comet-solar wind collision systems are essential. Based on an extensive interaction model, we will dicuss here the most important consequences of the charge exchange model to identify prime experimental and observational needs.

ELECTRON CAPTURE EXPERIMENTS

Electron capture reactions can be studied by a large variety of experimental methods that focus either on the projectile or the neutral electron donor, each offering its own particular advantages. The experimental technique used at the KVI, Photon Emission Spectroscopy (PES) allows for the measurement of state selective cross sections by the interpretation of line emission spectra following charge exchange reactions [18].

We will demonstrate the complexity of charge exchange reactions by briefly discussing recent experimental results on collisions between fully stripped helium and water molecules (for details see [19,20]).

Helium is very abundant in cometary atmospheres, whereas water dominates the inner regions of comets. Although this collision system is clearly of great importance

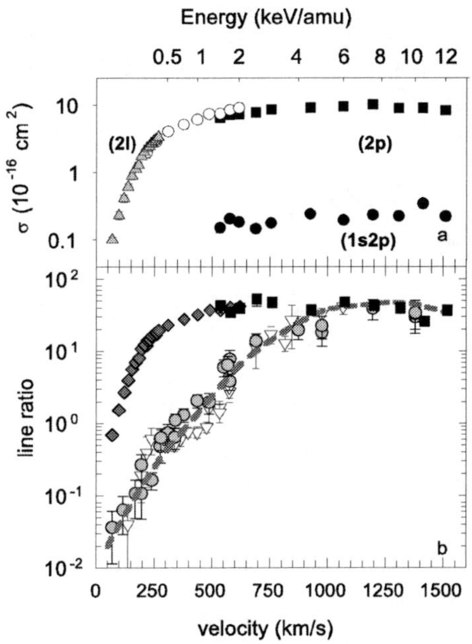

FIGURE 2. Dependency of line emission on the projectile velocity. He^{2+} colliding on H_2O. Present data for state selective cross sections for HeII*(2p)* are indicated by black squares, state selective cross sections for HeII*(2l)* by open circles and grey triangles [20]. KVI PES data for HeI*(1s2p)* are indicated by black circles. **B.** Experimental line ratios between 30.4 nm and 58.4 nm emission for He^{2+} colliding on H_2 (white triangles), CO (grey circles) [17] and H_2O (measured - black squares, estimated – grey diamonds). At low velocities, the HeI cross section is estimated by extrapolation, see text. Lines are drawn to guide the eye. Figure taken from [19].

to comet-solar wind interactions, no theoretical or studies exist for this system thus far. When fully stripped helium ions collide with water molecules, they can capture one or two electrons. These processes lead to two emission lines in the FUV, the HeII*(1s-2p)* 30.4 nm and the HeI$(1s^2$-$1s2p)$ 58.4 nm lines for single- and double electron capture, respectively.

A summary of experimental results for collisions between He^{2+} and water molecules is shown in Figure 2. Single electron capture into the $n=2$ shell is the dominant channel at high velocities, but decreases rapidly for velocities below 500 km/s. At lower velocities, single electron capture is dominated by capture into the ground state, a process that is accompanied with a considerable amount of electronic access energy, which in case of molecular targets can be absorbed and released in dissociative processes [20-23]. Electron capture into the HeI*(1s2p)* state appears to be more or less independent of the ion's velocity.

In Figure 2b, the He^{2+}-H_2O line ratio is compared to the line ratios obtained with H_2 and CO targets [17]. While the line ratios for collisions with H_2 and CO molecules show overall the same behaviour, the line ratio for collisions with water shows a rather different behaviour over the whole range of typical solar wind velocities.

As is illustrated by the example of electron capture by helium ions, the variables in charge exchange mechanisms are manifold making it a complex reaction to study, both theoretically and in the laboratory. For the interaction between highly charged ions and neutrals, in particular H_2O and atomic O, velocity dependent cross sectional data is basically lacking [24].

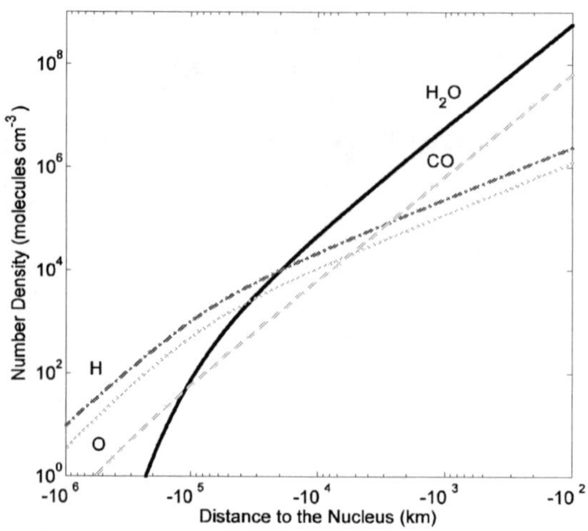

FIGURE 3. Neutral partical densities in the outflowing gas of a Halley type comet ($Q = 10^{29}$ of which 10% CO) at a distance of 1A.U. to the Sun. Beyond 20,000 km from the nucleus, the cometary atmosphere is dominated by the atomic water dissociation products H and O.

COMET CHARGE EXCHANGE MODEL

Several models are now available to simulate the observed X-ray emission resulting from the solar wind-coma interaction. These models can be roughly qualified as either macro- or microscopic. Macroscopic models study the magnetohydrodynamical interaction, using rather simplified atomic physics. Their goal is to model the global morphological characteristics of the solar wind-coma interaction and interpret the observed emission structure as parameters of the wind flux and the physical properties of the coma [9,14,25]. The resulting models reveal a very characteristic shell-shaped morphology in front of the nucleus, perpendicular to the comet-sun axis, in good agreement with the observations.

Microscopic models on the other simulate the X-ray spectrum of comets by following the excitation of the ions as it cascades to the ground state. These models invariably use detailed data on the atomic processes but simplify the environment in which these processes occur [26-28]. Their goal is to understand in detail the microscopic processes characterizing the interaction between solar wind ions and cometary neutral species and to relate the observed spectra to the physical characteristics of the Solar wind and/or the coma.

From recent observations with XMM, Chandra and FUSE it has become clear that cometary XUV emission depends on properties of both the comet (gas production rate, composition, distance to the Sun) and the solar wind (speed, composition). We briefly describe our model [17] here to identify the experimental and observational needs for the interpretation of cometary XUV emission.

We assume that a comet with a production rate Q has a spherically expanding neutral coma. This coma interacts with solar wind ions, penetrating from the Sunward side following straight line trajectories. The charge exchange processes between solar wind ions and coma neutrals is explicitly followed both in the change of the ionization state of the solar wind ions and in the relaxation cascade of the excited ions.

The density distribution of cometary molecules and their dissociation products is calculated using a standard Haser model with outstreaming gas velocity $v = 0.85/\sqrt{R}$, where R is the heliocentric distance [29]. Lifetimes of the outflowing neutrals are limited by photo dissociation or photo ionization in the solar radiation field [30]. For a Halley-like comet, the distribution of neutrals along the nucleus-Sun axis is shown in Figure 3, assuming a gas production rate of $7*10^{29}$ molecules per second, of which 10% CO and a heliocentric distance of 1.0 AU [31]. Up to a distance of more than 20,000 km from the nucleus, the atmosphere is dominated by water molecules, while the outer regions consist mainly of water dissociation products (H and O).

Next, the evolution of the charge state distribution of the solar wind ions is followed as they capture electrons from cometary neutrals. We assumed typical solar wind abundances from spaceborne mass spectroscopy [26] and combined this with velocity dependent total single electron capture cross sections from Phaneuf *et al.* [32] to calculate charge state distributions for $C^{5\text{-}6+}$, $N^{6\text{-}7+}$ and $O^{7\text{-}8+}$.

The third step is to model emission spectra of the ions involved. Cascade spectra are calculated for all the ions tracked in the. The initial population of excited states *(nl)* following charge exchange is strongly dependent on the collision velocity. For collisions between highly charged ions and cometary molecules (water, CO) only some scattered data is available [24]. Here, we have been careful to use velocity dependent, state selective electron capture cross sections for collisions between bare, hydrogen-like ions with atomic hydrogen [33-35]. Forbidden lines are included, assuming a triplet/singlet ratio of 3:1. Multiple electron capture is not included in the model.

A 3-D integration assuming cylindrical symmetry around the comet-Sun axis finally yields the absolute intensity of cometary emission lines. Effects due to the observational geometry are included in the model at this step. Effects of the Chandra instrument response are modeled by assuming a spectrometer resolution of $\sigma = 25$ eV.

RESULTS

A rough distinction in two types of wind is made: fast winds with velocities up to 1000 km/s from large, deep polar coronal holes and slow winds with velocities below 400 km/s, originating from smaller coronal holes close to the equator [36]. These two types of winds also differ in the charge distribution of their ions. We have simulated the charge exchange interaction with a Halley-like comet for a slow wind of 200 km/s and a fast wind of 600 km/s. In addition, we consider a 'decellerated' wind, i.e. a slow wind with the composition of the fast wind.

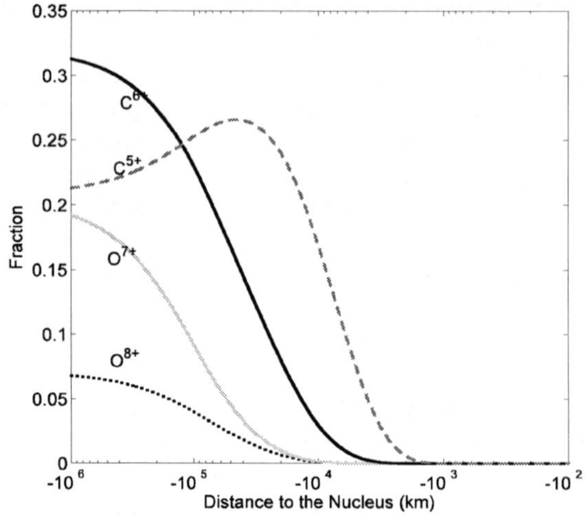

FIGURE 4. Charge state distribution of slow solar wind carbon and oxygen ions flying along the comet-Sun axis through a Halley type comet ($Q = 10^{29}$ of which 10% CO) at a distance of 1A.U. to the Sun. From this Figure, penetration depths are 55,000 km (O^{8+}), 80,000 km (O^{7+}), 30,000 km (C^{6+}) and 5,000 km (C^{5+}).

The charge state distribution of oxygen and carbon ions flying along the nucleus-Sun axis are given in Figure 4. We find penetration depths (the distance from the nucleus where $\tau_{CXE} \sim 1$) for charge exchange between 5,000 km (C^{5+}) to 80,000 km (O^{7+}) from the nucleus. The C^{5+} population behaves most interestingly, as it can be seen to increase until after 40,000 km from the nucleus, it starts to decrease. This effects is caused by the small cross section of the C^{5+} electron capture cross section; the population is fed from the C^{6+} population, but drained only slowly.

Synthetic X-ray spectra have been produced for slow, fast and a decelerated fast wind due to a Halley-like comet at 1 A.U. from the Sun. The corresponding spectra are shown in Figure 5. The strongest and most easily resolved emission is due to the OVII charge exchange line at 566 eV, the OVIII at 660 eV, and the CVI and CV charge exchange lines at 368 and 308 eV, respectively, as was observed in the spectra of Linear S4 and McNaught-Hartley [7,37]. Spectra due to a fast wind are dominated by carbon emission, regardless whether the interaction with the cometary atmosphere decelerates the wind or not.

The slow wind contains more ions of higher charged states than the fast wind, resulting in a 'harder' X-ray spectrum. The most important features of a slow wind spectra are therefore the OVII and OVIII features above 500 eV.

IMPLICATIONS FOR EXPERIMENTS AND OBSERVATIONS

As was illustrated in Figure 3, the spatial extent and elemental distribution of the cometary cloud of neutrals is determined primarily by the solar radiation field. H_2O and OH have very short lifetimes compared with CO, H and O so that, whereas the inner regions of the coma consist mainly of water vapor, the outer regions are populated with its dissociation products. At larger distances to the Sun, the water density extends further from the comet, but due to the fact that the solar heating is reduced outgassing rates are lower. Consequently densities become lower and the comet becomes both collisionally thin to charge exchange and difficult to observe. The penetration depth of solar wind ions is mainly determined by the comet's mass loss rate, its heliocentric distance and the electron capture cross section of the ion involved.

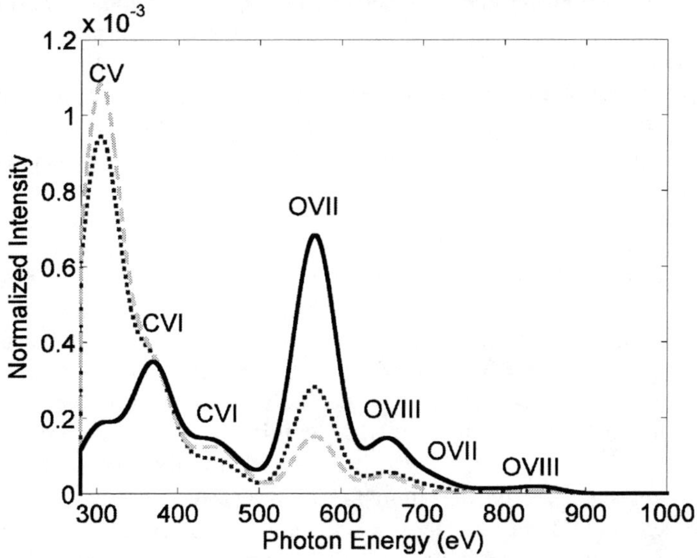

FIGURE 5. Synthetic spectrum for a Halley type comet (Q = 10^{29} of which 10% CO) at a distance of 1A.U. to the Sun. Spectra were modeled for three different types of solar wind. The solid black line indicates a spectrum due to slow wind, the dashed grey line is for fast wind and the dotted line is for a decellerated wind.

Laboratory experiments

From the previous sections, it is concluded that X-ray emission from comets is mainly driven by the interaction of fully-stripped and hydrogen-like oxygen, carbon, neon and nitrogen ions with H, O, H_2O and CO. In large comets close to the Sun, atoms will be the main electron donors for the solar wind ions, but for smaller comets the situation is more complicated as the solar wind ions reach deep into the cometary atmospheres.

The collision energies of relevance are 200-1000 km/s (or 0.2–6 keV/amu), but lower collision energies are also of relevance since the solar wind can be significantly decellerated due to the interaction with the cometary plasma [10,25,38].

In order to model cometary emission, both total- and state electron capture cross sections are required. For the line emission, *n*- and *nl*-selective cross sections are needed and the whole subsequent decay cascade needs to be included to model the Lyman or K-line series of the different ions [15]. In collisions on multi-electron targets the modelling is complicated by the fact that two- or more-electron capture followed by autoionization may add substantially to the Lyman emission and in particular to the K-α line emission [12,39].

Results

Our models predict that the consequences of the charge exchange interaction are manifold. Clear spectral and morphological effects should be detectable for different comet-solar wind encounters.

The modelling shows a clear spatial effect in large comets that could be exploited observationally. Because oxygen ions have a much larger cross section, oxygen is depleted outside the field of view, greatly enhancing the relative importance of the carbon emission features.

In the collisionally thin cometary atmospheres of comets with a mass loss rate around $Q=10^{28}$/s, the limiting factor is not the number of incoming solar wind minor ions, but the scarce neutral gas coma target molecules. Solar wind ions are able to fly through the cometary atmosphere relatively unhindered. The emission morphology will thus not show the characteristic crescent shape of the more active comets. Instead, the model source function is a diffuse cloud with a maximum around the nucleus.

Most of the comets observed thusfar have similar spectra, because theyhave similar physical conditions. In particular, the limited sensitivity urges astronomers to focus on larger comets ($Q>10^{29}$/s) within a limited distance to the Sun (<3 A.U.), and such comets that also approach the Earth within 1 A.U. are (fortunately) rare. Clearly, the launch of more sensitive X-ray telescopes will not only allow for a better spectral resolution, but will also provide us with observations of different comet-solar wind systems as no longer the rare occasion of a very large comet has to occur.

CONCLUSIONS

Cometary XUV emission offers a new window on the complex interaction between comets and the solar wind, because charge exchange reactions of highly charged ions are extremely sensitive to velocity and electron donor species. To make full use of this potentially powerful diagnostics, an interdisciplinary effort is therefore called for to turn the next generation cometary XUV observations into valuable, remote diagnostics for studying solar system plasmas.

ACKNOWLEDGMENTS

This work is supported within the framework of the FOM-EURATOM association agreement. The authors thank C. Lisse and S. Wolk for many fruitful discussions.

REFERENCES

1. Lisse, C. M., Dennerl, K., Englhauser, J., Harden, M., Marshall, F. E., Mumma, M. J., Petre, R., Pye, J. P., Ricketts, M. J., Schmitt, J., Trumper, J., West, R. G., *Science* **5285**, pp. 205 (1996)
2. Dennerl, K., Astron.Astroph. 394, pp.1119-1128 (2002)
3. McCammon, D., et al., *Ap.J.* **576**, pp. 188-203 (2002)
4. Pepino, R., Kharchenko, V., Dalgarno, A., Lallement, R., *Ap.J.* **617**, pp. 1347-1352 (2004)
5. Krasnopolsky, Vladimir A., Greenwood, Jason B., Stancil, Phillip C., *SSRv* **113**, pp. 271-374 (2004).
6. Lisse, C.M., Cravens, T.E., Dennerl, K., "X-ray and Extreme Ultraviolet Emission from Comets" in *Comets II*, edited by Festou, M.C. et al., University of Arizona Press (in press)
7. Lisse, C. M., Christian, D. J., Dennerl, K., Meech, K. J., Petre, R., Weaver, H. A., Wolk, S. J., *Science* **292**, pp. 1343 (2001)
8. Cravens, T.E., Geophys. Res. Lett. **24**, pp. 105 (1997)
9. Häberli, R. M., Gombosi, T. I., DeZeeuw, D. L., Combi, M. R., Powell, K. G. *Science* **276**, pp. 939-942 (1997)
10. Cravens, T.E., *Science* **296**, pp. 1042 (2002)
11. Janev, R. K. and Winter, H. P., *Phys. Rep.* **117**, pp. 265 (1985)
12. Hoekstra, R., Ciric, D., de Heer, F. J. and Morgenstern, R., *Phys.Scr.* **28**, pp. T81 (1989).
13. Fritsch, W. and Lin, C. D., *Phys. Rep.* **202**, pp. 1 (1991).
14. Wegmann, R., Dennerl, K., Lisse, C.M., *Astron. Astrophys*. **428**, pp. 647 - 661 (2004)
15. Kharchenko, V., Rigazio, M., Dalgarno, A., Krasnopolsky, V. A., *Ap.J.* **585**, pp. L73-L75 (2003)
16. Beiersdorfer, P., Boyce, K. R., Brown, G. V., Chen, H., Kahn, S. M., Kelley, R. L., May, M., Olson, R. E., Porter, F. S., Stahle, C. K., Tillotson, W. A., *Science* **300**, pp. 1558-1560 (2003)
17. Bodewits, D., Juhász, Z., Hoekstra, R., Tielens, A. G. G. M., *Ap.J.* **606**, pp. L81-L84 (2004)
18. Lubinski, G., Juhász, Z., Morgenstern, R., Hoekstra, R., *Phys.Rev.Lett.* **86**, pp. 616-619 (2001)
19. Bodewits, D., Tielens, A.G.G.M., Morgenstern, R., Hoekstra, R., *NIMB* (in press)
20. Seredyuk, B., McCullough, R.W., Tawara, H., Gilbody, H.B., Bodewits, D. Hoekstra, R., Tielens, A.G.G.M., Sobocinski, P., Pesic, D., Hellhammer, R., Sulik, B., Stolterfoht, N, Abu-Haija, O., Kamber, E.Y., *Phys. Rev. A* (in press)
21. Hodgkinson, J. M., McLaughlin, T. K., McCullough, R. W., Geddes, J., Gilbody, H. B., *J.Phys.B* **28** pp. L393-L397 (1995).
22. Hoekstra, R., De Heer, F.J., and Morgenstern, R., *J. Phys. B* **24,** pp. 4025 (1991)
23. Kearns, D. M., Gillen, D. R., Voulot, D., Greenwood, J. B., McCullough, R. W., Gilbody, H. B., *J.Phys.B* **34,** pp. 3401-3406 (2001)
24. Bodewits, D., McCullough, R.W., Tielens, A.G.G.M., Hoekstra, R., *Phys.Scr.C* **70** pp. 17 (2004)
25. Wegmann, R., Schmidt, H. U., Lisse, C. M., Dennerl, K., Englhauser, J., *P&SS* **46**, pp. 603-612 (1998)
26. Schwadron, N. A., Cravens, T. E., *Ap.J.* **544**, pp. 558-566 (2000)
27. Kharchenko, V., Dalgarno, A., *J. Geoph.Res.* **105**, pp. 18351 (2000)
28. Kharchenko, V., Dalgarno, A., J., *Ap.J.* **554**, pp. L99-L102 (2001)
29. Ip, W.-H, *Ap.J.* **343**, pp. 946 (1989)
30. Huebner, W. F., Keady, J. J., Lyon, S. P., *Ap&SS* **195**, pp. 1-294 (1992)
31. Fuselier, S. A., Shelley, E. G., Goldstein, B. E., Goldstein, R., Neugebauer, M., Ip, W.-H., Balsiger, H., Reme, H., *Ap.J.* **379** pp. 734-740 (1991)
32. Phaneuf, R.A., Janev, R.K., Pindzola, M.S., *"Collisions of Carbon and Oxygen Ions with Electrons, H, H_2 and He"*, Oak Ridge, Oak Ridge National Laboratory (1987)
33. Fritsch, W., and Lin, C. D., *Phys. Rev. A* **29**, pp.3039-3051 (1984)
34. Shipsey, E. J., Green, T. A., Browne, J. C., *Phys. Rev. A* **27**, pp.821-832 (1983)
35. Green, T. A., Shipsey, E. J., Browne, J. C., *Phys. Rev. A* **25**, pp.1364-1373 (1982)
36. Neugebauer, M., Forsyth, R. J., Galvin, A. B., Harvey, K. L., Hoeksema, J. T., Lazarus, A. J., Lepping, R. P., Linker, J. A., Mikic, Z., Steinberg, J. T., von Steiger, R., Wang, Y.-M., Wimmer-Schweingruber, R. F., *J. Geoph.Res.* **103** A7, pp. 14587-14600 (1998)
37. Krasnopolsky, V. A., Christian, D. J., Kharchenko, V., Dalgarno, A., Wolk, S. J., Lisse, C. M., Stern, S. A, *Icarus* **160**, pp. 437-447 (2002)
38. Goldstein, B. E., Neugebauer, M., Balsiger, H., Drake, J., Fuselier, S. A., Goldstein, R., Ip, W. H., Rettenmund, U., Rosenbauer, H., Schwenn, R., Shelley, E. G., *Astron.Astroph.* **187,** pp. 174 (1987)
39. Hasan, A. A., Eissa, F., Ali, R., Schultz, D. R. and Stancil, P. C., *Ap.J.* **560**, pp. L201 (2001).

Charge-Exchange Mechanism of X-ray Emission

V. Kharchenko

ITAMP, Harvard-Smithsonian Center for Astrophysics, Cambridge, MA 02138

Abstract. The charge-exchange mechanism of X-ray emissions induced by the solar wind ions from cometary and planetary atmospheres and interstellar gas is discussed. The X-ray photon spectra of individual ions, such as highly charged O^{q+}, C^{q+}, N^{q+}, Ne^{q+} and other ions, are described in detail. Cascading spectra of X-ray and EUV emissions of heavy solar wind ions are calculated for different ion velocities and compared with the results of laboratory measurements. Relative intensities of different spectral lines of the charge-exchange X-ray/EUV emission are used for the remote diagnostic of the compositions and velocities of the solar wind plasma. Synthetic X-ray spectra, computed for the slow and fast solar winds, are applied for the analysis of observed cometary and planetary X-rays and predictions of X-ray spectra from interstellar gas. The contribution of charge-transfer emission to the X-ray diffuse background is discussed. Brightness maps of the charge-exchange source of X-ray in the heliosphere are constructed taking into account the anisotropy of the solar wind plasma and a realistic distribution of the H and He gas.

Keywords: cometary and heliospheric X-rays, highly charged solar wind ions, charge-exchange collisions
PACS: 95.30.Dr, 96.50.Ci, 96.50.Gn

INTRODUCTION

Spectra of X-ray emission induced by highly charged ions in collisions with neutral atoms and molecules in laboratory plasma experiments have been investigated for decades [1, 2]. Unique information on the temperature, density, and plasma composition has been inferred from detected intensities of different UV, EUV, and X-ray spectral lines. Discovery and observations of X-ray emission from the comets [3, 4, 5, 6, 7], and the successful explanation of the features of cometary X-rays by Charge-Exchange (CX) collisions between the Solar Wind (SW) ions and cometary gas [8, 9, 10] have stimulated the interest of astrophysical community in the CX mechanism of X-ray emission. Since the first detection of the cometary X-rays [3], more than twenty comets have been identified as bright X-ray sources inside the heliosphere [11, 12], and cometary X-rays have been already used as a probe for a diagnostic of solar wind compositions [13, 14, 15, 16, 17, 18].

Very soon after the discovery of cometary X-rays, a new idea on the origin of the local diffuse X-ray background was formulated: The soft variable part of X-ray background is produced in charge-exchange collisions between heavy solar wind ions and interstellar gas following into the heliosphere [19, 20]. Observations with the Chandra and XMM X-ray telescopes and analysis of ROSAT archive data on the X-ray background emission have shown strong correlations between the solar wind flux and intensities of the soft X-ray background with photon energies below 1 keV [21, 22, 23, 24, 25]. Local background X-rays are also induced in CX collisions of heavy SW ions with neutral atoms from the geocorona, and at least half of X-ray background intensity may be at-

CP774, *X-Ray Diagnostics of Astrophysical Plasmas,* edited by Randall K. Smith

tributed to the charge-exchange mechanism. Spectra of the geocoronal CX X-rays have been detected recently in the "dark Moon" observations done with the Chandra X-ray telescope [26].

Planetary atmospheres interacting with the solar wind produce X-rays and EUV emissions, but total X-ray fluxes from planets are formed mostly by scattering and fluorescence of solar X-ray photons. The CX emission can be distinguished from solar X-rays because of its specific spectral and spatial distribution. Direct observation of CX X-rays from the Martian atmosphere [27] and the Earth [26] have been done with the Chandra X-ray telescope.

The specific role of the CX mechanism has been suggested as an explanation of the high intensities of X-ray fluxes observed from the polar regions of the Jovian atmosphere: The energetic oxygen and sulfur ions, precipitating into the atmosphere, lose and capture electrons in stripping and CX collisions with the atmospheric H and He atoms and H_2 molecules [28, 29, 30, 31]. Initial charges of precipitating ions are low, but their energies may be very high, up to 10-20 MeV [32]. As these ions slow down by collisions, each ion radiates a very large number of X-ray photons. This provides a high radiative yield per precipitating ion and so, a high intensity of X-ray emission. The scenario of X-ray production in the Jovian polar regions is different from the heliospheric and cometary X-ray mechanism, where a single heavy SW ion induces a few photons.

X-RAY SPECTRA INDUCED BY INDIVIDUAL SW IONS

The solar wind, which is a flux of plasma moving with velocities of 100 - 1000 km/s, consists of electrons e , protons H^+, alpha particles He^{2+}, and a small fraction of heavy highly charged ions Z^{q+}. The kinetic, thermal, and magnetic features of the SW plasma are completely governed by e and H^+, but only heavy ions Z^{q+}($C^{5,6+}$, $N^{6,7+}$, $O^{6,7,8+}$, $Ne^{7,8,9+}$...) are capable of producing EUV and X-ray photons. The mechanism of X-ray emission in the CX collisions involves two steps:

- the selective population of excited electronic states of highly charged ions $Z^{*(q-1)+}$;
- the radiative cascade from populated excited states to the ground state.

The selective population of excited ion states depends on the collision energy and the electronic structure of the neutral atoms or molecules. Populations of excited states with different principal, orbital, and spin quantum numbers $\{n, L, S\}$ in collisions with highly charged SW ions

$$Z^{q+} + A \rightarrow Z^{*(q-1)+}(n, L, S) + A^+ \tag{1}$$

have been extensively studied in laboratory plasma and ion-beam experiments. The majority of highly charged SW ions capture electrons in the set of states with principal quantum numbers n = 3 -6 depending on the collisional energy and the nature of the neutral target atoms [1, 2, 33, 34]. The captured electron relaxes down to the ion ground state radiating several photons. The spectra of emitted photons depends upon the initial population of excited states produced in the collision. Cascading spectra are shown in Fig.1a for O^{*q+} (q= 4-7) ions excited in CX collisions of $O^{(q+1)+}$ with water molecules

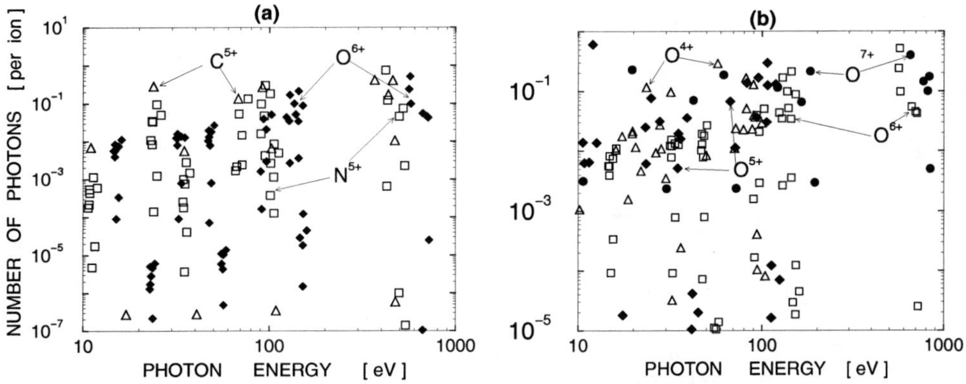

FIGURE 1. Spectra of individual solar wind ions Z^{q+} induced in a single CX collision with H_2O molecules. The numbers of cascading photons per single Z^{q+} ion are shown [a] for C^{5+} (triangles), N^{5+} (squares), and O^{6+} (diamonds) ions and [b] for different charge states of excited oxygen ions O^{4+} (triangles), O^{5+} (diamonds), O^{6+} (squares), and O^{7+} (solid circles).

H_2O. Spectra have been calculated for a collisional velocity of 400 km/s, corresponding to the average velocity of the slow solar wind. The cascading emission spectra are shown in Fig.1b for important ions of different heavy elements of the SW plasma. The cascade process produces simultaneously UV, EUV, and X-ray photons. Soft photons, which are emitted in the transition between ion excited states, provide unique information on state populations in CX collisions. Measurements of cascading UV and EUV photons allow to determine intensities of X-ray photons emitted in the same collisions.

Populations of ion states in CX collisions at different velocities have been comprehensively studied in laboratory experiments. There are two major methods of investigation: detections of the cascading photons and measurements of the energy losses in CX collisions. Both methods provide reliable information on state selective cross sections of electron-capture collisions, and these laboratory data are used for modeling cometary and heliospheric X-rays induced in the charge-exchange collisions. In Fig.2a, the theoretical X-ray emission spectra of hydrogenic Ne^{*9+} ions (solid curve) excited in CX collisions are compared with the data of the recent laboratory measurements (circles) [35, 36]. The theoretical and experimental spectra are in good agreement. The thin curves in Fig.2a are theoretical predictions for the high resolution spectra, and the insert figure shows the intensity of soft cascading photons with energies below 400 eV. X-ray spectra of O^{*7+} ions detected in the ion-beam O^{*8+} + He collisional experiments [34] by *Greenwood et al.* (2001) are compared in Fig.2b with theoretical spectra computed without any adjustable parameters. The predicted spectra of soft cascading photons with energies below 200 eV are shown by the thin solid and dashed curves. Simultaneous detection of the soft and hard X-ray photons of cascading CX spectra could provide a new accurate method for diagnosing laboratory and astrophysical plasmas.

The majority of heavy solar wind ions have a complicated level structure, reflecting the specific angular and spin quantum numbers of ion core electrons. CX emission spec-

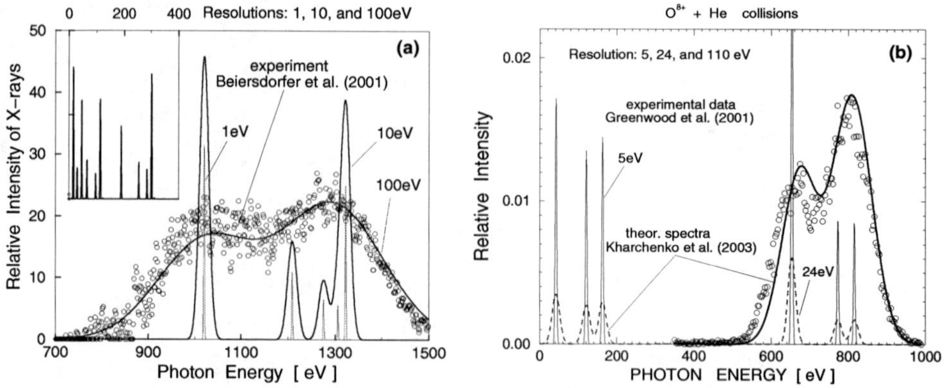

FIGURE 2. Soft and hard cascading photons in CX emission: (**a**) Theoretical (solid curve) and experimental (circles) spectra of EUV and X-rays induced in Ne^{10+} + Ne collisions [35, 36]. The predicted soft cascading photons are shown in the inserted figure. (**b**) Emission spectra of O^{*7+} ions excited in O^{8+} + He collisions. Theoretical spectra are computed for the photon energy resolutions of 5, 24, and 110 eV.

tra are sensitive to the selection rules for radiative cascading in multi-electron ions. In Fig.3 the experimental spectra of oxygen helium-like ions O^{*6+} induced in O^{7+} + He collisions [34] are compared with the results of our theoretical evaluations. Excited O^{*6+} ions are formed in the singlet (S=0) or triplet (S=1) electronic states. The singlet states generate a set of hard X-ray photons in allowed transitions, similar to transitions in the hydrogenic ions. The resonance $2^1P \rightarrow 1^1S$ transitions with a photon energy of 574 eV produce the brightest line of the singlet spectra. The triplet states decay to the ground state only via spin-forbidden transitions from the metastable state $2^3S \rightarrow 1^1S$ (561 eV) or intercombination transitions $2^3P \rightarrow 1^1S$ (568 eV) . There are no reliable experimental and theoretical data on relative populations of the singlet and triplet states in CX collisions. The relative intensities of 561, 568, and 574 eV lines can be calculated using an adjustable parameter - the ratio R of the triplet to singlet state populations. The laboratory spectra in Fig.3a include only allowed and intercombination transitions. The emission photons from the metastable state can not be detected in ion-beam experiments due to a long life-time $\sim$ 0.001s of $O^{*6+}(2^3S)$ metastable ions. Metastable ions with a velocity of 400 km/s move about 400 m without any emission and decay outside the detection area. In the heliospheric and cometary environment a distance of 400m is always available for an uninterrupted emission process and the 561 eV line should be a brightest feature of O^{*6+} X-ray spectra. Observations of cometary X-rays support this prediction [7, 12]. For comparison with the results of beam experiments, the theoretical spectra of O^{*6+} emission for R = 1.3, 3, and 5.8 are plotted in Fig.3a without the contribution of 561 eV line. The best fitting R= 5.8 value is about two times larger than statistical one. The ratio of intensities $\eta = [I(2^3S) + I(2^3P)]/I(2^1P)$ has been used for the determination of temperatures in the hot plasma model [37]. The CX mechanism predicts η-values to be much larger than η for the hot plasma. This difference may be used to test various mechanisms of X-ray emission from astrophysical plasmas.

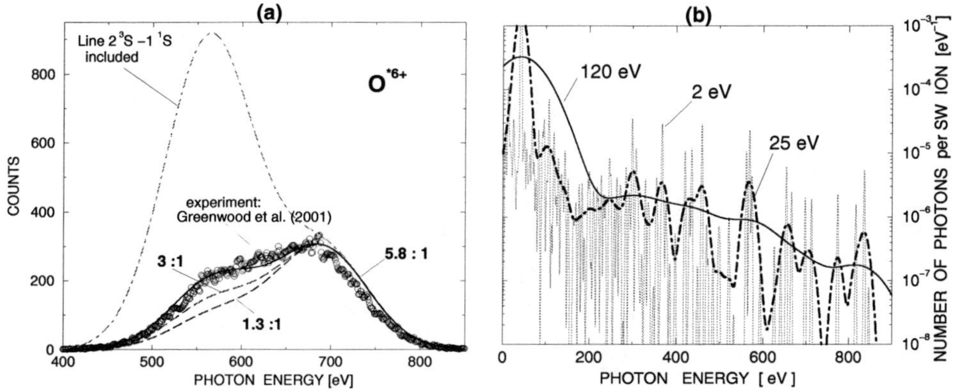

FIGURE 3. **(a)** The ion-beam experimental spectra of O^{*6+} X-ray emission (circles) are shown together with theoretical curves for different ratios R= 1.3, 3, and 5.8 of the triplet/singlet populations in O^{7+} + He collisions. The upper dot-dashed curve is the CX spectra including transitions from the metastable state. **(b)** The computed total EUV and X-ray spectra induced by the slow SW ions are shown for the different photon energy resolutions of 2, 25, and 120eV.

SPECTRA OF COMETARY EUV AND X-RAY EMISSION

Theoretical and experimental data on CX emission spectra of individual SW ions can be applied to the analysis of detected spectra of cometary X-rays. The volume density of the photon emission $N_{ph}(\hbar\omega)$ [cm^{-3} s^{-1} eV^{-1}] depends on the local density of heavy solar wind ions n_i $(i \equiv C^{q+}, N^{q+}, O^{q+}, Ne^{q+}, Fe^{q+}, Mg^{q+}...)$, the density of the neutral gas atoms and molecules ρ_k $(k \equiv H_2O, H, O, OH...)$, the collisional velocities associated with the SW ion velocities v_i, the cross sections of CX collisions $\sigma_{i,k}(v_i)$, and the photon yield function $P_{i,k}(\hbar\omega)$

$$N_{ph}(\hbar\omega) = \sum_{i,k} n_i \rho_k v_i \sigma_{i,k}(v_i) P_{i,k}(\hbar\omega), \tag{2}$$

where $P_{i,k}$ is the probability of inducing a selected photon $\hbar\omega$ from the "i"-th excited ion in a CX collision. $P_{i,k}$ are functions of collisional velocities, and these dependencies have been derived from CX emission spectra of individual ions, discussed in the previous section. The total spectrum $N_{ph}(\hbar\omega)$ of the cometary X-ray emission induced by the slow solar wind from a cometary atmosphere is shown in Fig.3b. A large number of individual lines can be seen at the photon energy resolution of 2 eV, but the spectrum at a low resolution of 120 eV is represented by the smooth curve, which resembles electron-bremsstrahlung emission. Spectra of cometary X-rays at the intermediate resolution of 25 eV include sharp features composed of several spectral lines. These features could provide valuable information on heavy ion abundances, if accurate data on cascading CX spectra could be applied for analysis [15, 16].

Spectra of cometary X-rays depend on the composition and velocity of the SW plasma. The fast and slow solar winds are described by different velocities and compo-

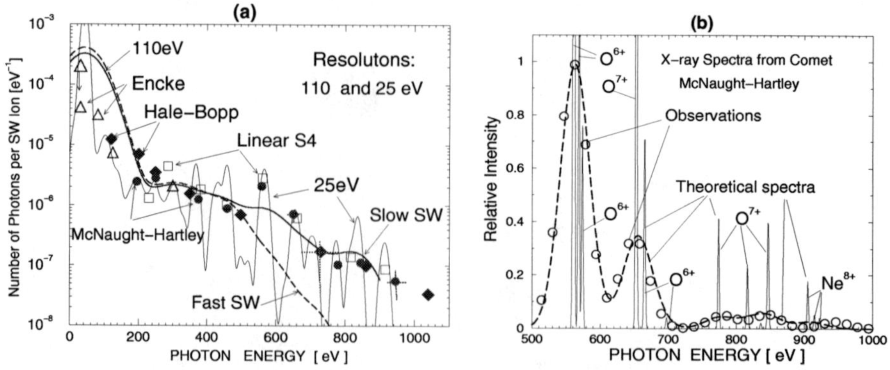

FIGURE 4. **(a)** Comparisons between the theoretical spectra of the CX mechanism and observational data on cometary X-rays. Data on EUV and X-ray spectra, observed from different comets, are: circles - McNaught-Hartley [15, 12]; squares -Linear S4 [7, 18, 12]; diamonds - Hale-Bopp [38, 39]; triangles -Encke [40]. Theoretical spectra have been calculated for the slow solar wind (solid curve) and the fast solar wind (dashed curve) for a photon energy resolution of 110 eV. The thin curve is the slow SW spectra with resolution of 25 eV. **(b)** Determination of the SW composition from the cometary X-ray emission spectra. The computed X-ray spectra (dashed curve) are compared with the X-ray spectra of Comet McNaught-Hartley (circles) detected with the Chandra X-ray telescope [15, 16]. The relative charge abundances of the SW ions $[O^{7+}]$: $[O^{8+}]$: $[Ne^{9+}]$ = 1: 0.34 : 0.02 provide the best agreement between theoretical spectra and observations.

sitions, and the CX spectra induced by these winds should be different. Cometary EUV and X-ray emission spectra computed for the typical compositions of the fast (dashed curve) and slow (solid thick curve) solar winds are compared in Fig.4a with spectra of cometary X-rays detected with different X-ray telescopes from different comets at different times. The observed spectra are well described by emission from the slow SW ions. This is consistent with observational conditions: the majority of comets has been detected near the solar ecliptic plane, where the SW plasma is presented by the slow solar wind. Spectra from the fast and slow solar winds are different at the photon energies above 500 eV. The slow solar wind produces more energetic photons because it has more abundant fractions of highly charged ions, such as O^{7+} and O^{8+}. These theoretical spectra do not involve any adjustable parameters and variations of the SW ion charge and elemental composition could provide a better agreement between theory and observations.

Cometary X-ray emission have been already used as a probe for a diagnostic of the solar wind composition. In Fig.4b, the X-ray emission spectra from Comet McNaught-Hartley with photon energies above 500 eV are compared with theoretical spectra computed for a variable composition of the SW heavy ions. We found abundances of oxygen and neon ions $[O^{7+}]$: $[O^{8+}]$: $[Ne^{9+}]$ = 1: 0.34 : 0.02 provide the best agreement between theoretical and observational spectra [16]. This illustrates a diagnostic capacity of the X-ray emission induced in charge-exchange collisions.

The spectra of CX emission provide a new opportunity to investigate the interaction

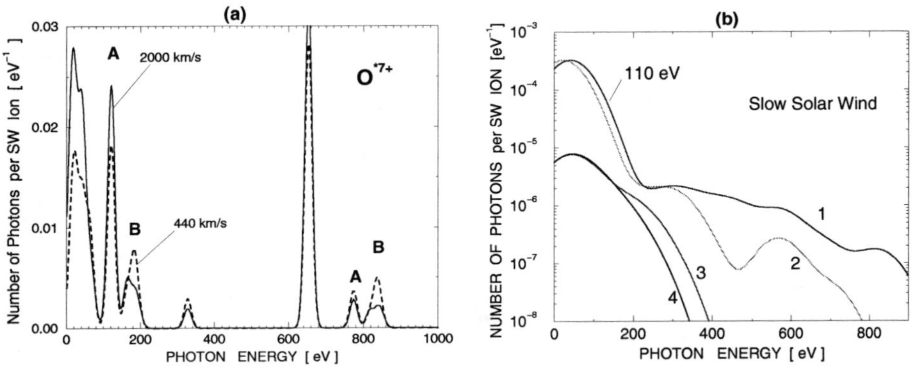

FIGURE 5. (a) CX emission spectra of O^{*7+} ions at velocities of 440 km/s (dashed curve) and 2000 km/s (solid curve). The features **A** and **B** show the strong velocity dependence of the intensity ratio N^A_{ph}/N^B_{ph}. (b) CX emission spectra induced by the slow SW from atmospheric regions with different collisional depth ξ, with a photon energy resolution of 110 eV. Numbers near curves indicate the ξ-values for CX collisions: ξ = 1,2,3, and 4 collisions.

between the SW plasma and planetary/cometary atmospheres. The sensitivity of the CX emission spectra to the velocity of individual ions could be used to determine the SW plasma velocities. In Fig.5a cascading spectra of CX emission of O^{*7+} ions are shown for the SW velocity of 440 and 2000 km/s. The higher velocity corresponds to the plasma parameter in the coronal mass ejections, and the CX spectrum for this velocity is shown by the solid curve. The intensity ratio N^A_{ph}/N^B_{ph} of the different spectral features **A** and **B** depends on ion velocity, and its calibrated value could be used for a velocity determination.

Charge composition of heavy SW ions can be changed in sequential collisions with atoms or molecules, if a neutral gas is collisionally thick. The spectra of CX emission induced from atmospheric regions with different collisional depths are shown in Fig.5b. The numbers along the spectral curves indicate collisional depth.

X-RAY EMISSION FROM HELIOSPHERIC GAS

Interstellar gas, mostly H and He, penetrates deeply inside the solar system. Heliospheric distributions of neutral atoms are governed by the photoionization and gravitational attraction of the Sun. Spatial distributions of the neutral H gas have been investigated in observations of the Ly-α absorption and have been modeled in plasma-hydrodynamic calculations [24, 42]. The homogeneous mixture of H and He in the interstellar gas is spatially separated inside the heliosphere. The local maximum of He density in downwind direction, the helium focusing cone [41], provides a remarkable opportunity to investigate independently the emission from the heliospheric He and H gases. CX collisions of the heavy SW ions with the neutral H and He produce X-rays, which are considered as a local component of the diffuse X-ray background [21, 22, 23, 24, 25].

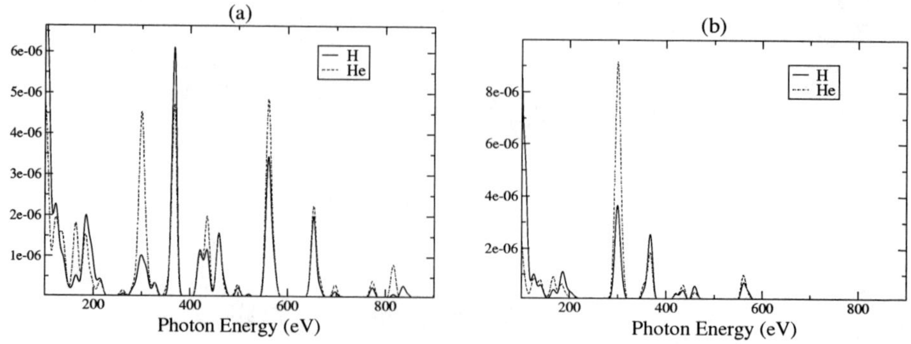

FIGURE 6. **(a)** Spectra of X-ray emission induced by the slow solar wind from the heliospheric H (solid curve) and He (dashed curve) gases. The spectra are normalized to unity of the solar wind flux. **(b)** X-ray spectra induced from the interstellar H and He gas by the fast solar wind.

In Fig.6a the CX X-ray emission spectrum induced by the slow SW from the hydrogen and helium gases are shown. This spectrum is normalized per single SW ion, including protons and alpha particles, or per unity of the SW flux. The spectrum induced by the fast SW is depicted in Fig.6b. The differences between the spectra of the fast and slow winds arises due to ion velocities and compositions. The slow solar wind generates a larger number of energetic photons because of the more abundant fraction of ions in highest charge states [43]. The anisotropy of the solar wind plasma could be observed by detecting spectra of CX heliospheric emission in different directions. Observations of the background X-rays near the ecliptic heliospheric regions should yield CX spectra of the slow solar wind (Fig.6a) and observations in directions of the heliospheric poles should yield mostly X-ray spectra of the fast wind (Fig.6b).

Investigations of the soft X-ray background have been carried out with different X-ray satellites [19, 20, 21, 22, 23, 24, 25, 26]. The contribution of the local CX X-ray sources has been evaluated as 50% of the total intensities of X-rays photons between 0.2 - 1 keV. Directional intensities of the CX emission obtained in line of site observations with X-ray telescopes are sensitive to the morphology of X-ray sources produced by the CX mechanism. In Fig.7 the distribution of the X-ray power density of CX emission $W[\text{ eV}^{-1}\text{ cm}^{-3}\text{ s}^{-1}]$ induced in the interaction between the solar wind and heliospheric hydrogen (Fig.7a) and helium (Fig.7b) gas are shown in the ecliptic plane. Distances are measured in AU units with the Sun at the origin. Interstellar gas is traveling from the left to the right at an angle of 6° with respect to the ecliptic plane. The crescent shape in Fig.7a is similar to the X-ray image observed from comets. The morphology of CX X-rays induced from He is different: The helium focusing cone is clearly seen in the downside of the Sun. The power density of CX X-rays, generated from both hydrogen and helium in the ecliptic plane, is presented in Fig.8a. The anisotropy of CX emission sources is illustrated in Fig.8b, where the map of the power density is shown in the plane defined by the heliospheric polar axis and interstellar wind direction. The contrast between the emission from the regions of the fast and slow solar winds is seen in this figure.

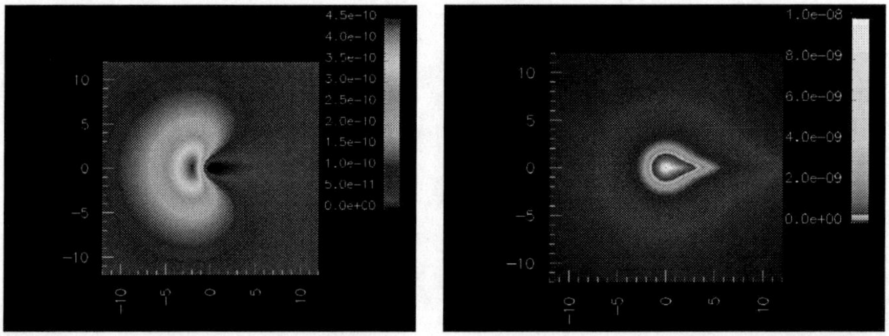

FIGURE 7. (**a**) X-ray power density [eV cm^{-3} s^{-1}] generated by SW ions from the H interstellar gas in the ecliptic plane . (**b**) Power density generated by SW ions from the He gas.

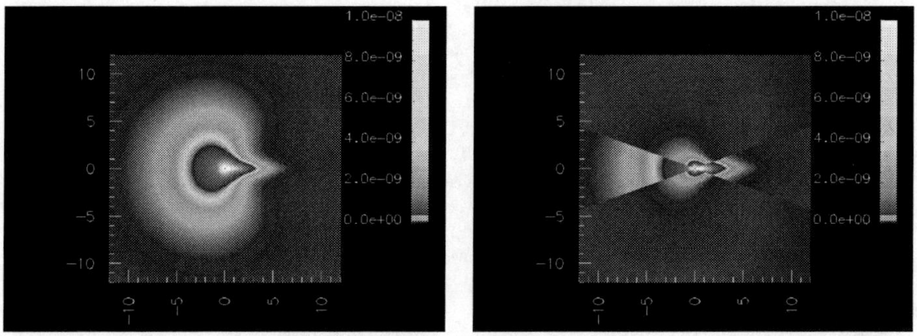

FIGURE 8. (**a**) Total power density of X-ray CX emission from H and He in the ecliptic plane. (**b**) X-ray power density in the plane defined by the heliospheric polar axis and the direction of the incoming interstellar gas .

CONCLUSIONS

Cascading photons induced by heavy SW ions in CX collisions with neutral atoms and molecules describe well all properties of observed cometary X-ray emission. Cometary X-ray emissions have been used as a probe of the SW ion compositions.

The CX mechanism provides about 50% of the soft X-ray background below 1 keV. X-ray emissions induced in CX collisions with H and He heliospheric gases have different spectra and may be distinguished by their different spatial distributions. The helium focusing cone should be a bright source of the charge-exchange emission with the predicted spectra of EUV and X-ray photons. Spectra and intensities of the heliospheric CX X-rays provide information on the parameters of the SW plasma and may be used for a remote diagnostic of the solar wind.

The data on the spectra and morphology of the heliospheric X-rays can be used for analysis of X-rays induced by stellar winds from astrospheres [44].

ACKNOWLEDGMENTS

I am grateful to A. Dalgarno and all my collaborators in the projects on the charge-exchange mechanism of X-rays. This work has been supported by the NASA grants NNG04GD57G and NAG5-13331.

REFERENCES

1. R. K. Janev & H. Winter, *Phys. Rep.*, **117**, 265 (1985)
2. Janev, R. K. (ed.), *Atomic and molecular processes in fusion edge plasmas*, Plenum, New York, 1995.
3. C. M. Lisse et al., *Science*, **274**, 205 (1996)
4. K. Dennerl, J. Englhauser, & J. Trümper, *Science*, **277**, 1625 (1997)
5. M. J. Mumma, V. A. Krasnopolsky, and M. J. Abbott, *ApJ*, **491**, L125 (1997)
6. V. A. Krasnopolsky & M. J. Mumma, 2001, *ApJ*, **549**, 629 (2001)
7. C. M. Lisse et al., *Science*, **292**, 1343 (2001)
8. T. E. Cravens, Geophys. Res. Lett., **24**, 105 (1997)
9. R. M. Häberli et al. , *Science*, **276**, 939 (1997)
10. V. Kharchenko & A. Dalgarno, *J. Geophys. Res.*, **105**, 18351 (2000)
11. T. E. Cravens, Science, **296**, 1042 (2002)
12. V. A. Krasnopolsky, J. B. Greenwood, and P. C. Stancil, *Space Sci. Rev.*, **113**, 271 (2004)
13. N. A. Schwadron & T. E. Cravens , *ApJ*, **544**, 558 (2000)
14. V. Kharchenko and A. Dalgarno, *ApJ*, **554**, L99 (2001)
15. V. A. Krasnopolsky et al., *Icarus*, **160**, 437 (2002)
16. V. Kharchenko, M. Rigazio, A. Dalgarno, and V. A. Krasnopolsky, *ApJ*, **585**, L73 (2003)
17. P. Beiersdorfer et al., *Science*, **300**, 1558 (2003)
18. V. A. Krasnopolsky, *Icarus*, **167**, 417 (2004)
19. D. P. Cox, Modeling the local bubble, in *The Local Bubble and Beyond*, edited by D. Breitschwerdt, M. J. Freyberg, & J. Trümper, p. 121, Springer-Verlag, New York, 1998
20. T. E. Cravens, *ApJ*, **532**, L153 (2000)
21. T. E. Cravens, I. P. Robertson, & S. L. Snowden, *J. Geophys. Res.*, **106**, 24883 (2001)
22. I. P. Robertson, I. P., & T. E. Cravens, *J. Geophys. Res.*, **108**, 8031 (2003)
23. I. P. Robertson, I. P., & T. E. Cravens, *Geophys. Res. Lett.*, **30**, 8 (2003)
24. R. Lallement, *A&A*, **418**, 143 (2004)
25. S. L Snowden, M. R. Collier, and K. D. Kuntz, *ApJ*, **610**, 1182 (2004)
26. B. J. Wargelin et al., *ApJ*, **607**, 596 (2004)
27. K. Dennerl, *A&A*, **394**, 1119 (2002)
28. G. R. Gladstone et al., *Nature*, **415**, 1000 (2002)
29. T. E. Cravens et al., *J. Geophys. Res.*, **100**, 17153 (1995)
30. V. Kharchenko, W. Liu, & A. Dalgarno, *J. Geophys. Res.*, **103**, 26687 (1998)
31. G. Branduardi-Raymont et al., *A&A*, **424**, 331 (2004)
32. T. E. Cravens et al., *J. Geophys. Res.*, **108**, Art.1465 (2003)
33. J.B. Greenwood et al., *ApJ*, **533**, L175 (2000)
34. J. B. Greenwood et al. , *Phys. Rev. A*, **63**, 062707 (2001)
35. P. Beiersdorfer et al., *ApJ*, **549**, L147 (2001)
36. M. Rigazio, V. Kharchenko, & A. Dalgarno, *Phys. Rev. A.*, **66**, 064701 (2002)
37. R. K. Smith et al., *ApJ*, **556**, L91 (2001)
38. A. Owens et al., *ApJ*, **493**, L47 (1998)
39. C. M. Lisse, et al., *Earth, Moon, Planets*, **77**, 283 (1999)
40. C. M. Lisse et al. , *Icarus*, **141**, 316 (1999)
41. J. G. Michels et al., *ApJ*, **568**, 385 (2002)
42. G. P. Zank, *Space Sci. Rev.*, **89**, 413 (1999)
43. R. Pepino, V. Kharchenko, A. Dalgarno, and R. Lallement, *ApJ*, **617** , 1347 (2004)
44. B. J. Wargelin and J. J. Drake, *ApJ*, **546**, L57 (2001)

Soft X-ray and Optical Laboratory Spectra to Simulate the Solar Wind on Comets: O^{5+} + CO

T. Ehrenreich*, K. Miller*, P. Gee*, Q. Kessel*, E. Pollack*, W.W. Smith*, N. Djuric+, J. Lozano+, S.J. Smith+ and A. Chutjian+

** Physics Dept. and Institute of Materials Science, University of Connecticut, Storrs, CT 06269*
and
+ Jet Propulsion Laboratory, California Institute of Technology, Pasadena, CA 91109

Abstract. UV/soft x-ray and optical measurements of relative line emission cross sections from collisions of O^{5+} with CO vapor have been made, as part of a laboratory study of interactions between minority solar-wind ions and gases found in comets. The interaction of O^{5+} with CO yields prototypical information applicable to other highly-charged solar wind ion interactions. The data are at least approximately consistent with a simple classical Coulomb "over-the-barrier" model prediction: strong XUV emission (~13-25 nm range) is observed in collisions in the slow and fast solar wind velocity range. We attribute this light to electron capture from the target molecule by the projectile ion into an excited n=4 level, plus cascade into n=3, resulting in emission lines from $(O^{4+})^*$. The optical spectra, taken at higher velocities, show a lack of line emission from projectile states in the visible range (400-850nm); the visible lines observed can be attributed primarily to excitation of atoms from the molecular target following dissociation.

Keywords: comets, solar wind, charge-transfer collisions, electron capture
PACS: 34.10.+x, 34.70.+e, 52.20.Hv, 95.30.Dr, 95.30.Ky, 95.85.Mt, 96.50.Ek

INTRODUCTION

We describe prototypical measurements of UV/x-ray emission spectra from O^{5+} + CO beam-gas collisions, as part of a program to study interactions of highly-charged minority ions found in the solar wind with cometary molecular gases. The goal is a detailed understanding of space-based soft x-ray emission spectra seen when comets approach the sun. Although O^{5+} is not a major solar wind ion, this initial study can be extended to other solar-wind ion interactions. The XUV data were obtained at the NASA/Caltech Jet Propulsion Laboratory highly-charged ion facility.[1] The visible emission collision spectra were obtained at the University of Connecticut Van de Graaff Accelerator Laboratory. Although the data so-far are not extensive, they are generally consistent with theoretical models of XUV emission observed from the vicinity of comets as they approach the Sun, assuming the charge-transfer mechanism.[2]

---

[1] See A. Chutjian, J.B. Greenwood, and S.J. Smith, in *Applications of Accelerators in Research and Industry,* J.L. Duggan and I.L. Morgan, editors, Am. Inst. Of Physics **CP 475,** New York, 1999, pp. 881-886; see also J.B. Greenwood, et al., Phys. Rev. A **63,** 062707 (2001).
[2] T.E. Cravens, *Geophysical Research Letters* **24**, 105 (1997); see also Häberli, et al., *Science* **276**, 939 (1997).

CP774, *X-Ray Diagnostics of Astrophysical Plasmas,* edited by Randall K. Smith

DISCUSSION OF OBSERVED SPECTRA

Figure 1 below shows XUV line spectra optained in the 10-21 nm range with 2.2 keV/amu O^{5+} ion beam collisions with a CO gas target, where the ion velocity has been chosen to simulate the fast solar wind velocity ~700 m/s. Additional lines were observed just below and above this wavelength range, with provisionally identified transitions shown in the Grotrian diagram of Figure 2. The strongest lines in this figure appear to originate mainly from excited states of O^{4+} ions created when the projectile

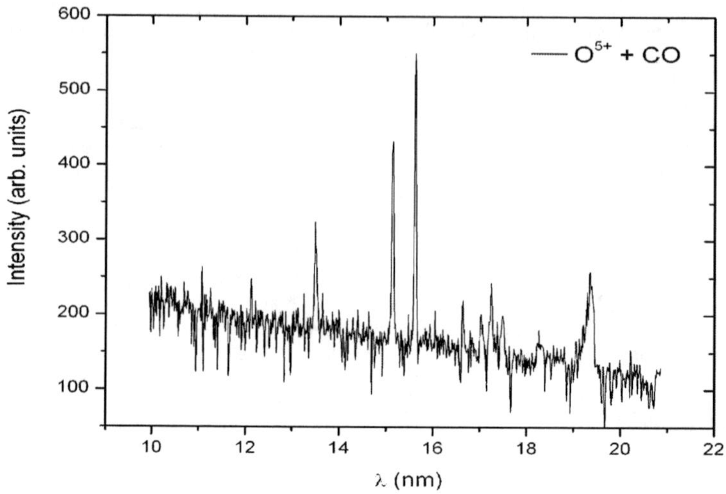

FIGURE 1. Line intensity vs. wavelength for 2.2 keV/amu O^{5+} collisions with CO in the XUV range 10-21 nm (sloping background is an artifact). Analogous spectra were obtained up to 28 nm for O^{5+}, for O^{6+} and from ~1.5 nm with O^{7+} projectiles. For probable line identifications, see Figure 2.

O^{5+} beam ions capture electrons from the molecular target gas, in this case CO. The wavelengths in the Grotrian diagram of Figure 2 below correspond to line identifications from the NIST tables[3] for O^{4+} (O V spectrum), and generally agree with the measured wavelengths in the spectra, within 1-2 times the experimental resolution of ~0.1 nm (1 Å) FWHM. The spectra were taken at JPL with a Minuteman/McPherson Model 310G 1.0 m grazing incidence vacuum monochromator equipped with an XUV CCD camera. (Calibration errors of ~0.1nm can occur.) Future work will involve extending spectra to shorter wavelengths down to ~2 nm and observing highly charged ion collisions with other gases deemed to be important in understanding the comet observations. Spectra in the visible range were also taken, with O^{3+} and O^{+} on CO at 26-128 keV/amu using the Van de Graaff: these were all remarkably similar in showing many atomic lines due to excitation of atoms from the

[3] NIST Atomic Spectra Database (Standard Reference Database 78), available online at www.nist.gov.

dissociating target gas.[4] [Only one weak visible line, at 421.3 nm, was attributed to the O V spectrum.]

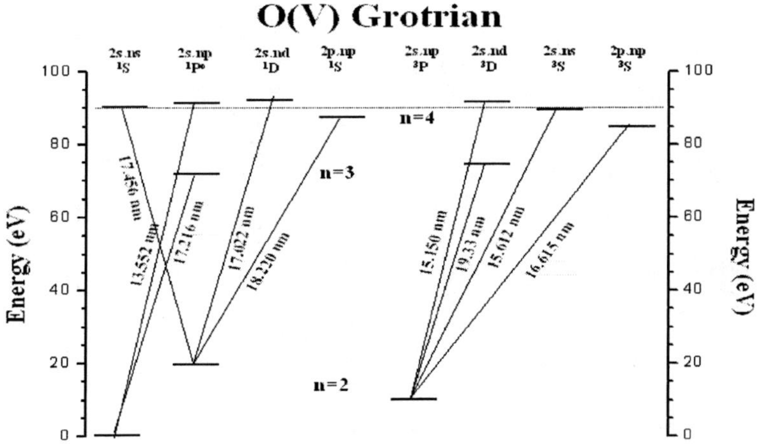

FIGURE 2. Grotrian Diagram showing probable identification of the prominent transitions observed.

INTERPRETATION

The Coulomb "over-the-barrier" model (COB) is quite useful in describing one-electron capture to a highly-charged projectile from an atomic (or, by extension, molecular) target, using an effective potential for the active electron in the combined field of the projectile ion and the target.[5] As Figure 2 shows, the initial state of the observed XUV transitions (some of them in Figure 1) involves one excited electron in O^{4+} radiating mainly from n=4 (n=3 may be populated by IR cascade). The strong lines seen are in rough agreement with the predictions of the simple COB model for electron capture from a molecular target into an excited state of the highly-charged projectile ion.[6] [7]

---

[4] Visible spectra at lower energies and higher charge states could be taken in the future to better represent solar wind-comet gas interactions. The optical spectral region appears not to be as sensitive to charge transfer processes from high charge states on molecules as the XUV region.

[5] See, R. Hoekstra and R. Morgenstern, "Electron Capture by Slow Multicharged Ions from Atoms", Chapter 14 in: *Trapping Highly Charged Ions: Fundamentals and Applications*, John Gillaspy, ed., Nova Science Publishers, Huntington, NY, 2001.

[6] These data are preliminary and further work is in progress, as well as comparisons with synthetic spectra.

[7] The research at the University of Connecticut has been sponsored by NASA EPSCOR Grant NCC5-601 and that at JPL/Caltech through agreement with NASA. We acknowledge helpful discussions with V. Kharchenko, T.E. Cravens and D. Bodewits.

VI: X-RAY DIAGNOSTICS FOR NEUTRON STARS

Atomic Data for the Atmospheres of Strongly Magnetized Neutron Stars

Matthias Klews*, Günter Wunner† and Klaus Werner*

*Institut für Astronomie uud Astrophysik. Univ. Tübingen, Sand 1, D-72076 Tübingen, Germany
†Institut für Theoretische Physik 1, Universität Stuttgart, D-70550 Stuttgart, Germany

Abstract. We have developed a new method for the fast computation of wavelengths and oscillator strengths for medium-Z atoms and ions, up to iron, at neutron star magnetic field strengths. The method is a parallelized multi-configuration Hartree-Fock approach based on finite-element and B-spline techniques. It turns out that typically 15 finite elements are sufficient to calculate energies to within a relative accuracy of 10^{-5} in 4 or 5 iteration steps using B-splines of 6th order, with parallelization speed-ups of 20 on a 26-processor machine. Results have been obtained for the energies of the ground states and excited levels and for the transition strengths of astrophysically relevant atoms and ions in the range $Z = 2 \ldots 26$ in different ionization stages.

Keywords: neutron stars, magnetic fields, atomic data, synthetic spectra
PACS: 31.15.Ne, 32.60.+i, 95.30.Ky, 97.60.Jd

Calculations of synthetic spectra [1] of strongly magnetized neutron stars are bedeviled by the lack of data for atoms in intense magnetic fields. While the behavior of hydrogen and helium has been investigated in detail (see, e.g., [2]), complete and reliable data for heavier elements, in particular iron, are still missing. Since neutron stars are formed by the collapse of the iron cores of massive stars, it may be assumed that their atmospheres contain an iron plasma. Our objective is to fill the gap and calculate as comprehensively as possible energies, wavelengths, and oscillator strengths of medium-Z atoms and ions up to $Z = 26$ in neutron star magnetic field strengths.

Obviously, the method for achieving this goal must be highly efficient since for the calculation of synthetic spectra data of many thousands or even millions of atomic transitions are required. As in previous work on the problem ([3],[4],[5],[6],[7]) we exploit the fact that a strong magnetic field results in an approximate decoupling of the dynamics of the electrons parallel and perpendicular to the field.

In the calculations presented here we restrict ourselves to the adiabatic approximation, in which the single-particle wavefunctions take the form: $u_i(\rho, \varphi, z) = \phi_{nm}(\rho, \varphi) \cdot P_{nm\nu}(z)$. Here, $\phi_{nm}(\rho, \varphi)$ are Landau wave functions, describing the motion perpendicular to the field, and the $P_{nm\nu}(z)$ are the longitudinal wave functions to be determined. The spins of the electrons are all aligned antiparallel to the magnetic field and need not be accounted for explicitly. The total N-electron wave function is constructed as a Slater determinant of the single-particle wave functions, and the unknown longitudinal wave functions are determined from the Hartree-Fock equations, which follow from inserting the total N-electron wave function into Schrödinger's variational principle for the total energy (for an explicit form of the equations see, e.g. [5], [8]).

The novel feature of our approach is to use finite-element and B-spline techniques to solve the Hartree-Fock equations for atoms in strong magnetic fields. This is accom-

CP774, *X-Ray Diagnostics of Astrophysical Plasmas,* edited by Randall K. Smith

plished through the following steps: 1) decomposition of the z-axis into finite elements with quadratically widening element borders; 2) third-order B-spline expansion of the single-particle wave functions on the individual finite elements; 3) formulation of the variational principle equivalent to the Hartree-Fock equations in terms of the expansion coefficients. This leads to a simple system of linear equations for the expansion coefficients, which is solved numerically, and, since the direct and exchange interaction potential terms depend on the wave functions, in a self-consistent way. The iteration procedure is initialized by distributing the electrons on magnetic sublevels according to the level scheme of the hydrogen atom in intense magnetic fields.

We have calculated *energies and oscillator strengths of the elements He up to Fe in intense magnetic fields*. Here, we can only present sample results (see Table 1 and Figure 1). A detailed compilation can be found in [8].

TABLE 1. Ground state energies of iron in all ionization stages from helium-like to neutral for different values of the magnetic field strength (in adiabatic approximation, magnetic field in Tesla, energies in – eV; in brackets: number of electrons not occupying a single-particle m ground state).

num. e^-	$1 \cdot 10^8$	$5 \cdot 10^8$	$1 \cdot 10^9$	$2 \cdot 10^9$	$5 \cdot 10^9$
2	16667	32163	42210	54984	77045
3	21465	41738	54967	71861	101192
4	25409	49668	65571	85943	121447
5	28897[1]	56460	74677	98067	138952
6	32117[1]	62393	82644	108697	154343
7	34909[1]	67636	89697	118122	168021
8	37362[1]	72306	95986	126539	180259
9	39558[2]	76486	101623	134090	191258
10	41620[2]	80239	106688	140884	201167
11	43449[2]	83614	111247	147003	210106
12	45075[2]	86782	115349	152516	218168
13	46521[2]	89633[1]	119038	157477	225433
14	47869[3]	92182[1]	122348	161932	231965
15	49066[3]	94454[1]	125308	165920	237820
16	50121[3]	96472[1]	127944	169476	243045
17	51047[3]	98254[1]	130278	172627	247684
18	51865[4]	99817[1]	132331	175401	251774
19	52594[4]	101175[1]	134118	177821	255349
20	53216[4]	102358[2]	135671[1]	179910	258443
21	53737[4]	103386[2]	137030[1]	181689	261084
22	54166[4]	104240[2]	138164[1]	183174	263291
23	54508[4]	104931[2]	139087[1]	184396	265126
24	54783[5]	105469[2]	139813[1]	185351	266578
25	54982[5]	105867[2]	140356[1]	186003	267686
26	55109[5]	106134[2]	140728[1]	186595	268446

To speed up the calculations, we parallelized our computations. Our parallelization strategy is: a) each processor calculates one or several electrons, depending on the total number of processors, b) single-particle wave functions are broadcast from each processor to every other processor. As the coefficient vectors in the B-spline basis are

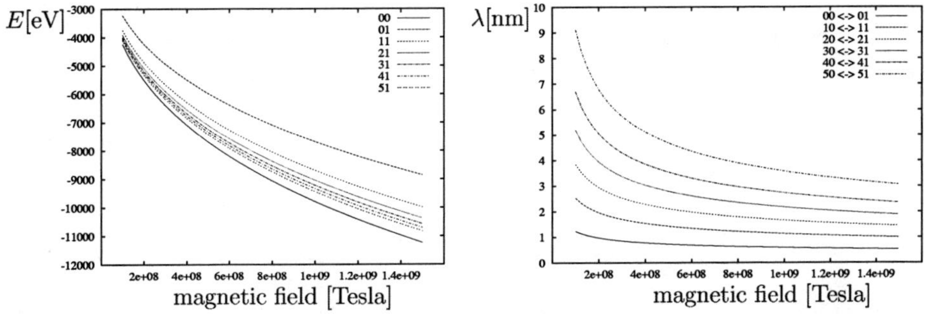

FIGURE 1. Left: Energies of the ground state $(0,0)$ and excited states of carbon as a function of the magnetic field (legend: quantum numbers $(-m, \nu)$ of the excited electron). Right: Wavelengths of transitions between the states of carbon shown in the left-hand figure as a function of the magnetic field.

small ($dim \approx 20$), there is only little communication between the nodes. Typical speed-ups by a factor of 20 were obtained on a 26-processor Athlon XP1600+ cluster (i.e., from 1200 to 60 CPU seconds in the calculation of the ground state energy of iron in a magnetic field of 5×10^8 Tesla).

We are currently working on the following problems to render a sound interpretation of observed spectra in physical terms possible: 1) inclusion of excited Landau levels to calculate electrons close to the nucleus to even higher accuracy, and to extend the range of the magnetic fields to smaller fields; 2) automatization of the selection of the initial wave functions to calculate many possible configurations in large ranges of parameters without human interference; 3) calculation of comprehensive databases for energies and oscillator strengths in strong magnetic fields; 4) calculation of synthetic spectra using the new database.

This work was supported by Deutsche Forschungsgemeinschaft, SFB 382.

REFERENCES

1. K. Werner, and S. Dreizler, "The Classical Stellar Atmosphere Problem" in *Computational Astrophysics,* edited by H. Riffert, K. Werner, *J. of Computational and Applied Mathematics*, Elsevier (1998).
2. H. Ruder, G. Wunner, H. Herold, and F. Geyer, *Atoms in Strong Magnetic Fields*, Springer-Verlag, Heidelberg, (1999).
3. P. B. Jones, *MNRAS*, **216**, 503 (1985).
4. D. Neuhauser, K. Langanke, and S. E. Koonin *Phys. Rev. A*, **33**, 2084–2086 (1986).
5. M. C. Miller, and D. Neuhauser, *MNRAS*, **253**, 107 (1991).
6. M. V. Ivanov, and P. Schmelcher, *Phys. Rev. A*, **61**, 022505-1–13 (2000).
7. K. Mori, and C. J. Hailey, *ApJ*, **564**, 914–929 (2002).
8. M. Klews, "Discretization Methods for the Investigation of Atoms in Time-dependent Electric Fields, and in Extremely Strong Magnetic Fields," *Doctoral Thesis*, University of Tübingen, (in German), (2003). `http://www.theo1.physik.uni-stuttgart.de/forschung/sfb382a15/klews_2003.ps.gz`

Surface Atomic Spectral Lines from Weakly Magnetic Rotating Neutron Stars

Sudip Bhattacharyya*, M. Coleman Miller† and Frederick K. Lamb**

*University of Maryland at College Park; sudip@astro.umd.edu
†University of Maryland at College Park; miller@astro.umd.edu
**University of Illinois at Urbana-Champaign; fkl@uiuc.edu

Abstract. The profiles of atomic spectral lines from the surfaces of neutron stars can provide important information about stellar structure and the emission geometry, which in turn may help constrain models of the high density cold matter in the cores of neutron stars. An observed surface line will be affected by several physical effects, such as Doppler shifts (due to stellar rotation), special relativistic beaming, gravitational redshifts, light-bending, and frame-dragging. We have computed theoretical models for such lines, taking all these effects into account in the context of weakly magnetized, rapidly rotating neutron stars in low-mass X-ray binaries. We find that the stellar radius-to-mass ratio can be inferred from surface line profiles to better than 5%, even for a broad and asymmetric observed line. Our results also indicate that a signature of frame-dragging may be detected from future detailed studies of the shapes of surface atomic lines.

INTRODUCTION

The recent report (Cottam, Paerels, & Méndez 2002) of atomic resonant scattering lines from the surface of the neutron star in the low mass X-ray binary (LMXB) EXO 0748−676 has stimulated new work on the physics of surface spectral lines from neutron stars (see, for example, Özel & Psaltis 2003; Loeb 2003; Bildsten, Paerels, & Chang 2003; Villarreal & Strohmayer 2004). This is because the identification of such a line provides the gravitational redshift at the neutron star surface, and hence the stellar radius-to-mass ratio, a key parameter that constrains the equation of state of dense cold matter.

Cottam et al. (2002) detected three surface lines (Fe XXV Hα, Fe XXV 2–3, and possibly O VIII Lyα) using the burst data from XMM-Newton. The corresponding surface gravitational redshift value for all three lines is ~ 0.35, which gives a dimensionless stellar radius-to-mass ratio (R/M) of 4.4. The narrowness of the observed iron lines is consistent with the slow rotation rate of the neutron star (spin frequency ~ 45 Hz; Villarreal & Strohmayer 2004), but such a narrow line provides information only about R/M, and does not help us constrain other system parameters. Therefore, while the slow stellar rotation for this particular source was an important reason behind the detection of these lines with XMM-Newton (because the rotational Doppler broadening was small, and hence the depth of the line was large enough for detection), it will be difficult to constrain other system parameters by future observation of surface lines from this source with better instruments. However, as the typical spin frequency of an accreting neutron star in an LMXB is $200-600$ Hz (Lamb & Yu 2004), the rotational Doppler broadening of the observed line may enable us to study the detailed shape of the line, and hence

CP774, *X-Ray Diagnostics of Astrophysical Plasmas,* edited by Randall K. Smith

constrain several system parameters. Therefore, future observation of such broad lines by larger area detectors may be able to determine more than one stellar structure parameter with enough precision to obtain tight constraints on models of the equation of state of dense matter.

We have computed theoretical models of observed line profiles, using realistic neutron star equation of state models, and considering the effects of stellar spin and special and general relativity on the lines. We demonstrate that it is possible to determine the stellar surface gravitational redshift value (and hence R/M) with high precision, even if the observed line is broad and asymmetric. We also show that the properties of the two expected flux minima in a line may be used to detect frame-dragging.

MODEL COMPUTATION

We assume the X-ray emitting portion of the surface to be a belt that is symmetric around the spin axis of the neutron star. This is justified, because during a typical integration time (order of seconds) for spectroscopic measurements, and for a stellar spin frequency > 10 Hz, any hot spot on the stellar surface will be effectively smeared into an axisymmetric belt. In our calculation, we consider the following physical effects on the observed line profile structure: (1) Doppler shift due to stellar rotation, (2) special relativistic beaming, (3) gravitational redshift, (4) light-bending, and (5) frame-dragging (Özel & Psaltis 2003 considered the first four of these effects). To include the effects of light-bending, we trace back the paths of the photons (numerically, in the Kerr spacetime) from the observer to the source using the method described in Bhattacharyya, Bhattacharya, & Thampan (2001).

For a given neutron star matter equation of state, we have several source parameters. These are two stellar structure parameters (radius-to-mass ratio and spin frequency), one binary parameter (observer's inclination angle), two emitting-belt parameters (polar angle position and angular width of the belt), and a parameter n describing the emitted specific intensity distribution (in the corotating frame) of the form $I(\alpha) \propto \cos^n(\alpha)$, where α is the emission angle measured from surface normal. To find the values of other stellar structure parameters (mass and specific angular momentum) required for line profile calculations, we compute (for a given equation of state model) the structure of the rotating neutron star, using the formalism given by Cook, Shapiro, & Teukolsky (1994; see also Bhattacharyya et al. 2000, Bhattacharyya 2002).

RESULTS AND DISCUSSIONS

We obtain two main results (Bhattacharyya, Miller, & Lamb 2004):

1) If the observed line profile is broad and asymmetric, the surface gravitational redshift measured from the energy of the deepest minimum (or, highest maximum for emission lines) will typically have a large error (Özel & Psaltis 2003). Even if we use the centroid energy of the line profile obtained from the arithmetic mean of the low-energy and high-energy edges of the observed line, the fractional error in the inferred redshift value, and hence in the stellar R/M, will typically be large. We find, however, that if we

calculate the line centroid using the geometric mean of the low-energy and high-energy edges of the observed line, the percentage error in the inferred R/M value is in general less than 5% (Bhattacharyya, Miller, & Lamb 2004). This is the accuracy needed for strong constraints on neutron star equation of state models.

2) We find that the ratio of the flux deficit in the low-energy minimum to that in the high-energy minimum is increased by frame-dragging. This ratio, therefore, contains an observable signature of frame-dragging (Bhattacharyya, Miller, & Lamb 2004).

In a more detailed study we will relate the observable line profile parameters (such as line width, ratio of flux deficits in the flux minima, etc.) to the full set of source parameters. Future observations of surface spectral lines with larger detector area instruments in proposed X-ray space missions (e.g., Constellation-X) will help constrain stellar parameters, and will provide a way to detect frame-dragging.

ACKNOWLEDGMENTS

We thank Arun V. Thampan for providing the code used to compute the structure of rapidly spinning neutron stars and Geoff Ravenhall for providing tabulations of the A18+δv+UIX equation of state. This work was supported in part by NSF grant AST 0098436 at Maryland, by NSF grant AST 0098399 and NASA grants NAG 5-12030 and NAG 5-8740 at Illinois, and by the funds of the Fortner Endowed Chair at the University of Illinois.

REFERENCES

1. Bhattacharyya, S. 2002, A&A, 383, 524.
2. Bhattacharyya, S., Bhattacharya, D., & Thampan, A.V. 2001, MNRAS, 325, 989
3. Bhattacharyya, S., Miller, M.C., & Lamb, F.K. 2004, submitted to ApJ Letters (astro-ph/0412107).
4. Bhattacharyya, S., Thampan, A. V., Misra, R., & Datta, B. 2000, ApJ, 542, 473
5. Bildsten, L., Chang, P., & Paerels, F. 2003, ApJ, 591, L29.
6. Cook, G.B., Shapiro, S.L., & Teukolsky, S.A. 1994, ApJ, 424, 823
7. Cottam, J., Paerels, F., & Méndez, M. 2002, Nature, 420, 51.
8. Lamb, F.K., & Yu, W. 2004, in Binary Radio Pulsars, ed. F. A. Rasio & I.H. Stairs (ASP Conf. Series), in press
9. Loeb, A. 2003, Phys. Rev. Lett., 91, 071103
10. Özel, F., & Psaltis, D. 2003, ApJ, 582, L31
11. Villarreal, A.R., & Strohmayer, T.E. 2004, ApJ, 614, L121.

VII: DIELECTRONIC RECOMBINATION AND OTHER PLASMA PROCESSES

Ionization and Recombination with Electrons: Laboratory Measurements and Observational Consequences

Daniel Wolf Savin

Columbia Astrophysics Laboratory, Columbia University, New York, NY 10027, USA

Abstract.
Reliably interpreting spectra from both photoionized and electron-ionized cosmic plasmas requires an accurate understanding of the ionization balance of the source and the underlying atomic physics which produces the ionization structure. Uncertainties in ionization balance calculations hamper our ability to model cosmic sources and to infer reliably their properties. Of particular importance are accurate rate coefficients for electron impact ionization (EII) and dielectronic recombination (DR). Here I briefly review the astrophysical implications due to current uncertainties in EII and DR and discuss our ongoing laboratory work aimed to remove these uncertainties.

Keywords: electron-ion ionization; electron-ion recombination; elementary processes in plasmas
PACS: 34.80.Kw; 34.80.Lx; 52.20.-j; 95.30.Dr

INTRODUCTION

Spectroscopic observations of astrophysical plasmas are used to infer the properties of the cosmos. The aim of laboratory astrophysics is to reduce atomic physics uncertainties to the point where discrepancies between spectral observations and models tells us something about the observed sources and cannot be attributed to errors in atomic data.

As an example of the role atomic data play in astrophysics, relative abundances of cosmic plasmas can be inferred using

$$\frac{n_{\rm A}}{n_{\rm H}} \propto \frac{1}{\frac{n_q}{n_{\rm A}}} \frac{I_{line}}{\alpha_{line}} \tag{1}$$

where $n_{\rm A}$ is the particle density of element A, $n_{\rm H}$ the density of H, n_q the density of charge state q for element A, I_{line} the intensity of a line from ion A^{q+}, and α_{line} the rate coefficient for producing this line. This equation clearly shows that accurate ionization balance calculations are needed for reliable relative abundance determinations. This in turn requires accurate ionization and recombination data. Our research aims to improve the electron impact ionization (EII) and electron-ion recombination data used for modeling astrophysical plasmas, thereby improving our understanding of the cosmos.

ASTROPHYSICAL PLASMAS

Cosmic plasmas can be divided into two broad classes: electron-ionized and photoionized. Electron-ionized plasmas are formed in stellar atmospheres, supernova remnants,

CP774, *X-Ray Diagnostics of Astrophysical Plasmas,* edited by Randall K. Smith

the interstellar medium, galaxies, and clusters of galaxies. In collisional ionization equilibrium (CIE), an ion typically forms at an electron temperature $T_e \sim I_p/2$ where I_p is the ionization potential of the ion. In these plasmas the ionization balance is determined largely by the balance of EII and high temperature dielectronic recombination (DR).

Photoionized plasmas form in the environs of planetary nebulae, H II regions, the intergalactic medium (IGM), X-ray binaries (XRBs), and active galactic nuclei (AGNs). Here the ionization is due to photons and the resulting photoelectrons. In photoionization equilibrium (PIE), an ion forms typically at $T_e \sim I_p/20$. In photoionized plasmas low temperature DR is the dominant recombination mechanism for many ions.

Equation 1 demonstrates the importance of reliable atomic collision data for both electron-ionized and photoionized plasmas. We note here another example applicable to electron-ionized plasmas. For a given element in CIE we can write

$$\frac{n_{q+1}}{n_q} = \frac{C_q}{\alpha_{q+1}} \tag{2}$$

where C_q is the EII rate coefficient of charge state q forming charge state $q+1$ and α_{q+1} the electron-ion recombination rate coefficient of charge state $q+1$ forming charge state q. It is readily apparent that predicted line ratios involving charge states q and $q+1$ will be affected by uncertainties in the ionization and recombination data used for modeling.

In the astrophysics community, the most commonly used ionization balance calculations for electron-ionized plasmas are those of Arnaud and Rothenflug [1], Arnaud and Raymond [2], and Mazzotta et al. [3]. Two commonly used codes for modeling photoionized plasmas include those of Ferland et al. (CLOUDY [4]) and Kallman and Bautista (XSTAR [5]). The reader can find more information about the atomic physics relevant to electron-ionized and photoionized plasmas in these publications and the works cited therein.

ELECTRON IMPACT IONIZATION (EII)

EII occurs when an electron collides with an atom or ion and knocks out one or more electrons, thereby resulting in two or more outgoing electrons. In astrophysics, the most commonly used EII rate coefficients are from the 1985 paper of Arnaud and Rothenflug [1]. These were updated in 1992 for iron by Arnaud and Raymond [2]. Common wisdom in the astrophysics community has held that these data are reliable (e.g., Ref. [6]). Partly as a result there has been no significant attempt to update the EII data base in over 15 years. However, the common widsom is an incorrect perception. Our review of the EII data base has turned up large uncertainties which call into question the reliability of the EII data used for plasma modeling.

Another compilation of recommended EII data from the late 80s is that of Lennon et al. [7]. These data use essentially the same scant set of laboratory measurements and theoretical calculations as do Arnaud and her collaborators. However, as shown in Fig. 1, the two recommended EII rate coefficients can differ from one another by a factor of 2 or more at temperatures relevant for CIE plasmas. We find similar errors in many other astrophysically relevant isoelectronic sequence.

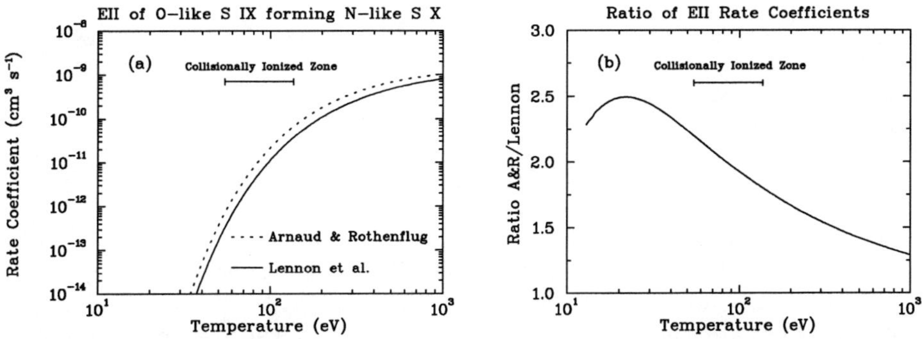

FIGURE 1. Shown for O-like S^{8+} are (a) the recommended EII rate coefficients of Arnaud & Rothenflug [1] and Lennon et al. [7] and (b) the ratio of the two. The horizontal bars in (a) and (b) shows the temperature range over which this ion is predicted to form in CIE [3].

EII studies are theoretical and experimentally difficult. Calculations are computationally challenging as they need to take into account an infinite number of final states. Experimental results are often compromised by unknown quantities of metastable state ions in the ion beam. These two difficulties are part of the reason why there has been no significant improvement to the EII data base in the last 15 or so years.

Research Plan

In order to address the EII data needs of the astrophysics community we have initiated a new series of EII measurements using the Oak Ridge National Laboratory (ORNL) electron-ion crossed-beams apparatus [8]. Recently we have added a gas attenuation cell to the ion beam. This is installed between the ion source and the electron-ion collision region. Using this cell we can determine the metastable fraction of our ion beam. Ground state and metastable state ion can have significantly different cross sections for electron capture (EC) from an attenuating gas. Under these conditions, one of the two states will be preferentially lost as the attenuating gas density is increased. A plot of the logarithm of ion current versus gas density will be bi-linear with the slopes proportional to the EC cross section. Such a plot can be used to infer the relative populations of ground and metastable state ions (e.g., Refs. [9, 10]).

Additionally, we plan on using the attenuation cell to control the metastable fraction of the ion beam. We will carry out EII measurements for two different fractions of metastable ions in the beam (f_1 and f_2). This will yield as a function of beam energy

$$\begin{aligned} \sigma_1(E) &= (1-f_1)\sigma_g(E) + f_1\sigma_m \\ \sigma_2(E) &= (1-f_2)\sigma_g(E) + f_2\sigma_m \end{aligned} \quad (3)$$

where σ_g and σ_m are the EII cross sections for the ground and metastable states, respectively. From measurements of σ_1, σ_2, f_1, and f_2, we will extract the values for σ_g and

σ_m. Initial measurements are planned for ions which have long-lived metastable states and for which reliable ground and metastable state EII cross sections measurements have not previously been achieved.

DIELECTRONIC RECOMBINATION (DR)

DR is a two step recombination process which begins when an electron collisionally excites a core electron of an ion and is simultaneously captured into a Rydberg level. The core electron excitation can be labeled $Nl_j \rightarrow N'l'_{j'}$ where N is the principal quantum number, l its orbital angular momentum, and j its total angular momentum. The energy of this recombined system lies in the continuum and the complex may autoionize. The DR process is complete when the system emits a photon which reduces the total energy of the recombined system to below its ionization threshold.

The first step of the DR process is commonly called dielectronic capture (DC). Energy conservation requires that the excitation energy ΔE of the core electron is balanced by the sum of the kinetic energy E_k of the incident electron plus the binding energy E_b of the level into which it is captured, i.e., $\Delta E = E_k + E_b$. Because both ΔE and E_b are quantized, E_k is also quantized. Hence, DC and hence DR are resonance processes.

Theoretical DR rate coefficients for K-shell ions are believed to be well understood. This is partially due to a sustained series of electron beam ion trap (EBIT) and ion storage ring measurements over the past 15 years. These measurements have provided important benchmarks for the development of modern theoretical methods for calculating DR. Reviews of experimental and theoretical K-shell DR are given in Refs. [11, 12]. Overall agreement between theory and experiment is on the order of 20%.

The situation for L- and M-shell ions, though, has not been so good. DR theory for these many-electron systems are theoretically and computationally challenging and the DR database for these ions has been known to highly unreliable for a number of years. The implications of these DR uncertainties have been explored for solar spectra [12, 13], the IGM [14, 15], and X-ray photoionized plasmas such as are formed in the environs of XRBs and AGN [16, 17].

The importance of reliable M-shell Fe DR data has also been shown by the analyses of recent Chandra and XMM AGN spectra. These spectra show a broad unresolved transition array (UTA) which is attributed to absorption by M-shell Fe ions. Models are able to match AGN spectral features from abundant second row elements. However, these models over-predict the average Fe charge stage. This is believe to be due to the theoretical underestimate of the relevant low temperature DR rate coefficients for M-shell Fe [18, 19].

Research Program

In order to provide reliable DR data for plasma modeling, we are carrying out storage ring measurements for a wide range of L- and M-shell ions. Our work has focused primarily on DR measurements for iron because it is the highest Z cosmically abundant element and plays an important role in cosmic plasmas in regimes of parameter space

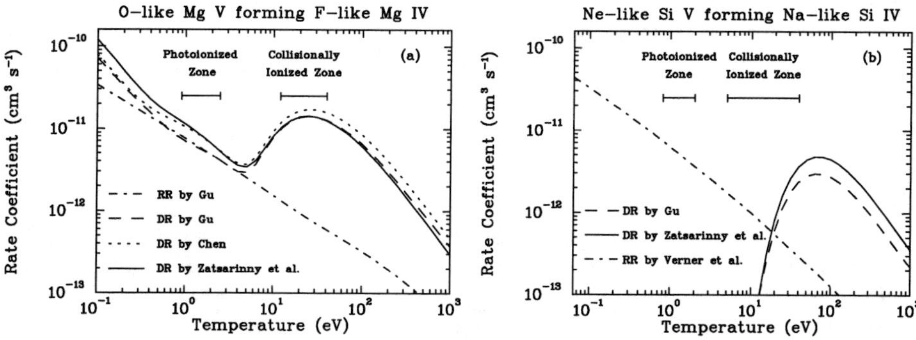

FIGURE 2. Shown are various state-of-the-art DR calculations for (a) O-like Mg^{4+} forming F-like Mg^{3+} from Chen [29], Gu [30], and Zatsarinny et al. [31] and (b) Ne-like Si^{4+} forming Na-like Si^{3+} from Gu [30] and Zatsarinny et al. [32]. The radiative recombination data in (a) are from Ref. [33] and in (b) from Ref. [34]. The horizontal bars in (a) and (b) shows the temperature range over which each ion is predicted to form in PIE [5] and CIE [3].

where other ions have already been fully stripped. Our work also provides important benchmarks for modern DR theory. These tested methods can then be used to calculated the needed DR data for other ions in the measured isoelectronic sequences. However, as we discuss below, we are finding that even if the various state-of-the-art DR calculations are in good agreement with experiment for a given iron ion, that does not guarantee that they will be in good agreement with one another for isoelectronic ions at lower atomic number Z.

We have carried out measurements of DR via $\Delta N = 0$ core excitations for Al-like Fe^{13+} (in preparation), Mg-like Fe^{14+} (in preparation), Na like Fe^{15+} [20, 21], F-like Fe^{17+} [16, 22], O-like Fe^{18+} [16, 23], N-like Fe^{19+} [24], C-like Fe^{20+} [25], B-like Fe^{21+} [25], and Be-like Fe^{22+} (in preparation). We have also carried out $\Delta N = 1$ measurements for Na-like Fe^{15+} [20, 21], O-like Fe^{18+} [16, 23], and Be-like Fe^{22+} (in preparation). Additionally, we note that there have been $\Delta N = 0$ and $\Delta N = 1$ DR measurements for Li-like Si^{11+} [26] and Cu^{26+} [27], and also for F-like Se^{35+} [28]. To summarize, storage ring $\Delta N = 0$ DR measurements exist for all L-shell ions and a few of the M-shell ions. The situation is far spottier for $\Delta N = 1$ DR and we are unaware of any storage ring measurements for isoelectronic sequences not mentioned above.

A detailed discussion of typical DR data collection techniques and DR resonance spectra are given in the references above and the papers cited therein. The point which we wish to make here is that even with all the benchmarking that has been going on, for some ions the various state-of-the-art DR calculations have still not converged. This can readily be seen in Fig. 2(a) where the modern DR results differ from one another by a factor of 2.5 at temperatures relevant for photoionized plasmas. This is attributed to the uncertainties in the calculated theoretical energies of the low energy DR resonances which dominate the low temperate DR rate coefficient. Interesting also are the theoretical results for Ne-like ions shown in Fig. 2(b). Though these calculations have yet to be benchmark by storage ring results, one can see that the two published modern DR results differ by a factor of nearly 1.7. This is surprisingly large considering the typical 20%

estimates for the uncertainties in theoretical $\Delta N = 1$ DR rate coefficients. This large difference is likely due to one of the calculations not having included enough states.

CONCLUSIONS

Our review of EII has turned up surprisingly large uncertainties in the EII data being used for plasma modeling. Though some of these issues were first noted by Kato et al. [35], these results did not filter into the astrophysics community. Clearly much theoretical and experimental work needs to be done to improve the EII data base. Our laboratory work will help to address this issue.

The situation is somewhat better for DR of L-shell ions. There has been a significant improvement to the DR data base for these ions thanks to a combined experimental and theoretical efforts over the past few years. However, a number of issues remain to be resolved and further experimental work is needed.

For high temperature DR, there is in general good agreement between theory and experiment for the few L-shell isoelectronic sequences which have been measured. But there are some surprising discrepancies between various modern DR calculations for systems which have not yet been benchmarked.

For low temperature DR, state-of-the-art theoretical methods are often still not able to calculate reliable DR rate coefficients. This can be traced directly back to the accuracy of the atomic structure codes used to calculate the energies of the low-lying doubly-excited autoionizing states that are important for low temperature DR. For many-electron ions with partially-filled shells, there has been no significant advance in atomic structure methods for over 30 years. This stands as possibly the greatest challenge for producing reliable low temperature DR rate coefficients. Given the limited laboratory resources which can be devoted to this problem, it will not be possible to measure all the astrophysically relevant low energy DR resonances. Theoretical calculations must be used for all the unmeasured systems. But this cannot be achieved without any advances in atomic structure methods.

The situation is probably even worse for M-shell ions. Few benchmark measurements exist to guide the theory. And any theoretical and computational challenges experienced for L-shell ions will be even greater for M-shell ions.

It is probably fair to summarize this overview by saying that it will take many more years of combined theoretical and experimental efforts before the EII and DR data bases are truly reliable.

ACKNOWLEDGMENTS

The author thanks his collaborators M. Bannister, E. Landi, J. M. Laming, W. Mitthumsiri, A. Müller, M. Schnell, S. Schippers, and A. Wolf for many stimulating discussions. We also thank T. Kato for providing the EII fit parameters of Lennon et al. in electronic form and for M. Chen for providing his corrected DR fit parameters for Mg^{4+}. DWS supported in part by the NASA Space Astrophysics Research and Analysis program and the NASA Solar Physics Research, Analysis, and Suborbital program.

REFERENCES

1. M. Arnaud and R. Rothenflug, *Astron. Astrophys. Suppl. Ser.* **60**, 425–457 (1985).
2. M. Arnaud and J. Raymond, *Astrophys. J.* **398**, 394–406 (1992).
3. P. Mazzotta, G. Mazzitelli, S. Colafrancesco, and N. Vittorio, *Astron. Astrophys. Suppl. Ser.* **133**, 403–409 (1998).
4. G. J. Ferland, K. T. Korista, D. A. Verner, J. W. Ferguson, J. B. Kingdon, and E. M. Verner, *Publ. Astron. Soc. Pacific* **110**, 761-778 (1998).
5. T. Kallman and M. Bautista, *Astrophys. J. Suppl. Ser.* **133**, 221–253 (2001).
6. J. C. Raymond, "Radiation from Hot, Thin Plasma," in *Hot Thin Plasmas in Astrophysics*, edited by R. Pallavicini, Kluwer Academic Publishers, Dordrecht, 1988, pp. 3–20.
7. M. A. Lennon, K. L. Bell, H. B. Gilbody, J. G. Hughes, A. E. Kingston, M. J. Murray, and F. J. Smith, *J. Chem. Phys. Ref. Data* **17**, 1285–1363 (1988).
8. M. E. Bannister, *Phys. Rev. A* **54**, 1435–1444 (1996).
9. M. Zuo, S. J. Smith, A. Chutjian, I. D. Williams, S. S. Tayal, and B. M. McLaughlin, *Astrophys. J.* **440**, 421–429 (1995).
10. D. B. Reisenfeld, L. D. Gardner, P. H. Janzen, D. W. Savin, and J. L. Kohl, *Phys. Rev. A* **60**, 1153–1164 (1999).
11. A. Müller, *At. Plasma Material Interactions* **6**, 59–100. (1995).
12. D. W. Savin and J. M. Laming, *Astrophys. J.* **566**, 1166–1177 (2002).
13. N. S. Brickhouse, "EUV Spectroscopy of Stellar Coronae," in *Atomic Processes in Plasmas Tenth Topical Conference*, edited by A. L. Osterheld and W. H. Goldstein, AIP Conference Proceedings 381, American Institute of Physics, Woodbury, New York, 1996, pp 31–38.
14. D. W. Savin, *Astrophys. J.* **533**, 106–112 (2000).
15. A. Aguirre, J. Schaye, T. S. Kim, T. Theuns, M. Rauch, and W. L. Sargent, *Astrophys. J.* **602**, 38–50 (2004).
16. D. W. Savin, et al., *Astrophys. J. Suppl. Ser.* **123**, 687–702 (1999).
17. D. W. Savin, et al., "Measurements of Low Temperature Dielectronic Recombination in L-shell Iron for Modeling X-ray Photoionized Cosmic Plasmas," in *Atomic Processes in Plasmas Twelfth Topical Conference*, edited by R. C. Mancini and R. A. Phaneuf, AIP Conference Proceedings 547, American Institute of Physics, Woodbury, New York, 2000, pp. 267–278.
18. H. Netzer, *Astrophys. J.* **604**, 551–555 (2004).
19. S. B. Kraemer, G. J. Ferland, and J. R. Gabel *Astrophys. J.* **604**, 556-561 (2004).
20. J. Linkemann, et al., *Nucl. Instrum. Methods B* **98**, 154–157 (1995).
21. A. Müller, *Int. J. Mass Spectrom.* **192**, 9–22 (1999).
22. D. W. Savin, et al., *Astrophys. J. Lett.* **489**, L115–118 (1997).
23. D. W. Savin, et al., *Astrophys. J.* **576**, 1098–1107 (2002).
24. D. W. Savin, et al., *Astrophys. J. Suppl. Ser.* **138**, 337–370 (2002).
25. D. W. Savin, et al., *Astrophys. J. Suppl. Ser.* **147**, 421–435 (2003).
26. J. Kenntner, et al., *Nucl. Instrum. Methods B* **98**, 142–145 (1995).
27. G. Kilgus, D. Habs, D. Schwalm, A. Wolf, N. R. Badnell, and A. Müller, *Phys. Rev. A* **46**, 5730–5740 (1992).
28. A. Lampert, A. Wolf, D. Habs, G. Kilgus, D. Schwalm, M. S. Pindzola, and N. R. Badnell, *Phys. Rev. A* **53**, 1413–1423 (1996).
29. M. H. Chen, *Phys. Rev. A* **66**, 052715 (2002).
30. M. F. Gu, *Astrophys. J.* **590**, 1131-1140 (2003).
31. O. Zatsarinny, T. W. Gorczyca, K. T. Korista, N. R. Badnell, and D. W. Savin, *Astron. Astrophys.* **412**, 587–595 (2003).
32. O. Zatsarinny, T. W. Gorczyca, K. T. Korista, N. R. Badnell, and D. W. Savin, *Astron. Astrophys.* **426**, 699-705 (2004).
33. M. F. Gu, *Astrophys. J.* **589**, 1085-1088 (2003).
34. D. A. Verner and G. J. Ferland *Astrophys. J. Suppl. Ser.* **103**, 467–473 (1996).
35. T. Kato, K. Masai, and M. Arnaud, "Comparison of Ionization Rate Coefficients of Ions from Hydrogen through Nickel," *National Institute for Fusion Science Report, Nagoya, Japan, NIFS-DATA-14* (1991).

Drs. Daniel Savin and Paola Testa at a break in the workshop.

Dielectronic Recombination, Photoionization and Opacities

N. R. Badnell*, M. J. Seaton† and H. P. Summers*

*Department of Physics, University of Strathclyde, Glasgow G4 0NG, UK
†Department of Physics and Astronomy, University College London, London WC1E 6BT, UK

Abstract. We review methodologies for describing dielectronic recombination (DR). We look briefly at the state-of-play of theory vs experiment. We then discuss how to deliver DR data in a form suitable for modelling and we review the status of the partial and total DR database being assembled for the modelling of dynamic finite-density plasmas. We comment on the role of fields and how modelling l-changing collisions approximately models the dynamic part of the plasma microfield.

We also consider the inverse process of photoionization and discuss how the same approach to DR is being used for inner-shell photoionization and the determination of opacities in stellar interiors. In particular, we pay attention to the role of opacity in connection with the discrepancy between the helioseismology observations and stellar evolution models for the position of the base of the solar convective zone.

Keywords: atomic data – atomic processes – stars: interiors
PACS: 34.80.Kw, 96.60.Jw, 98.38.Bn

DIELECTRONIC RECOMBINATION

Dielectronic recombination (DR) is the dominant electron–ion recombination process in many astrophysical and laboratory plasmas. It plays an important role in determining both the level populations and the ionization balance of both high- and low-temperature non-LTE plasmas over a wide range of electron densities and its accurate description is key to spectral analysis [1], [2]. Dielectronic recombination can be viewed as a two-step process. Firstly, a free electron excites an electron in an ion X^{+z}, say, and in the process transfers sufficient of its energy that it is captured into an autoionizing state of the ion X^{+z-1}. The process is reversible since the total energy of the system remains conserved. However, if an electron (either the captured electron or an electron in the parent core) then makes a spontaneous radiative transition which leaves the ion in a non-autoionizing (bound) state then the recombination can be viewed as complete, at least in the low-density (coronal) limit. Dielectronic recombination can take place via many intermediate autoionizing states — indeed, entire Rydberg series. It was the significance of the effective statistical weight of so many available states that led Burgess [3] to recognize the importance of dielectronic recombination in the first instance. A consequence of this is that a huge population structure model is required in principle for further progress. In astrophysics and fusion, the problem has usually been made manageable by simply summing over all final states so as to produce a total dielectronic recombination rate coefficient. In combination with radiative recombination, this is the effective recombination coefficient from the point of view of an ionization state only —

CP774, *X-Ray Diagnostics of Astrophysical Plasmas,* edited by Randall K. Smith

the so-called 'coronal picture'. In this picture, excited-state populations are depleted exclusively by spontaneous radiative transitions and are small compared to those of ground states — with which they are in quasi-static equilibrium. Collisional processes are negligible, except with ground state targets. In turn, the ionization state is determined by balancing the effective recombination rate of (the ground state of) X^{+z} against the collisional ionization rate of (the ground state of) X^{+z-1}.

For dynamic finite-density plasmas, there are therefore two critical limitations to the coronal approximation which must be addressed. Both require a re-appraisal of the strategy for computing dielectronic recombination data. Firstly, from the atomic point of view, a dynamic plasma is one for which the timescale of change in plasma parameters (especially electron temperature, T_e, and electron density, N_e) is comparable with the lifetime of the metastable populations of its constituent ions. This situation is a concern, for example, for impurities near plasma contacted surfaces in fusion [4] and for active solar events [5]. The population of an ionization stage may no longer be assumed to be concentrated in the ground state of each ion. A significant population is found in the metastables, and they may not be in quasi-static equilibrium with the ground state. This means that such metastables are the starting point for recombination events, and so the time-evolution of their populations must be tracked in the same manner as for the ground states. We call this the generalized collisional–radiative (GCR) picture [6]. Even if the metastable populations are in quasi-static equilibrium with the ground state, they are significant and so are still a source term for dielectronic recombination which cannot be ignored.

The second issue is that of finite-density effects. The coronal (zero-density limit) picture assumes that, following the two-step dielectronic recombination process, the resultant (non-autoionizing) excited-state (electron) radiatively cascades back down to the ground state without collisional disruption. At finite electron densities, this radiative cascade can be interrupted by further electron collisions which redistribute the population — in particular, ionization (possibly stepwise) out of excited states, which reduces the effective dielectronic recombination rate. Collisional–radiative modelling removes the limitations of the coronal model, but at the cost of much more elaborate excited-state population calculations. These in turn require much more detailed dielectronic recombination data, and in an easily accessible form. The onset of these density effects on the 'post-DR' population structure depends markedly on ion charge, but can be significant even at electron densities as low as $N_e \gtrsim 10^8 \text{cm}^{-3}$, typical of the solar corona. At much higher densities, $N_e \sim 10^{14} \text{cm}^{-3}$, redistributive collisions can interrupt the two-step dielectronic process itself.

Therefore, we require dielectronic recombination from the metastable states as well as the ground. Secondly, we require final-state resolved data, i.e. we need to know the specific level that each two-step recombination ends-up in — the generalized collisional–radiative population rate equations govern the subsequent time-evolution of these states. As we have already indicated, there are very many possible final states accessible to the dielectronic recombination process which would seem to make for unacceptably large tabulations. However, it is not only impracticable, but unnecessary, to treat each final state in the same manner for the purposes of collisional–radiative modelling. The techniques of matrix condensation and projection reduce the effective number of high-lying states by the progressive bundling of representative states and project the full influence of the high-lying states down onto a fully-resolved low-lying set [6]. (It is the low-level set which is the focus of detailed spectral analysis.) With this in mind, we can tailor our dielectronic recombination tabulations to reflect this situation [7].

Work is in progress [7] to calculate multi-configuration intermediate coupling (*level 2*) dielectronic recombination rate coefficients from the (ground plus) metastable levels of an ion to all possible final states, resolved by level, and/or bundling, appropriate for generalized collisional–radiative modelling. Coverage is of all elements applicable to astrophysics and magnetic fusion viz. He, Li, Be, B, C, N, O, F, Ne, Na, Mg, Al, Si, P, S, Cl, Ar, Ca, Ti, Cr, Fe, Ni, Zn, Kr, Mo and Xe. Sequences covered and published so far include O-like [8], Be-like [9], C-like [10], Li-like [11], B-like [12], Ne-like [13] and N-like [14]. LS-coupling (*level 1*) data for many elements of these sequences was calculated by Badnell (1991–92, unpublished) and incorporated into the atomic database part of the Atomic Data and Analysis Structure (ADAS) and is routinely drawn into the generalized collisional–radiative part of ADAS [15].

Dielectronic recombination rate coefficient modelling

Recombination rate coefficients include contributions from three-body, radiative and dielectronic recombination. In ADAS, three-body recombination rate coefficients are obtained from electron-impact ionization rate coefficients, via detailed balance. This also ensures that the correct Saha–Boltzman limit is reached at high electron densities. Since three-body recombination is separate from dielectronic and radiative recombination, it is not necessary to consider it further. However, quantum mechanically, dielectronic and radiative recombination are indistinguishable processes which interfere with each other. In practice [16], this interference is a very small effect and can safely be neglected for our purposes. This is the independent processes (IP) approximation whereby dielectronic and radiative recombination can be considered separately and is the approach taken by the database aspect of ADAS. Separate data files exist for dielectronic (*adf09*) and radiative recombination (*adf08*) and they can be updated independently. Our current focus is dielectronic recombination but we are also in the process of systematically updating our RR database so as to mirror DR, and this is implemented within a new RR dataset specification *adf48* which is almost identical in structure to that of *adf09*.

In the isolated resonance (IR) approximation, the partial dielectronic recombination rate coefficient α_{iv}^{z+1} from an initial metastable state v of an ion X^{+z+1} into a resolved final state i of an ion X^{+z} is given by [3]

$$\alpha_{iv}^{z+1} = \left(\frac{4\pi a_0^2 I_{\rm H}}{k_{\rm B}T_{\rm e}}\right)^{3/2} \sum_j \frac{\omega_j}{2\omega_v} \times \frac{\sum_l A^{\rm a}_{j\to v,E_c l} A^{\rm r}_{j\to i}}{\sum_h A^{\rm r}_{j\to h} + \sum_{m,l} A^{\rm a}_{j\to m,E_c l}} \, {\rm e}^{-E_{\rm c}/(k_{\rm B}T_{\rm e})}, \qquad (1)$$

where ω_j is the statistical weight of the $(N+1)$-electron doubly-excited resonance state j, ω_v is the statistical weight of the N-electron target state (so, $z = Z - N - 1$) and the autoionization ($A^{\rm a}$) and radiative ($A^{\rm r}$) rates are in inverse seconds. Here, $E_{\rm c}$ is the energy of the continuum electron (with orbital angular momentum l), which is fixed by the position of the resonances, and $I_{\rm H}$ is the ionization potential energy of the hydrogen atom (both in the same units of energy), k_B is the Boltzman constant, T_e the electron temperature and $(4\pi a_0^2)^{3/2} = 6.6011 \times 10^{-24}$ cm^3.

The effect of interacting resonances on dielectronic recombination has been investigated by [16] and can safely be neglected, at least in the absence of external electric and magnetic fields (see below). While autoionization rates can be determined (within the isolated resonance approximation) via the fitting of resonances calculated in a close-coupling approximation, or via the extrapolation of threshold close-coupling collision strengths using the correspondence principle, it is usual now to introduce a further approximation — that of using distorted waves (DW), i.e. the autoionization rates are calculated via perturbation theory using the Golden Rule. This is the only approximation that we have made so far that may need to be reconsidered in certain cases. In low-charge ions, a perturbative distorted wave calculation may give inaccurate autoionization rates compared to those calculated in a close-coupling approximation. However, this only has a direct effect on the partial dielectronic recombination rate if the autoionization rates do not 'cancel-out' between the numerator and denominator of (1) — typically, autoionization rates are orders of magnitude larger than radiative rates.

We use the code AUTOSTRUCTURE [17] to calculate multi-configuration intermediate coupling energy levels and rates within the IPIRDW approximation. The code can make use both of non-relativistic and semi-relativistic wavefunctions. The low-n problem is no different from the one of computing atomic structure. The high-n problem ($n \gtrsim 25$) is solved using the correspondence principle. The Rydberg nl radial function is approximated by a zero-energy continuum. This becomes 'exact 'in the limit $n \to \infty$. Diagonalization of N- and $(N+1)$-electron Hamiltonians is maintained for all n. No extrapolation or further approximation is required.

Electric fields and doubly-excited states

Ion and electron l-redistributive cross sections for doubly-excited states are evaluated following the method of Pengelly & Seaton [18]. This is to be viewed as a very simplified treatment of the dynamic part of the plasma microfield associated with ions which move closer than their neighbours to the target. The quasi-static part of the ion microfield, which is approximately 1/3 of the total, is ignored at this level of analysis. We use the Burgess–Bethe General Program (BBGP) to determine the population structure of the l-subshells of the representative set of n-shells. The ratio of the sum over l-substates at a given density to zero density yields a BBGP finite-density adjustment factor of the total n-shell capture at zero-density. We then use this adjustment factor on the detailed *level 1* and *level 2* DR data so as to assess the size of the effect of dynamic part of the plasma microfield on DR.

The energy differences are critical to the problem of doubly-excited-state redistribution. They are small, tending to zero at large l. Thus, the redistributive cross-sections are very large, remaining finite in the degenerate level limit only because of the finite radiative lifetime of the target (which is short for resonant states) or through screening of the projectile by nearest neighbours. For the $l = 0, 1, 2$ waves of the n-shell spectator electron, quantum defect expansions of the form $\mu_{nl} = a_0 + a_1/n^2$ are usually available.

For the higher $l(>2)$, the large number of energies required can be estimated more economically from the dipole polarizabilities, α_p^{pol}, of the parents, following Edlen [19].

Experiment, DR and the role of electric fields

The last decade has seen an enormous amount of experimental activity in the area of dielectronic recombination. In particular, heavy-ion storage rings coupled with electron-coolers have provided a wealth of data for partial dielectronic recombination. (The partial here is by the intermediate resonance state rather than the final state.) Most data is of the 'bundled-n' form, but some l-resolution is possible for very low-lying states. The iso-electronic sequences studied range almost exclusively from H-like through to Mg-like and nuclear charges have ranged between Z=2 and 92. In all cases, in the absence of external fields, there is rarely any significant disagreement with theory, i.e. outside of the experimental uncertainty. In Fig. 1, we show representative comparisons of dielectronic recombination data for $O^{5+} + e^- \rightarrow O^{4+}$[20] and $Fe^{18+} + e^- \rightarrow Fe^{17+}$[21], calculated in the IPIRDW approximation with AUTOSTRUCTURE and which illustrates the level of accuracy that can be expected of the theoretical data. For further comparisons, see the experimental review by Savin elsewhere in this volume.

One major area of uncertainty is the role of external fields on dielectronic recombination, and it is this more than anything that renders pointless efforts to compute (zero-density) field-free data to an accuracy of better than $\simeq$20% , say. It has long been known that the high Rydberg states that frequently dominate the dielectronic recombination process can be Stark-mixed by weak electric fields [22], in particular the plasma microfield [23], and so increase the partial rate coefficients by factors of 2, or 3, or more, over a wide range of n. Recently, the picture has been further complicated by the discovery that magnetic fields, when crossed with an electric field, strongly affect the electric field enhancement — by reducing it in most cases [24], [25]. While this suppression of the electric field enhancement is advantageous towards the use of field-free dielectronic recombination data, it is disadvantageous in terms of trying to compute field-dependent data for plasma modelling.

Previously, it appeared that a reasonable approach would be to use the values of the plasma microfield (which in turn depends on the plasma density) for the electric field strength for use in the generation of field dependent (i.e. density dependent) data as input to collisional–radiative modelling. This in itself ignored any further (e.g. external) electric fields that might be present in the plasma environment, beyond the plasma microfield. The recognition of the importance of magnetic fields as well makes a comprehensive solution to dielectronic recombination in a plasma a distant goal and partial data accurate to $\simeq$20% as meaningful as necessary. Furthermore, field enhancement is sensitive to interacting resonances as well [26] unlike the field-free case. We do note again that high Rydberg states in a finite density plasma are brought into LTE by (electron) collisions. A preliminary study [27] showed that the effect of the plasma microfield on the density-dependent effective recombination rate coefficient was suppressed by collisions driving high Rydberg states into LTE — larger values for the microfield, which lead to larger enhancements of the zero-density rate coefficient, corresponds to denser

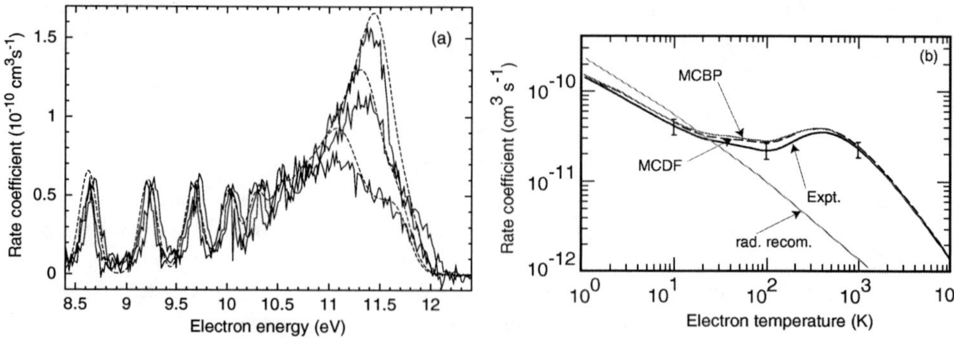

FIGURE 1. **a)** A theoretical reconstruction is shown of the observed dielectronic recombination resonances for the light ion $O^{5+} + e^- \rightarrow O^{4+}$. The measurement is from the CRYRING heavy-ion storage ring [20]. Note that the dielectronic recombination data from the methods described in section have been convoluted with the experimental velocity distribution, which has the units of a rate coefficient. The three pairs of curves correspond to three different beam energies which results-in three different cut-offs in the maximum principal quantum number detected. The experimental and theoretical data agree quite nicely. **b)** The Maxwellian dielectronic recombination rate coefficient for $Fe^{18+} + e^- \rightarrow Fe^{17+}$ is shown. The measurement (solid curve) is from the Heidelberg storage ring [21] with an experimental error assessed as $\lesssim 20\%$. The theoretical IPIRDW results include the $1 \rightarrow 2$, $2 \rightarrow 2$ and $2 \rightarrow 3$ core excitations in the multi-configuration Breit–Pauli (MCBP, dotted curve) and Dirac–Fock data (MCDF, dashed curve) with $n_{\max} = \infty$. The worst deviation from experiments is $\lesssim 30\%$ (for the $3l3l'$ resonances) with a typical uncertainty $\lesssim 20\%$ for the direct state-selective coefficients to individual levels and for the total dielectronic rate coefficient. There is excellent agreement between the Breit–Pauli and Dirac–Fock results. The Breit–Pauli approach is used for our mass data production.

plasmas for which collisions drive more states into LTE.

PHOTOIONIZATION AND OPACITIES

Photoionization

Dielectronic recombination is the time-reversed process of photo-excitation–autoionization while radiative recombination is the time-reversed process of direct photoionization. Consequently, the whole machinery developed for photorecombination can be applied to photoionization. Another review in this volume (Palmeri) covers description of the K-edge in X-ray spectra. Here we focus on the application to opacity determination.

Opacities

Energy is produced by nuclear reactions at the centre of a star, at a temperature of a few times 10^7 K, and escapes at the stellar surface. The structure of a star is determined by the equations for conservation of mass and of energy, an equation for

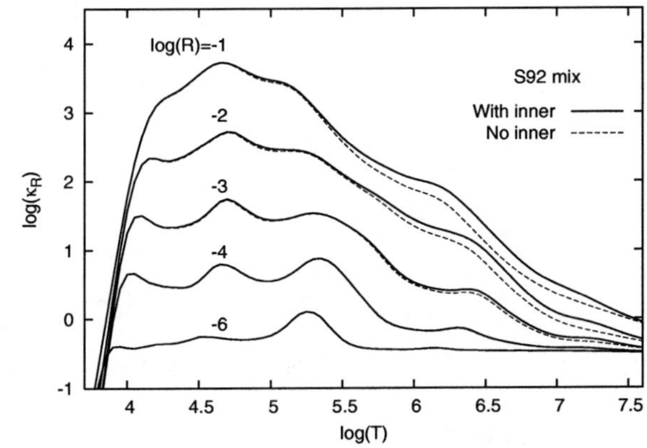

FIGURE 2. Rosseland-mean opacities from OP for S92 mix, with and without inner-shell contributions.

hydrostatic equilibrium, and by the temperature gradient (see e.g. [28]). In regions for which convection does not occur, the temperature gradient is determined by the *Rosseland-mean opacity* ($\kappa_{\rm R}$).

First results for opacities from the Opacity Project (OP) were published by Seaton *et al.* [29] and were in good general agreement with those from the OPAL project [30] except for regions of high densities and temperatures. In a more recent paper, Badnell and Seaton [31] confirmed a suggestion made by Iglesias and Rogers [32] that the discrepancies in those regions were due to the omission of important inner-shell processes in the OP work. Further comparisons between OPAL results and those from OP with inclusion of all important inner-shell processes are given by Seaton and Badnell [33] for the case of the 6-element mix (H, He, C, O, S and Fe) of [32]. In [31] and [33], the required inner-shell atomic data were computed using the code AUTOSTRUCTURE [17]. Similar computations of the required inner-shell atomic data for all cosmically-abundant elements (He, C, N, O, Ne, Na, Mg, Al, Si, S, Ar, Ca, Cr, Mn, Fe and Ni) have now been made by Badnell *et al.* [34]. We discuss some of the results from [34].

Contributions of inner-shell transitions to Rosseland-means

We use the variable $R = \rho/T_6^3$ introduced in the OPAL work, where ρ is mass–density and $T_6 = 10^{-6} \times T$ with T in K. The behaviour of R in some stellar models is shown on Figs. 1 and 2 of [29].

Fig. 2 shows, for the S92 solar mix, $\log(\kappa_{\rm R})$ against $\log(T)$ for $\log(R) = -1, -2, -3, -4, -5$ and -6, both with and without inclusion of inner-shell transitions. It is seen that the inner-shell transitions contribute very little to the means at the lower densities, say $\log(R) \lesssim -4$. The reason is that, at a lower density, ions in a given ionization stage will have maximum abundance only at a lower temperature: and at the lower temperatures features in the monochromatic opacities due to inner-shell

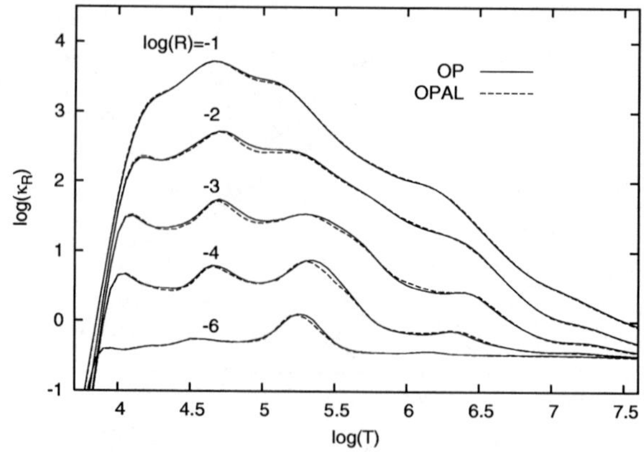

FIGURE 3. Rosseland-mean opacities from OP and OPAL for the S92 mix.

transitions will appear at larger values of the frequency variable corresponding to regions where its weighting function within the Rosseland-mean is small.

Comparisons with OPAL

OPAL data can be obtained from the OPAL web site[1] for any required composition. Fig. 3 shows $\log(\kappa_R)$ against $\log(T)$ for $\log(R) = -1, -2, -3, -4$ and -6, from both OP and OPAL. The over-all agreement is seen to be good. The feature at $\log(T) \simeq 5.2$, usually known as the 'Z-bump', is mainly due to transitions in iron ions with $N = 13$ to 18. It is seen that, compared to OPAL, the OP feature is shifted to slightly higher temperatures.

Use of the logarithmic scale for $\log(\kappa_R)$, as in Fig. 2, does not allow one to see the finer details of the level of agreement between the two calculations. The general level of agreement is actually in the region of 5 to 10%. An excursion to larger differences at $\log(T) \simeq 5.5$, approaching 30% at $\log(R) = -4.5$, is found in the high-temperature wing of the Z-bump, as shown on Fig. 3.

The solar radiative interior

Helioseismology provides remarkably accurate values for the depth, R_{CZ}, of the solar convection zone (CZ). With the earlier estimates of solar element abundances, such as S92 or GN93 (see [34]), values of R_{CZ} and other data from helioseismology were found to be in good agreement with results from solar models computed using OPAL opacities. However, with the new abundances it is found in two recent papers to be necessary to

[1] www-phys.llnl.gov/Research/OPAL/

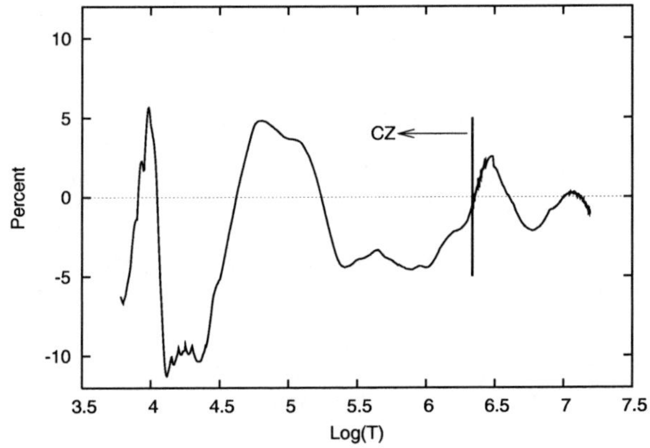

FIGURE 4. Percentage differences, (OP–OPAL), between opacities for a solar model BP04 [38]. The convection zone is the region with $\log(T) \lesssim 6.34$.

increase opacities in the vicinity of R_{CZ}: by 19% according to Basu and Antia [35]; and by 21% according to Bahcall *et al.* [36]. In a further detailed study of the problem, Bahcall *et al.* [37] show that there are also discrepancies for the solar profiles of sound speed and density and of helium abundance. They argue that all such discrepancies would be removed if the opacities were larger than those from OPAL by about 10% in the region of $2 \times 10^6\text{K} \lesssim T \lesssim 5 \times 10^6\text{K}$ (0.7 to $0.4R_{\text{sun}}$).

In the solar model of Bahcall and Pinsonneault [38], referred to as BP04, the base of the convection zone is at $\log(T) \simeq 6.34$ ($r \simeq 0.715R_{\text{sun}}$). In the CZ the abundances are independent of depth and the values used for the model were similar to those of the S92 mix (i.e. not taking account of recently proposed revisions). Below the convection zone, $\log(T) \gtrsim 6.34$ ($r \lesssim 0.715R_{\text{sun}}$), abundances depend on the nuclear reactions which have taken place and vary with depth.

Fig. 4 shows percentage differences, (OP–OPAL), between opacities from OP and those used for BP04. In the region below R_{CZ} the relative abundances of metals are assumed to be independent of depth but the mass-fractions are variable. At the lowest temperatures (say $\log(T) \leq 4.5$) κ_R is varying very rapidly as a function of T and the differences (OP–OPAL) are as large as 10%. Throughout the rest of the CZ the differences are never larger than about 5%. The problem of 'What is wrong with the sun?' [39] remains.

ACKNOWLEDGMENTS

The contributions of Z. Altun, M. A, Bautista, K. Butler, J. Colgan, F. Delahaye, T. W. Gorczyca, K. Korista, S. D. Loch, C. Mendoza, D. M. Mitnik, M. G. O'Mullane, P. Palmeri, M. S. Pindzola, D. W. Savin, A. D. Whiteford, A. Yumak, O. Zatsarinny, and C. J. Zeippen are gratefully acknowledged.

REFERENCES

1. P. G. Judge, V. Hubeny, and J. C. Brown, ApJ, **475**, 275 (1997).
2. A. C. Lanzafame, D. H. Brooks, J. Lang, *et al.* A&A, **384**, 242 (2002).
3. A. Burgess, ApJ, **139**, 776 (1964).
4. H. P. Summers, N. R. Badnell, M. G. O'Mullane, *et al.* , *Plasma Phys. Control. Fusion*, **44**, B323 (2002).
5. A. F. Lanza, D. Spadaro, A. C. Lanzafame, *et al.* , ApJ, **547**, 1116 (2001).
6. H. P. Summers, and M. B. Hooper, *Plasma Physics*, **25**, 1311 (1983).
7. N. R. Badnell, M. G. O'Mullane, H. P. Summers, *et al.* , A&A, **406**, 1151 (2003).
8. O. Zatsarinny, T. W. Gorczyca, K. T. Korista, *et al.* , A&A, **412**, 587 (2003).
9. J. Colgan, M. S. Pindzola, A. D. Whiteford and N. R. Badnell, A&A, **412**, 597 (2003).
10. O. Zatsarinny, T. W. Gorczyca, K. T. Korista, *et al.* , A&A, **417**, 1173 (2004).
11. J. Colgan, M. S. Pindzola and N. R. Badnell, A&A, **417**, 1183 (2004).
12. Z Altun, A. Yumak, N. R. Badnell, *et al.* , A&A, **420**, 775 (2004).
13. O. Zatsarinny, T. W. Gorczyca, K. T. Korista, *et al.* , A&A, **426**, 699 (2004).
14. D. M. Mitnik and N. R Badnell , A&A, **425**, 1153 (2004).
15. H. P. Summers, *Atomic Data and Analysis Structure User Manual, Version 2.6* (2003). webpage *http://adas.phys.strath.ac.uk*
16. M. S. Pindzola, N. R. Badnell, and D. C. Griffin, *Phys. Rev. A*, **46**, 5725 (1992).
17. N. R. Badnell, *J. Phys. B*, **19**, 3827 (1986).
18. R. M. Pengelly, and M. J. Seaton, MNRAS, **127**, 165 (1964).
19. B. Edlèn, *Handbuch der Physik* **27**, Springer, 1964.
20. S. Böhm, S. Schippers, W. Shi, *et al.* , *Phys. Rev. A*, **65**, 052728 (2002).
21. D. W. Savin, S. M. Kahn, J. Linkemann, et al. ApJ, **576**, 1098 (2002).
22. A. Burgess, and H. P. Summers, ApJ, **157**, 1007 (1969).
23. V. L. Jacobs, J. Davis, and P. C. Kepple, *Phys. Rev. Lett.*, **37**, 1390 (1976).
24. F. Robicheaux, and M. S. Pindzola, *Phys. Rev. Lett.*, **79**, 2237 (1997).
25. S. Böhm, S. Schippers, W. Shi, *et al.* , *Phys. Rev. A*, **64**, 032707 (2001).
26. F. Robicheaux, M. S. Pindzola, D. C. and Griffin, *Phys. Rev. Lett.*, **80**, 1402 (1998).
27. N. R. Badnell, M. S. Pindzola, W. J. Dickson, *et al.* , ApJ, **407**, L91 (1993).
28. G. W. Collins, *Fundamentals of stellar structure*, Freeman, 1989.
29. M. J. Seaton, Yu Yan, D. Mihalas, and A. K. Pradhan, MNRAS, **266**, 805 (1994).
30. F. J. Rogers and C. A. Iglesias, ApJS, **79**, 507 (1992).
31. N. R. Badnell, and M. J. Seaton, *J. Phys. B*, **36**, 4367 (2003).
32. C. A. Iglesias, and F. J. Rogers, ApJ, **443**, 460 (1995).
33. M. J. Seaton, and N. R. Badnell, MNRAS, **354**, 457 (2004).
34. N. R. Badnell, M. A. Bautista, K. Butler, *et al.* , MNRAS, At Press (astro-ph/0417044)
35. S. Basu, and H. M. Antia, ApJ, **606**, L85 (2004).
36. J. N. Bahcall, A. M. Serenelli, and M. Pinsonneault, ApJ, **614**, 464 (2004).
37. J. N. Bahcall, S. Basu, M. Pinsonneault, and A. M. Serenelli, ApJ, submitted (astro-ph/0407060)
38. J. N. Bahcall, and M. Pinsonneault, *Phys. Rev. Lett.*, **92**, 121301 (2004).
39. J. N. Bahcall, Physics World, Submitted.

VIII: X-RAY SPECTRAL ANALYSIS OF INDIVIDUAL SOURCES

What is the origin of the soft excess in AGN?

Małgorzata Sobolewska* and Chris Done†

*Copernicus Astronomical Center, Warsaw, Poland
†University of Durham, Durham, UK

Abstract. We investigate the nature of the soft excess below 1 keV observed in AGN. We use the XMM-Newton data of the low redshift, optically bright quasar, PG 1211+143, and we compare it with the Narrow Line Seyfert 1 galaxy, 1H 0707-495, which has one of the strongest soft excesses seen. We test various ideas for the origin of the soft X-ray excess, including a separate spectral component (for example a low temperature Comptonized component), a reflection-dominated model, or a complex absorption model. All three can give good fits to the data, and χ^2 fitting criteria are not sufficient to discriminate among them. Instead, we favor the complex absorption model on the grounds that it is the most physically plausible.

Keywords: accretion, accretion disks – galaxies:active – X-rays:galaxies – galaxies:individual:PG 1211+143, 1H 0707-495 – atomic processes
PACS: 32.80.Fb, 97.10.Gz, 98.54.Cm, 98.62.Ra

INTRODUCTION

Several models are considered in the literature to explain the origin of the soft excess below 1 keV in AGN, which cannot be explained by either of the two emission components usually considered in accreting black holes, namely the accretion disk and the high energy power law Comptonized emission. Spectra with a strong soft excess also often show a strong deficit at ~ 7 keV which again has no obvious identification.

Partial covering models allow for a spatially non-uniform cold absorber attenuating only a fraction of intrinsic radiation. They can explain the sharp features around 7 keV. However, they require an enormous overabundance of iron (5-30 times the Solar value [1]; 5 times the Solar value [2]), and they cannot simultaneously explain the soft X-ray emission. Thus, we do not consider partial covering models in our analysis as they only provide a method to explain the sharp feature around 7 keV.

The soft excess may originate in a distinct unknown physical process which manifests itself as **a separate spectral component** (Model 1). It can be modeled e.g. with low temperature Comptonization models [3]). The apparent temperature of this component is rather similar ($\sim 0.1 - 0.3$ keV) across a diverse selection of AGN (e.g. [4]).

An obvious way to produce a fixed soft excess energy is to relate it to atomic processes rather than continuum emission. In reflection models, the dramatic strength of the soft excess requires that the source is **reflection dominated** (Model 2). This might happen if the disc fragments at high accretion rates, hiding the hard X-ray source among many clumps. Such models can successfully fit the spectra, but need a large range in ionization states of the reflecting material, and supersolar abundances ([5]).

Alternatively, the strong jump in opacity for partially ionized material could result in a soft excess from **absorption** (Model 3, [6]). The underlying continuum may be modeled

CP774, *X-Ray Diagnostics of Astrophysical Plasmas,* edited by Randall K. Smith

by only one component. The soft excess and hardening of the spectrum at high energies would be only artifacts of absorption acting at 1-2 keV, while the H-like iron K alpha resonance absorption from the same material could produce the deficit at ~ 7 keV.

RESULTS

The best fit models (and their components) together with the data to model ratios are shown in Figure 1. The models are additionally modified by galactic absorption, cold absorption at the redshift of the source, and one (in 1H 0707-495) or two (in PG 1211+143) warm absorbers modeling the narrow absorption features. Thus, the only thing that differentiates the three models is the description of the origin of the soft excess.

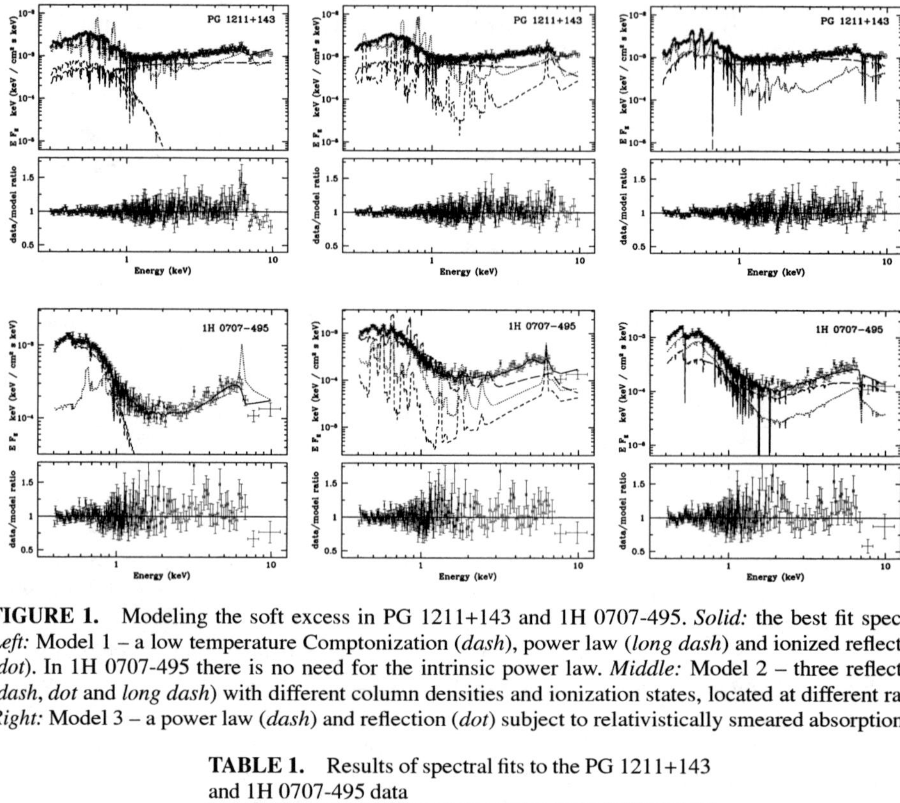

FIGURE 1. Modeling the soft excess in PG 1211+143 and 1H 0707-495. *Solid:* the best fit spectra. *Left:* Model 1 – a low temperature Comptonization (*dash*), power law (*long dash*) and ionized reflection (*dot*). In 1H 0707-495 there is no need for the intrinsic power law. *Middle:* Model 2 – three reflectors (*dash*, *dot* and *long dash*) with different column densities and ionization states, located at different radii. *Right:* Model 3 – a power law (*dash*) and reflection (*dot*) subject to relativistically smeared absorption.

TABLE 1. Results of spectral fits to the PG 1211+143 and 1H 0707-495 data

Object	Model 1	Model 2 χ^2 (*d.o.f.*)	Model 3
PG 1211+143	989 (939)	972 (937)	998 (939)
1H 0707-495	311 (288)	292 (283)	318 (287)

In PG 1211+143 the three models give virtually the same quality fits in terms of the reduced $\chi^2 = 1.04$–1.06. In 1H 0707-495 the model of complex absorption results in a fit comparable to that of the model with a distinct spectral component (with reduced χ^2 of 1.10 and 1.08, respectively). The best fit is obtained with the model of complex reflection (with iron abundance set to 2 times the Solar abundance) with reduced χ^2 of 1.03. The values of χ^2 and number of degrees of freedom in fits are presented in Table 1.

DISCUSSION AND CONCLUSIONS

Based only on the χ^2 criterion it is not possible to uniquely distinguish between the three kinds of models as each of them provides an adequate description. Moreover, many **model** uncertainties can contribute to χ^2, e.g. in both the reflection and absorption scenarios we expect an (unknown!) range of ionization states to be present. Hence direct comparison of χ^2 can be misleading when the models are known to be incomplete. Instead, we should be guided as well by physical plausibility.

The model with a distinct component involves an emission from an unknown physical process, with unknown process fixing its typical energy, which does not seem to have an analogue in the spectra of galactic black holes (GBHs) [7], and cannot simultaneously explain the 7 keV feature.

By contrast, both reflection and absorption are much more plausible, as they give a physical reason for the fixed energy of the soft excess and can reproduce the structure around iron K. However, the reflection models require quite strong supersolar abundances, and there is no evidence for reflection dominated spectra in the GBH systems.

The absorption model seems to reproduce the strong soft excesses well with moderate column densities ($\sim 10^{22-23}$ cm^{-2}) of solar abundance material. Such material would not be seen in the GBHs because it would be completely ionized by the much higher accretion disc temperatures seen in the stellar mass black hole systems. Thus the complex absorption model provides an interesting alternative because (1) it does not require a separate soft component in the spectra, (2) the hard radiation slope resulting from the fits is similar to that of GBHs in the high/soft state, (3) it describes the spectra of AGNs in terms of atomic processes, explaining the universal shape of the soft excess.

ACKNOWLEDGMENTS

This work was partially supported by the Polish Committee for Scientific Research through grants 1P03D01626 and 2P03D00322.

REFERENCES

1. Tanaka, Y., Boller, T., Gallo, L., Keil, R., & Ueda, Y. 2004, *PASJ*, **56**, L9
2. Gallo, L. C., Tanaka, Y., Boller, T., et al., W. N. 2004, *MNRAS*, **353**, 1064
3. Magdziarz, P., Blaes, O. M., Zdziarski, A. A., Johnson, W. N., & Smith, D. A. 1998, *MNRAS*, **301**, 179
4. Czerny, B., Nikołajuk, M., Różańska, A., et al. 2003, *A&A*, **412**, 317
5. Fabian, A. C., Ballantyne, D. R., Merloni, A., et al. 2002, *MNRAS*, **331**, L35
6. Gierliński, M. & Done, C. 2004, *MNRAS*, **349**, L7
7. Gierliński, M. & Done, C. 2004, *MNRAS*, **347**, 885

Chandra-LETGS observation of Mrk 279: modeling the AGN internal structure

E. Costantini[*,†], J.S. Kaastra[*], K.C. Steenbrugge[*], N. Arav[**], J.R. Gabel[**] and G. Kriss[‡]

[*]*SRON, Sorbonnelaan, 2, 8534CA Utrecht, The Netherlands*
[†]*Astronomical Institute, Utrecht University, P.O. Box 80000, 3508TA Utrecht, The Netherlands*
[**]*CASA, University of Colorado, 389 UCB, Boulder, CO, USA*
[‡]*STSI, 3700 San Martin Drive, Baltimore, MD 21218, USA*

Abstract. We present the *Chandra*-LETGS analysis of Mrk 279, a bright Seyfert 1 galaxy. The spectrum shows a variety of features arising from different physical environments in the vicinity of the black hole. We detect three absorption components, located at the redshift of the source. One of them is likely to arise from the host galaxy. An additional component due to absorption by a collisionally ionized gas at z=0 is also observed. Such an absorption can be produced in the outskirts of our Galaxy or in the local group. The emission spectrum is rich in narrow and broad emission features. We tested the hypothesis that the broad emission lines originate in the Broad Line Region (BLR), as studied in the UV band. We find that at least 2 components with different ionization parameters are necessary to simultaneously interpret the broad emission line measured in the UV (by HST-STIS and FUSE) and in the X-rays.

Keywords: X-ray spectra, active galaxies, photoionization of atoms and ions
PACS: 32.30.Rj, 98.54.Cm, 32.80.Fb

INTRODUCTION

The high spectral resolution now available in the X-ray band led to a deeper understanding of the physical processes occurring in active galaxies. Mrk 279 is a type 1 Seyfert galaxy. For these objects, according to the standard model [e.g. 1], our line of sight should intercept an outflowing, highly ionized gas. It is now recognized that many components, differing in ionization level, column density, and outflowing velocity, are actually constituting this outflow [e.g. 6, 9, 13] and that the UV and the X-ray band share at least some of these components [e.g. 13, 10]. Yet, the geometry, the distance, and therefore the density of the X-ray outflow is not yet clear [e.g. 4, 13, 10]. The observation angle for Seyfert 1 objects allows to investigate the innermost parts of the AGN, from the broad line region [BLR, see e.g. 3], at a distance of few light days from the black hole, down to the accretion disk. The bulk of the X-rays are produced here [e.g. 11]. The BLR is likely to produce X-ray radiation, but up to now this radiation was never convincingly observed in X-ray spectra [13, 12].

CP774, *X-Ray Diagnostics of Astrophysical Plasmas,* edited by Randall K. Smith

RESULTS

We processed the *Chandra*-LETGS data of Mrk 279 following the approach described in [7]. The residuals to a basic continuum model show significant excess emission, especially in the 18-24 Å wavelength range, where the main oxygen line features are (O VII triplet and O VIII Lyα). We interpret these emission features as broad lines, possibly arising from the BLR. The best fit shows that the excess can be modeled by Gaussian profiles for O VII, O VIII, N VII, and C VI. Evidence of weaker broad excesses is found for N VI and C V (Fig. 1) and possibly neon (see next paragraph for the modeling). The inclusion of broad emission lines is required to properly fit the absorbed spectrum. The best fit requires indeed absorption by two components of ionized gas. For the lower ionized component we find: $N_H \sim 1.6 \times 10^{22} cm^{-2}$, and $\log\xi \sim 0.57$, while for the high ionization component the parameters are: $N_H \sim 4 \times 10^{22} cm^{-2}$ and $\log\xi \sim 2.5$. The two absorbers seem to belong to systems with different outflow velocity, -220^{+50}_{-90} and -570^{+100}_{-70} km s^{-1} for the low and high ionization gas, respectively. Another component, whose main features are O I and O II has been modeled by a collisionally ionized gas at a temperature $t \sim 8$ eV. Mrk 279 shows also ionized absorption taking place at z=0 [8]. In the LETGS band the deepest features of this collisionally ionized component are O VII and C VI.

The Broad Line Region

Thanks to the simultaneous FUSE and HST observation [2], we could compare the X-ray line luminosity with the ones measured for the UV broad emission lines. We used Cloudy [5] for the modeling. The input ionizing spectral continuum was evaluated directly from the HST, FUSE and *Chandra* data. We chose a grid of physical parameters for the BLR clouds and then chose the best fit to our data through χ^2 minimization. In Fig. 2 we display the intrinsic integrated luminosity measured for the X-ray and UV lines (star points) *vs* the rest frame wavelength of the lines, together with our best fit. Two ionization components for the emitting medium are certainly needed in order to explain the UV and the lower ionization X-ray lines (labeled as "cold" and "warm" in Fig. 2). This result is based mainly on the more precise UV measurements. The higher ionization X-ray lines, such as C VI, N VII, O VIII, Ne IX, and Ne X may require the contribution of a higher ionization component ("hot" in Fig.2). The ionization parameters of these components are roughly $\log\xi \sim$ -1.5, 1.6 and 3.6 for the "cold", "warm", and "hot" components, respectively. The distance from the ionizing source of the emitting cloud is around 10^{16-17} cm, and the density ranges between $10^{9-12} cm^{-3}$. The ionization parameter is given by: $\xi = L/nr^2$. Here only the ionizing luminosity L is directly measured. Density effects on the emission may help in determining the parameters in the nr^2 term. However, as a caveat, the r and n values are still coupled.

REFERENCES

1. R. Antonucci, 1993, ARA&A , 31, 473

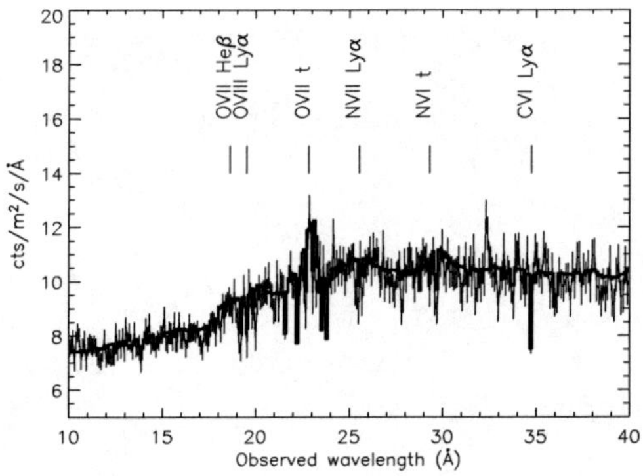

FIGURE 1. Best fit model for the LETGS data of Mrk 279. The labels refer to the broad lines added in the model.

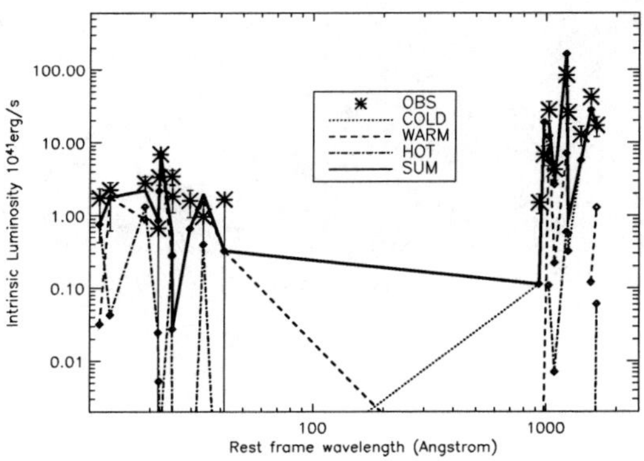

FIGURE 2. Cloudy prediction of the ionization state of the BLR emitting clouds (see text).

2. N. Arav, J. S. Kaastra, G. A. Kriss, K. T. Korista, J. Gabel, & D. Proga, 2004, astro-ph/0406200
3. J. A. Baldwin, 1997, ASP Conf. Ser. 113: IAU Colloq. 159: Emission Lines in Active Galaxies: New Methods and Techniques, 80
4. E. Behar et al. 2003, ApJ, 598, 232
5. G. J. Ferland, et al. 1998, PASP, 110, 761
6. J. S. Kaastra et al. 2000, A&A, 354, 84
7. J. S. Kaastra et al. 2002, A&A, 386, 427
8. J. S. Kaastra, N. Arav, E. Costantini et al. 2005, in prep.
9. S. Kaspi et al. 2001, ApJ, 554, 216
10. Y. Krongold, F. Nicastro, N. S. Brickhouse, et al. 2003, ApJ, 597, 832
11. K. Nandra, & K. A. Pounds, 1994, MNRAS , 268, 405
12. P. M. Ogle, K. O. Mason, M. J. Page, et al. 2004, ApJ, 606, 151
13. K. C. Steenbrugge et al., 2004, A&A, accepted.

Drs. Frits Paerels and Duane Liedahl, holding the youngest attendee, Celeste Paerels.

Diagnostics on the Location and Structure of Seyfert Warm Absorbers

Yair Krongold*, Nancy S. Brickhouse†, Martin Elvis†, Duane A. Liedahl**, Fabrizio Nicastro† and Smita Mathur‡

*IA-UNAM
†Harvard-Smithsonian Center for Astrophysics
**Lawrence Livermore National Laboratory
‡Ohio State University

Abstract. At least half of all Active Galaxies show the presence of strong absorption features produced by highly ionized material. These features are blueshifted, implying outflowing winds. The location of these winds is of great importance to understanding the structure of quasars. Variability of the absorption features in response to flux variations of the central source is one of the best diagnostics to constrain the location and structure of AGN winds. Here we show that the opacity in the absorber in NGC 3783 clearly changes (at a level $> 10\sigma$) in a timescale of 31 days responding to a factor of ~ 2 change in the ionizing continuum. There is also evidence for variability in the O VI K edge (at $\sim 3\sigma$). The observed changes are consistent with the gas producing these absorption features being close to photoionization equilibrium. The observed change in opacity implies a density $> 1 \times 10^4$ cm^{-3}, and so locates the gas within 6 pc of the X-ray source. The scenario in which the gas is composed of a continuous radial range of ionization structures is ruled out, as in such scenario, no opacity variations are expected. Rather, the structure of the absorber is likely composed of heavily clumped gas.

Keywords: galaxies: absorption lines – galaxies: Seyferts – galaxies: active – galaxies: X-ray

INTRODUCTION

The X-ray and UV spectra of at least half of all Active Galactic Nuclei (AGNs) show the presence of prominent features produced by highly ionized material in our line of sight (Crenshaw, Kraemer and George 2003). The study of these warm absorbers has been intensified lately, due to the improvement in the quality of UV and X-ray spectra from STIS, FUSE, and the grating spectrometers on *Chandra* (HETGS, LETGS) and XMM-Newton (RGS). Both UV and X-ray spectra show absorption lines with moderate velocity blueshifts ($\sim$1000 km s^{-1}), implying outflowing winds. The mass loss rate from these outflows can be a substantial fraction of the accretion rate needed to power the AGN continuum, and so these 'quasar winds' are likely to be dynamically important to AGNs (Mathur, Elvis & Wilkes 1995). Understanding the physical state or states of this gas, and hence the properties of AGN winds would be a valuable advance in the understanding of quasars.

In particular, the location of these winds is of great importance to constrain the structure of these objects, but is currently a matter of great debate. If the wind originates from the accretion disk, widely thought to be the origin of the AGN continuum, then all the emission and absorption features in AGN spectra, including the broad emission

CP774, *X-Ray Diagnostics of Astrophysical Plasmas,* edited by Randall K. Smith

lines (BELs), could arise from this wind (Elvis 2000). If instead the wind arises from the inner edge of the obscuring torus (Krolik & Kriss 2001) then several disconnected gaseous components are present in AGNs, and the mass flux in the wind is large (Netzer et al. 2003, Behar et al. 2003, Ogle et al. 2004). A large scale wind, with a continuous radial decrease in ionization parameter (Kinkhabwala et al. 2002) would tie together type 1 and type 2 AGNs, as high ionization kpc scale 'cones' are seen around several type 2 AGNs in Chandra images (Ogle et al. 2003 and references therein), and these would produce high ionization absorbers if viewed along the cone axis toward the nucleus.

These are quite different views of the same observations, and better diagnostics are needed to decide between these scenarios. Variability of the X-ray absorption features can provide a direct test of these models, as rapid variability requires clumped gas, while in continuous radial structures of ionization parameters the variability effects should be washed out by the average absorption in the flow. Variability of the features can also shed light on the location of the absorber. Studies of X-ray absorber variability in AGNs have been undertaken for a decade or so. The results from these early papers, based on low resolution X-ray spectra, suggest variations. However these variations are hard to interpret since high resolution X-ray spectra from Chandra and XMM-Newton now show that the absorbers are far more complex than the simple models used in those papers. The high resolution X-ray spectra though, typically have insufficient signal-to-noise to subdivide in time bins. A rare exception is the integrated 900ks exposure on NGC 3783 with the Chandra gratings.

THE PROMINENT WARM ABSORBER IN NGC 3783

NGC 3783 is an extensively studied bright Seyfert galaxy, at a redshift 0.00976 ± 0.00009 (2926 ± 28 km s^{-1},de Vaucouleurs et al. 1991). NGC 3783 has been the subject of extensive monitoring by the High Energy Transmission Grating Spectrometer (HETGS) on board the *Chandra X-ray Observatory* with a total exposure of $\approx$ 900 ksec in six observations (Kaspi et al. 2002). The first observation was carried out on January 2000 and had a duration of 56 ksec. Between February and June of 2001, the following five observations were taken, each one with an exposure time of 170 ksec. The resulting spectrum covers the 0.5-10 keV energy range, and is the best obtained so far for an ionized absorber in a Seyfert galaxy, with more than 2000 counts per resolution element at 7 Å (for a detailed explanation of the data reduction see Krongold et al. 2003). This spectrum showed the presence of more than one hundred absorption lines from a wide range of ionized species. A few weak emission lines, mainly from O and N, were also detected. The spectrum further revealed a wide absorption feature in the $16-17$ Å range. Kaspi et al. (2002) identified this feature as an unresolved transition array (UTA), arising from numerous Fe M-shell inner-shell $2p-3d$ transitions. The spectrum is presented in Figure 1.

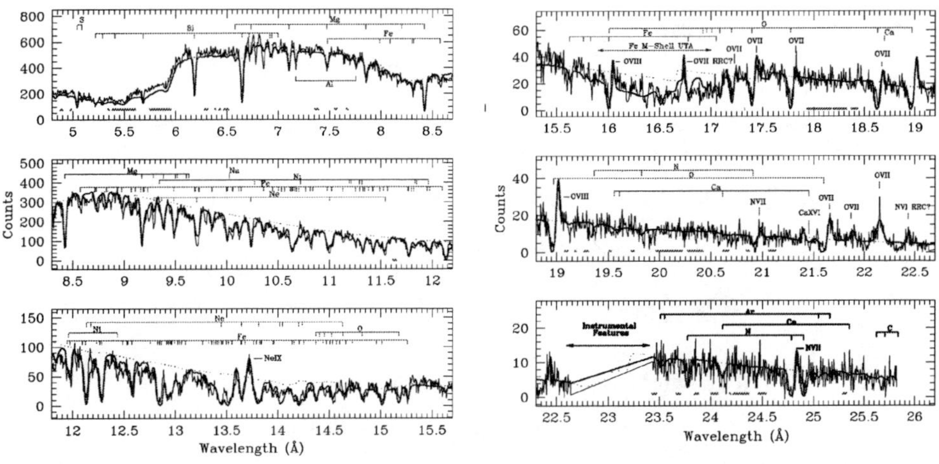

FIGURE 1. Two phase absorber model plotted against the first-order MEG spectrum of NGC 3783. Absorption lines predicted are marked in the top. Single labels stand for emission lines. Line free zones are indicated at the bottom of each panel. The continuum level (including edge continuum absorption) is overplotted for comparison (dotted line).

Modelling the Spectra

With the aim of modelling in a self-consistent way all the absorption features observed in the spectra of UV and X-ray sources, we have developed PHASE (Krongold et al. 2003), a code that calculates absorption due to an ionized plasma. The parameters of the code are (1) the ionization parameter (defined over the entire Lyman continuum, as $U = Q(H)/4\pi r^2 n(H)c$ with $Q(H)$ the number of H ionizing photons per second, r the distance to the source, $n(H)$ the H density, and c the speed of light.), (2) the equivalent hydrogen column density, (3) the outflow velocity, and (4) the internal micro-turbulent velocity of each one of the absorbers.

PHASE draws the ionization balance from CLOUDY (v.90.04, Ferland 1997), and includes ~ 4000 lines arising from the ground level. The line list was mainly extracted from the Astrophysical Plasma Emission Database (APED v. 1.3.0, Smith et al. 2001), but we also added nearly 200 features from the "Atomic Data for Permitted Resonance Lines" by Verner et al. (1996), as well as data for inner shell transitions extracted from Behar et al. (2001).

We performed spectral fitting of the NGC 3783 data using PHASE integrated into the Sherpa package in CIAO.

TABLE 1. Two Phase Absorber Model for the full (900 ks) *Chandra* observation of NGC 3783 (Krongold et al. 2003)

Parameter	High-Ionization Component	Low-Ionization Component
Log U*	0.76±0.1	-0.78±0.13
Log N_H (cm^{-2})*	22.20±.22	21.61±0.14
V_{Turb} (km s^{-1})	300	300
V_{Out} (km s^{-1})*	788±138	750±138
T (K)[†]	$9.52 \pm 0.44 \times 10^5$	$2.58 \pm 0.39 \times 10^4$
Log T/U (∝P)**	5.22±0.12	5.19±0.20

* Free parameters of the model.

[†] Derived from the column density and ionization parameter, assuming photoionization equilibrium.

** The pressure $P \propto n_e T$. Assuming that both phases lie at the same distance from the central source $n_e \propto 1/U$, and $P \propto U/T$.

The Warm Absorber

The spectrum of NGC 3783 could not be reproduced by a single absorbing component. Therefore, we included a second one with different ionization. Our results are listed in Table 1. The plot with the model is presented in Fig. 1. The 0.5-10 keV intrinsic continuum of the source was well represented by a single power law ($\Gamma = 1.53$) and a soft black-body component (kT $\sim$ 0.1 keV).

We find no evidence for cosmic dust or anomalous abundances in any of the absorbers. It is noteworthy that the outflow of both components is consistent with a single velocity. While kinematically indistinguishable and with column densities a factor 4 apart, our two components have dramatically different ionization parameters: the high ionization phase (HIP) has a value $\approx$ 35 times larger than the low ionization phase (LIP). The HIP, gives rise to the absorption by N VII, O VIII, Ne IX-X, Mg X-XII, Al XII-XIII, Si XI-XIV, S XIII-XVI, Ar XII-XVII, Ca XIII-XVIII, Fe XVII-XXII, and Ni XIX-XXI. On the other hand, the LIP produces absorption by C VI, N VI, O VII, Ne V-VIII, Mg VII-IX, Si VII-IX, and the Fe M-shell inner transitions forming the UTA (Fe VII-XII).

The model presents a surprisingly simple picture: two gas outflows with quite different temperatures. The model also suggests that no third component with high H column density and intermediate U is consistent with the data. Six free parameters are sufficient to model > 100 absorption features. An important result is that while O VIII and Fe XVII-XXII are the main features of the hotter phase, the signature of the cooler phase can be attributed to the UTA and the O VII lines. This is noteworthy, since traditionally, the O VII absorption has been associated with that of O VIII and warm absorbers. In our model, O VII is consistent with a cooler medium producing significant amounts of O VI and other lower ionization species. Without the constraints introduced by the UTA this distinction could not be realized (for more details see Krongold et al. 2003).

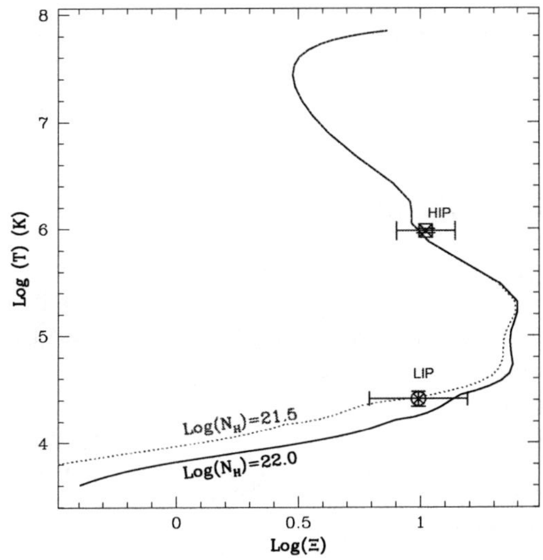

FIGURE 2. Curve of thermal stability derived from our analysis. Pressure equilibrium between the HIP and the LIP can be observed.

Two Phases of the Same Medium?

The presence of a two phase medium is suggested by the fact that the two absorbing components are in pressure equilibrium: the gas pressure (P) of each phase is proportional to the electron density (n_e) and the temperature (T); therefore, with the adopted definition of the ionization parameter, $P \propto U/T$. Under the assumption that the two components lie at the same distance from the central source, as is plausible given their identical kinematics, the two phases turn out to have the same pressure (see Table 1). To further illustrate this point, we show in Figure 2 the thermal equilibrium curve (S curve) derived for our model. The pressure equilibrium is evident. We notice that the HIP lies in the intermediate part of the curve, in a region where the slope is negative. Regions of the curve with negative slope are thermally unstable. However, the shape of the thermal equilibrium curve strongly depends on (1) the SED, particularly on the unobservable region between 100 Å and 912 Å, and (2) on the metallicity of the material. Considering different shapes for the SED in this unknown gap as well as different metalicities can give rise to thermal curves with an intermediate stable region in the region occupied by the HIP.

It is important to notice that a two phase medium, with the components in pressure equilibrium has also been found in the warm absorber present in NGC 985 (Krongold et al. 2004).

OPACITY VARIATIONS OF THE WARM ABSORBER IN NGC 3783

The High and Low State Spectra

To study the temporal variability of the absorber in NGC 3783, we proceeded in the following way: Two of the six observations (obsids 2090 and 2091) are separated by only 3 days and have extremely similar count rates (0.38 cts s^{-1}, 5-25 Å), but lower fluxes than the rest of the observations (see Netzer et al. 2003 for detailed light curves). Then, we coadded these two observations to obtain a low state spectrum (hereafter LS). The observation with the highest count rate (obsid 2093; 0.74 cts s^{-1} i.e., twice the count rate of the LS) began ~31 days after the end of the LS observation. We will refer to this observation as the "high state" (hereafter HS). This grouping represents the best configuration to study variations in the ionization state of the absorbing gas, in response to flux variations. The other 3 observations in the 900 ks exposure have flux levels intermediate between the HS and LS, and are therefore, not suitable to detect variability in the absorber (see Krongold et al. 2005 for further details).

Variability of the Ionized Absorber

In Figure 3 we present a comparison of the LS and HS spectra between 15 and 18 Å, the region of the UTA. The data has been grouped in bins of size 0.25 Å to increase the S/N ratio of the spectra. Although such binning washes out the signature of isolated narrow absorption lines, it is better for the study variability in broad blended features, such as the UTA. The LS has been scaled by a factor 3.0 to match the flux level of the high state. However, we note that the results presented here are not sensitive to this factor. There is a clear shift of ~ 0.2 Å in the position of the UTA toward shorter wavelengths in the HS compared to the LS. This shift is highly statistically significant and is far larger than wavelength calibration uncertainties. The change can only be attributed to a physical change in the absorber.

To compare the observed opacity variation with theoretical predictions from photoionization models, we have generated HS and a LS absorption models with PHASE. We used the best (5-25 Å) fitting model for the low ionization component described in the last section (and presented in Table 1) as an "average state" of the gas producing the UTA, as this model was based on the averaged spectrum of the six observations. Since the count rate from the LS to the HS changed by a factor ~ 2, we have selected values for the ionization parameter U of the high and low states also separated by a factor of 2. We have used the equivalent H column density, and the turbulent and outflow velocities of the absorber derived from the 900 ks model (Table 1).

Figure 3b presents a comparison of these high and low ionization models. There is a striking similarity between the changes observed in the spectra and the changes predicted by the models. In particular, the spectra clearly shows that the shape of the UTA (along with the O VII K edge located at 16.7 Å, a feature heavily blended with the UTA) is varying as expected for gas close to photoionization equilibrium. Furthermore, the O VI

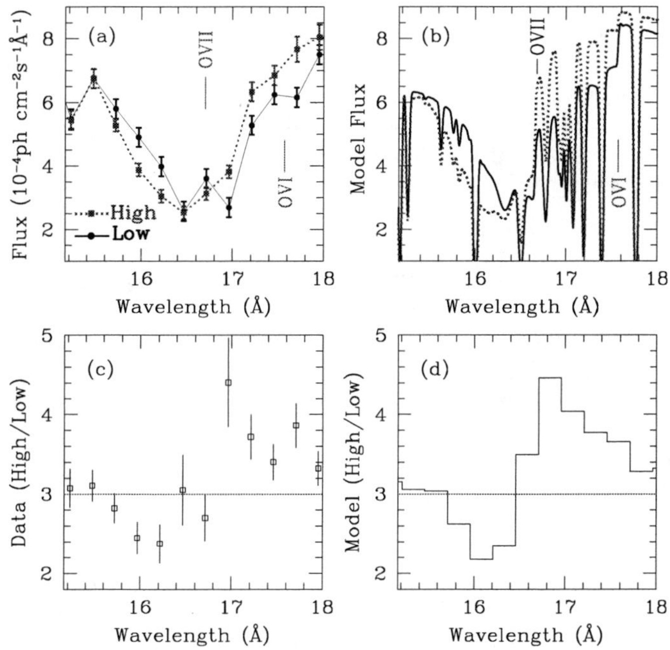

FIGURE 3. (a) *Chandra* HETGS spectra of NGC 3783 binned in channels of 0.25 Å in size. The low state is scaled up by a factor of 3. A clear variation in the position of the UTA can be observed, as well as a change in the O VI K edge (17.6 Å). (b) Photoionization models for the expected opacity variation by gas in photoionization equilibrium, as a response to a change by a factor of 2 in flux of the central source. (c) Ratio between high and low state data. (d) Ratio between high and low state model, binned at 0.25 Å. The significance in the variation is $> 10\sigma$.

K edge located at 17.6 Å also appears to be varying as predicted by the model.

In panel (c) of Figure 3 the ratio between the observed HS and LS spectra is presented. The errors in this figure have been calculated in quadrature from the errors in each spectrum. The variation in the ratio is again highly significant, at a level $> 10\sigma$ for the UTA, and $\sim 3\sigma$ for the O VI edge. Figure 3d, shows the ratio between the high state and low state models, and for clarity we have applied the same binning factor used for the data. Both panels show striking similarities. We conclude that the LIP, responsible for the absorption by the UTA, is responding to the ionization continuum and is possibly in photoionization equilibrium.

Using the UTA to Constrain the Changes in the Ionization State of the Gas

It is important to note that the UTA sets tight constraints in the ionization state of the absorber because (1) this feature is strongly dependent on the charge states of Fe producing it, since several charge states of the same element are observed at the same time, and (2) because it is a broad feature, and therefore its detection involves a large number of detector channels which allows for a heavy binning of the spectra to consider large number of counts (see also Behar et al. 2003). We stress, then, that the observed change in the position of the UTA is a highly robust measurement of the opacity variation in the ionized absorber in NGC 3783.

The Structure of Ionized Absorbers

The observed variability of the UTA helps in understanding the structure of ionized absorbers in AGN. A popular model for the structure is that of a continuous range of ionization structures spanning a large location in the nuclear environment (rather than a localized absorber), and spanning more than two orders of magnitude in ionization parameter (e.g. Ogle et al. 2004, and references therein). In such a model many ions of the same element have comparable column densities. Thus no variability is expected in the opacity of the absorber with moderate flux variations (by a factor < 3), because the average column density of each ion remains constant in the flow. This is because the amount of material in the line of sight changing from charge state i to charge state $i+1$, as the result of a continuum variation, is almost equally replaced by the amount of material transformed into charge state i from charge state $i-1$. Thus variation is only expected at the boundaries of the flow. In this case the observed UTA represents the "averaged absorption" from the dominant Fe charge states in all shells and will remain constant to moderate continuum variations. The observed variation of the UTA rules then out a continuous radial range of ionization stages, and implies that the absorbing gas is highly clumped.

The Location of the Absorber

The change in the opacity of ionized plasma depends not only on the change of flux, but also on the change in the spectral energy distribution (SED) of the central continuum source. This should be kept in mind because there is a larger change in the soft X-ray region of the spectra than in the hard region between the LS and HS observations (Netzer et al. 2003). However, the data does not allow us to measure spectral variations beyond 25 Å, which is the most critical part of the spectrum for the ionization state of the LIP. This means that the soft excess might extend up to the Lyman limit, or might rather be a localized feature, fading out at relatively large energies (i.e. $E < 0.5$ keV). Evidently these two possibilities of SED would produce dramatically different effects on the opacity of the ionized gas. Additional complexity is added because at least part of the changes observed in the soft band are not intrinsic to the continuum source, but produced by the change in opacity of the absorber. All these effects make an exact determination of the expected variation of the gas in photoionization equilibrium impossible to achieve. Such determination would require a self consistent global model, over time resolved broad band data. However, we stress again, that the change in the shape of the UTA is consistent with a change in ionization parameter by a factor of 2. Since the observed total flux changed also by this factor, there is support for a scenario where the LIP is close to photoionization equilibrium.

Assuming photoionization equilibrium applies, and using the temperature and ionization parameter deduced from our model (Table 1), and the recombination times from the leading charge states of Fe producing the UTA (Shull & van Steenberg 1982), we can set limits on the density and location of the absorber. Considering the elapsed time between the LS and HS ($\sim$ 31 days) as an upper limit for the destruction timescale of a single ion (see Nicastro et al. 1999) we find $n_e > 1 \times 10^4$ cm^{-3} and D$<$ 5.7 pc.

Using an XMM observation, Behar et al. (2003) reported a slight shift in the position of the UTA in NGC 3783 during a change of flux by a factor of 2 in a timescale $\sim$ 1.5 days, but rejected photoionization equilibrium arguing that the change in the position of the UTA would have had to be 0.4 Å with such change in flux. However, this calculation considers only the change in opacity for Fe. Including absorption from all the ions in the LIP reduces the expected difference to 0.2 Å, because of the presence of bound-free transitions (particularly the O VII and O VI K edges, see Fig. 3b). Therefore, the variation observed by Behar et al. (2003) might still be consistent with photoionization equilibrium. Furthermore, the S/N ratio of their low and high state data might not be enough to detect such variation in a significant way. Deciding this is beyond the scope of the present paper, however we note the following: (1) If the XMM S/N ratio is indeed sufficient to detect the expected opacity variation, and this variation is not observed, then the gas should be located between 5.7 and 1.3 pc from the central source excluding the Broad Line Region, as suggested by Behar et al. (2003). (2) If the expected variation is detected, the gas should be closer than 1.3 pc. (3) If the S/N ratio is not sufficient to detect the expected opacity variation, then the gas can lie anywhere within 5.7 pc.

In these last two cases the location of the absorber could thus, well be at subparsec scales, as suggested by Reeves et al. (2004, Distance $<$ 0.1 pc) for the Fe K-shell absorption. This is also consistent with measurements for ionized absorbers in other objects, which suggest that the location of the absorber might be similar to that of the broad emission line region, e.g. NGC 4051 (Nicastro et al. 1999), NGC 3516 (Netzer et al. 2002), and NGC 7469 (Blustin et al. 2003).

In any case, the detected variability presented here rules out the presence of an absorber extending tens or hundreds of parsecs outside the nucleus (and a continuous range of ionization stages), and rather suggests a localized absorber, as has also been suggested by the evidence of transverse motion of the absorbing gas with respect to our line of sight in several objects (see Crenshaw, Kraemer and George 2003, and references therein). Given these properties, it is clear that the absorber (at least the LIP) in the nuclear environment of NGC 3783 is very different from the extended X-ray emission gas observed is Seyfert 2 galaxies.

CONCLUSIONS

We present a self consistent model of the 900 ks *Chandra* HETGS spectrum of NGC 3783, the best X-ray spectrum available so far from a Seyfert galaxy. The warm absorber model leads to a simple picture consisting of a two phase wind. The model can reproduce more than 100 features with only 6 parameters. The two absorbing components are in pressure equilibrium, and are consistent with a single outflow. We unambiguously ($> 10\sigma$) detect opacity variations in the Fe M-shell unresolved transition array (UTA), in response to an ionizing continuum change by a factor of 2. The UTA is the best diagnostic for detecting opacity changes in ionized plasma. The changes are as expected for gas close to photoionization equilibrium implying a density $n_e > 1 \times 10^4$ cm^{-3}. This result locates the absorbing material within 6 pc of the central source. The strong change in ionization state rules out a picture where the absorber extends radially in a continuous flow over hundreds of parsecs, implying that this gas is different than the one observed in emission in Seyfert 2 galaxies. The observations also imply that the gas is heavily clumped, since in a continuous distribution of ionization stages no variability is expected with moderate flux variations.

ACKNOWLEDGMENTS

This research has been partly supported by CONACyT grant 40096-F, NASA Contract NAS8-39073 (*Chandra* X-ray Center), NASA grant NAS G02-3122A, and Chandra General Observer Program TM3-4006A.

REFERENCES

1. Behar, E., Rasmussen, A. P., Blustin, A. J., Sako, M., Kahn, S. M., Kaastra, J. S., Branduardi-Raymont, G., & Steenbrugge, K. C. 2003, ApJ, 598, 232
2. Behar, E., Sako, M., & Kahn, S. M. 2001, ApJ, 563, 497
3. Blustin, A. J., et al. 2003, A&A, 403, 481
4. Crenshaw, D. M., Kraemer, S. B., & George, I. M. 2003, ARA&A, 41, 117
5. de Vaucouleurs, G., de Vaucouleurs, A., Corwin, H. G., Buta, R. J., Paturel, G.,
6. Elvis, M. 2000, ApJ, 545, 63
7. Ferland G.J. 1997, CLOUDY Version 90.04
8. Kaspi, S. et al. 2002, ApJ, 574, 643
9. Kinkhabwala, A., et al. 2002, ApJ, 575, 732
10. Krolik, J. H. & Kriss, G. A. 2001, ApJ, 561, 684
11. Krongold, Y., Nicastro, F., Brickhouse, N.S., Elvis, M., Liedahl D.A. & Mathur, S. 2003, ApJ, 597, 832
12. Krongold, Y., Nicastro, F., Elvis, M., Brickhouse, N.S., Mathur, S. & Zezas, A. 2004, ApJ, in press
13. Krongold, Y., Nicastro, F., Elvis, M., Brickhouse, N.S. & Mathur, S. 2005, ApJ, in press
14. Mathur, S., Elvis, M., & Wilkes, B. 1995, ApJ, 452, 230
15. Netzer, H., Chelouche, D., George, I. M., Turner, T. J., Crenshaw, D. M., Kraemer, S. B., & Nandra, K. 2002, ApJ, 571, 256
16. Netzer, H., et al. 2003, ApJ, 599, 933
17. Nicastro, F., Fiore, F., & Matt, G. 1999, ApJ, 517, 108
18. Ogle, P. M., Mason, K. O., Page, M. J., Salvi, N. J., Cordova, F. A., McHardy, I. M., & Priedhorsky, W. C. 2004, ApJ, 606, 151
19. Ogle, P. M., Brookings, T., Canizares, C. R., Lee, J. C., & Marshall, H. L. 2003, A&A, 402, 849
20. Reeves, J. N., Nandra, K., George, I. M., Pounds, K. A., Turner, T. J., & Yaqoob, T. 2004, ApJ, 602, 648
21. Shull, J. M. & van Steenberg, M. 1982, ApJS, 48, 95
22. Smith, R. K., Brickhouse, N. S., Liedahl, D. A., & Raymond, J. C. 2001, ApJL, 556, L91
23. Verner, D. A., Verner, E. M., & Ferland, G. J. 1996, Bulletin of the American Astronomical Society, 28, 907

Outflows, Accretion Disks & the IGM: High Resolution Chandra & XMM Spectroscopy of Active Galactic Nuclei

Tahir Yaqoob[*,†]

Department of Physics & Astronomy, Johns Hopkins University, 3400 N. Charles St., Baltimore, MD 21218, USA
NASA/GSFC, Code 662, Greenbelt, MD 20771, USA

Abstract. We give a brief review of what has been learned from high spectral resolution X-ray grating spectroscopy of type 1 active galaxies since the launch of *Chandra* and *XMM-Newton*.

Keywords: X-rays, Active Galaxies, Spectroscopy, Emission Lines, Absorption Lines, IGM
PACS: 90

PHOTOIONIZED OUTFLOWS

A thorough review (in 2003) of our observational and theoretical understanding of the so-called 'warm absorbers', or photoionized outflows, prevalent in over $\sim 50\%$ of Seyfert 1 galaxies, was given in [1]. Since then, there are now enough grating observations of Seyfert 1 galaxies to begin to perform collective studies in order to deduce some general properties of the photoionized outflows. At 0.5 keV the *Chandra* Medium Energy Grating (MEG) has a FWHM velocity resolution is ~ 300 km s^{-1}, a factor of ~ 1000 better than CCD resolution at the same energy. At 2 keV the MEG FWHM resolution is ~ 1100 km s^{-1}. The *Chandra* Low Energy Grating (LEG) and *XMM-Newton* Reflection Grating Spectrometers (RGS) both have spectral resolutions which are about half that of the MEG. Blustin et al. [2] have collected together results from both *Chandra* and *XMM-Newton* for 23 sources, and details of references for works on the individual AGN are given in that paper. McKernan et al. [3] have collected a sample of 15 Seyfert 1 galaxies observed with the *Chandra* MEG and performed a uniform analysis in order to deduce overall properties of the sample. These studies have been complemented by very detailed modeling of NGC 3783 by several research groups, using the deepest exposure with the *Chandra* MEG (~ 900 ks) of a Seyfert galaxy to date [4, 5, 6]. *XMM-Newton* RGS data have also been extensively studied and modeled [7]. UV observations, some simultaneous with the X-ray observations, have been used to study the relation between the UV and X-ray absorbers in NGC 3783 (see [8] and references therein) and other AGN (see [1, 9] and references therein).

We summarize the findings of the above studies below.

Photoionization. It has been clear from the beginning that the dominant ionization mechanism is photoionization: a range of highly ionized species are present (He-like and H-like ions of the lighter, abundant elements like N, O, and Ne are very common) at relatively low temperatures (the latter indicated by certain line ratios and/or recombi-

CP774, *X-Ray Diagnostics of Astrophysical Plasmas,* edited by Randall K. Smith

nation continua).
Blueshifted Absorption. The soft X-ray absorption lines are blueshifted with respect to the systemic velocity of the galaxy by $\sim 0-1000$ km s^{-1} (the lower end of this range being most common). These blueshifts indicate outflowing gas. Evidence for much faster outflows has been found in several quasars by *XMM-Newton* from absorption features due to highly ionized Fe (e.g. [10, 11, 12]). Outflow velocities in the range $\sim 20000-60000$ km s^{-1} have been reported, with associated absorbing column densities typically exceeding 10^{23} cm^{-2} (but see 'Intervening Absorption' below).
Line Widths/Kinematics. The X-ray absorption lines are typically unresolved, even at the highest spectral resolution afforded by *Chandra*. In a few tentative cases resolved absorption lines have been found (e.g. see [3]) and in some cases, such as NGC 3783, it has even been possible to deduce the presence of more than one kinematic component to the absorber from the absorption line profile (e.g. [4]).
Multiple Components. It has been found that invariably, more than one ionized absorber component is required to account for the data. The different components can have very different column densities and ionization parameters. Typically the column densities of the warm absorber components are in the range $\sim 10^{21-22}$ cm^{-2}, and the ionization parameter ($\xi \equiv L_{\rm ion}/n_e r^2$) in the range $\sim 10^{1-3}$ [2, 3]. If one calculates the pressure of the different components of the absorber, it is often found that in the same source the pressure appears to be the same for different components which may have very different ionization parameters and column densities. This has suggested that the warm absorber components may be on the near-vertical unstable part of the heating/cooling curve and may therefore exist in the form of a multi-temperature wind [13]. However, recent time-resolved MEG spectroscopy of NGC 3783 argues against such a model [14].
Emission Lines. Emission lines in type 1 AGN in the soft X-ray band are much less common than absorption lines (see [2, 3] and references therein). Unresolved emission lines have been clearly detected in a few sources (notably He-like and/or H-like lines from the lighter elements). In addition, broadened, soft X-ray relativistic emission lines have been reported [15], but there is ambiguity in the process of their deconvolution from the rest of the complex spectrum and their existence remains rather controversial [16, 17].
Key Physical Parameters. Unfortunately, some of the key physical properties of the photoionized outflows in AGN are still elusive. In particular, the physical location of the absorbing material, its covering factor, its volume filling factor, and its density cannot be directly extracted from the current data without making some assumptions. For example, to obtain the distance of the absorber from the central X-ray source, r, requires a detailed variability study, but only a handful of sources even have more than one observation at high spectral resolution so there is insufficient data to address the question. Estimates range from tenths of a parsec to tens of parsecs (e.g. [6, 14, 18]). Blustin et. al. [2] derived r for a sample of objects, without the use of variability data, based on the assumption that the measured velocity must be greater than the escape velocity. They derived distances which would put the outflow in the vicinity of the putative obscuring torus that is required by AGN unification schemes. If this is correct, torus-wind models of the photoionized outflows would be favored. However, the basic assumption may not be valid. We do not know what drives the outflows and whether they are gravitationally bound to the central mass or not. It is also important to bear in mind that the warm

absorbers may only correspond to that part of the outflow which is *observable*. For example, there may a part of the outflow closer to the central engine which is completely ionized and therefore rendered unobservable.

Whilst the covering factor may be crudely estimated simply from the frequency of occurrence of warm absorbers in AGN, the volume filling factor cannot be easily estimated. Blustin et. al. [2] estimate that the filling factor is $< 8\%$, but this is based on the assumption that the momentum carried by the photoionizing radiation is equal to the momentum carried by the outflow. However, this assumption is not necessarily valid. We do not know whether the outflow is actually driven by the radiation. For example, the outflow may originate close to the accretion disk and be influenced by dynamical effects. The density of the absorber cannot be deduced from the model-fitting of time-averaged spectra. There are insufficient data to estimate it with useful precision from variability studies (but see [6]). Although the density can be estimated from certain emission-line ratios, it is not known whether the emission lines originate in the same material as that responsible for the absorption.

If one cannot measure the filling factor of the outflows, or their density, it is not possible to calculate absolute mass outflow rates. Blustin et. al. [2] calculated mass outflow rates for a sample of AGN based on their assumptions about the filling factor. Even with the small filling factors assumed by [2], the mass outflow rates are comparable to the accretion rates in over half of the sample. Although [3] did not calculate filling factors, they concluded that the mass outflow rates for their MEG sample could easily be comparable to the accretion rates. This result is somewhat puzzling. However, [2] concluded that since the kinetic energy in the outflows is of the order of only $\sim 1\%$ of the bolometric luminosity, the outflows are energetically unimportant to the AGN, and may not even be related to the AGN at all. Such a conclusion may be premature: by analogy, even the accretion efficiency in an AGN may be $\ll 1$, we know that the accretion certainly does influence profoundly the physics of the AGN.

Correlations. Using the *Chandra* MEG sample, McKernan et al. [3] searched for correlations between various physical quantities associated with the outflows, as well as between outflow parameters and other properties of the AGN. No correlations were found between the ratio of ionizing luminosity to the Eddington ratio ($L_{\rm ion}/L_{\rm Edd}$) and the column densities of the absorbers or the ionization parameters. Nor was any correlation found between $\dot{m}/C$ (where $\dot{m}$ is the mass accretion, and C is the product of the filling factor and covering factor), and the black-hole mass. On the other hand, a tentative correlation was found between $\dot{m}/C$ and $L_{\rm ion}/L_{\rm Edd}$. There was also a tentative anti-correlation between the outflow velocity and $L_{\rm ion}/L_{\rm Edd}$. The latter two correlations, if true, would imply that the less efficient accretors have *faster* outflows. This might be expected if accreting matter that experiences a greater 'barrier' to accretion that results in the same matter 'bouncing' back and ejected in a faster outflow. Further, the outflow velocity amongst sources in the sample appeared to fall into a bimodal distribution: $v < 300$ km s^{-1} and $v > 600$ km s^{-1}. Finally, no correlation was found between any of the properties of the photoionized outflows and the properties of the core of the Fe-K emission line at ~ 6.4 keV, although the statistical errors on the equivalent width (EW) of the Fe-K line tend to be large.

Intervening Absorption and 'WHIGM'. In a recent study [19] searched for absorption features specifically due to O VII and O VIII at $z = 0$ in the same sample of low-redshift

AGN observed with the *Chandra* MEG as [3, 20]. Evidence for absorption in local filaments in about half of the lines-of-sight was found. Curiously, it was also found that the absorption features in the Fe K band found by *XMM-Newton* in PG1211+143 and in PDS 456 may also be due to WHIGM, although the very high column densities of the absorption lines cannot easily be reconciled with such an interpretation. Finally, [21] were fortunate to catch the blazar Mkn 421 in a very high flux state and thus detected several absorption lines due to intervening systems, with high signal-to-noise. This yielded a measurement of the mass-density consistent, within the uncertainties, with the mass density of the missing baryons which had thus far remained elusive.

FE K EMISSION LINES

Line Core. *ASCA* found that broad Fe K emission lines in Seyfert 1 galaxies were common [22, 23], and the broadening has been attributed to relativistic effects in an accretion disk rotating around a black hole ([24] and references therein). A major limitation of the ASCA (CCD) data was that, with a velocity resolution of only $\sim$ 7000 km s^{-1} FWHM in the Fe K band, it was not possible to disentangle any Fe Kα emission-line component that might originate in matter more distant from the central X-ray source than the disk (such as the BLR, NLR, or the putative obscuring torus). *Chandra* HEG observations have improved this situation, since the HEG has a FWHM resolution of $\sim$ 1800 km s^{-1} at 6.4 keV. The core of the Fe Kα emission line (which carries a significant fraction of the total line luminosity) was studied with the HEG in a sample of 15 low-redshift Seyfert 1 galaxies and the results reported in [20]. In all except one object (Akn 564, for which the signal-to-noise was poor) a line peak at 6.4 keV was found (mean energy 6.404 ± 0.005 keV), corresponding to fluorescence in neutral or lowly-ionized Fe. The mean width for the Fe Kα line core was measured to be 2380 ± 760 km s^{-1} FWHM. The EWs measured for the line core were in the range $\sim 50 - 100$ eV in individual sources. Even in cases in which the line core was unresolved by the *Chandra* HEG, more than one location for the line emission remains a possibility (emission lines from the BLR, torus, and NLR are expected to have FWHM of $\sim 2000 - 5000$ km s^{-1}, $\sim 700 - 1000$ km s^{-1}, and a few hundred km s^{-1} respectively).

Notable Chandra HEG Results. Some of the most interesting results that have come from measurements in Seyfert 1 galaxies of time-averaged spectra of the Fe K region using the highest spectral resolution available are the following. (1) Resolution of a broad Fe K line in NGC 7314 (as measured by *ASCA*) into three distinct Fe K lines (Fe I-XVII Kα, Fe XXV, and Fe XXVI Lyα), unresolved by the *Chandra* HEG [25]. (2) A double-peaked line in IC 4329A, one peak at $\sim$ 6.4 keV, and the other at $\sim$ 6.9 keV [26]. (3) A possible 'Compton-shoulder' to the red side of the main peak due to Fe Kα in NGC 3783 [4]. A Compton shoulder is not a unique description of the data: a disk-line profile can also account for the data [27].

Blue Wings. In parallel with the high-resolution *Chandra* grating measurements, *XMM-Newton* has made advances in measurements of the Fe K line profiles in AGN over *ASCA* with EPIC, its high-throughput CCD detectors (see [28] and references therein). Notably, in several AGN, a line profile characterized by a peak at $\sim$ 6.4 keV, with a prominent *blue* wing has been found in several AGN (Mkn 509 [29, 30, 31]; NGC 5506

[32]; NGC 7213 [33]).

Broad vs. Narrow Lines. Unprecedented quality of data has been obtained for the archetypal 'broad Fe K line' AGN, MCG −6-30-15 with *XMM-Newton* and the data studied by several research groups ([34] and references therein). However, in most cases, *XMM-Newton* does not find evidence of the 'classic' relativistic, skewed Fe K line profile of the kind observed in MCG −6-30-15. This is actually not inconsistent with *ASCA* results, since a *ASCA* found a variety of line shapes [23]. We find in fact that the broad Fe K *Chandra* HEG line profiles are consistent with the *ASCA* data for the 15 AGN in the HEG sample [35]. The key is that one must be careful to distinguish the case that a broad, relativistic line is *not required* by the data, and the case that a broad, relativistic line is *ruled out*. A relativistic disk line with an EW of only $\sim 50-100$ eV, that originated in a disk with a low inclination angle and/or flat radial emissivity (flatter than $\sim r^{-2}$) would be very hard for *XMM-Newton* to detect, as would a disk line which is *extremely* broadened (stretching over several keV). Typically, when upper limits on the EW of a possible disk line are quoted for *XMM-Newton* data, certain disk line parameters such as the inclination angle and radial emissivity index are fixed at presumed values. The above two scenarios are explicitly not included in the modeling. To make matters worse, whether the presence of a broad Fe K line is inferred or not often depends on how one models the continuum. If one is not looking for a disk line which is several keV wide, one may not find it, and instead that line may be modeled as part of the continuum. The converse is also true: a complex continuum may masquerade as a broad disk line.

High-Z Fe K Lines. Although Fe K lines at high redshift are rare (and *broad* lines even rarer), they have been observed. The clearest example of a broad Fe K line at redshift higher than 0.1 is that of the quasar E1821+643 at $z = 0.297$ from a *Chandra* HEG observation [36]. Although Fe K lines have been observed in quasars with redshift as high as 2.55 [37], they are unresolved and the signal-to-noise is poor. There are no clear examples of a broad Fe K line at a redshift higher than 0.297, except for those in the stacked spectra from *XMM-Newton* deep fields [38], in which the AGN distribution redshift peaks at ~ 1.7.

Variability. Aside from gross variability of the Fe K line profile (e.g. see [34] and references therein), there have been increasing reports, for several sources, of narrow emission features appearing at unusual energies, generally in the 5–6.4 keV range, in the AGN rest frame (see [39] and references therein). These have been interpreted as Fe K emission from localized 'hot-spots' on an accretion disk. If this is the correct interpretation such features will be potentially powerful model-independent diagnostics of the black-hole mass and spin (see [40, 41] and references therein) since the energy of these narrow features (as a function of time) does not depend on the emissivity distribution over the disk. First attempts at time-resolved spectroscopy of these features seem to support the current interpretation of them [42].

ACKNOWLEDGMENTS

This work was supported by NASA grants NAG5-10769, NNG04GB78A, and *Chandra* grant AR4-5009X. The author thanks Barry McKernan and Urmila Padmanabhan for their substantial contributions to some of the work described in this paper.

REFERENCES

1. Crenshaw, D. M., Kramer, S. B., & George, I. M. 2003, ARA&A, 41, 117
2. Blustin, A. J., Page, M. J., Fuerst, S. V., Branduardi-Raymont, G., & Ashton, C. E. 2005, MNRAS, in press (0411297)
3. McKernan, B., Yaqoob, T., & Reynolds, C. S. 2005, MNRAS, submitted
4. Kaspi, S., *et al.* 2002, ApJ, 574, 643
5. Krongold, Y., Nicastro, F., Elvis, M., Brickhouse, N. S., Liedahl D., & Mathur, S. 2003, ApJ, 597, 832
6. Netzer, H., *et al.* 2003, ApJ, 599, 933
7. Behar, E., Rasmussen, A. P., Blustin, A. J., Sako, M., Kahn, S. M., Kaastra, J. S., Branduardi-Raymont, G., & Steenbrugge, K. C. 2003, ApJ, 598, 232
8. Gabel, J. R., Kramer, S. B., & Crenshaw, D. M. 2005, ApJ, in press (astro-ph/0405524)
9. Arav, N., Kaastra, J., Gerard, K. A., Korista, K. T., Gabel, J. D. 2005, ApJ, in press (astro-ph/0406200)
10. Pounds, K. A., King, A. R., Page, K. L., & O'Brien, P. T. 2003, MNRAS, 346, 1025
11. Pounds, K. A., Reeves, J. N., King, A. R., Page, K. L., O'Brien, P. T., & Turner, M. J. L. 2003, MNRAS, 345, 705
12. Reeves, J. N., O'Brien, P. T., & Ward, M. J. 2003, ApJ, 593, L65
13. Krolik, J. H., & Kriss, G. A. 2001, ApJ, 561, 684
14. Krongold, Y., Nicastro, F., Brickhouse, N. S., Elvis, M., & Mathur, S. 2005, ApJ, in press (astro-ph/0411554)
15. Sako, M., et al. 2003, ApJ, 596, 114
16. Lee, J. C., Ogle, P. M., Canizares, C. R., Marshall, H. L., Schulz, N. S., Morales, R., Fabian, A. C., & Iwasawa, K. 2001, ApJ, 554, L13
17. Turner, A. K., Fabian, A. C., Lee, J. C., Vaughan, S. 2004, MNRAS, 353, 319
18. Netzer, H., Chelouche, D., George, I. M., Turner, T. J., Crenshaw, D. M., Kramer, S. B., & Nandra, K. 2002, ApJ, 571, 256
19. McKernan, B., Yaqoob, T., & Reynolds, C. S. 2004, ApJ, 617, 232
20. Yaqoob, T., & Padmanabhan, U. 2004, ApJ, 604, 63
21. Nicastro, F. 2005, Nature, in press (astro-ph/0412378)
22. Nandra, K., George, I. M., Mushotzky, R. F., Turner, T. J., & Yaqoob, T. 1997, ApJ, 477, 602
23. Yaqoob, T., Padmanabhan, U., Dotani, T., & Nandra, K. 2002, ApJ, 589, 487
24. Reynolds, C. S., & Nowak, M. A. 2003, Phys. Rep., 377, 389
25. Yaqoob, T., George, I. M., Kallman, T. R., Padmanabhan, U., Weaver. K. A., & Turner, T. J. 2003, ApJ, 596, 85
26. McKernan, B., & Yaqoob, T. 2004, ApJ, 608, 157
27. Yaqoob, T., et. al. 2005, in preparation
28. Reeves, J. N. 2003, in ASP Conf. Ser., Active Galactic Nuclei, from Central Engine to Host Galaxy, ed. S Collin, F. Combes & I. Shlosman, (San Francisco: ASP), Vol. 290, 35
29. Pounds, K. A., Reeves, J. N., O'Brien, P., Page, K., Turner, M. J. L., Nayakshin, S. 2001, ApJ, 559, 181
30. Page, K. L., O'Brien, P. T., Reeves, J. N., & Turner, M. J. L. 2003, MNRAS, 347, 316
31. Page, M. J., Davis, S. W., & Salvi, N. J. 2003, MNRAS, 343, 1241
32. Matt, G., Guainazzi, M., Perola, G. C., Fiore, F., Nicastro, F., Cappi, M., & Piro, L. 2001, A&A, 377, L31
33. Bianchi, S., Matt, G., Balestra, I., & Perola, G. C. 2003, A&A, 407, L21
34. Miniutti, G., Fabian, A. C., Goyder, R., & Lasenby, A. N. 2003, MNRAS, 344, L22
35. Yaqoob, T., Padmanabhan, U., Reeves, J. N., Markowitz, A., & Serlemitsos, P. J. 2005, ApJ, submitted
36. Yaqoob, T., & Serlemitsos, P. J. 2005, ApJ, in press
37. Chartas, G.; Eracleous, M.; Agol, E.; Gallagher, S. C. 2004, ApJ, 606, 78
38. Streblyanska, A., Hasinger, G., Finoguenov, A., Barcons, X., Mateos, S., Fabian, A. C., A&A, submitted (astro-ph/0411340)
39. Turner, T. J., Kramer, S. B., & Reeves, J. N. 2004, ApJ, 603, 62
40. Yaqoob, T. 2001, in ASP Conf. Ser., New Century of X-ray Astronomy, ed. H. Inoue, & H. Kunieda (astro-ph/0108032)
41. Dovčiak, M., Karas, V., Yaqoob, T. 2004, ApJS, 153, 205
42. Iwasawa, K., Miniutti, G., & Fabian, A. C. 2004, MNRAS, 355, 1073

A study of spectra of Cyg X-3 observed by BeppoSAX

A. Szostek* and A. A. Zdziarski*

*Centrum Astronomiczne im. M. Kopernika, Bartycka 18, Warszawa, Poland

Abstract. We model the ~1–200 keV spectra of Cygnus X-3 observed by *BeppoSAX*. The continuum, modeled by Comptonization in a hybrid plasma, is modified by the strongly ionized plasma of the stellar wind of the Wolf-Rayet companion star. Discrete absorption and emission spectral features are modeled with XSTAR. The model has been applied to phase-resolved spectra in the hard and soft spectral states.

Keywords: X-rays: binaries, X-rays: individual (Cygnus X-3), stars: Wolf-Rayet, accretion, accretion-disks, line: identification,
PACS: 32.30.Rj, 97.10.Me, 97.30.Eh, 97.0.Jp, 98.62.Mw, 98.70.Qy

INTRODUCTION

Cygnus X-3, a high-mass X-ray binary, contains a compact object (either a black hole or a neutron star) and a Wolf-Rayet star. The orbital period is very short, 4.8 h. A strong, dense, stellar wind from the companion strongly modifies the intrinsic continuum by absorption and emission.

High-resolution spectroscopy provides valuable diagnostic tools to examine the discrete absorption and emission features of the wind. However, the relatively narrow energy band of *Chandra* does not allow us to model the intrinsic broad-band continuum. On the other hand, the low spectral resolution and the lack of soft X-ray coverage of *RXTE* do not allow us to study the discrete emission and absorption features imprinted by the wind. Thus, here we take advantage of the data from *BeppoSAX*, featuring broad-band coverage and medium energy resolution.

OBSERVATIONS AND MODEL

We analyze 9 out of 10 observations of Cyg X-3 performed by *BeppoSAX*, using the MECS2 (1.7–10.5 keV) and PDS (15–120 keV) data. Each observation contained more than four orbits. We use the parabolic ephemeris [6] to divide the data into intervals of 1/6 of the period, and then add together the spectra from a given observation corresponding to the same phase interval. The minima and maxima correspond to the phases of −0.08–0.08 and 0.42–0.58, respectively.

All the spectra have been fitted with a physically-motivated model including Comptonization of blackbody emission by hybrid (thermal and nonthermal) electrons [1], [3], Compton reflection from an ionized medium [5], emission and absorption in the photoionized wind calculated with the XSTAR photoionization code [4], neutral absorption,

CP774, *X-Ray Diagnostics of Astrophysical Plasmas,* edited by Randall K. Smith

and a relativistically broadened Fe Kα line [2].

RESULTS AND CONCLUSIONS

The softest and the hardest spectra at the maximum of the orbital modulation are plotted in Fig. 1. The soft-state spectrum is dominated by a soft blackbody-like component and a high-energy tail due to emission of nonthermal electrons. The hardest spectrum observed by *BeppoSAX* is well modeled by thermal Comptonization, with no nonthermal electrons required by the fit, and weak Compton reflection. This spectrum is still softer than the hard spectra observed by *RXTE*, similar to the average spectrum 2 in the classification of [7]. Both soft and hard state spectra are significantly absorbed and contain numerous discrete spectral features.

The models fitted to all analyzed spectra in the maxima and minima are shown in Fig. 2(*a*, *b*). We clearly see that whereas the spectra below $\sim$10 keV show a continuity of flux levels, the spectra above $\sim$20 keV show a bimodal distribution. Thus, we base our state classification here on the latter, with the soft and hard state corresponding to the low and high, respectively, level of the high-energy flux. (This classification may differ from some other ones used for X-ray binaries.) Interestingly, all the (soft-state) spectra with the weak tail correspond to an almost unique spectrum below 10 keV, forming a thick line in Fig. 2(a). The state transition (i.e., the decline of the tail amplitude) takes place only when the soft X-ray flux reaches its very maximum. The soft and hard state spectra intersect at the pivot energy of $\sim$10 keV.

The amplitude of the orbital modulation, see Fig. 2(*c*, *d*), below $\sim$10 keV is comparable with that of the aperiodic variability, see Fig. 2(*a*, *b*). On the other hand, X-rays above $\sim$ 20 keV change much less. We find that the flux of the Fe Kα line does not vary significantly during either the orbital modulation or state transition. This is suggestive of the origin of this emission from an extended region. Naturally, the equivalent width is the highest at the minimum of the continuum flux, which correspond to the phase minimum of the hard state.

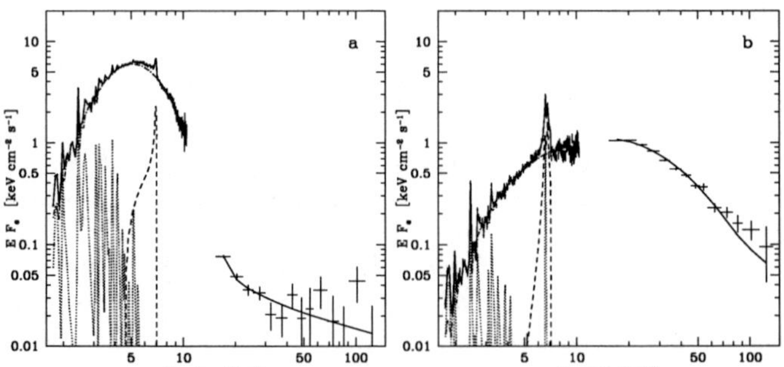

FIGURE 1. Results of spectral fits. (*a*) The softest and (*b*) the hardest observed spectra at the maximum of the orbital modulation plotted together with the model (*solid curves*) consisting of Comptonization continuum (*dot-dashed curves*) soft X-ray line emission (*dotted curves*) and the Fe K line (*dashed curves*).

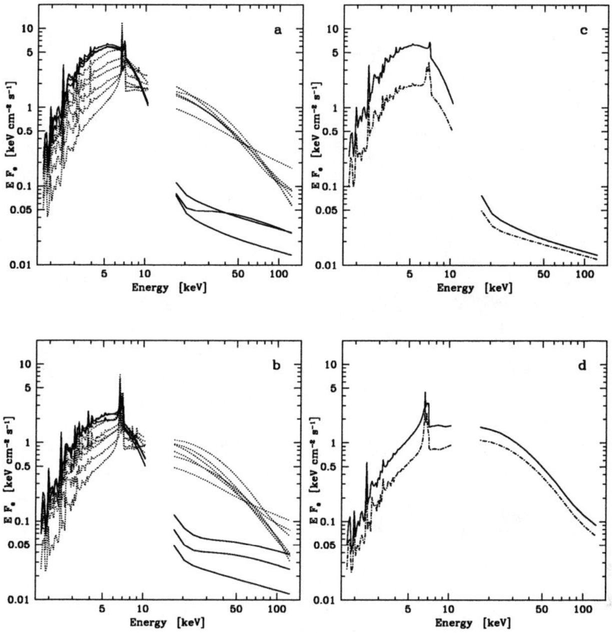

FIGURE 2. The models fitted to the phase (*a*) maxima and (*b*) minima in the soft (*solid curves*) and hard (*dotted curves*) states, and the models fitted to (*c*) the softest and (*d*) the hardest spectra in the phase minimum (*dot-dashed curves*) and the maximum (*solid curves*).

Although the energy resolution of *BeppoSAX* does not allow for precise separation of adjacent spectral features in soft X-rays or precise measurement of their equivalent width, we still have been able to identify several lines and recombination continua such as: Si XII, Si XIV Ly α, RRC Si XIII, S XVI Ly α, Si XIV, Ar XVII, Ar XVIII Ly α, RRC S XVI, and Ca XIX.

ACKNOWLEDGMENTS

We thank L. Hjalmarsdotter for valuable comments on this paper. This research has been supported by KBN grants 1P03D01827, 4T12E04727 and PBZ-KBN-054/P03/2001.

REFERENCES

1. P. S. Coppi, *ASP Cons. Ser. Vol.*, 1999, 161, 375.
2. A. C. Fabian, M. J. Rees, L. Stella, N. E. White, *MNRAS*, 1989, 238, 729.
3. M. Gierlinski, et al., *MNRAS*, 1999, 309, 496.
4. T. R. Kallman, R. McCray, *ApJS*, 1999, 50, 263.
5. P. Magdziarz, A. A. Zdziarski, *MNRAS*, 1995, 273, 837.
6. N. S. Singh, et al., *A&A*, 2002, 392, 161.
7. A. Szostek, A. A. Zdziarski, astro-ph/0401265.

Turning cool star X-ray spectra upside down

K. Werner* and J. J. Drake†

*Institute for Astronomy and Astrophysics, University of Tübingen, 72076 Tübingen, Germany
†Harvard-Smithsonian Center for Astrophysics, MS 3, 60 Garden St., Cambridge, MA 02138, USA

Abstract. H1504+65 is a young white dwarf with $T_{\rm eff}$=200 000 K and is the hottest post-AGB star ever analysed with detailed model atmospheres. *Chandra* LETG+HRC-S spectra have revealed the richest X-ray absorption line spectrum recorded from a stellar photosphere to date. The line forming regions in this extremely hot photosphere produce many transitions in absorption that are also observed in emission in cool star coronae. We have performed a detailed comparison of *Chandra* spectra of H1504+65 with those of Procyon and α Cen A and B. NLTE model spectra for the white dwarf have enabled us to identify a wealth of absorption lines from highly ionised O, Ne and Mg. In turn, these features have allowed us to identify coronal lines whose origins were hitherto unknown.

Keywords: stars: atmospheres – stars: coronae – X-rays: stars
PACS: 97.10.Ex

High-resolution X-ray spectroscopy performed with *Chandra* and XMM-Newton allows very detailed studies of coronae about cool stars. While many individual emission lines were detected for the first time in stellar spectra by the Extreme Ultraviolet Explorer Satellite (EUVE; see, e.g., [1]), the resolving power of $\lambda/\Delta\lambda \sim 200$ of the EUVE spectrographs was a quite modest compared with that of present day X-ray observatories. In particular, the unprecedented resolution capabilities of the *Chandra* X-ray Observatory Low Energy Transmission Grating Spectrograph (LETG) in the 30-170 Å range ($\lambda/\Delta\lambda \sim 1000$) that overlaps with the EUVE Short Wavelength spectrometer (70-170 Å), have revealed many more weak spectral lines. Owing largely to a historical lack of attention to the 30-170 Å range, a large fraction of these remain unidentified. Identification of these features is desirable because they could be used as a spectroscopic diagnostics, because they potentially contribute to the flux of diagnostic lines currently employed, and because they contribute to the overall plasma radiative loss.

Two nearby stars that have illuminated the forest of lines in the 30-170 Å range are α Cen (G2V+K1V) and Procyon (F5IV). These stars exhibit classical solar-like X-ray emitting coronae. Indeed, analogues of the relatively X-ray faint Sun are difficult to observe because they become unreachable with current instrumentation beyond a few parsecs, and α Cen and Procyon represent the nearest and brightest coronal sources with solar-like activity. Only a small fraction of the multitude of lines between 30-170 Å seen their *Chandra* LETG spectra could be identified based on current radiative loss models [2,3]. Drake et al. (in prep.) have estimated that these models underestimate the true line flux in the range 30-70 Å by factors of up to 5 or so.

The "missing lines" are predominantly transitions involving $n = 2$ ground states in abundant elements such as Ne, Mg, Si, S and Ar—the analogous transitions to the Fe "L-shell" lines between $\sim$ 8-18 Å, together with Fe $n = 3$ (the "M-shell") transitions [4,5]. Some of these lines have been identified based on Electron Beam Ion Trap experiments

CP774, *X-Ray Diagnostics of Astrophysical Plasmas,* edited by Randall K. Smith

[6,7,8]. In the present paper we approach this problem from a new perspective, namely through a *Chandra* observation of the photosphere of the hottest white dwarf (WD) known, H1504+65, and its quantitative analysis by means of detailed non-LTE model atmospheres.

H1504+65 has an effective temperature of 200 000 K. It belongs to the PG1159 spectral class, which are hot, hydrogen-deficient (pre-) white dwarfs. Their surface chemistry (typical abundances: He=33%, C=48%, O=17%, Ne=2%, mass fractions) suggests that they exhibit matter from the helium-buffer layer between the H- and He-burning shells in the progenitor AGB star [9]. This is likely because the PG1159 stars have suffered a late He-shell flash, a phenomenon that drives the fast evolutionary rates of such famous stars like FG Sge and Sakurai's object. H1504+65 is in fact a peculiar member of this class, because it is also helium-deficient. Its atmosphere is mainly composed of carbon and oxygen plus neon and magnesium (C=48%, O=48%, Ne=2%, O=2%, mass fractions). H1504+65 is a unique object, considering its high $T_{\rm eff}$ and chemical surface composition, and we have speculated that it represents the naked C/O core of a former red giant [10].

Chandra LETG+HRC-S spectra from H1504+65 have revealed the richest X-ray absorption line spectrum recorded from a stellar photosphere to date. We have recently performed a detailed analysis of this spectrum [10] and we use here the photospheric spectrum of H1504+65 together with an appropriate model atmosphere to identify a number of emission lines in the coronae of α Cen A, α Cen B, and Procyon. The difference in particle densities in the WD photosphere and in the coronae amounts to many orders of magnitude (roughly $n_e=10^{10}$ and $10^{13} - 10^{18}\,{\rm cm}^{-3}$, respectively). However, the temperature in the line forming regions of the WD (up to 300 000 K) is comparable to the low-temperature component of multi-temperature fits to coronal spectra that is required to account for the lines of low-ionisation stages (e.g. 630 000 K for Procyon [2]). As a consequence, numerous lines from O VI, Ne VI-VIII and Mg VI-IX are visible in the soft X-ray spectra of both the cool star coronae (in emission) and the hot WD photosphere (in absorption). Lines from higher ionisation stages are formed in the high-temperature regions of the coronae (T of the order 1–2.5 million K for the stars studied in this paper), hence, their respective absorption line counterparts cannot be formed in the WD photosphere.

We have performed a detailed comparison of the H1504+65 photospheric absorption line spectrum with the coronal emission line spectra of α Cen A, α Cen B, and Procyon. We have also used the model spectrum of H1504+65 for this purpose. The use of our synthetic spectrum helps considerably to identify lines in the coronal spectra. Lines from 65 multiplets of O VI, Ne VI-VIII, and Mg VI-VIII are identified in both H1504+65 (or its model) and in at least one of the considered coronae. Many of these were already identified by Raassen et al. [2,3], but the majority represents new identifications. An example how the spectra compare is shown in Fig. 1. It displays some new line identifications in the coronal spectra, see for example the 87.46Å resonance line of Ne VII in α Cen A.

To conclude, with the help of a detailed model spectrum for H1504+65 we have found that a large number of lines from multiplets of O, Ne, and Mg are present in both the photospheric absorption line spectrum and the coronal emission line spectra. In the coronal spectra we have newly identified lines from about 40 multiplets of O VI,

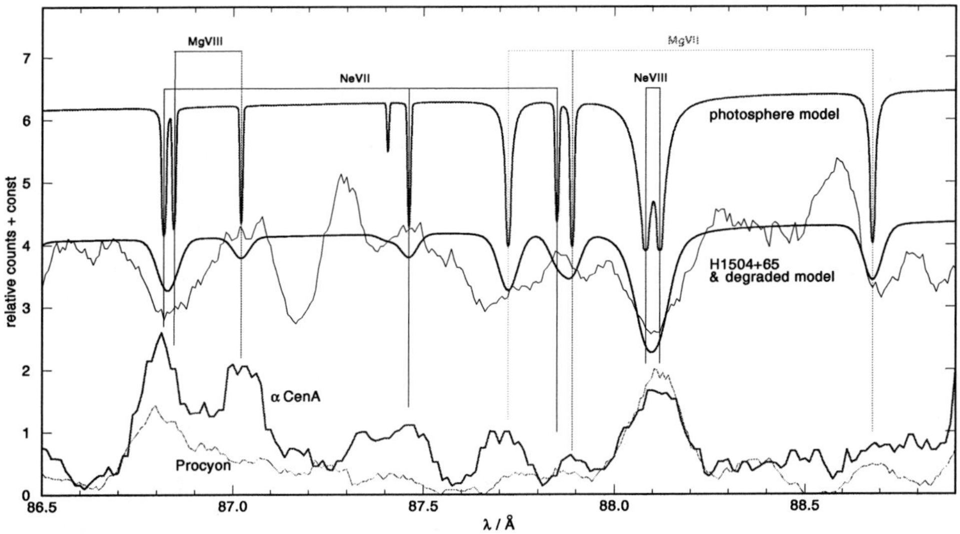

FIGURE 1. Comparison of *Chandra* spectra of H1504+65 with Procyon and α Cen A. Lines are in absorption in the white dwarf photosphere H1504+65 and in emission in the cool star coronae. Top: photosphere model for H1504+65 with line identifications for Mg VII and Ne VII. Middle: Degraded model spectrum plotted over H1504+65 observation. Bottom: Procyon and α Cen A spectra. All lines from highly ionised Ne and Mg are identified for the first time in a cool star corona, except for Mg VIII 86.85/87.02Å and the Ne VIII doublet at 88.1Å, which were identified by [2,3]).

Ne VI-VII, and Mg VI-VIII. Some of these lines are blends with previously known lines, which are in use for diagnostic purposes, hence, their contribution to the line flux must be considered in detailed spectral analyses. A detailed version of this paper will be published in Astronomy & Astrophysics.

Analysis of X-ray data in Tübingen is supported by the DLR under grant 50 OR 0201. JJD was supported by NASA contract NAS8-39073 to the *Chandra X-ray Center*.

REFERENCES

1. Drake J.J., Laming J.M., Widing K.G. 1996, ApJ, 443, 393
2. Raassen A.J.J., Mewe R., Audard M., et al. 2002, A&A 389, 228
3. Raassen A.J.J., Ness J.-U., Mewe R., et al. 2003, A&A, 400, 671
4. Drake J.J. 1996, in Cool Stars; Stellar Systems; and the Sun: 9, ed. R. Pallavicini, A.K. Dupree, ASP Conference Series, 109, 203
5. Jordan C. 1996, in Astrophysics in the Extreme Ultraviolet, IAU Coll. 152, ed. S. Bowyer, R.F. Malina, Dordrecht: Kluwer Academic Publ., p. 81
6. Beiersdorfer P., Lepson J.K., Brown G.V., et al. 1999, ApJL, 519, 185
7. Lepson J.K., Beiersdorfer P., Brown G.V., et al. 2002, ApJ, 578, 648
8. Lepson J.K., Beiersdorfer P., Behar E., Kahn S.M. 2003, ApJ, 590, 604
9. Werner K. 2001, in Low Mass Wolf-Rayet Stars: Origin and Evolution, ed. T. Blöcker, L.B.F.M. Waters, A.A. Zijlstra, Ap&SS, 275, 27
10. Werner K., Rauch T., Barstow M.A., Kruk J.W. 2004, A&A, 421, 1169

Soft X-ray emission lines of Fe XV in spectra of the Sun and Capella

J. J. Drake*, F. P. Keenan†, S. Chung*, N.S. Brickhouse*, K.M. Aggarwal†, A.Z. Msezane**, R.S.I. Ryans† and D.S. Bloomfield†

*Smithsonian Astrophysical Observatory (SAO), MS 3, 60 Garden Street, Cambridge, MA 02138
†Department of Pure and Applied Physics, Queen's University Belfast, Belfast, BT7 1NN, Northern Ireland, U.K.
**Center for Theoretical Studies of Physical Systems, Clark Atlanta University, Atlanta, GA 30304

Abstract. Recent calculations of atomic data for Fe XV have been used to generate theoretical line ratios involving $n = 3$–4 transitions in the soft X-ray spectral region ($\sim$52–83 Å) for a wide range of electron temperatures and densities applicable to solar and stellar coronal plasmas. The line ratio calculations are compared with solar flare observations from a rocket-borne spectrograph (XSST) and with a co-added spectra of Capella obtained with the *Chandra* LETGS, representing the highest signal-to-noise observation achieved for a stellar source in the 30-80 Å soft X-ray region. Results are promising for use of Fe XV as astrophysical plasma diagnostics: Agreement between theory and observation is generally good, after particular account is taken of line blending. The 82.76 Å emission line in the XSST spectrum is identified, for the first time to our knowledge in an astrophysical source, as the $3s3d\ ^3D_3$–$3s4p\ ^3P_2$ transition of Fe XV.

Keywords: atomic data, stellar coronae, x-ray spectroscopy

INTRODUCTION

Since the potential of EUV lines of Fe XV as diagnostics for hot, coronal plasma was first noted[3] many authors have calculated emission line intensities for this ion and compared these with solar observations (for example, ref.[6] and references therein). However, Fe XV also shows a rich emission line spectrum in the relatively unexplored soft X-ray region, $\sim$50–80 Å, arising from $n = 3$–4 transitions. In contrast to the EUV lines, there has been little work on the soft X-ray transitions in the astrophysical literature, probably as a result of the limited availability of high quality solar spectra for this wavelength region.

Here, we use the most recent atomic physics calculations for Fe XV to generate soft X-ray line ratios for a wide range of electron temperatures and densities applicable to coronal plasmas. We compare these in detail with XSST observations of a solar flare[1], and with a *Chandra* spectrum of Capella, constructed from multiple observations, which hence has one of the highest signal-to-noise ratios ever achieved for an astrophysical emission line source in this soft X-ray region. This allows us to perform, for the first time to our knowledge, an investigation of the Fe XV $n = 3$–4 lines in an astrophysical source other than the Sun.

CP774, *X-Ray Diagnostics of Astrophysical Plasmas,* edited by Randall K. Smith

TABLE 1. Fe XV transitions and predicted and observed line ratios considered in this study.

λ(Å)	Transition	Ratio	Solar Flare I(λ)/I(59.40 Å)	Theory	Capella I(λ)/I(59.40 Å)	Theory
52.91	$3s^2\ ^1S_0$–$3s4p\ ^1P_1$	R_1	0.78 ± 0.23	0.58 ± 0.12	0.55 ± 0.06	0.58 ± 0.12
53.11	$3s^2\ ^1S_0$–$3s4p\ ^3P_1$	R_2	0.36 ± 0.11	0.18 ± 0.04	0.16 ± 0.03	0.18 ± 0.04
55.78	$3s3p\ ^3P_1$–$3s4d\ ^3D_{1,2}$	R_3	0.33 ± 0.10	0.13 ± 0.03	…	…
56.17	$3s3p\ ^3P_2$–$3s4d\ ^3D_3$	R_4	0.54 ± 0.16	0.23 ± 0.05	0.26 ± 0.06	0.21 ± 0.04
59.40	$3s3p\ ^1P_1$–$3s4d\ ^1D_2$	…	…	…	…	…
63.97	$3p^2\ ^1D_2$–$3s4f\ ^1F_3$	R_5	0.35 ± 0.08	0.40 ± 0.08	0.39 ± 0.06	0.40 ± 0.08
66.25	$3s3p\ ^3P_2$–$3s4s\ ^3S_1$	R_6	2.6 ± 0.8	0.23 ± 0.05	…	…
69.65	$3s3p\ ^1P_1$–$3s4s\ ^1S_0$	R_7	3.0 ± 0.9	3.6 ± 0.7	2.9 ± 0.2	3.6 ± 0.7
69.93	$3s3d\ ^3D_1$–$3s4f\ ^3F_2$	R_8	0.28 ± 0.08	0.16 ± 0.03	0.28 ± 0.06	0.15 ± 0.03
69.98	$3s3d\ ^3D_2$–$3s4f\ ^3F_{2,3}$	R_9	0.33 ± 0.10	0.25 ± 0.05	0.35 ± 0.07	0.24 ± 0.05
70.05	$3s3d\ ^3D_3$–$3s4f\ ^3F_{3,4}$	R_{10}	0.73 ± 0.22	0.46 ± 0.09	0.49 ± 0.06	0.44 ± 0.09
73.47	$3s3d\ ^1D_2$–$3s4f\ ^1F_3$	R_{11}	0.75 ± 0.23	0.50 ± 0.10	…	…
82.76	$3s3d\ ^3D_3$–$3s4p\ ^3P_2$	R_{12}	1.4 ± 0.4	1.2 ± 0.2	1.3 ± 0.1	1.2 ± 0.2

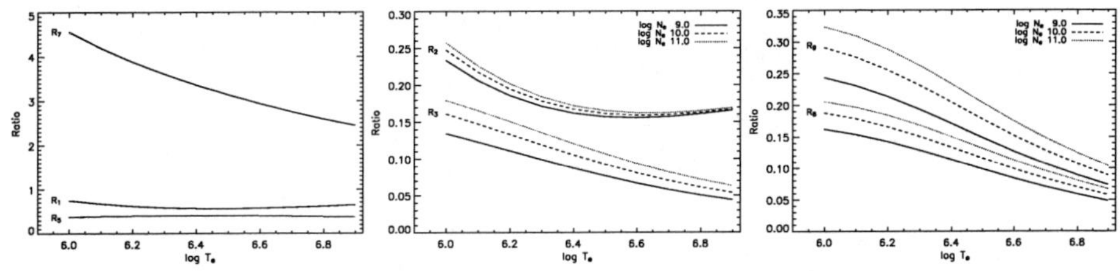

FIGURE 1. The theoretical Fe XV emission line intensity ratios (photon units) plotted as a function of logarithmic electron temperature for different values of electron density ($N_e = 10^{10}$ cm^{-3} unless otherwise noted)

THEORETICAL RATIOS

Our model Fe XV ion consisted of the energetically lowest 53 fine-structure levels belonging to the $3s^2$, $3s3p$, $3p^2$, $3s3d$, $3p3d$, $3d^2$, $3s4s$, $3s4p$, $3s4d$, $3p4s$ and $3s4f$ configurations. Electron impact excitation rates were taken from ref.[2]; Einstein A-coefficients are from [5]. Proton impact excitation is only important for transitions among $3s3p$ ^{3}P levels of Fe XV; rates were taken from [7]. The transitions studied here are listed in Table 1. Theoretical emission line intensity ratios are illustrated as a function of temperature in Figure 1; lines are ratioed to 59.40 Å—the cleanest and most reliably detected Fe XV line in the soft X-ray spectrum. The accuracy of our theoretical line ratio calculations is estimated to be $\pm 20\%$.

OBSERVATIONS AND RESULTS

Solar line strengths are from an M-class flare, recorded with 54s and 145s exposures on Kodak 101–07 emulsion by an X-ray spectrograph (XSST) with 0.02 Å resolution covering 11–97 Å during a rocket flight on 1982 July 13[1]. Capella (G1III+G8III) line fluxes were measured from co-added observations obtained over several years by the *Chandra* LETG+HRC-S. The resulting spectrum has wavelength coverage 2-170 Å with a resolution of 0.05 Å (Figure 2). Spectral lines were measured using the PINTofALE IDL software suite.[1] Observed solar flare and Capella emission line ratios are compared with theoretical values in Table 1.

We find very good agreement between XSST observations and theory for ratios R_5 and R_7; 63.97 and 69.65 Å lines are well-detected and relatively unblended. Identification of the 63.97 Å feature as Fe XV $3p^2$ ^{1}D–$3s4f$ ^{1}F[4] and not Al VIII $2s^22p^2$ $^3P_{1,2}$–$2s2p^23p$ $^3D_{2,3}$[1] is confirmed, as is Fe XV $3s3p$ ^{1}P–$3s4s$ ^{1}S.

The observed ratios R_1, R_9, R_{10}, R_{11} and R_{12} are only slightly larger than theory. We note that our classification of 82.76 Å as Fe XV $3s3d$ $^3D_3$–$3s4p$ $^3P_2$ is first time this transition is identified in an astrophysical source.

---

[1] Freely available from http://hea-www.harvard.edu/PINTofALE

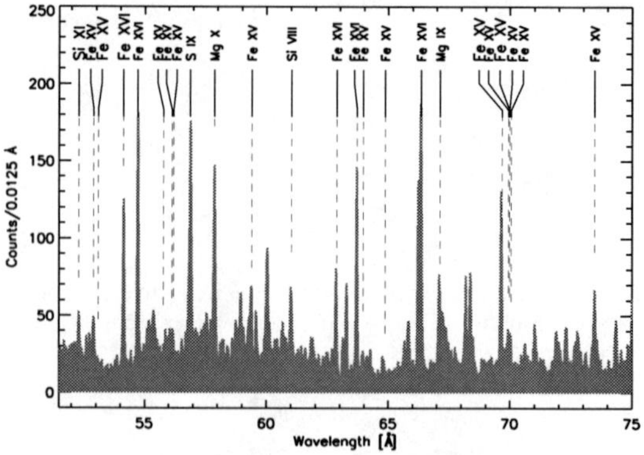

FIGURE 2. The LETG+HRC-S spectrum of Capella in the 50–75 Å range. Lines of Fe XV analysed in the present paper are identified, together with some other prominent transitions of Fe, Mg, Si and S. The pseudo-continuum is comprised of a superposition of weak lines from the $n = 2$ levels of abundant elements, mostly Mg, Si, S and Ar, but also including contributions from the less abundant elements Na and Al.

The observed values of ratios R_2, R_3, R_4, R_6 and R_8 in the XSST spectrum are all *significantly larger* than our theoretical predictions, implying that Fe XV 53.11, 55.78, 56.17, 66.25 and 69.93 Å are blended to varying degrees.

Agreement between theory and experiment for Capella is generally good, and somewhat better than for the solar flare spectrum. Discrepancies, such as R_8, are similar to the XSST results, with the observed ratio being greater than the theoretical one, indicating some degree of blending. In most instances, the better agreement with theory we find for the Capella line ratios appears to be due to a lesser degree of contamination by blends: much of the blending in this spectral region is caused by species formed at temperatures too low to have significant populations in the hotter Capella coronae.

ACKNOWLEDGMENTS

JJD, NSB and SC acknowledge support by NASA contract NAS8-39073 to the *Chandra X-ray Center*. FPK is grateful to AWE for the award of a William Penney Fellowship.

REFERENCES

1. Acton, L. W., Bruner, M. E., Brown, W. A., Fawcett, B. C., Schweizer, W., & Speer, R. J. 1985, ApJ, 291, 865
2. Aggarwal, K. M., Keenan, F. P., & Msezane, A. Z. 2003, A&A, 410, 349
3. Bely, O., & Blaha, M. 1968, Sol.Phys., 3, 563
4. Bhatia, A. K., Mason, H. E., & Blancard, C. 1997, At. Data Nucl. Data Tables, 66, 83
5. Deb, N. C., Aggarwal, K. M., & Msezane, A. Z. 1999, ApJS, 121, 265
6. Kastner S.O., Bhatia A.K., 2001, ApJ, 553, 421
7. Landman, D. A., & Brown, T. 1979, ApJ, 232, 636

Chandra HETGS Observations of B Supergiants

Wayne L. Waldron

L-3 Communications Government Services, Inc., 1801 McCormick Dr., Suite 170, Largo, MD 20774, wayne.waldron@L-3com.com

Abstract. Preliminary results from the first HETGS observation of an early B supergiant, ε Orionis (B0 Ia), are presented. The observed emission line spectrum and X-ray emission line characteristics are compared to those of previous O-star studies. In addition, the significance of the observed stellar wind distribution of X-ray temperatures, as determined from several OB stars, is discussed.

Keywords: <Enter Keywords here>
PACS: 95.85.Nv, 97.10.Me, 97.20.Ec, 97.20.Pm

INTRODUCTION

The X-ray emission from OB stars is believed to arise from a distribution of stellar wind shocks. However, analyses of *Chandra* High Energy Transmission Grating Spectrometer (HETGS) data from OB stars indicate that there are some problems with this interpretation [1, 2, 3, 4]. Observations of B supergiants extend our understanding of the UV flux dependence of He-like forbidden to intercombination emission line ratio (f/i) by providing measurements at lower UV flux levels. The diversity in B supergiant wind properties as compared to other OB stars allows us to probe different distributed X-ray source models and test (over a broader range in OB stellar effective temperatures) the developing concept that the observed He-like X-ray emission lines from OB stars are emerging from their associated X-ray continuum optical depth unity radii.

HETGS OBSERVATION AND X-RAY EMISSION LINE PROPERTIES

The B supergiant, ε Ori (B0 Ia), was observed with the HETGS on 12 DEC 2003 for a total exposure time of 95.65 ks. The High Energy Grating (HEG) and Medium Energy Grating (MEG) $\pm 1^{st}$ order source count rates are respectively 0.0155 and 0.0755. The MEG spectrum is shown in Figure 1. Although the overall distribution of X-ray emission lines are comparable to those seen in O-star spectra, differences are seen at higher energies. The Mg XII line (8.42 Å) is significantly weaker and the Si XIV line (6.18 Å) is absent (i.e., the maximum X-ray temperature in ε Ori is considerably lower).

The X-ray emission line properties are determined by assuming that all lines can be represented by Gaussian line profiles superimposed on an underlying bremsstrahlung continuum. The resultant line fits are found to be in very good agreement with the observed line profiles. The results for ε Ori indicate minimal blue-shifted lines and no evidence of any line asymmetry (using the line asymmetry diagnostic suggested by

CP774, *X-Ray Diagnostics of Astrophysical Plasmas,* edited by Randall K. Smith

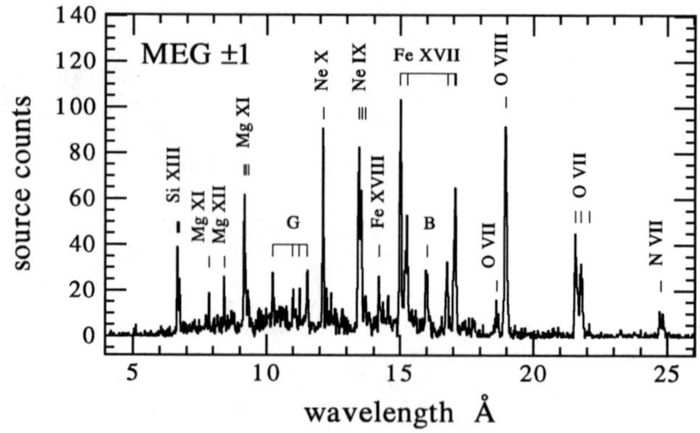

FIGURE 1. The MEG $\pm 1^{st}$ order source count spectra at a resolution of 0.02 Å. The most likely line identifications are indicated. The line group labeled G consists of Ne X (10.24 Å), Fe XXIII (10.98 Å), Fe XVII (11.24 Å), and Ne IX (11.54 Å). The line blend labeled B consists of O VIII and Fe XVIII (16.02 Å).

[4]) which is consistent with the majority of other O-star analyses. The average line shift of the ε Ori emission lines is $-230\,\mathrm{km\,s^{-1}}$ and the average line width (HWHM) is $630\,\mathrm{km\,s^{-1}}$. Surprisingly this HWHM is $\sim 33\%$ smaller than the HWHM of O-stars with comparable mass loss rates. The significance of this result is currently being investigated.

One of the key findings from other O-star analyses is that the *fir*-inferred radii as determined from He-like f/i line emission ratios are well correlated with their respective X-ray continuum optical depth unity radii [2, 3, 4]. The derived *fir*-inferred radii from ε Ori provide additional confirmation of this observed dependence. The most likely interpretation of this dependence is discussed by [1, 2].

WIND DISTRIBUTION OF X-RAY TEMPERATURES

The X-ray temperature distribution in OB stellar winds can be determined by using the *fir*-inferred radii and their associated X-ray temperatures, T_X, as determine from H-like to He-like line emission ratios [3, 4]. It is found that the results for ε Ori are in agreement with other O-star results, i.e., the largest T_X occur at the deepest parts of the wind. This dependence for several OB supergiants is shown in Figure 2. There is an important disagreement with shock theory that can best be illustrated by using the standard relation between shock velocity and T_X, $T_X\ (MK) = 13(V_S/1000)^2$ [5]. The shock velocity is defined as a fraction η of the local wind speed, $V_S = \eta V(r)$, and η is a scaling parameter < 1 with a typical value of 0.5. As evident in Figure 2, the shock model does *not* work for any of the observed T_X above 7 MK! Furthermore, there is no evidence for any high T_X at large radii (> 3) which is surprising since shock model simulations

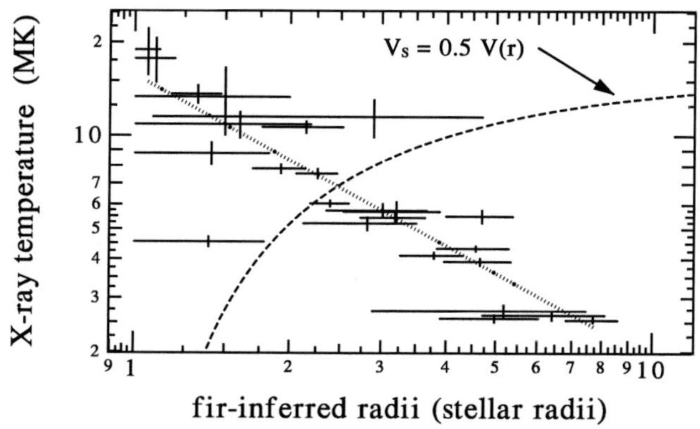

FIGURE 2. The observed radial distribution of X-ray temperatures in OB supergiant stellar winds. The radial location of shock produced X-ray temperatures are restricted to lie below the dashed line. A best fit to the data (dotted line) indicates that $T_X\ (MK) \sim 16(r/r_*)^{-1}$.

predict stronger shocks in this region [6]. Although the observed X-ray emission below 7 MK appears to be consistent with *weak* wind shocks, the deeply embedded high X-ray temperatures *cannot* be explained by shock theory. This suggests that the hot plasma near the star is confined in some other way, perhaps in magnetic loop structures as is commonly the case for cool stars.

ACKNOWLEDGMENTS

Support for this work was provided by NASA through Chandra Award Number GO3-4013X issued by the Chandra X-ray Observatory Center, which is operated by the Smithsonian Astrophysical Observatory for and on behalf of NASA under contract NAS8-03060.

REFERENCES

1. W. L. Waldron, and J. P. Cassinelli, *Astrophys. J.*, **548**, L45 (2001).
2. J. P. Cassinelli, N. A. Miller, W. L. Waldron, J. J. MacFarlane, and D. H. Cohen, *Astrophys. J.*, **554**, L55 (2001).
3. N. A. Miller, J. P. Cassinelli, W. L. Waldron, J. J. MacFarlane, and D. H. Cohen, *Astrophys. J.*, **577**, 951 (2002).
4. W. L. Waldron, J. P. Cassinelli, N. A. Miller, J. J. MacFarlane, and J. C. Reiter, *Astrophys. J.*, **616**, 542 (2004).
5. J. J. MacFarlane, and J. P. Cassinelli, *Astrophys. J.*, **347**, 1090 (1989).
6. A. Feldmeier, *Astron. Astrophys.*, **299**, 523 (1995).

Modeling CHANDRA Low Energy Transmission Grating Spectrometer Observations of Classical Novae with PHOENIX

A. Petz*, P. H. Hauschildt*, J. U. Ness† and S. Starrfield**

*Hamburger Sternwarte, Gojenbergsweg 112, 21029 Hamburg, Germany; [apetz,phauschildt]@hs.uni-hamburg.de
†Oxford University, Theoretical Physics, 1 Keble Road, Oxford, OX1 3NP, UK; ness@thphys.ox.ac.uk
**Dept of Physics and Astronomy, ASU, P. O. Box 871504, Tempe, AZ 85287-1504; sumner.starrfield@asu.edu

Abstract. We use the PHOENIX code package to model the X–ray spectrum of Nova V4743 Sagittarii observed with the LETGS onboard the Chandra satellite on 19 March 2003. To analyze nova atmospheres and related systems with an underlying nuclear burning envelope at X–ray wavelengths, it was necessary to update the code with new microphysics. We demonstrate that the X–ray emission is dominated by thermal bremsstrahlung and that the hard X–rays are dominated by Fe and N absorption. Preliminary models are calculated assuming solar abundances. It is shown that the models can be used to determine element abundances in the nova ejecta by increasing the absorption in the shell.

Keywords: Stellar atmospheres, Novae, CHANDRA, V4743Sgr
PACS: 97.10.Ex

INTRODUCTION

We have modeled X–ray spectra of classical novae (CNe) with the PHOENIX–code version 13 [1]. With our models it is possible to determine the effective temperature of the nova atmosphere and the formation of X–ray emission in the nova shell. We show that it is possible to determine the abundances of the ejecta by modelling X–ray spectra of CNe and present our first results for the abundances of C, N, and O.

For the comparison of our model with observations we use an X–ray spectrum of Nova V4743 Sagittarii (Sgr) observed with the LETGS onboard the Chandra satellite in March 2003 [2].

MODELING NOVA ATMOSPHERES IN X–RAYS WITH PHOENIX

Our model atmospheres are 1D spherical symmetric, expanding, and time independent. The radiative equilibrium is solved in the comoving frame and the radiative transfer is treated as special relativistic for an atmosphere in full NLTE with line blanketing of 8534 atomic levels from H, He, C, N, O, Ne, Mg, and Fe. We use an exponential density law ($\rho(r) \propto r^{-n}$) with a gradient of $n = 3$ and a standard velocity field ($v(r) = \dot{M}/4\pi r^2 \rho(r)$, $\dot{M}$ = const) with an outer velocity of $v_{\rm out} = 2500\,{\rm km\,s^{-1}}$.

CP774, *X-Ray Diagnostics of Astrophysical Plasmas,* edited by Randall K. Smith

The earlier versions of PHOENIX used atomic data from CHIANTI Version 3 (CHIANTI3) and the line lists of Kurucz. For this work, we have implemented two new atomic databases, CHIANTI Version 4 (CHIANTI4) [3] and APED[1], because the old databases did not provide enough data for the X–ray energy range. Using the new databases, we have extended PHOENIX to use many new spectral lines in the X–ray waveband down to 1 Å, improved data for electron collision rates, new data for proton collision rates, and better data for thermal bremsstrahlung.

X–RAY OBSERVATIONS OF NOVA V4743 SGR

The solid curve of Fig. 1 shows the observed X–ray spectrum of Nova V4743 Sgr on March 19, 2003 with the LETGS onboard the CHANDRA satellite [2]. At wavelengths greater than ≈ 55 Å, it is dominated by second and higher dispersion orders, and these wavelengths will not be considered in our analysis. The effective areas used to convert from ct s^{-1} to flux are determined with the CIAO software package[2], version 3.0.

An examination of the spectrum shows that it is not a black–body but resembles a stellar atmosphere with deep absorption features and, possibly, some weak emission lines. The strongest lines are from the two highest ionisation stages of C, N, and O. An extensive analysis of the observation has been carried out by [2].

MODEL WITH SOLAR ABUNDANCES

The best fit with solar abundances to the spectrum of Nova V4743 Sgr is shown in Fig. 1, left panel. The model has an effective temperature of $T_{\rm eff} = 5.8 \times 10^5$ K and a bolometric luminosity of $L_{\rm bol} = 50,000 L_\odot$. To get the correct slope for the pseudo–continuum, a value of $n_h = 4.0 \times 10^{21}\,\mathrm{cm}^{-2}$ for the hydrogen column density has to be used. The quality of the fit is independent of the luminosity of the model. This was already found for earlier solar models in other wavelength ranges [4].

Close inspection of the measured spectrum reveals that some spectral lines are not reproduced well or are missing in the model spectrum. This is because we have used only solar abundances in this model and have not increased abundances of, for example the CNO elements, as is generally observed in novae and predicted by theory [5]. Furthermore all absorption lines are too weak and there is too much emission around $\lambda \sim 24$ Å. Increasing the abundances should increase the absorption and the fit should improve with solar abundances.

The X–ray emission is dominated by thermal bremsstrahlung from the atmosphere surrounding the WD and the hard spectral range of $\lambda \lesssim 29$ Å is dominated by iron and nitrogen absorption. The atmosphere is in strong departure from LTE and is very extended with the highest ionization stages of elements in the outermost layers.

---

[1] http://cxc.harvard.edu/atomdb/
[2] http://cxc.harvard.edu/ciao/

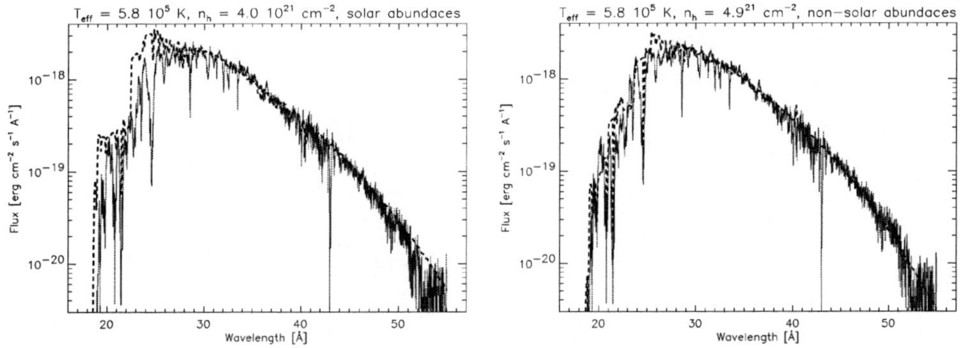

FIGURE 1. Model atmosphere of Nova V4743 Sgr with solar (left) and non–solar abundances (right). The solid curve is the observed spectrum.

MODEL WITH NON–SOLAR ABUNDANCES

A model with non–solar abundances (Fig. 1, right panel) produces a spectrum which fits the observation much better than with solar abundances. It was calculated with a 22 times solar abundance of nitrogen and oxygen. Accordingly the nitrogen to oxygen ratio is equal to the solar value. The abundance of carbon with 1.25 times solar is only slightly higher than in solar material.

The strengths of some spectral lines now fit much better to the observation. For example, the fits of the N VII line at $\lambda \sim 24.8$ Å and the O VII line at $\lambda \sim 21.6$ Å. Around $\lambda \sim 24$ Å there is still too much emission.

ACKNOWLEDGMENTS

Some of the calculations presented here were performed at the Höchstleistungs Rechenzentrum Nord (HLRN) and at the National Energy Research Supercomputer Center (NERSC), supported by the U.S. DOE. We thank all these institutions for a generous allocation of computer time. Part of this work was supported by the DFG (*Deutsche Forschungsgemeinschaft*), project number HA 3457/2–1. S. Starrfield was partially supported by grants from NASA–CHANDRA, NASA–Theory, and NSF to ASU.

REFERENCES

1. Petz, A., Hauschildt, P. H., Ness, J. U., & Starrfield, S. 2004, AstroPh-0410370
2. Ness, J. U., Starrfield, S., Burwitz, V., et al. 2003, ApJ, 594, L127
3. Young, P. R., Zanna, G. D., Landi, E., et al. 2003, ApJS, 144, 135
4. Hauschildt, P. H. & Starrfield, S. 1995, ApJ, 447, 829
5. Starrfield, S., Truran, J. W., Wiescher, M. C., & Sparks, W. M. 1998, MNRAS, 33, 804

Analysis of a XMM-Newton Spectrum of the Extremely Hot White Dwarf in Nova V4743 Sgr

T. Rauch*,†, K. Werner† and M. Orio**,‡

*Dr.-Remeis-Sternwarte, Bamberg, University of Erlangen-Nürnberg, Germany
† Institute for Astronomy and Astrophysics, University of Tübingen, 72076 Tübingen, Germany
**Istituto Nazionale di Astrofisica (INAF), Osservatorio Astronomico di Torino, Italy
‡ Department of Astronomy, University of Wisconsin, Madison, USA

Abstract. Classical novae occur in close binary systems (main-sequence star + white dwarf). V4743 Sgr is a very fast nova exhibiting a X-ray spectrum which is consistent with a view on the optically thick atmosphere of an extremely hot white dwarf (WD). It exhibits strong absorption features of the C V, C VI, N VI, N VII, and O VII resonance lines indicating an effective temperature of about 610 kK.

We present preliminary results of a spectral analysis by means of line-blanketed non-LTE atmosphere models of a XMM-Newton RGS-1 spectrum of V4743 Sgr, taken about half a year after its outburst.

Keywords: Stars: AGB and post-AGB – Stars: atmospheres – Stars: individual: V 4743 Sgr – Stars: novae, cataclysmic variables – X-rays: stars
PACS: 97.30.Qt

V4743 Sgr has been discovered as a nova on Sep 20, 2002 (Haseda et al. 2002). Until Mar 2003, it evolved into the brightest super-soft X-ray source in the sky. A first CHANDRA spectrum shows a flux maximum around 30 Å and strong C V, N VI, and N VII (Fig. 1) absorption features which suggest temperatures of 1 – 2 MK (Ness et al. 2003). These lines are blue shifted (-2400 km/sec) – most likely we see expanding gas which consists out of CNO-processed material.

However, in the phase after the outburst, i.e. when the H burning is still on-going (for years), the surface composition of the WD is poorly known. In order to make some progress, we decided to use our static NLTE models to investigate basic parameters like effective temperature (T_{eff}) and surface composition. Although we neglect any effect of the velocity field, our aim is at least a qualitative modeling of the XMM RGS-1 spectrum of V4743 Sgr, obtained on Apr 4, 2003 with an exposure time of about 10 h. The EPIC-pn count rate in timing mode was 1348 cts/sec (Orio et al. 2003).

For a preliminary analysis, we employed plane-parallel, static models which are calculated with TMAP, the Tübingen NLTE Model Atmosphere Package (Werner et al. 2003, Rauch & Deetjen 2003). The first set of models is composed of H+He+C+N+O, then Ne+Mg+S are added subsequently in a line-formation calculation (Tab. 1). However, in general TMAP can treat all elements from hydrogen to the iron group simultaneously (Rauch 1997, 2003).

Small grids of model atmospheres were calculated in order to match the relevant T_{eff} range (0.5 – 1 MK). 208 levels of H+He+C+N+O+Ne+Mg+S are treated in NLTE with 332 individual line transitions (Tab. 1).

CP774, *X-Ray Diagnostics of Astrophysical Plasmas,* edited by Randall K. Smith

TABLE 1. Statistics of the model atoms used in our calculations of the NLTE model atmospheres. The notation is: NLTE = levels treated in NLTE, LTE = LTE levels, RBB = radiative bound-bound transitions

atom	ion	NLTE	LTE	RBB	atom	ion	NLTE	LTE	RBB
H	I	5	11	10	Ne	VII	10	50	12
	II	1	-	-		VIII	8	18	15
He	I	1	21	0		IX	5	6	3
	II	10	22	45		X	1	0	0
	III	1	-	-	Mg	VIII	1	19	0
C	V	29	21	60		IX	3	23	1
	VI	15	21	26		X	2	3	1
	VII	1	0	-		XI	5	6	1
N	V	5	15	6		XII	1	0	0
	VI	17	7	33	S	X	1	7	0
	VII	15	21	30		XI	5	29	1
	VIII	1	0	-		XII	10	12	15
O	VI	1	34	0		XIII	9	21	10
	VII	19	7	21		XIV	9	1	15
	VIII	15	30	30		XV	1	0	0
	IX	1	0	-					
						total	208	409	332

A comparison with the spectrum of V4743 Sgr via XSPEC yields a $T_{\rm eff}$ of about 705 kK, however, the flux at wavelengths smaller than 26Å is too low. Decreasing the C abundance by a factor of 0.01, results in a much better fit in this region (Fig. 1).

A crucial point for the understanding of processes during the nova outburst may be the time-dependent prediction of surface abundances as well as the spectral analysis of high-resolution X-ray spectra taken from outburst to the end of surface H burning. The inspection of available XMM-Newton spectra of V4743 Sgr, taken on a half-year time scale has shown that about weekly spectra are highly desirable. This would also allow to investigate the velocity reliably.

Starrfield et al. (2000) calculated the yields of thermonuclear runaways of O/Ne/Mg WD. We adopted the average of their models' abundance ratios and calculated a grid of test models. Their fluxes do not reproduce the observation better than our simple C-reduced models. Thus, this issue will be a challenge for both theory and spectral analysis in the future.

The comparison of our preliminary model atmosphere fluxes with the Apr 2003 XMM-Newton RGS spectrum of V4743 Sgr, however, has shown that at solar abundance ratios we cannot fit the overall flux distribution. Although the fit is not perfect yet, it is obvious that the C/N abundance ratio is much smaller (about a hundred times) than the solar value, typical for CNO processed material. The fit to the observation based on our first test models is best at a $T_{\rm eff}$ of about 610 kK and $\log N_{\rm H} = 20.54\ \left[{\rm cm}^{-2}\right]$.

Analysis of X-ray data in Tübingen is supported by the DLR under grant 50 OR 0201.

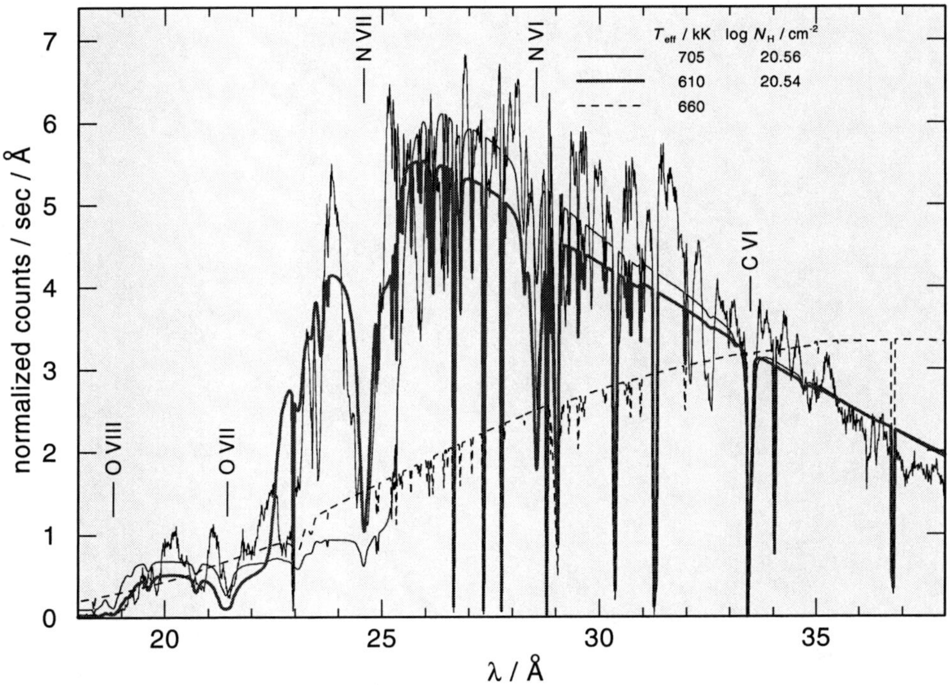

FIGURE 1. Comparison of the XMM-Newton RGS-1 spectrum with NLTE model fluxes (thick: H:He:C:N:O = 18.9:75.1:0.0006:0.016:5.96, thin: solar). $T_{\rm eff}$ and $N_{\rm H}$ are determined with XSPEC. A black body (dashed) is shown for comparison only and to identify instrumental features of XMM-Newton. The synthetic spectra are shifted by -2400 km/sec

REFERENCES

1. Haseda, K., West, D., Yamaoka, H., Masi, G. 2002, IAU Circ. 7975
2. Kato, T., Hishikura, T., West, J.D., et al. 2002, IAU Circ. 7976
3. Ness, J.-U., Starrfield, S., Burwitz, V., et al. 2003, ApJ 594, L127
4. Orio, M., Leibowitz, E., Rodriguez, P., et al. 2003, IAU Circ. 8131
5. Rauch, T. 1997, A&A 320, 237
6. Rauch, T. 2003, A&A 403, 709
7. Rauch, T., Deetjen, J.L. 2003, in: Workshop on Stellar Atmosphere Modeling, eds. I. Hubeny, D. Mihalas, K. Werner, The ASP Conference Series (San Francisco: ASP), 288, 103
8. Starrfield, S., Sparks, W.M., Truran J.W., Wiescher, M.C. 2000, ApJS 127, 485
9. Werner, K., Dreizler, S., Deetjen, J.L., Nagel, T., Rauch, T., Schuh, S.L. 2003, in: Workshop on Stellar Atmosphere Modeling, eds. I. Hubeny, D. Mihalas, K. Werner, The ASP Conference Series (San Francisco: ASP), 288, 31

Cataclysmic Variable Grating Emission Lines

E. M. Schlegel*, V. Rana, K. P. Singh† and P. Barrett**

*SAO, Cambridge, MA
†TIFR, Mumbai, India
**STScI, Baltimore, MD

Abstract. We describe work-in-progress to analyse the available HETG spectra on sixteen spectra from ten cataclysmic variables. We approach the analysis from dual directions, using globally-fit, multi-temperature components and via a line-by-line delineation of emission line properties. As an ultimate science goal, we intend to compare the results of the two approaches for consistency.

Keywords: cataclysmic variables, X-ray, spectroscopy
PACS: 95.85.Nv, 97.80.Gm, 95.75.Fg

INTRODUCTION

The goal of X-ray spectroscopy of cataclysmic variables (CVs) is similar to that in the optical: determine the dynamics and geometry of the emission region(s). Prior to *Chandra* and *XMM*, at most a handful of all astrophysical objects had been observed at high spectral resolution, hampering the detailed understanding of the innermost accretion disk or magnetic accretion regions.

We briefly discuss a summary of a work-in-progress to analyze the available *Chandra* high-energy grating (HETG) data on CVs. A total of 16 spectra for 10 CVs are currently available (10/2004). The CVs include 4 dwarf novae (U Gem×2, SS Cyg×3, WX Hyi, SU UMa), 4 intermediate polars (EX Hya, AO Psc, V1223 Sgr, GK Per), one old nova (V603 Aql), and one Z Cam star (V426 Oph).

For our science goal, we intend to compare the results from global spectral fits to the results inferred from a line-by-line analysis. We do not include the low-energy grating (LETG) data, nor the sources detected with the *XMM* reflection gratings at this time.

ANALYSIS

The HEG and MEG data were extracted and the ± orders combined. To test the quality of new fitting codes, we compared spectral fits using the Mekal code (Kaastra 1992; Mewe, Gronenschild, & van den Oord 1985; Mewe et al. 1986; Liedahl et al. 1995 [2, 5, 6, 3] and as available in `xspec`, (Arnaud 1996) [1]) with the APEC code (Smith et al. 2001) [7]. Figure1 shows a best-fit result for the summed EX Hya MEG spectra using multi-temperature Mekal and APEC components. Generally the two codes yield similar results for the HEG fits because of the easier inclusion of all K-shell features. However, the 15-17 Å region shows many residuals using the Mekal model, including several strong transitions. The difference between the Mekal and APEC models clearly reflects the improved number of lines available to the APEC code, particularly for the Fe

CP774, *X-Ray Diagnostics of Astrophysical Plasmas,* edited by Randall K. Smith

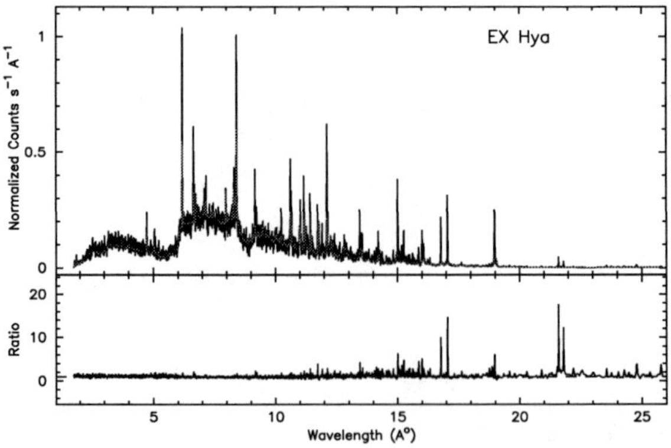

FIGURE 1. Comparison of models using (top) Mekal and (bottom) APEC codes. Note the large residuals in the 15-17 Å range for the Mekal code. Note that both codes are missing lines at ~22 Å.

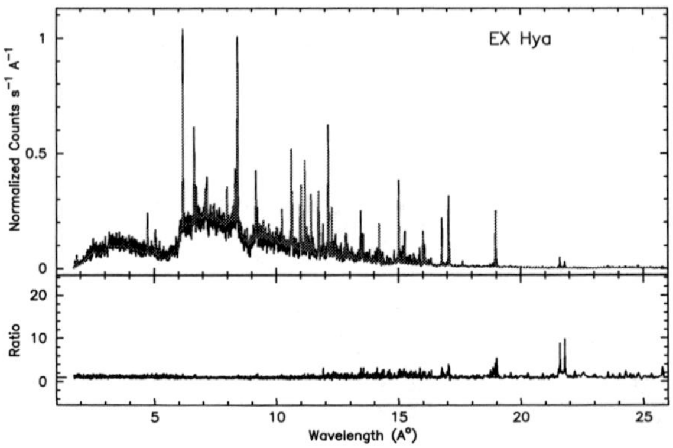

FIGURE 2. APEC model fit.

L region. The χ^2/ν is 0.84 and 0.75 respectively for the HEG fit, but 2.04 for the Mekal model and 1.37 for the APEC model when applied to the MEG data.

For the line-by-line fits, we adopted unresolved (= zero-width) gaussians, fitting for line position and strength. A bremsstrahlung component accounted for the pseudo-continuum generated by the blending of weak, unresolved lines. Approximately 330 lines are detected; another 230 upper limits are calculated assuming a zero-width gaussian fixed at the position of an expected line.

The line-by-line analysis is complicated by the unknown presence of velocity shifts, introducing some degree of "line confusion" into the line identification. A velocity shift of ~500 km s^{-1} corresponds to 0.023 Å at 15 Å, easily confusing the line's identity in crowded regions, eg, Fe L region in the 15 Å region.

Figure 3 shows the ratio of the Fe XVII 17.10 Å and Fe XVII 17.05 Å lines as calculated and as detected. Mauche et al. (2001) [4] first reported the lack of the 17.10 Å line in the EX Hya spectrum, indicating an electron density $>2\times10^{14}$ cm^{-3} for the

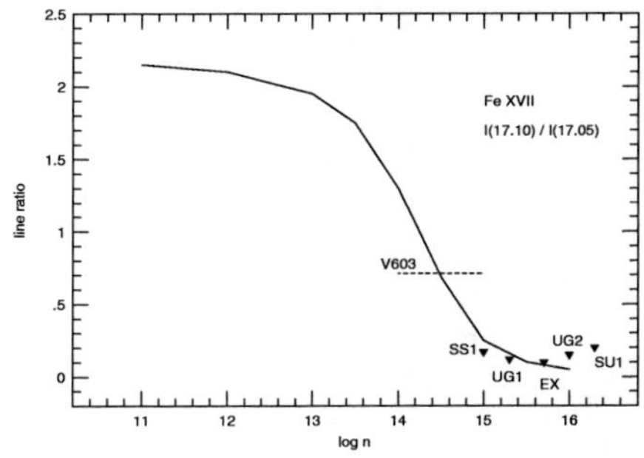

FIGURE 3. Comparison of values of the Fe XVII 17.10 Å / Fe XVII 17.05 Å ratio in the HETG data of CVs. Note that V603 Aql differs from the other CVs by ~2-5 in number density.

post-shock flow. Note that all of the CVs *except* V603 Aql show similar values for the ratio. The observed ratio in V603 Aql leads to a measure of the number density of $\sim 10^{14}$ cm^{-3}; the implied measure for the other CVs is higher, although photon statistics lead to some overlap.

The improved spectral resolution available with *Astro E2* will eliminate many or all of the line confusion issues associated with these spectra. Additional work on the other detected lines is on-going.

The research of EMS was supported by NASA contract NAS8-39073 to SAO for the Chandra X-ray Center.

REFERENCES

1. Arnaud, K. 1996, in ASP Conf. Ser 101, Astronomical Data Software and Systems V, ed. G. H. Jacoby & J. Barnes (San Francisco: ASP), 17
2. Kaastra, J.S. 1992, An X-Ray Spectral Code for Optically Thin Plasmas (Internal SRON-Leiden Report, updated version 2.0)
3. Liedahl, D.A., Osterheld, A.L., & Goldstein, W.H. 1995, ApJL, 438, 115
4. Mauche, C. W., Liedahl, D. A., & Fournier, K. B. 2001, ApJ, 560, 992
5. Mewe, R., Gronenschild, E.H.B.M., & van den Oord, G.H.J. 1985, A&AS, 62, 197
6. Mewe, R., Lemen, J.R., & van den Oord, G.H.J. 1986, A&AS, 65, 511
7. Smith, R. K., Brickhouse, N. S., Liedahl, D. A., Raymond, & J. C. 2001, ApJ, 556, L91

Modelling the X-ray spectra of high-velocity outflows from quasars.

S. A. Sim

Astrophysics Group, Imperial College London, Blackett Laboratory, Prince Consort Road, London, SW7 2AZ

Abstract. High-velocity outflows from supermassive black-holes have been invoked to explain the recent identification of strong absorption features in the hard X-ray spectra of several quasars. We present the results of new 2-D Monte Carlo radiative transfer calculations which synthesise X-ray spectra for such outflows. These show that simple, parametric bi-conical models with plausible choices for the wind parameters predict spectra that are in good qualitative agreement with observations. Particular attention is given to the bright quasar PG1211+143.

Keywords: radiative transfer – quasars: absorption lines – quasars: individual PG1211+143 – quasars: outflows – galaxies: active
PACS: 95.30.Jx, 98.54.Aj, 98.54.Cm

INTRODUCTION

Blueshifted narrow absorption lines have recently been reported in the hard X-ray spectra of several quasars (e.g. [1], [2], [3], [4]). Although there is some ambiguity (e.g. [5]) it is widely speculated that these features arise in relativistic outflows that are more highly ionized than those previously detected via ultraviolet features. Here, we present new radiative transfer calculations synthesizing hard X-ray spectra for a variety of simply parameterised bi-conical outflow models. These synthetic spectra can be contrasted with observations to test the outflow hypothesis and place constraints on the flow parameters. This initial study is focused on the nearby quasar PG1211+143 for which high quality data have been analysed by [3]. For full details of this investigation see [6].

METHOD

The hard (2 – 10 keV) X-ray spectra are computed using a new 2-D Monte Carlo radiative transfer code based on the 1-D code described by [7] which adopts the Macro Atom formalism developed by [8], [9]. The computations include radiative bound-free and bound-bound processes for 27 important ions using atomic data taken from the Xstar database [10]. Compton scattering of photons by free electrons is also included.

The ionization state is computed as a function of position in the flow by balancing photoionization rates obtained during the Monte Carlo calculations with radiative recombination rates.

CP774, *X-Ray Diagnostics of Astrophysical Plasmas,* edited by Randall K. Smith

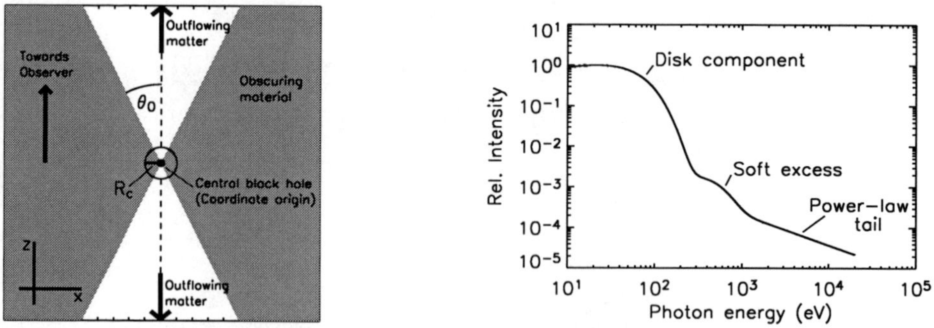

FIGURE 1. Left: cartoon showing the bi-conical outflow geometry adopted. Right: form of the input spectral energy distribution (the three components described in the text are identified).

Model geometry

A bi-conical outflow geometry as discussed by [3] and [11] is adopted (see Fig. 1, left panel). The wind is assumed to reach a terminal velocity ~ 0.1c (based on the observed line shifts from [3]) and to be launched at radius $R_c = 100R_s$ from the central black-hole, where R_s is the Schwarzschild radius of the black-hole. (The black-hole mass in PG1211+143 is taken as 4×10^7 $M_\odot$ [12]). The region outside the flow is assumed to contain obscuring material through which photons do not propagate.

Input radiation field

At the base of the flow the radiation field is assumed to consist of three components (see Fig. 1, right panel):

- A multi-colour black-body accretion disk spectrum. The total disk luminosity is fixed at the Eddington luminosity for the central black-hole.
- A hard X-ray power-law continuum. The slope and normalisation of this component are fixed to match the observed X-ray properties of PG1211+143.
- A soft excess. This is modelled as a black-body with temperature fixed at that fitted to the X-ray spectrum by [3].

RESULTS

Fig. 2 shows sample computed spectra for flows with differing mass-loss rates and opening angles. It is concluded that simple bi-conical outflow models can readily account for the absorption lines reported by e.g. [3]. Given that the observed spectral features are narrow, an interpretation in terms of a wind which subtends a fairly small solid angle as viewed from the black-hole is favoured [6].

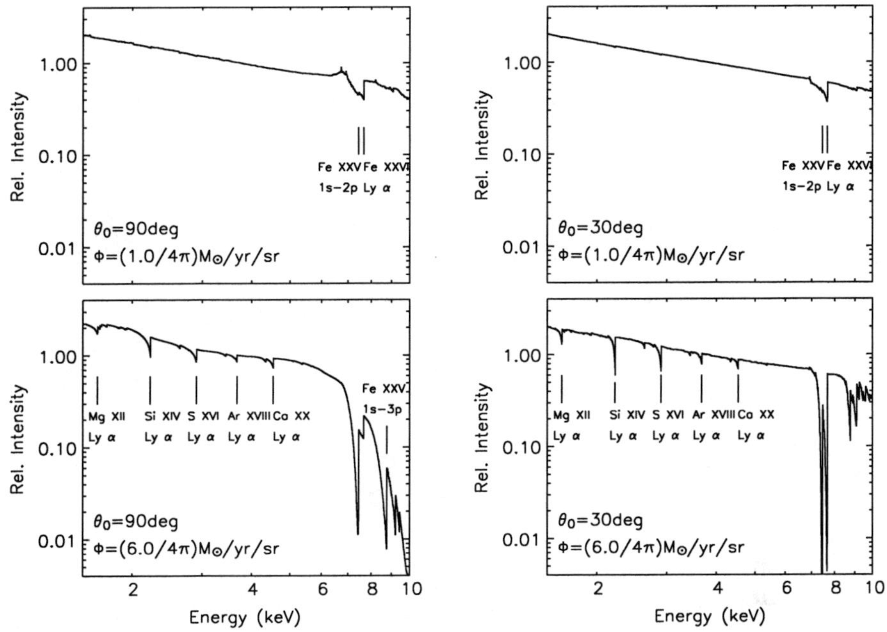

FIGURE 2. Computed spectra for flows with differing mass-loss rate per unit solid angle (Φ) and biconical half-opening angle (θ_0). Small opening angles produce features that are stronger and narrower than those obtained with wide opening angles. This is due to photon leakage through the conical boundaries which influences the ionization state [6].

ACKNOWLEDGMENTS

I wish to thank L. Lucy for many useful discussions and suggestions relating to this work. Thanks also to J. Drew, K. Nandra and A. L. Longinotti for their comments.

REFERENCES

1. G. Chartas, W. N. Brandt, S. C. Gallagher, and G. P. Garmire, *ApJ*, **579**, 169 (2002).
2. G. Chartas, W. N. Brandt, and S. C. Gallagher, *ApJ*, **595**, 85 (2003).
3. K. A. Pounds, J. N. Reeves, A. R. King, K. L. Page, P. T. O'Brien, and M. J. L. Turner, *MNRAS*, **345**, 705 (2003).
4. K. A. Pounds, A. R. King, K. L. Page, and P. T. O'Brien, *MNRAS*, **346**, 1025 (2003).
5. B. McKernan, T. Yaqoob, and C. S. Reynolds, *astro-ph*, **0408506** (2004).
6. S. A. Sim, *astro-ph*, **0410090** (2004).
7. S. A. Sim, *MNRAS*, **349**, 899 (2004).
8. L. B. Lucy, *A&A*, **384**, 725 (2002).
9. L. B. Lucy, *A&A*, **403**, 261 (2003).
10. M. A. Bautista, and T. R. Kallman, *ApJS*, **134**, 139 (2001).
11. A. R. King, and K. A. Pounds, *MNRAS*, **345**, 657 (2003).
12. S. Kaspi, P. S. Smith, H. Netzer, D. Maoz, B. T. Jannuzi, and U. Giveon, *ApJ*, **533**, 631 (2000).

Incorporating Atomic Data Errors in Stellar DEM Reconstruction

Hosung Kang*, David A. van Dyk†, Vinay L. Kashyap** and Alanna Connors‡

*Department of Statistics, Harvard University
†Department of Statistics, University of California at Irvine
**Harvard-Smithsonian Center for Astrophysics
‡Eureka Scientific

Abstract.
We develop a powerful new method to reconstruct stellar Differential Emission Measures (DEMs); its Bayesian framework allows us to incorporate atomic and calibration errors as prior information. For instance, known errors in the line locations, as well as lines missing from the atomic data base, can be included directly during fitting. Highly correlated systematic errors in the ion balance may be included as well, as a natural sequence during Monte Carlo sampling. Our method uses the statistical framework of data augmentation, where we treat photon counts in each level of a hierarchical structure as missing data. We demonstrate our method by fitting a selected subset of emission lines and continuum in Chandra and EUVE data of Capella to estimate the DEM that best describes the data, and simultaneously determine the element abundances. The Markov Chain Monte Carlo based method also naturally produces error estimates on the fit parameters.

Keywords: Differential Emission Measures, Emissivity Matrix, Atomic Data Error, Capella

INTRODUCTION

DEMs define the distribution of the amount of emitting material in coronae, as a function of temperature. DEMs summarize the structure of coronae on stars and provide a powerful tool for understanding the energetics in stellar atmospheres. Hence, it is an important determinant of the physical structure and evolutionary state of a star. Given the atomic line and continuum emissivities, and the observed photon counts spectrum, we aim to reconstruct the marginal distribution of the temperature of the corona.

SPECTRAL MODEL

We model the intensity of the *ideal* photon counts arriving at energy bin i without instrumental contamination, Y_i as independent Poisson variables with intensity

$$\lambda_i = \sum_{t=1}^{N_T} \left[\left\{ \Lambda_{it}^{C} + \sum_{\text{element } k} \gamma_k \left(\Lambda_{it}^{k,PC} + \Lambda_{it}^{k,L} \right) \right\} \delta_t \right], \tag{1}$$

where N_T is the number of temperature bins, the $I \times N_T$ matrix Λ^C represents the continuum emissivity matrix, the $I \times N_T$ matrices $\Lambda^{k,L}, \Lambda^{k,PC}$ represent the emission-line

CP774, *X-Ray Diagnostics of Astrophysical Plasmas,* edited by Randall K. Smith

and pseudo-continuum emissivity matrices, respectively for element k and γ_k are the element abundances for element k, and δ_t are the values of the DEM at temperature t.

Note the similarity in construction here to a spectral response matrix; we treat the emissivity matrix as a redistribution matrix to go from temperature space to wavelength space. The standard instrumental effective area and redistribution matrix are also then applied.

HIERARCHICAL MISSING DATA STRUCTURING

The model (1) has multiple components (e.g., background, continuum and emission lines) and complex structure due to large dimensional matrices. Data augmentation methods such as the EM [1] or DA algorithm [2] make model-fitting simpler especially in highly structured models. Thus, we impute the missing data at each level, treat them as if they were observed, and find the conditional posterior distribution of the next level missing data given the current level. This will eventually lead us to imputing the photon counts emitted from each temperature, where we treat the emissivity matrix as an RMF-like "response matrix" and trace the energy bin counts to the unobserved photon counts in a number of temperature bins.

ATOMIC DATA ERRORS

Even the best atomic emissivity databases have missing or misplaced lines and incorrect emissivities. Our method allows prior information on these issues to be directly incorporated into the analysis.

Error bars on the emissivity matrix: During each iteration of the MCMC sampling algorithm, we regenerate the emissivity matrix with an error of 10% for H-like and He-like lines and a 20% error on Fe XVII - Fe XX lines. Further corrections that include known differences between the predicted and observed fluxes in the EUV and X-ray regions (e.g., Fe XVIII in X-ray appears to be uniformly underpredicted by 30%; cf. Desai et al. 2004, this conference) can be included in a similar fashion.

Missing lines: The two main sources of missing lines in current emissivity database tables are DR lines associated with the weaker resonance lines and lines from high n ($n > 5$). We plan to include the former by suitably scaling a template derived from, e.g., the satellite lines associated with Fe XVII λ 15.02. The magnitude of the latter can be estimated by considering the asymptotic value of the fluxes due to lines from different n and locating them uniformly between the ionization limit and the last listed line. This has not yet been implemented.

Wavelength errors: Due to incomplete atomic data measurements as well as detector non-linearities, the observed locations of lines does not in general match the theoretical locations. We compensate for this effect by allowing the strongest lines in the spectral region to be shifted during the fit, and then move the remaining weaker lines accordingly.

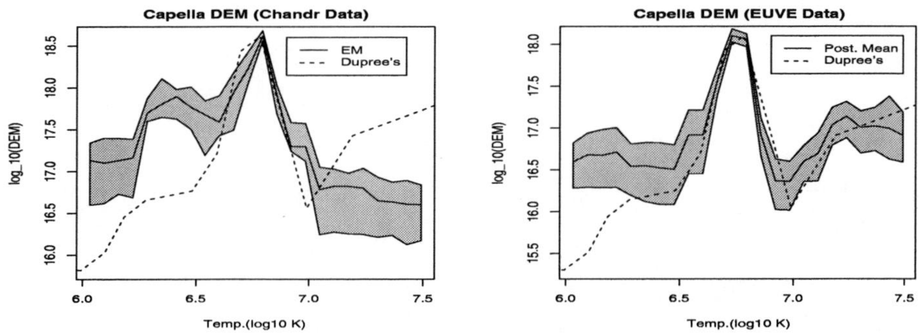

FIGURE 1. Error Bars for the Fitted DEMs from Chandra and EUVE data. The shaded area represents componentwise 95% posterior intervals for the DEM. The DEM reconstructed for the EUVE observation by Dupree et al. [3] is shown. The features in the DEMs are by and large consistent, though our reconstruction shows a significantly larger emission measure below logT=6.5. The second plot shows more consistent result because the two methods used the same data.

RESULTS

The raw spectrum of Capella collected from Chandra HRC-S/LETG data. We computed the maximum a posteriori estimate of the DEM using the EM algorithm starting from a flat DEM. We fit the model via MCMC to compute the posterior mean and componentwise 95% posterior intervals using the same starting values. See Figure 1.

ACKNOWLEDGMENTS

The authors gratefully acknowledge funding for this partially provided by NSF Grants DMS-01-04129, DMS-04-38240, and DMS-04-06085 and by NASA Contracts NAS8-39073 and NAS8-03060(Chandra X-ray Center). This work is a product of joint work with the California-Harvard astrostatistics collaboration (CHASC) whose members include J. Chiang, A. Connors, D. van Dyk, D. Esch, P. Freeman, H. Kang, V. L. Kashyap, X.-L. Meng, A. Siemiginowska, E. Sourlas, T. Park, Y. Yu, and A. Zezas. We thank Nancy Brickhouse for helpful comments.

REFERENCES

1. A. P. Dempster, N. M. Laird, and D. B. Rubin, *Journal of the Royal Statistical Society, Series B, Methodological*, **39**, 1–37 (1977).
2. M. A. Tanner, and W. H. Wong, *Journal of the American Statistical Association*, **82**, 528–550 (1987).
3. A. K. Dupree, N. S. Brickhouse, G. A. Doschek, J. C. Green, and J. C. Raymond, *Astrophysical Journal Letters*, **418**, 115–126 (1993).

IX: FUTURE MISSIONS, PLANS, AND PROSPECTS

The XRS Microcalorimeter on Astro-E2

J. Cottam*, K. R. Boyce*, G. V. Brown*,†, R. Fujimoto**, T. Furusho**, Y. Ishisaki‡, R. L. Kelley*, C. A. Kilbourne*, D. McCammon§, K. Mitsuda**, U. Morita‡, F. S. Porter*, T. Saab*, Y. Takai‡ and M. Yamamoto¶

*NASA/Goddard Space Flight Center
†Lawrence Livermore National Laboratory
**Institute for Space and Astronautical Science/JAXA
‡Tokyo Metropolitan University
§University of Wisconsin
¶Miyazaki University

Abstract. The XRS microcalorimeter will be launched in 2005 as part of the Astro-E2 mission. It will cover the energy band from 0.3 to 10 keV with a nearly constant energy resolution of 6.0 eV and a peak effective area of 200 cm^2 at 1.5 keV. The XRS will provide unprecedented throughput and resolving power, particularly at high energies. Detailed spectral features in the Fe K region will be resolved for the first time, providing access to spectroscopic diagnostics for a wide range of astrophysical objects. In this presentation we will describe the XRS instrument, details of its spectral performance, and how it compares to the current high-resolution instruments on the *Chandra* and *XMM-Newton* observatories.

Keywords: X-ray Instrumentation; X-ray Spectrometers; Microcalorimeters
PACS: 95.55.Ka; 95.55.Qf; 07.20.Fw

INTRODUCTION

The Astro-E2 observatory is a joint NASA and ISAS/JAXA mission scheduled for launch in the summer of 2005 [1]. The primary instrument onboard Astro-E2 is the high-resolution X-Ray Spectrometer (XRS). The XRS is a microcalorimeter array of 32 pixels mounted at the focal plane of one of five mirror modules. The energy of incident photons is converted to heat in HgTe pixel absorbers. Current-biased thermistors record the change in temperature as 'pulses', where the pulse height is roughly proportional to the photon energy. The XRS has a nearly constant energy resolution of $\Delta E_{\rm FWHM} \sim 6\,\rm eV$ across the operating band of 0.3 to 10 keV, and therefore a resolving power of $\mathscr{R} \sim 1000$ in the Fe K region. The high quantum efficiency of the detectors combined with the large collecting area of the telescopes results in an effective area of $A_{\rm eff} \sim 150\,\rm cm^2$ at 6 keV. The XRS will clearly provide unprecedented access to quantitative spectral diagnostics particularly at high energies. The XRS has undergone testing at the detector level and after full instrument integration. In the sections below we will briefly describe the energy scale, line spread function, effective area and background. The most current descriptions of the instrument characterization can be found on the Astro-E2 webpage [2].

CP774, *X-Ray Diagnostics of Astrophysical Plasmas,* edited by Randall K. Smith
2005 American Institute of Physics 0-7354-0259-0/05/$22.50

ENERGY SCALE

The energy scale of the XRS array is a complex, non-linear function that depends not only on the design parameters of the XRS array and the pulse detection algorithm, but also on the operating conditions including the thermal environment seen by the array and the photon flux. An analytic model of the XRS pulses is being developed. However, because of the inherent complexity, the energy scale is currently determined by empirically fitting measured pulse heights from known energy sources to a polynomial function. Because the energy scale depends on the specific operating conditions, a new energy scale will be generated in-flight using both onboard radioactive sources and astrophysical sources. An offset calibration pixel will be used to track the changes in the energy scale. The accuracy of the in-flight energy scale is expected to be $\Delta E_{\rm centroid} \sim 1.0$ to 2.0 eV (see Fig. 1).

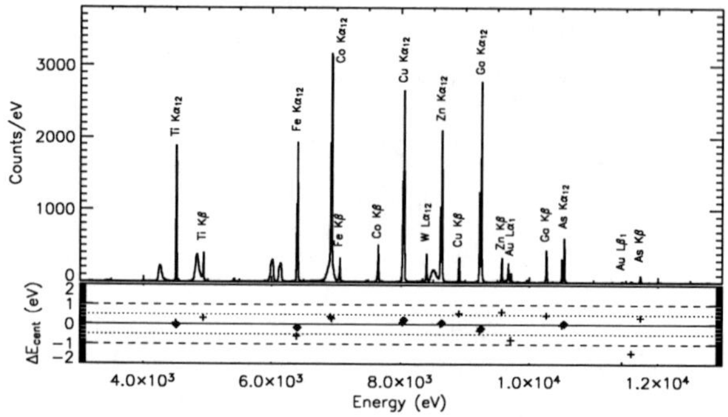

FIGURE 1. An energy-corrected spectrum obtained during ground calibration and the residuals between the measured and laboratory energies for the brightest lines. The filled diamonds represent the Kα lines that were used to generate the energy scale. The broad, unlabeled features are caused by Bragg reflection.

LINE SPREAD FUNCTION

The XRS Line Spread Function (LSF) consists of a Gaussian resolution kernel and a low energy tail. The resolution kernel is primarily determined by the temperature and voltage fluctuations caused by phonon and Johnson noise in the detector system. These produce a Gaussian distribution in the measured energy around the nominal value. The low-energy tail is caused by photon and electron escape events. It contains roughly $2-4\%$ of the total events. The LSF of the XRS has been measured by illuminating the array with a ^{55}Fe electron capture source and characteristic fluorescent emission lines (see Fig. 2). Except at low energies where the electron loss continuum begins to contribute, the XRS resolution kernel can be described by a single Gaussian that is only weakly dependent on energy. Each pixel is a unique detector with a unique energy resolution. The composite resolution, excluding two outlying pixels is $\Delta E_{\rm FWHM} \leq 6$ eV at 6 keV. This is not expected to change significantly in-flight.

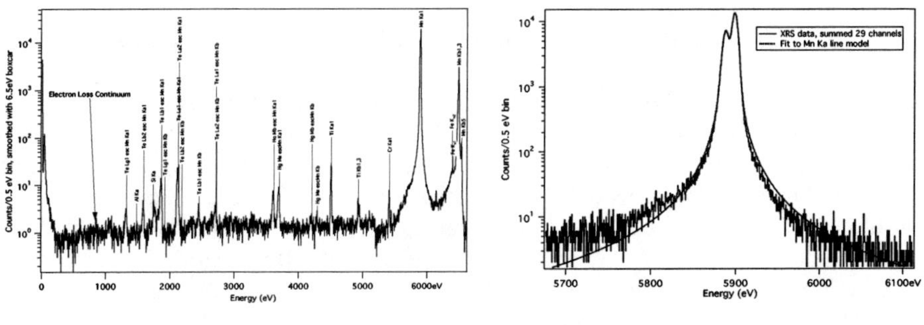

FIGURE 2. The measured broad-band spectrum from an ^{55}Fe source (Left) and a close-up of the Mn Kα lines (Right) with the modeled fit superimposed.

EFFECTIVE AREA

The effective area of the XRS is the product of the geometric area of the XRT telescopes and the efficiency of each component along the optical path. These include the reflection efficiency of the gold coated mirrors and the internal vignetting between the mirror shells, the overlap between the mirror PSF and the detector pixel absorbers, the transmission of five optical blocking filters, the detector quantum efficiency, and the electronic deadtimes. The full XRS effective area cannot be measured on the ground. However, the discrete structure in the effective area curve is dominated by the quantum efficiency of the HgTe absorbers at high energies and the transmission of the blocking filters at the low energies, both of which have been calibrated extensively. The global effective area curve must be calibrated in flight.

BACKGROUND

The XRS background comes primarily from direct interactions from cosmic-ray protons and from secondary electrons and photons generated in the detector housing. Minimum ionizing protons will deposit energies comparable to those of photon events. An anti-coincidence detector is positioned directly behind the calorimeter array for onboard rejection of these events. Modeling shows that only 0.1% of all protons incident on the array will miss the anti-coincidence detector and deposit energy less than 10 keV. Protons that hit the Si frame of the array can deposit enough energy to heat the array and generate small simultaneous pulses on multiple pixels. These events can be rejected by pixel-to-pixel coincidence, but they will contribute slightly to the overall instrument dead time. The dominant unrejected background comes from secondary events. To estimate these rates we have scaled the background spectrum observed with ASCA, since both satellites are in similar orbits. This gives a flat background with a rate of $\sim 7 \times 10^{-4}$ events/s over the array when the instrument is outside of the SAA.

COMPARISONS TO OTHER SPECTROMETERS

The XRS on Astro-E2 will complement the high-resolution spectrometers on the *XMM-Newton* and *Chandra* observatories [3, 4] (see Fig. 3). The RGS onboard the *XMM-Newton* and the HETGS and LETGS onboard *Chandra* are dispersive spectrometers. These are excellent for low energy point sources or mildly extended sources. The XRS, as a non-dispersive spectrometer, has a resolving power that increases with energy and is unaffected by the spatial extent of the incident light. It therefore provides the highest resolution spectra of high energy sources and spatially extended sources.

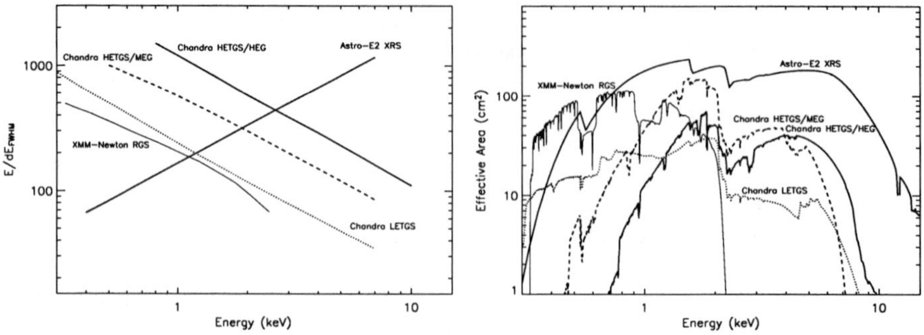

FIGURE 3. The resolving power (Left) and the effective area (Right) of the high-resolution spectrometers onboard the *XMM-Newton*, *Chandra*, and *Astro-E2* observatories.

REFERENCES

1. H. Inoue, *Proc. SPIE*, **4851**, 289 (2003).
2. http://heasarc.gsfc.nasa.gov/docs/astroe/astroegof.html
3. F. Jansen, D. Lumb, B. Altieri, J. Clavel, M. Ehle, C. Erd, et al., *A& A*, **365**, 1 (2001).
4. M. C. Weisskopf, B. Brinkman, C. Canizares, G. Garmire, S. Murray, L. P. Van Speybroeck, *PASP*, **114**, 1 (2002).

The Frontier in X-ray Spectroscopy: NASA's Constellation-X Mission

Ann E. Hornschemeier, Nicholas E. White

Laboratory for X-ray Astrophysics
NASA Goddard Space Flight Center

Harvey Tananbaum
Harvard-Smithsonian Center for Astrophysics

Abstract. This paper briefly describes the Constellation-X mission, giving its basic performance parameters such as effective area and spectral resolution, describes some example science that may be conducted, and gives a general update on the mission, due for launch in 2016. Constellation-X is one of the two flagship missions for the Beyond Einstein programand was designated as a high priority, second only to JWST, in the large space observatories category of the most recent National Academy decadal. Constellation-X will present a great improvement in spectroscopic sensitivity, enabling studies at sensitivities more than 100 times greater than the best high-resolution spectrometers observing in the 0.25-10 keV band during the current decade. The spectral resolution, R = E/ΔE, will range from 300 to 3000.

Keywords: Future missions.
PACS: 95.85.Nv

X-RAY ASTRONOMY'S PLACE IN NASA'S NEW STRUCTURE AND THE BEYOND EINSTEIN PROGAM

High-resolution X-ray spectroscopy in the Chandra and XMM era, with R (=E/ΔE) of approximately 300, has revolutionized our understanding of a number of fundamental issues [1] but due to relatively small mirror collecting areas at high spectral resolution (typically less than 200 cm^2), observations have been limited to extremely bright X-ray point sources and a small handful of diffuse astrophysical sources. What may the decade after the current one bring in terms of high-resolution X-ray spectroscopy? NASA's answer to this question is the upcoming Constellation-X mission, due for launch in 2016 to an L2 orbit. This paper briefly describes the technical capabilities that are planned in the 0.25-10 keV band. The mission also has a Hard X-ray Telescope (covering ~10-60 keV) with modest spectral resolving power (R=10) that is beyond the scope of this current paper. Additional details about the mission may be found on the official Constellation-X website (http://constellation.gsfc.nasa.gov).

CP774, *X-Ray Diagnostics of Astrophysical Plasmas,* edited by Randall K. Smith
2005 American Institute of Physics 0-7354-0259-0/05/$22.50

Constellation-X is one of the two flagship missions in NASA's Beyond Einstein program (for more details, see http://universe.nasa.gov). The three "Beyond Einstein questions" capture many of the biggest unsolved problems in astrophysics: What happens at the edge of a black hole? What is the mysterious dark energy that is observed to be pulling the Universe apart? What powered the Big Bang? Together with the gravitational wave mission LISA, Constellation-X will address the first two of these fundamental questions. Constellation-X is fully approved by NASA and is designated as a major priority by the Decadal Survey of the National Academy of Science. Constellation-X, via high-resolution X-ray spectroscopy, will detect inner accretion-disk X-ray emission and enable tests of General Relativity in the strong regime. Both missions are set to launch within the coming decade (LISA in 2012, Constellation-X in 2016).

PRACTICAL PROBLEMS IN HIGH-RESOLUTION X-RAY SPECTROSCOPY: THE CONSTELLATION-X MISSION

The universal task across the electromagnetic spectrum in building sensitive spectroscopic instruments is the need for large telescope apertures to produce high signal-to-noise spectra at ever increasing spectral resolution values. To illustrate the needs of X-ray astronomy, one may consider the "typical X-ray emitting source" characterized as the point source flux at which the vast majority of the diffuse X-ray background is resolved (for a recent review, see [2]). In the 0.5-2 keV band, where the majority of the X-ray background has now been resolved, approximately 70-80% of the X-ray background is resolved at 0.5-2 keV fluxes of 1×10^{-15} erg cm^{-2} s^{-1}, corresponding to incident photon fluxes of $\sim 6.5 \times 10^{-7}$ photons cm^{-2} s^{-1}.

Thus, to enable study of a large sample of "typical" X-ray emitting objects, this X-ray flux value is taken as a target goal for a typical observation (100 ks). Requiring 1000 counts per second, this leads to a collecting area requirement of 15,000 cm^{2}square centimeters in the 0.5-2 keV bandpass, which is the requirement of the Constellation-X mission at 1.25 keV.

Figure 1 shows the collecting area of Constellation-X compared to the X-ray spectroscopic workhorses of this decade: the Chandra High Energy Transmission Gratings (HETG) and Low Energy Transmission Gratings (LETG), the XMM-Newton Reflection Grating Spectrometer and the upcoming ASTRO-E2 calorimeter instrument. The figure shows the total combined effective area from both the +1 and −1 dispersion directions, but gives the first order effective area only for the LETG and HETG. Clearly Constellation-X will provide *two orders of magnitude increase in sensitivity* over the *best* performance of any of these high-resolution spectroscopy missions.

The effective area shown in Figure 1, which is for the two instruments operating in the 0.25-10 keV bandpass, is based upon existing demonstrated technology. Just like Chandra and XMM-Newton, the mission will have a gratings spectrometer. The baseline Con-X gratings will have a fixed $\Delta\lambda$ (~0.05 Å in the first order, half-power width) but gratings are under study that will exceed this value. The second component of high resolution X-ray spectroscopy on Constellation-X is provided by an X-ray calorimeter with a fixed ΔE value of 2-4 eV (FWHM), reaching R=3000 at Fe K energies (see Figure 2).

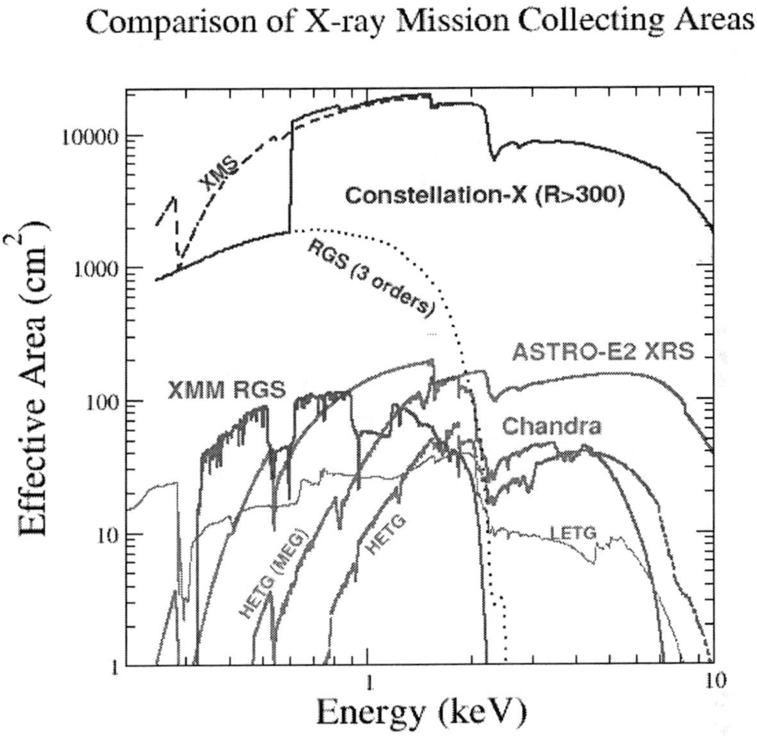

FIGURE 1. Effective area of Constellation-X in the 0.25-10 keV bandpass for both the calorimeter and gratings instruments where the resolution of either (or both) exceeds R=300. Note that the details of the XMS filter transmission below 1 keV will be strongly dependent upon the final mission configuration. The XMM and Chandra effective area curves are for the first order (both dispersion directions co-added).

What is not evident in Figure 1 is the tradeoff between collecting area, mirror figure (which affects the point-spread function; hereafter PSF) and the weight of the mirror assembly. If one compared the Chandra and XMM-Newton mirrors in terms of their effective area at 1 keV and the mass of the mirror assembly, one finds that the mass per square meter of effective area is 18,500 kg/m^2 for Chandra and 2300 kg/m^2 for XMM-Newton (Marcos Bavdaz, private comm.). The half-power diameter of the PSF, however, is 0.5" for Chandra and 14" for XMM-Newton. For the Chandra mission, great care was given to polishing the large amount of area over the four mirror shells whereas the XMM-Newton shells were fabricated in larger numbers, compromising mirror figure in the name of collecting area. Constellation-X has a requirement for a 15" half-power-diameter PSF (similar to XMM-Newton) with a goal of 5". Note that this spatial resolution goal is well-matched to a confusion limit at X-ray fluxes of a few times 10^{-16} erg cm^{-2} s^{-1}.

The spectral resolution requirement is a minimum of R=300, with higher requirements at Fe K (R=1500 from 6 keV to 10 keV). Figure 2 shows the spectral resolving power of the Constellation-X instruments over the 0.25-10 keV bandpass.

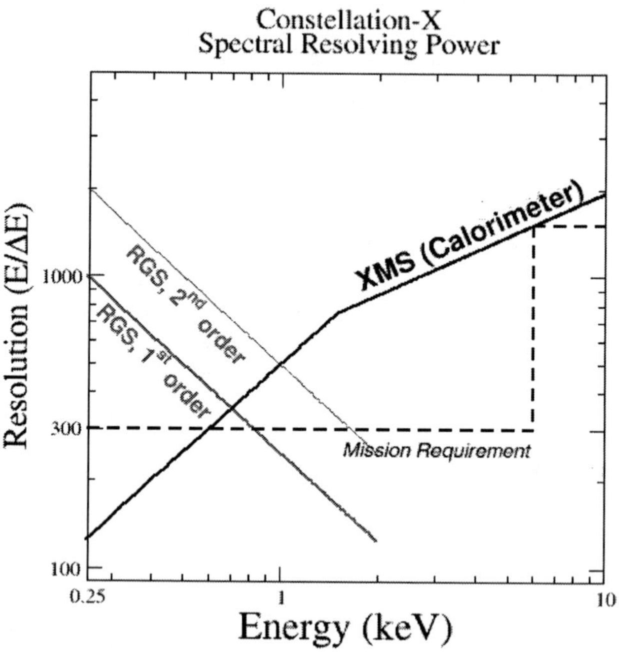

FIGURE 2. Spectral resolving power and requirements for the Constellation-X gratings and calorimeter instruments (as of February 2005). Note that calorimeters have a fixed ΔE (presented as FWHM) so the calorimeter curve reflects the current detector *requirements* for the calorimeter, not the actual performance of the calorimeter (e.g., the resolution curve might be built up with different types of calorimeter pixels in different locations of the array). These curves will change as the mission design evolves. Note that the gratings have a fixed Δλ (presented as HEW).

EXAMPLE CONSTELLATION-X SCIENCE

In this section we describe some of the significant science progress that is expected from the Constellation-X mission.

Strong gravity studies with Constellation-X

Ever since the detection of rapid X-ray variability over 20 years ago, it has been clear that X-ray observations of accreting black holes provide a window on the immediate vicinity of the black hole event horizon. Despite the advances with Chandra and XMM-Newton, progress is hampered by the fact that several different models (e.g., models with different black hole spin parameters) can fit the same line profile. Since the line profiles studied thus far are temporally averaged over at least a few orbital periods of the innermost disk, they do not contain sufficient information to uniquely constrain both the space-time geometry and the accretion disk physics.

Constellation-X, with its extremely high effective area and spectral resolution, will measure iron line profiles of the brightest AGN in a fraction of an orbit, allowing us to track individual orbiting substructures within the disk. Line profiles will be determined over the light-crossing time of the system for the more massive AGN systems, allowing searches for reverberation effects of dramatic X-ray flares across the disk surface. Such direct probing of the propagation of the X-ray signal provides a fundamental constraint on the nature of space-time around the black hole [3].

Dark Energy and Dark Matter Studies

The Universe today is dominated by Dark Energy, with ~70% of the energy density of the universe today is in the form of 'dark energy', 26% is in the form of dark matter and the rest is the sum of normal matter and neutrinos. Constellation-X will be able to help unravel the nature of both dark energy and dark matter through studies of clusters of galaxies. It is now clear that relaxed, simple clusters of galaxies can be used as 'standard candles' [4] to provide relative distances. The key is the theoretical prediction and observational verification that the baryonic mass fraction in rich clusters should be independent of redshift; X-ray observations are crucial since ~90% of all the baryons are in the hot X-ray emitting gas. Spectroscopic measurements with Constellation-X of the X-ray temperature and surface brightness of the gas in clusters, whose translation to gas mass depends on the absolute distance to the cluster will provide an independent measure of the expansion of the Universe. This will provide constraints on the equation of state of the dark energy w, and its derivative with time w'.

Evolution of Supermassive Black Holes Over Cosmic Time

Based upon the strong correlations observed between galaxy stellar mass and central black hole mass, it now seems likely that the development of black holes and galaxies are intimately connected, possibly due to accretion-related outflows that regulate star formation [5]. Thus understanding how supermassive black holes grow tells us how galaxies form. Indeed, one of the most important results of the last five years, which is an X-ray result, is that the growth of supermassive black holes over cosmic time has the most massive black holes forming first, a process often referred to as cosmic downsizing [6].

Constellation-X will study $z \approx 1$ active galactic nuclei (AGNs) with the detail only currently possible for the ~10 brightest local AGN (such as NGC 5548). Luminous $z > 6$ quasars will be more accessible to *Constellation-X* at high spectral resolution than $z > 1$ quasars are to *Chandra* and *XMM-Newton* at much lower spectral resolution. The large-scale AGN outflows (in absorption and emission) that likely regulate star formation in massive galaxies will be studied in detail, providing estimates of mass and energy outflow rates and chemical enrichment/heating of the IGM (see Figure 3).

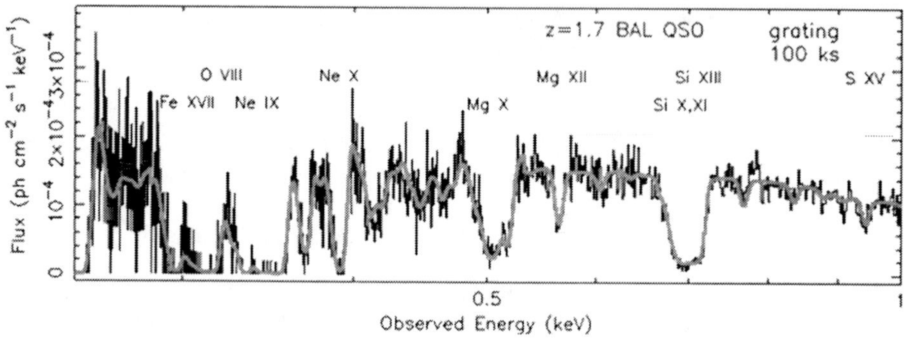

Figure 3: 100 ks *Constellation-X* simulation of a $z = 1.7$ quasar with an ionized, high velocity outflow Note the rich structure in the 0.2–1 keV band, including the resonance lines and edges of highly ionized O, Ne, Mg, and Si (rest frame E = 0.5–2.7 keV). These spectral features provide important constraints on the physical conditions in moderate column density gas (figure courtesy of S. Gallagher).

CURRENT STATUS OF THE CONSTELLATION-X PROGRAM

The Constellation-X mission baseline is to use four smaller spacecraft launched two at a time with two Atlas V or Delta IV launch vehicles. Each spacecraft has a 1.6m diameter spectroscopy X-ray telescope with a 10-m focal length, and three hard X-ray telescopes to extend the bandpass to 40 keV. Each of the soft (0.25-10 keV) X-ray telescopes will have a mass of <~250 kg (note that this translates to <200 kg per m^2 of effective area at 1 keV!) The mirrors will Wolter-type nested, grazing-incidence X-ray mirrors that will be manufactured using replication techniques similar to those used in the development of e.g., ASTRO-E2.

Alternate mission configurations are also under study including one large, monolithic X-ray mirror, with a longer 25 to 50m focal length. Such alternate configurations are now feasible due to the lower cost of the Delta IV "heavy" launch vehicle than was projected a decade ago. The mission will enter "Phase A" during ~2007, with an expected launch date of ~2016.

ACKNOWLEDGMENTS

We gratefully acknowledge the Constellation-X Facility Science Team and Science Panels who have contributed via advisory panels to the development of the Constellation-X science case. We thank Barb Mattson for helpful comments on a draft version. Particular thanks are given to Kathryn Flanagan, Paul Reid, Sarah Gallagher, Jean Cottam, Caroline Kilbourne and Richard Kelley for providing data for the figures.

REFERENCES

1. Paerels, F. B.S., and Kahn, S.M., *Annual Review of Astronomy &Astrophysics*, **41**, 291-342 (2003)
2. Brandt, W.N., and Hasinger, G., *Annual Reviews of Astronomy&Astrophysics*, in press (2005; astro-ph/0501058)
3. Young, A.J. and Reynolds, C.S, in *The Astrophysical Journal*, **529**, 101 (2000)
4. Allen, S. W., Schmidt, R. W., Ebeling, H., Fabian, A. C., van Speybroeck, L. et al., *Monthly Notices of the Royal Astronomical Society* **353**, 457
5. King, A.R. & Pounds, K.A., *Monthly Notices of the Royal Astronomical Society*, **345**, 657 (2003)
6. Cowie, L.L., Barger, A.J., Bautz, M.W., Brandt, W. N., and Garmire, G.P., *The Astrophysical Journal*, **584**, L57 (2003)
7. Barger, A. J., Cowie, L. L., Mushotzky, R. F., Yang, Y., Wang, W.-H., Steffen, A. T., Capak, P., *The Astronomical Journal,* **129**, 578 (2005)

Drs. Randall Smith, John Raymond, and Barham Smith proving that there are in fact two different Smiths working in X-ray spectroscopy with a single Raymond.

High Energy, High Resolution X-Ray Spectroscopy:
Microcalorimeters For Nuclear Line Astrophysics

E. Silver[a], H. W. Schnopper[a], C. Jones[a], W. Forman[a], S. Romaine[a], N. Madden[b], D. Landis[c], J. Beeman[c], E. E. Haller[c], M. Barbera[d], F. Christensen[e], B. Ramsey[f], S. Woosley[g] and R. Diehl[h]

[a]Smithsonian Astrophysical Observatory, [b]Lawrence Livermore NationalLaboratory, [c]Lawrence Berkeley National Laboratory, [d]Osservatorio Astronomico G.S. Vaiana, [e]Danish Space Research Institute, [f]George C. Marshall Space Flight Center, [g]University of California at Santa Cruz, [h]Max Planck Institut für Extraterrestriche Physik,

Abstract. We introduce focusing optics and microcalorimeter spectroscopy to nuclear line emission astrophysics with a balloon payload concept called, *B-MINE*. It is designed to probe the deepest regions of a supernova explosion by detecting ^{44}Ti emission at 68 keV with spatial and spectral resolutions that are sufficient to determine the velocity distribution of the ^{44}Ti emitting region. *B-MINE* has a thin plastic foil telescope multilayered to maximize the reflectivity in a 20 keV band centered at 68 keV and a microcalorimeter array optimized for the same energy band. This combination provides a reduced background, an energy resolution of 50 eV and a 3σ sensitivity in 10^6 s of 6×10^{-8} ph cm^{-2} s^{-1} at 68 keV. During the course of a long duration balloon flight, *B-MINE* could carry out a detailed study of the ^{44}Ti emission line centroids and width in CAS A. Details of the spectrometer and simulated results are presented.

INTRODUCTION

Nuclear line astrophysics provides a direct probe of the details of one of the most violent events in the universe -- a supernova explosion that expels the heavy elements into the ISM from the nuclear furnace in which they were created. Because of its production deep in the stellar core and 60 year half-life, ^{44}Ti provides a key diagnostic of supernova explosions. Although other elements are produced at high density and temperature in SN explosions, the key feature of ^{44}Ti is that its half life is longer than the few years required for the overlying strata to become optically thin at high energies and sufficiently short that the ^{44}Ti remains localized around the SN site while it emits intensely. Thus, the nuclear properties of ^{44}Ti make it unique among all elements of nature for probing deep into the cauldron of a SN explosion. Model nucleosynthesis yield calculations for Type Ia and Type II supernovae show that the conditions for ^{44}Ti production are significant only within the central cores of massive stars or in white dwarfs following a surface detonation. Hence, the mass in ^{44}Ti is an important diagnostic of the most extreme densities and temperatures.

Despite the importance of ^{44}Ti as a diagnostic of supernova explosions, observational evidence remains scarce. Only recently have instruments of sufficient sensitivity become available. COMPTEL detected ^{44}Ti emission from Cas A in the 1.156

CP774, *X-Ray Diagnostics of Astrophysical Plasmas*, edited by Randall K. Smith

MeV line (Figure 1) produced in the final decay of the ^{44}Ti chain as ^{44}Sc decays to stable ^{44}Ca [1] with an estimate of the line flux of (3.3 ±0.6) x 10^{-5} ph cm^{-2} s^{-1}. Using the Phoswich Detection System aboard BeppoSAX, Vink & Laming [2] measured the line strengths of the 67.9 keV and 78.4 keV lines emitted by ^{44}Sc in Cas A which is formed when the ^{44}Ti undergoes electron capture. The BeppoSax spectrum is shown in Figure 2. The flux in each line is (2.1 ±0.7) x 10^{-5} ph cm^{-2} s^{-1} which implies an initial ^{44}Ti mass of (0.8 – 2.5) x 10-4 $M_{\odot}$. ^{44}Ti emission also has been reported from GRO J0852-4642 [3], and possibly a young supernova remnant RX J0852.0-4622 [4] in the direction of the older Vela remnant.

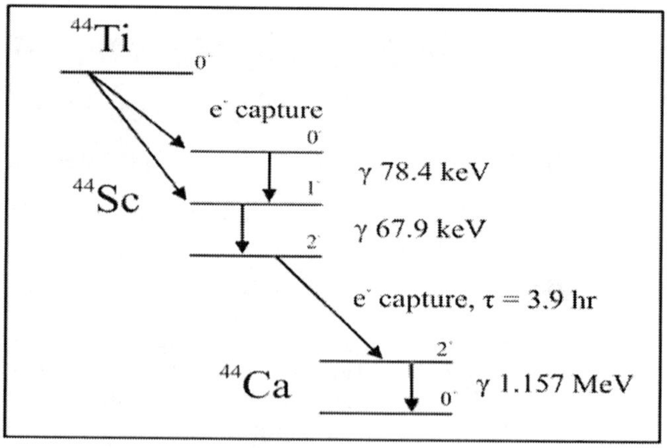

Figure 1. The ^{44}Ti decay chain.

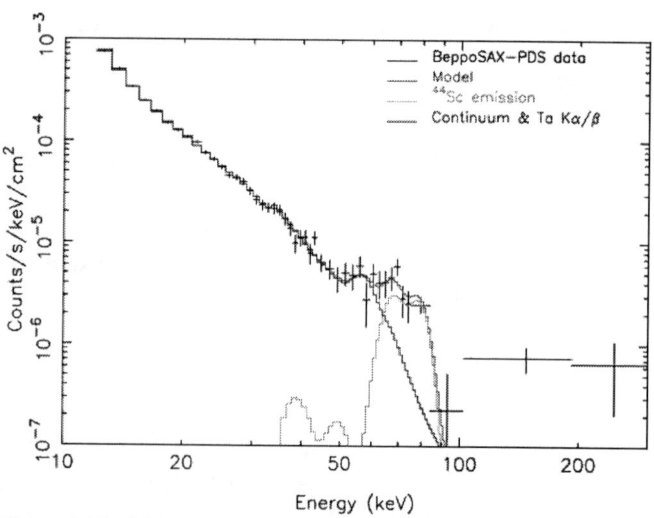

Figure 2. The high energy spectrum of CasA measured by the PSD on BeppoSAX [2].

We have developed a concept for a Balloon-borne Microcalorimeter Nuclear Line Explorer (*B-MINE*) to study the 68 keV ^{44}Ti emission from young supernova remnants. It is designed to study the structure and centroid of the ^{44}Ti line and its flux emitted by Cas A. We will also determine the mass of ^{44}Ti and the velocity of the ejecta, both critical to understanding the nucleosynthesis and following core collapse in a supernovae explosion.

B-MINE

B-MINE, the Balloon-Borne Microcalorimeter Nuclear Line Explorer, combines the high spectral resolution of a microcalorimeter array with modest focusing, multilayered optics to reduce the background and enhance the effective area in a balloon-borne experiment. The telescope has a focal length of 7.8 m and is a thin plastic foil, single reflection conical spiral design that is multilayered to maximize the reflectivity in a 20 keV band centered at 68 keV [5]. Tungsten/Silicon (W/Si) multilayers enhance the reflectivity at 68 keV resulting in a peak effective area of 60 cm^2 and a FOV that is 5 arc minutes in diameter.

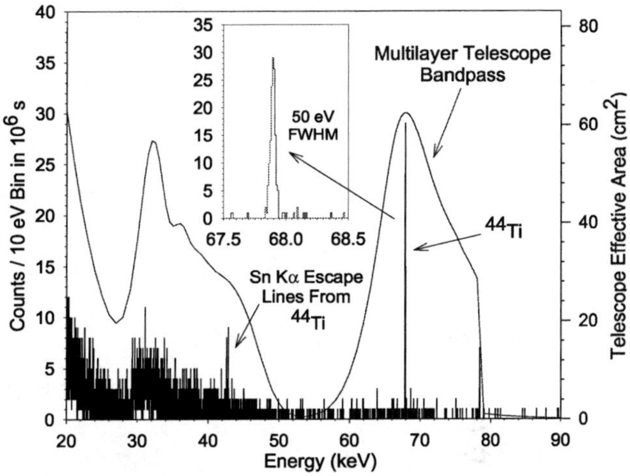

Figure 3. The blue curve shows effective area of the telescope (without detector).The black curve is a simulation of a 10^6 s observation of the Cas A supernova remnant. It includes the detector efficiency. The 68 and 78 keV (unredshifted) ^{44}Ti lines are both seen. The red curve in the inset is the 68 keV line broadened by the 50 eV detector resolution.

The primary energy range for *B-MINE*, centered at 68 keV, is defined by the distribution of outflow velocities of the ^{44}Ti ejecta. Velocities are presently unknown but could reach 4,000 km s^{-1} in Cas A. A secondary energy range of 30 - 50 keV includes the cyclotron resonance features previously observed in Her X - 1 and other binary pulsars. These two ranges can be achieved with a novel multilayer design. Figure 3 shows the effective area as a function of energy.

Novel Spiral Optic

The results from our laboratory tests demonstrate that a multi-shell plastic optic will meet our specifications for half energy width (HEW) and FWHM. Figure 4 shows the results of precision measurements of the scattering properties of the foil made at the Danish Space Research Institute. The folded scattering curve can be used to predict a minimum HEW of 34 arcsec for a single reflection telescope made from this plastic. This is the expected HEW for a perfectly built telescope, i.e., there are no figure errors introduced by the support structure or other mechanical and/or thermal sources. Since the microcalorimeter will obtain high resolution spectra that can be analyzed to reveal geometry of the ^{44}Ti emission from the Cas A SNR, the actual telescope requirement is relaxed to nothing more than a "*light bucket*" with a HEW ~ 5 arcmin.

The application of constant spacing spiral X-ray optics began in 1997 in our laboratory program.[6] An example of a spiral lens is shown in Figure 5. With excellent imaging results, (see Figure 6), we realized that the potential benefits of a conical spiral design could be applied to our *B-MINE* telescope.

Since a multilayer d-spacing of 10 Å can now be deposited, a single reflection from a graded multilayer spiral ribbon is all that is required to focus the hard x-rays in the two *B-MINE* energy bands. The optic will consist of a spiral ribbon lens with ~ 220 multilayered turns that range from 12 cm to 40 cm in diameter. The ribbon will be 50 mm wide. The multilayer d-spacing gradation is divided into two narrow ranges; a band centered at 68 keV and one for the 30 - 50 keV. We have carefully matched the range of d-spacings in the 68 keV band to accommodate the size of our primary science target (Cas-A) so that the effective FOV is about 5 arcmin.

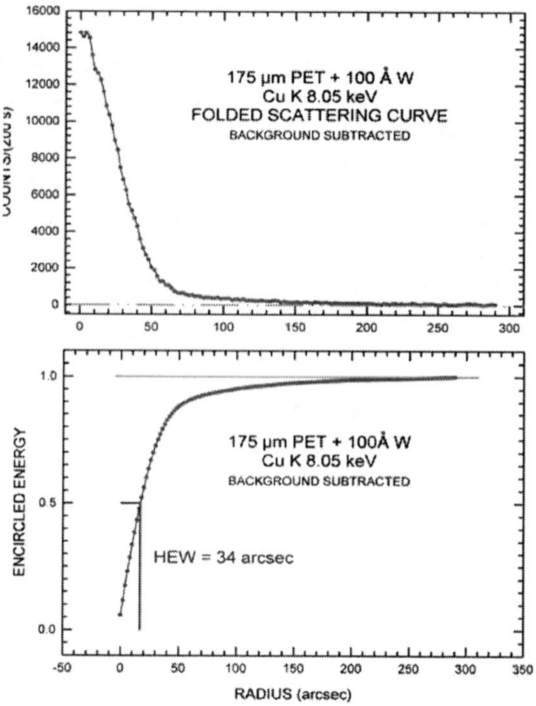

Figure. 4. The folded scattering curve(top) and the predicted HEW for a single reflection telescope.

Figure 5. The cylindrical spiral x-ray lens used for laboratory astrophysics and microanalysis.

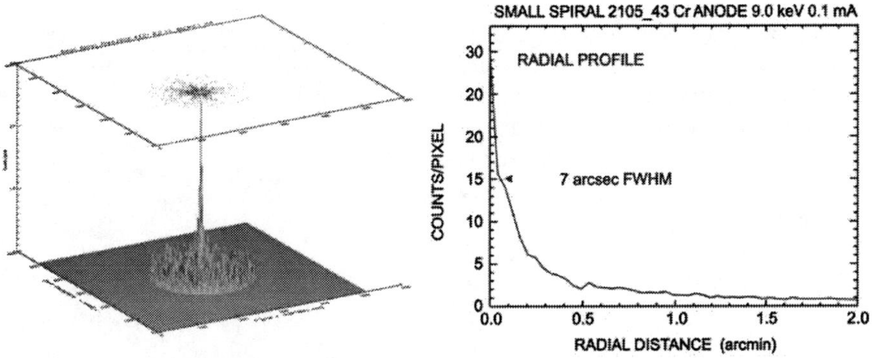

Figure 6. Imaging results using the small spiral lens shown in Figure 5.

Focal Plane Microcalorimeter Array

The B-MINE hard-X-ray microcalorimeter is based on a laboratory prototype that has a 25 μm thick tin X-ray absorber operating at a base temperature of 80 mk []. The prototype has a quantum efficiency (Q.E.) of 40% at 35 keV, 8% at 68 keV and an energy resolution of 50 eV. The resolving power at 68 keV is $E/\Delta E = 1360$. A typical X-ray spectrum using ^{241}Am and ^{55}Fe sources is shown in Figure 7.

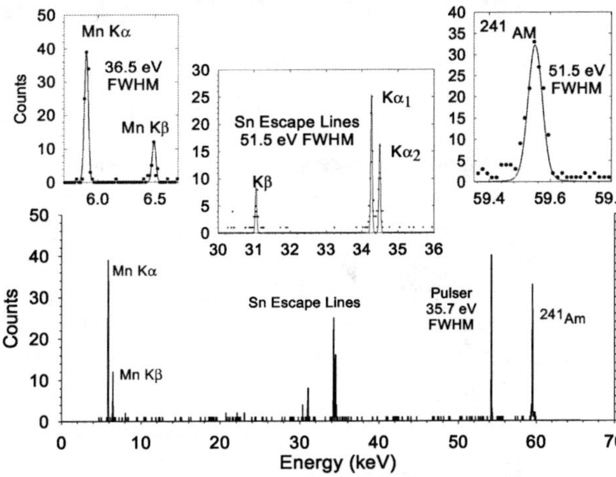

Figure 7. The spectrum of emissions from ^{241} Am and ^{55} Fe sources. The Np L lines between 13.7 keV and 20.8 keV produced in the ^{241} Am decay have been absorbed by a copper foil.

Recently, we have made several significant advances in detector design that have led to a reduction in low frequency (1/f) noise, improved stability, and resolution performance of 3.1 eV at 6 keV [13,14] and is now consistent with our theoretical model which also predicts sub 2 eV at 1 keV. The implications of this technical breakthrough for *B-MINE* are tantalizing because we expect improved performance in the 10 keV to 70 keV range as well. Based upon our recent advancements, our detector modeling codes have confirmed that the 50 eV energy resolution at 60 keV for our early detectors (figure 7) was dominated by front-end preamplifier noise. The resolution should have been 30 eV for the type of microcalorimeter components used at that time. The earlier microcalorimeter also was intrinsically less sensitive than the one used for the recent soft x-ray measurements. Thus, we can expect further improvement by adapting this more sensitive detector to this high energy application.

In Figure 8 we show the relationship between energy resolution and absorber quantum efficiency for three x-ray energies, 10 keV, 30 keV and 60 keV as determined from our computer model. By incorporating our recent detector advancements, we can choose to use a 350 μm thick tin absorber to achieve 80% quantum efficiency at 60 keV while maintaining an energy resolution of 14 eV. The potential increase in collecting area is 4-5 fold due to the improved detector efficiency. We can gain an additional factor of 1.6 by using the single reflection telescope described earlier. The microcalorimeter array is optimized for the same energy bands. An array of 20 × 20 pixels is required to cover the 5' × 5' angular size of Cas A with the 7.8 m *B-MINE* telescope. The details of the array may be found in the literature [7].

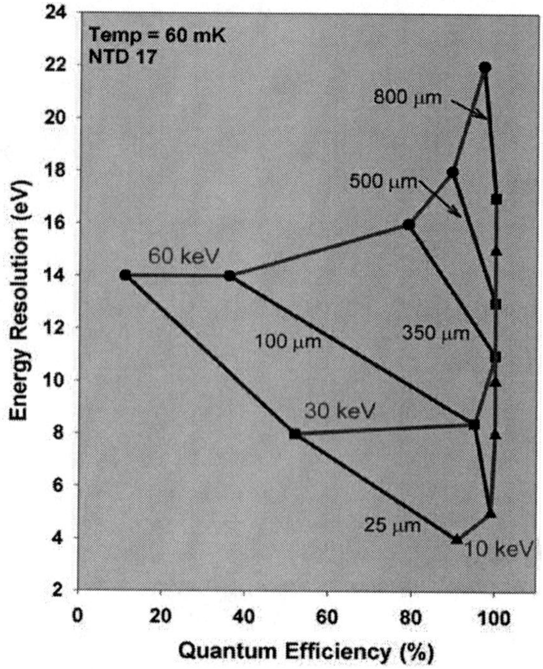

Figure 8. Predictions of energy resolution as a function of tin absorber quantum efficiency for three x-ray energies, 10 keV, 30 keV and 60 keV and 5 absorber thicknesses. The operating temperature is 60 mK

PREDICTED PERFORMANCE ON CAS A

Since a large area array of microcalorimeters is small compared to traditional hard x-ray detectors, the volume induced background is several orders of magnitude smaller than that in more conventional hard x-ray spectrometers. With an energy resolution of 50 eV at 68 keV, the sensitivity for nuclear line detection is unprecedented. Figure 9 shows the sensitivity to line detection for the baseline microcalorimeter (blue curves) with 50 eV resolution, 15% quantum efficiency at 68 keV (50 μm thick absorber) and a double reflection telescope; The sensitivity of the advanced microcalorimeter array with 80% absorption efficiency and a single reflection telescope is shown by the black curves. The

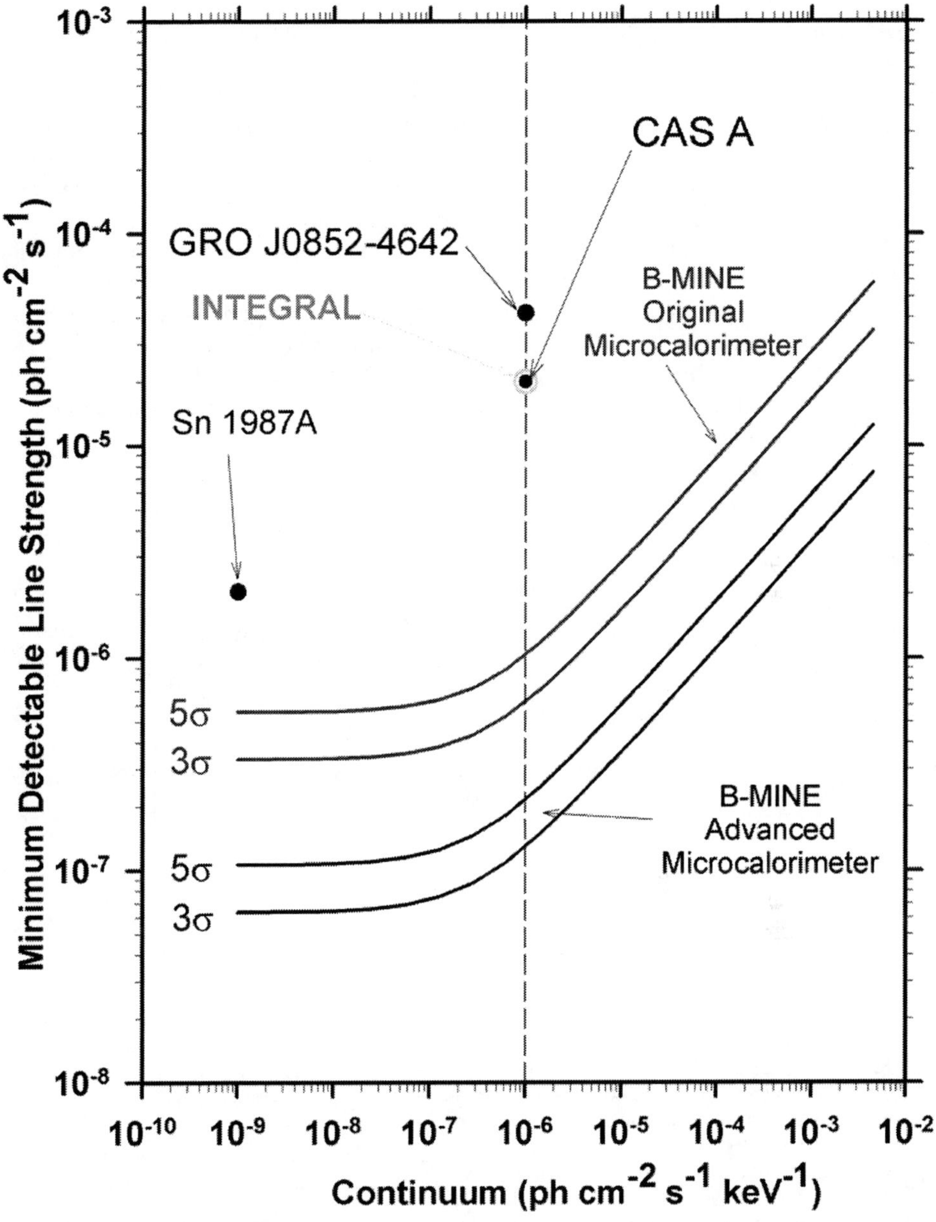

Figure 9. Minimum detectable line strength needed by *B-MINE* for 3σ and 5σ statistical precision.

3 σ sensitivity in 10^6 s for narrow line detection is 6×10^{-8} ph cm^{-2} s^{-1} at 68 keV for the advanced design.

The improved signal-to-background ratio of the microcalorimeter over past experiments on Cas A will provide a more precise measure of the flux, line shape, and centroid of ^{44}Ti emission. As a simple model one can assume a uniform outflow at a constant ejection velocity from the center of the supernova explosion. The radius of the ^{44}Ti shell is given by the time since the explosion and the ejecta velocity. Limb brightening will enhance the flux from the limb (zero velocity in the line-of-sight) compared with the flux from the center. Thus, the observed line emission will peak at zero velocity, with flux throughout the energy range corresponding to the maximum and minimum velocities. Assuming the baseline instrument design, Figure 10 shows spectra that would be obtained for Cas A if the ^{44}Ti ejecta had velocities of 300, 1000, or 4000 km s^{-1}. If ejecta have a range of velocities, the spectra would be a sum of these simulations,

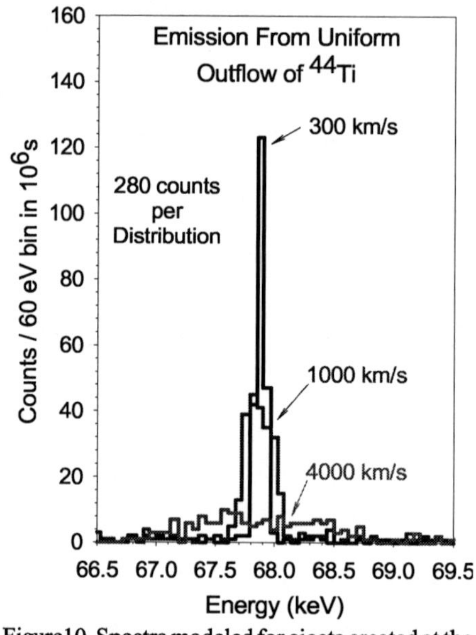

Figure10. Spectra modeled for ejecta created at the same time and moving with a constant velocity in all directions from the source center. X-ray emission comes from a shell of radius given by the velocity and the age of the remnant. Three different outflow velocities are modeled. The spectra are dominated by the zero redshifted emission from the source limb. The higher the velocity, the broader the distribution.

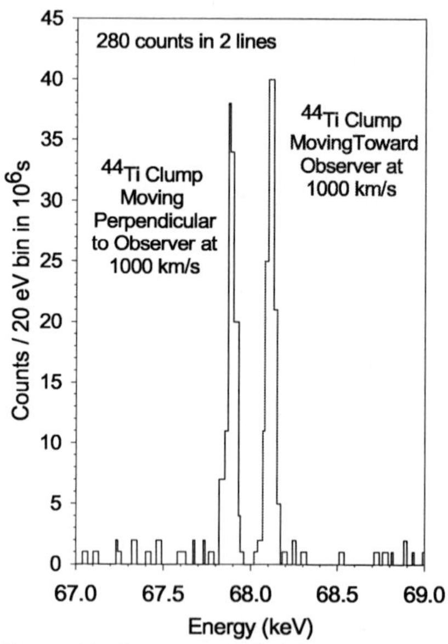

Figure 11. Spectra modeled for the case of directed emission from the center. In one case, the emission appears to come from the center of the source at maximum blue shift, in the other it comes from the limb at zero redshift. For the zero redshift case, emission from the entire limb cannot be distinguished spectroscopically from directed beams of particles striking the limb at one or more places. Imaging is required to resolve the emission site(s).

with the line broadened.[1]

With B-MINE's spectral resolution of 50 eV, one could detect broadening if the velocity of the expanding ^{44}Ti shell exceeds 300 km s^{-1}. Detecting velocity broadening would provide clear evidence for mixing of the ejecta layers, thus introducing new constraints on instabilities and transport mechanisms in massive stellar explosions.

If the ^{44}Ti source is composed of a few bright knots, the spectrum will have features that correspond to each knot. Again assuming the baseline instrument design, Figure 4 shows the simulated B-MINE spectrum of Cas A if the ^{44}Ti ejecta were in two knots, one moving toward the observer at 1000 km s^{-1} and the other moving perpendicular to the line of sight.

ACKNOWLEDGEMENTS

The authors wish to thank G. Austin, R. Ingram, G. Nystrom, J. Gomes , D. Boyd, and B. Podgorski for their technical input to the B-MINE concept. This work was performed in part by NASA Grant NAG5-5104.

REFERENCES

1.Iyudin, A. F. et al.,A&A,**284**, L1, 1994.
2.Vink & Laming , 2003, ApJ, **584**, 758.
3.Iyudin, A. F. et al., Astrop. Letters and Communications, **38**, 383,1999.
4. Aschenbach, B., Nature, **396**,141, 1998
5. H. W. Schnopper, et al. Proc. SPIE, Vol. 5537, 2004, 82 - 93.
6 Silver, E., Schnopper, H. and Ingram, R., US6094471: X-ray diagnostic system
7. Silver et al., SPIE, **4140**, 397, 2000

1

The simulations displayed in Figures 3, 10 and 11 are based on the baseline detector/telescope design. For the advanced design, one could expect as much as an 8-fold increase in total counts.

Benchmark Exercises for stellar X-ray Spectroscopy Testing (BEXST)

A. Maggio*, J. J. Drake†, F. Favata**, M. Güdel‡ and C. Jordan§

*INAF - Osservatorio Astronomico di Palermo, Italy
†Smithsonian Astrophysical Observatory, Cambridge, MA
**Astrophysics Division - Research and Space Science Department of ESA, ESTEC, Noordwijk, NL
‡Paul Scherrer Institut, Würenlingen and Villingen, Villingen, Switzerland
§Department of Physics (Theoretical Physics), University of Oxford, UK

Keywords: X-ray Spectroscopy, Stellar Coronae, Optically-thin Plasmas
PACS: 95.75.Fg, 95.75.Pq, 95.85.Nv, 97.10.Ex97.10.Tk, 95.30.Ky

INTRODUCTION

In recent years, different teams active in the field of stellar coronal physics have used different approaches for the reduction and analysis of spectra taken with Chandra and XMM-Newton. Various methods have been employed, some of which work on binned spectra while others use extracted line fluxes. These methods differ in their numerical inversion strategies or convergence techniques, their sample of emission lines used, their atomic physics databases, or the free parameters allowed in the underlying models. At present, no one has compared the results of all the different approaches in order to check whether these are compatible, and whether some (or all) of these lead to biases in the physical parameters derived.

This situation is made more complex by the growing indications that some of the usual assumptions in the above analyses are in fact invalid in the astrophysical environments of interest: in particular, stellar coronae are likely composed of independent structures with different pressures, rather than single structures in hydrostatic equilibrium, heating may be by transient impulsive events, and element abundances may not be uniform.

Following the results of the Splinter Session on the future of stellar X-ray spectroscopy [1], held at the 13th Cambridge Workshop on Cool Stars, Stellar Systems, and the Sun (Hamburg, July 2004), we are organizing a *benchmark network open to the whole community*, aimed to address the above issues. Here we present the rationale for this project, its organization, and current scheduling.

ASSESSMENT OF CURRENT UNCERTAINTIES

In order to describe the state of the art in stellar X-ray spectroscopy, we have considered several examples of results obtained with different methods and approaches. Comparison of these results, often based on the same data sets, provides a factual indication of the degree of agreement between them, and of the uncertainties we need to address.

CP774, *X-Ray Diagnostics of Astrophysical Plasmas,* edited by Randall K. Smith

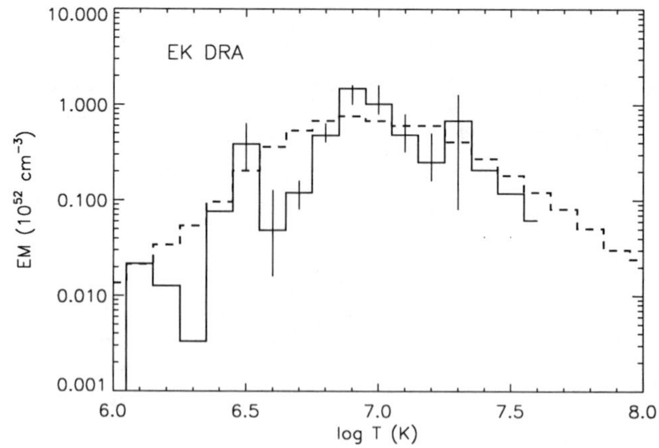

FIGURE 1. Plasma Emission Measure Distributions in the corona of EK Dra obtained with method *a* (dashed) and *b* (solid) (see text).

Emission Measure Distributions (EMDs)

Scelsi et al. [2] performed a comparison of EMDs vs. temperature, based on an XMM-Newton observation of the Hertzsprung gap giant 31 Com, obtained with the same method of reconstruction (the Markov-Chain Monte Carlo method by [3]), but different sets of emission lines and emissivities based on the CHIANTI database [4] or on the Astrophysical Plasma Emission Database (APED [5]). They showed that consistent results can be obtained with a careful selection of the emission lines employed for the MCMC reconstruction, in spite of substantial differences between the two databases.

Instead, Fig. 1 shows results obtained from the same XMM-Newton data set (an observation of the young solar-type analog EK Dra with the Reflection Grating Spectrometer) and the same APED emissivities, but different methods: (a) an analysis based on few selected Fe XVII–XXV lines, the O VIII/O VII flux ratio, and the continuum level [6], (b) the MCMC method applied to all the emission lines with sufficiently high S/N ratio, well-identified, and density-independent [7].

Abundances

The plots in Fig. 2 show the dispersions of different determinations of the Fe abundance and three abundance ratios for selected targets, reported in recent works (2001–2004; see [8]). The scatter is to be compared with nominal error ranges of 10–50% for individual element abundances, in most cases (not including uncertainties in the atomic physics, calibration problems, or coronal inhomogeneities).

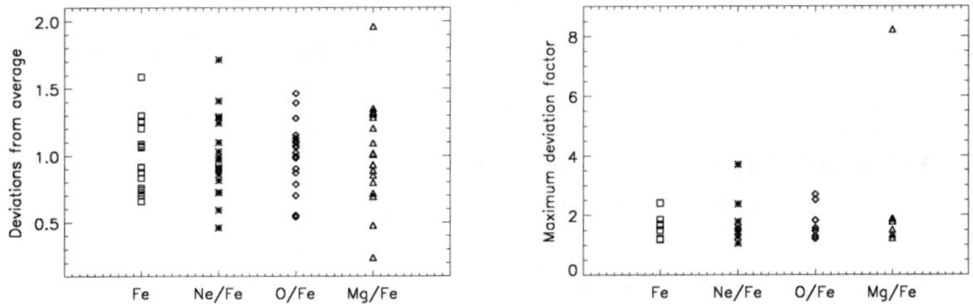

FIGURE 2. *Right:* Ratios of individual measurements to the average value (for each star and quantity), for 7 late-type stars with multiple determinations. *Left:* Ratios of maximum to minimum measured values, obtained for the same data sample.

Densities

The state of the art in the determination of coronal plasma densities from He-like triplets is illustrated in Fig. 3, where electron densities derived from O VII line ratios for 8 stars in common to the recent works by [9] and [10], are compared.

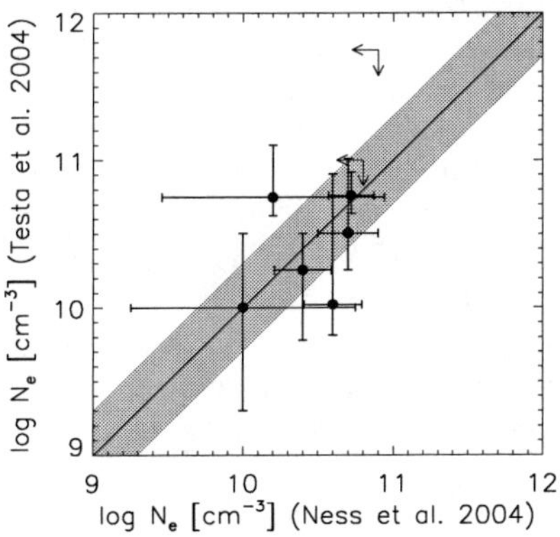

FIGURE 3. Comparison of electron densities derived from O VII line ratios by [9] and [10]. All measurements based on *Chandra*/MEG spectra. Deviations of factors ≤ 2 indicated by the shaded area.

HINTS OF COMPLEXITY

There are a number of results which point to complexity in the temperature and pressure structure of coronae. For example, the temperatures derived from the temperature-sensitive G(T) ratios ($(i+f)/r$, with i, f, and r denoting the fluxes of the intercombination, forbidden, and resonance lines of He-like triplets) are significantly lower than the ionization temperatures, and also with respect to the expectations based on the reconstructed EMDs [10].

BEXST ORGANIZATION AND TIMING

The above examples indicate overall agreement on the characteristic EMDs and relative abundances, but there is some freedom for variation in the details. One of the aims of BEXST is to understand how to gain physical insight on the complex coronal environment from the EMDs and density measurements; a second goal is how far we can trust "details" in EMDs or abundance patterns, which may be relevant for the physical interpretation of the results.

The above reflections suggest the need for an end-to-end comparison of data analysis methods and interpretation approaches. In order to make further progress in stellar coronal physics we need to base the physical deductions on results which other scientists are able to reproduce, even with different methodological approaches. Reproducibility of the results is at the base of good science. To this aim several of the participants to the Splinter Session on the future of stellar X-ray spectroscopy, held at the 13th Cool Stars Workshop (CS13), have agreed to pursue a "hare and hounds" exercise: one group will take care of producing simulated high-resolution X-ray spectra, based on reasonable (not necessarily simple) physical assumptions; other groups will independently analyze the simulated data set(s) and then meet to compare the results. This is not intended to be a contest, but a way to assess the robustness or limits of different modeling approaches.

At the time of CS13, several groups already expressed interest in this project, including groups at OAPA (Palermo, Italy), SAO (Cambridge, USA), ESA/ESTEC (Noordwjik, NL), PSI (Villigen, Switzerland), Oxford University (UK), Columbia University (New York, USA), and SRON (Utrecht, NL).

Simulated data will be made available on a dedicated WWW site (`http://bexst.astropa.unipa.it`), which will host also the BEXST results once available. We have foreseen also a small workshop to discuss the outcomes of the project and to issue a set of recommendations for future works.

In conclusion, BEXST will help the community to discuss/compare/improve our approaches and data bases, and hence to improve our confidence on stellar X-ray spectroscopy, after understanding its current limits. Reproducibility of the results is one of the major goals.

REFERENCES

1. A. Maggio, J. J. Drake, F. Favata, and M. Güdel, "What future for stellar X-ray Spectroscopy?," in *Cool Stars, Stellar Systems, and the Sun. XIII Cambridge Workshop*, edited by F. Favata, 2004, vol. in press of *ESA SP*.
2. L. Scelsi, A. Maggio, G. Peres, and P. Gondoin, *A&A*, **413**, 643–655 (2004).
3. V. Kashyap, and J. J. Drake, *ApJ*, **503**, 450–+ (1998).
4. K. P. Dere, E. Landi, P. R. Young, and G. Del Zanna, *ApJS*, **134**, 331–354 (2001).
5. R. K. Smith, N. S. Brickhouse, D. A. Liedahl, and J. C. Raymond, *ApJ*, **556**, L91–L95 (2001).
6. A. Telleschi, M. Güdel, K. Briggs, M. Audard, J.-U. Ness, and S. L. Skinner, *ApJ*, **in press** (2004).
7. L. Scelsi, A. Maggio, G. Peres, and R. Pallavicini, *A&A*, **in press** (2004).
8. M. Güdel, *Astr. Astroph. Review*, **12**, 71–237 (2004).
9. J.-U. Ness, M. Güdel, J. H. M. M. Schmitt, M. Audard, and A. Telleschi, *A&A*, **427**, 667–683 (2004).
10. P. Testa, J. J. Drake, and G. Peres, *ApJ*, **in press** (2004).

Progress and Plans for the Astrophysical Plasma Emission Code (APEC)

N.S. Brickhouse*, P. Desai*, R. Hoogerwerf*, D. A. Liedahl† and R. K. Smith**

*Harvard-Smithsonian Center for Astrophysics, 60 Garden Street, Cambridge, MA 02138
†Lawrence-Livermore National Laboratory, PO Box 808, L-41, Livermore, CA 94551
**NASA Goddard Space Flight Center/JHU, Code 662, Greenbelt, MD 20771

Abstract. APEC models for collisional ionization equilibrium are now available. These basic models have been benchmarked against astrophysical coronal spectra and show reasonable agreement; however, several atomic data issues need to be addressed. We plan to conduct an atomic data needs assessment and set priorities over the next year. New tools are under development to broaden the applicability of APEC models. We discuss the status of this development.

Keywords: atomic processes; X-rays
PACS: 32.30.Rj

COLLISIONAL IONIZATION EQUILIBRIUM MODELING

APEC[1] output models for plasmas in Collisional Ionization Equilibrium (CIE) have been publicly available for approximately four years [1] and are used in X-ray spectral fitting packages such as XSPEC [2] and Sherpa [3]. To date only low density APEC models are available in fitting packages, as a density-dependent table with a useful T_e and n_e grid is of an unmanageable size (4 to 10 GB). (Density-dependent models are available upon request and can be used in IDL.) We are working toward distributing the APEC code itself next fall. As a first application, we plan to guide users toward creating their own density-dependent table models for use in fitting. Users will also be able to change the ionization balance model to test for sensitivity.

ATOMIC DATA NEEDS ASSESSMENT

To date we have focused on the most critical atomic data needs for *Chandra* and *XMM-Newton*, using both compiled data from the literature and data produced within our collaboration (e.g. Liedahl's HULLAC calculations). Recent work has shown that theoretical methods can produce data of the quantity and, for the most part, accuracy needed; however, much work is needed to produce these data, critically assess them, and deliver them to the community via public plasma codes. A standardized atomic database, with associated tools, will bring some efficiency to this process. APED is a good

[1] http://cxc.harvard.edu/atomdb

CP774, *X-Ray Diagnostics of Astrophysical Plasmas,* edited by Randall K. Smith
2005 American Institute of Physics 0-7354-0259-0/05/$22.50

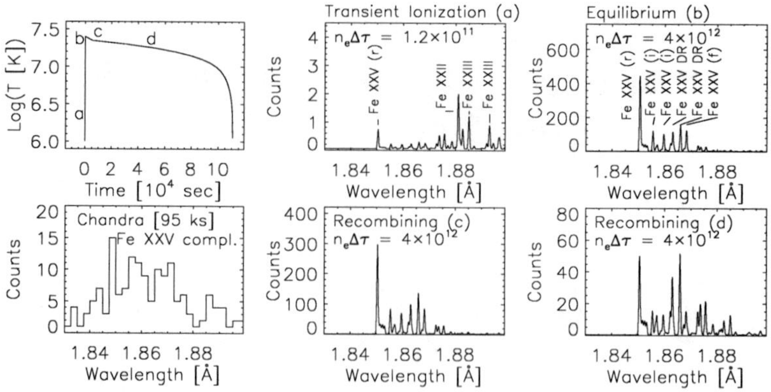

FIGURE 1. T_e vs time during a flare (*upper left*), simulated by arbitrarily imposing a high temperature ($\sim$ 25 MK) and allowing the gas first to ionize, and then to radiate, cool, and recombine under a constant $n_e = 10^9$ cm^{-3}. Scaling from *Chandra* HEG data for HR 1099 (*lower left*), four *Constellation-X* spectra for times **a-d** are shown. The signature feature during the ionizing phase **a** of the flare is Fe XXII λ1.881, with 20 counts in 120 s during ionization. Binsize is .00006 Å. Units of $n_e\Delta\tau$ are cm^{-3}s, with $\Delta\tau$ steps taken as 4 ks except in the ionizing case. For this figure, the time-dependent ionization state is calculated as described by [7], using the inner-shell excitation data of [8]. We thank R. Edgar for calculations.

starting point, but it needs to be generalized for new applications (e.g. photoionization modeling). Furthermore, tools to simplify the database (bundling lines together) should also increase the applicability and use of the data.

We plan to conduct an atomic data needs assessment during the upcoming year. This assessment will describe missing data as well as areas where we feel the data are not as accurate as necessary, either because processes are missing in the calculation (e.g. see Desai et al., this proceedings) or because the methods used are not optimal. We will also compare our models with other codes. We will produce a prioritized list of atomic data needed for APED, including the number of levels to meet the completeness goal, the accuracy desired, and the parameter range (e.g. T_e) needed.

Laboratory data will also be considered in the atomic data assessment. In addition to laboratory wavelengths, which are directly added to the database, line ratios from tokamaks or ebits (electron beam ion trap) can serve to validate (or not) the models, since they are, in principle, generated under known plasma conditions. However, current experimental results show some discrepancies that need to be understood before the results can be applied to astrophysics [4, 5]. Measurements of cross sections, made at selected energies, are particularly useful in conjunction with theoretical models that cover the broad parameter space [6].

Data compilation is a daunting task, but we are working toward a well-defined goal for completeness of the atomic database. We expect the database eventually to include approximately 5 million transitions (a huge increase over the 1200 lines in Raymond-Smith). For all ions relevant to EUV through X-ray spectra, we require fine-structure levels up to at least principal quantum number $n = 5$, along with the collisional and radiative rates among these levels. APED includes only approximate data for N, Al, and

Ni, and does not yet include many of the odd-Z elements. Satellite lines are also needed. For H- and He-like ions APED contains nearly 500 dielectronic recombination satellites; yet for many ions, there are no data existing in the public domain. The APED structure allows the inclusion of uncertainty estimates for atomic rates.

NON-EQUILIBRIUM IONIZATION MODELING

Although APEC can calculate the ionization balance directly (and thus handle ionizing and recombining conditions), APED does not yet contain all the necessary ionization and recombination rate coefficients nor does it include inner-shell excitation data. Unlike many NEI codes, APEC will self-consistently calculate the time-dependent level populations. Figure 1 shows an application.

PHOTOIONIZED PLASMA MODELING

The Photoionization and Absorption Spectral Engine (PHASE) calculates absorption due to an ionized plasma based on an ionization balance from CLOUDY [9], the radiative data in APED, and additional inner-shell transitions culled from the literature [10]. The calculations assume a simple geometry that consists of a central source that emits an ionizing continuum and clouds of gas intercepting our line of sight. Given an appropriate spectral energy distribution, the absorption is calculated using Voigt profiles to properly account for line saturation. The first public release of PHASE is due soon.

Accurate data are needed, particularly for inner-shell transitions, including fine structure resolution. To date, only abbreviated data or selected strong lines are publicly available (e.g. [11]). Additional processes, such as collisional excitation for low-lying levels, may also need to be considered.

ACKNOWLEDGMENTS

This project is supported by NASA contract NAS8-39073 to the Smithsonian Astrophysical Observatory for the Chandra X-ray Center and by NASA LTSA NAG5-3559.

REFERENCES

1. Smith, R. K., Brickhouse, N. S., Raymond, J. C., & Liedahl, D. A. 2001, ApJ, 556, L91
2. Dorman, B. & Arnaud, K. 2001, ASP Conf Proc, 238, 415
3. Freeman, P., Doe, S., & Siemiginowska, A. 2001, SPIE, 4477, 76
4. Laming, J. M. et al. 2000, ApJ, 545, L161
5. Beiersdorfer, P. et al. 2002, ApJ, 576, L169
6. Savin, D. W. et al. 2002, ApJ, 576, 1098
7. Esser, R., Edgar, R. J., & Brickhouse, N. S. 1998, ApJ, 498, 448
8. Phillips, K. et al. 1983, ApJ, 265, 1120
9. Ferland, G. 2000 ???
10. Krongold, Y. et al. 2003 ???
11. Behar, E., Sako, M., & Kahn, S. 2001, ApJ, 563, 497

Recent developments of the CHIANTI database in the X-ray wavelength range

E. Landi*, K.P.Dere†, P.R.Young**, G.Del Zanna‡, H.E.Mason§ and M. Landini¶

*Artep, Inc., Ellicott City, MD 21042 and Naval Research Laboratory, Washington DC, USA
†George Mason University, Fairfax, USA
**Rutherford Appleton Laboratory, Chilton, UK
‡Mullard Space Science Laboratory, UK
§University of Cambridge, Cambridge, UK
¶Universita' degli Studi di Firenze, Largo E. Fermi 2, 50125, Firenze, Italy

Abstract. CHIANTI is a database of atomic data and transition probabilities that allows the calculation of synthetic spectra and plasma diagnostics of optically thin plasmas. It is primarily suited for wavelengths shorter than 2000 Å but in principle can also be used at longer wavelengths. In the present paper we illustrate the latest development of the CHIANTI database, that include a huge expansion of the database in the X-ray wavelength range, the renewal of several data in the EUV range, and the inclusion of ionization and recombination in the level population calculation. This enhanced database will constitute the Version 5 of CHIANTI.

Keywords: X-ray spectra – X-ray diagnostics – Atomic databases
PACS: 32.30.Rj

INTRODUCTION

This paper describes the latest extension of the CHIANTI database (Dere et al. 1997), that will constitute version 5.0. Several changes have been made over Version 4.0, in order to take into account the new and accurate data produced since the last release, and to improve and expand the existing database. The main changes have been made in two different wavelength ranges: X-rays (8-17 Å) and UV (700-1500 Å). The changes in the X-ray wavelength range have been driven by the recent observations obtained with the XMM and Chandra, that have provided high-quality, high-resolution spectra that require large atomic models to be successfully reproduced by spectral codes. Also, observations with RHESSI and with older solar missions like P78-1, SMM and Yohkoh have shown that CHIANTI Version 4 required some improvements in the 1.5-8 Å range as well. The changes in the UV range have been triggered by the identification of a number of transitions between excited $n = 3$ configurations of coronal ions of Magnesium to Argon emitted by the solar corona. Such transitions are excellent diagnostic tools for the measurement of plasma temperature, emission measure and velocities in virtually all possible solar coronal plasmas, but the data necessary to calculate their emissivities were not available in the literature. Data for a few more ions were also added, and new, improved data for Fe XII were included, that allow to solve some discrepancies between Fe XII predicted emissivities and observations noted in the past.

The Version 5.0 of the CHIANTI database represents a big improvement over the

CP774, *X-Ray Diagnostics of Astrophysical Plasmas*, edited by Randall K. Smith

previous version, and it will be released for public use in late 2004/early 2005 at the website

http://wwwsolar.nrl.navy.mil/chianti.html

NEW DATA FOR FE XVII TO FE XXIII

The X-ray spectrum of the Sun and solar-like stars presents a number of strong spectral lines in the 8-12 Å wavelength range, that are emitted by high-energy configurations with principal quantum number $n \geq 4$ in Fe XVII to Fe XXIII. When compared to lines emitted by lower energy configurations from the same ion, or to lines from other ions of the same element, these transitions can provide excellent line pairs for the measurement of the plasma temperature. No data was publicly available in the literature to calculate the emissivities of these lines, so the Version 4.0 of CHIANTI did not include them. Recently, Landi & Gu (2004) calculated energy levels, radiative transition rates and collisional transition rates for high-energy configurations of Fe XVII to Fe XXIII, with the aim of providing the astrophysical community with a complete set of data for the calculation of these lines. Landi & Gu (2004) used the relativistic, distorted wave Flexible Atomic Code (FAC) by Gu (2003), that allows to calculate energy levels, A values and collision strengths as a function of incident electron energy. Resonances can be taken into account explicitly by FAC, and were included by Landi & Gu for the $n = 3$ levels. The configurations included by Landi & Gu in their calculations are as follow:

Fe XVII	$n = 3$ to $n = 7$	267 levels
Fe XVIII	$n = 3$ to $n = 5$	336 "
Fe XIX	$n = 3$ to $n = 5$	630 "
Fe XX	$n = 3$ to $n = 5$	725 "
Fe XXI	$n = 3$ to $n = 5$	590 "
Fe XXII	$n = 3$ to $n = 5$	513 "
Fe XXIII	$n = 3$ to $n = 5$	168 "

These data allow to reproduce a number of spectral lines neglected in earlier versions of the CHIANTI database and in other spectral codes, as shown in Figure 1.

NEW DATA FOR K-SHELL LINES

A large amount of data to predict the emissivities of K-shell transitions in the Fe spectra and also in several other ions have been added. The K-shell transitions provide important lines observed below 2 Å and can provide excellent temperature diagnostics when used in combination with direct excitation lines. We have included large numbers of lines for Si XIII, S XV and Ca XIX (see Landi et al., this issue) and for all the ions of Fe, from the work of Bautista et al. (2004) and references therein.

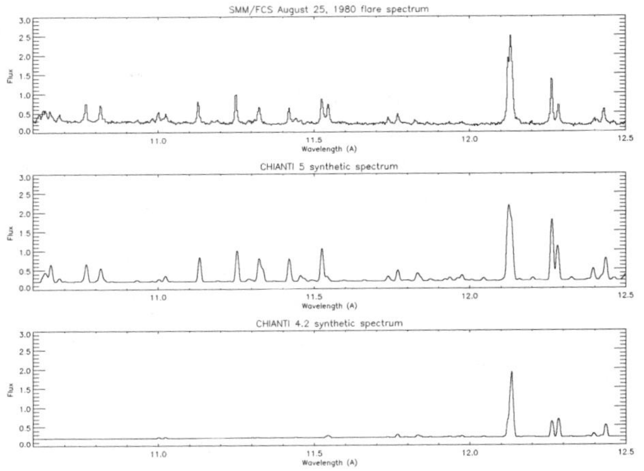

FIGURE 1. Observed and synthetic spectrum of a solar flare in the 11.6-12.5 Å wavelength range. Top: observed spectrum; middle: synthetic spectrum with the new Fe XVII-XXIII data; bottom: synthetic spectrum with CHIANTI 4.2.

NEW DATA IN THE EUV AND UV RANGES

Kink & Engström (1997) and Kink et al. (1997) indentified several $n = 3 \rightarrow n = 3$ lines in the solar UV spectrum observed by SUMER in the 500-1600 Å range, that could be used for diagnostics but for which no transition rates and atomic data were available in the literature. Since then, several other transitions of this kind from other ions have been identified in the solar spectrum (Curdt et al. 2004). As part of the CHIANTI project, we have undertaken the calculation of such data for the O-like and N-like sequence, in an effort to provide complete datasets that allow the prediction and the diagnostic use of such lines. So far, data for Ne III, Mg V,VI, Si VII,VIII, S IX,X,XI, Ar XII, Ca XIV, Ni XXI have been produced and published by E.Landi and Dr. A.K.Bhatia. More calculations will be published in the future dealing with the rest of the ions in those sequences.

The calculations include the $n = 2$ and $n = 3$ levels and provide atomic data and transition rates for all the transitions in the sequence, and have been calculated using the Distorted Wave approximation.

In addition to $n = 3$ data, O-like ions have been improved by using the semi-relativistic calculations by Zhang & Sampson (2002). This dataset includes all the necessary data for the calculation of the emissivities for the $n = 2$ sequence and fills in several gaps in the existing literature. Radiative data have been taken from a variety of different sources. These data replace old data, in many cases based on interpolation.

Several other ions have been updated in CHIANTI Version 5.0, and a few more have been included for the first time. These ions are listed in Table 1. Of particular importance is Fe XII, since this ion provides some of the most intense lines in the EUV solar and stellar spectrum at around 190 Å and 350 Å; these lines have provided diagnostic ratios routinely used to measure the plasma density, and are used as prime lines for the development of narrow-band imagers. However, atomic physics problems in this ion were found (i.e Binello et al 2001). As part of the CHIANTI project and of the Iron Project, some new calculations were carried out that solve these problems.

Also, we have calculated and inserted in CHIANTI complete datasets for Ca VII and Ca VIII. These ions, neglected in the literature, have been routinely observed by CDS and SUMER, and are extremely useful for element abundance diagnostics in the quiet Sun and active regions, especially when studied together with Ne VI and VII.

TABLE 1. Data for ions outside the C-like, N-like and O-like sequence, included in Version 5.0.

He II	New R-Matrix calculation
C II	New radiative data
C III	New radiative data
N III	New radiative data
O III	New radiative data
O IV	New radiative data
O V	New radiative data
Ne VI	New R-Matrix calculation
Al II	New ion
S IV	New radiative data
Cl II,XVI,XVII	New ions
K V,XVIII,XIX	New ions
Ca VII,VIII	New ions
Fe VII	New calculation
Fe IX	New R-Matrix calculation
Fe XII	New R-Matrix calculation
Fe XV	New calculation

IONIZATION AND RECOMBINATION

Ionization and recombination processes were recognized to be of crucial importance for the calculation of line emissivities in the X-ray range by Gu (2003). This work emphasizes the fact that the level populations of the $n = 3$ levels in Fe XVII-XXIII were in many cases strongly influenced by direct recombination or cascades following recombination from higher levels. Such influences are directly translated into changes in the line emissivities in the X-ray range.

In Version 5.0 of CHIANTI, we have included ionization and recombination into level populations making use of the Gu (2003) data for Fe XVII-XXIII, and using an approximate treatment of these process. The CHIANTI model for ionization and recombination assumes that the plasma can be described under the Coronal Model Approximation, where the total population of the excited levels of an ion is negligible compared to the population of the ground level. In this case, recombination and ionization processes can be included in a relatively straightforward way, since they can be treated as a correction to the case where population are calculated neglecting them.

$$\left(\frac{N_j}{N_g}\right)_{ion+rec} = \left(1 + \frac{n_{q-1}CI + n_{q+1}REC}{n_q CE}\right) \times \left(\frac{N_j}{N_g}\right)_{no\ ion+rec} \tag{1}$$

where n_{q-1}, n_q, n_{q+1} are the ion fractions for the ion (n_q) and its adjacent states of ionization, CI is the total collisional ionization coefficient to level j (plus cascades), REC is the total recombination coefficient to level j (plus cascades) and CE is the total excitation rate to level j.

The correction $\aleph$ is temperature sensitive and can be very large. It has significant effects on intensities of observed X-ray lines. The correction $\aleph$ causes the emissivity of several observed lines to change significantly, although for them the correction is normally less than a factor 2.

The only limitation of this approach lies in the breakdown of the coronal model approximation at high densities for a few ions. In particular, the greater problems arise for Fe XIX,XX and XXI when densities reach 10^{12} cm^{-3}, since by then the populations of the metastable levels is significant and invalidates the Equations 1 and 2. However, such densities are reached only in strong flares. The other ions either start to have problems at 10^{13} cm^{-3} (Fe XXII) or higher, or can always be described by the coronal model approximation (Fe XVII, Fe XXIV).

RE-ASSESSMENT OF ENERGY LEVELS

In order to properly match observed energy levels to the levels in the atomic models for Fe XVII-XXIII in CHIANTI we have carried out a new *ab-initio* line identification study using CHIANTI predicted emissivities and published observed spectra from the Sun and the laboratory. Such identification work is extremely important since incorrect line identifications or energy level assignment can lead to serious errors in the synthetic spectra.

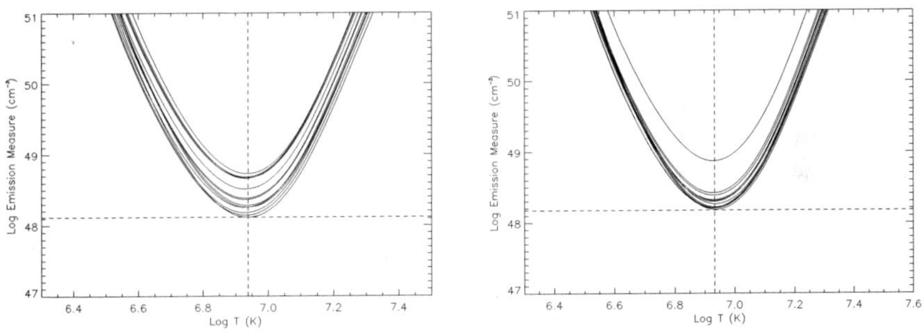

FIGURE 2. EM loci obtained from observed Fe XIX spectral lines. Left: case with NIST energy levels without assessment; right: case with re-assessed energy levels.

In the past, a large body of work has been carried out to match observed and theoretical energy levels, but in only a few cases those results are compatible with the new, extensive atomic models used in CHIANTI. The problem is that identifications in the past were carried out using different atomic models than the ones used in the present version of CHIANTI. Such atomic models, usually more limited due to computer memory limitations, in nearly all cases used the LS-coupling notation to describe energy levels. The inclusion of more configurations, the use of more sophisticated approximations in the calculations, and the inadequacy of the LS-Coupling notation to describe the $n = 3$ configurations in highly ionized iron ions, lead to different LS-coupling notation assignment to the same level in different calculations. For this reason, LS-coupling identifications made correctly with one atomic model might not be compatible with the LS-coupling notation of another model.

In order to provide the right assignment to each observed line we have analyzed a few observed spectra, identifying spectral lines by matching their wavelengths *and* intensity to predictions made using Version 5.0 of the CHIANTI database. We have used two X-ray high-resolution spectra acquired by SMM/FCS on August 25, 1980 and July 2, 1985 to identify spectral lines of Fe XVII to XX and a few more lines from Fe XXI to XXIV, and the laboratory spectrum of Brown et al. (2002) to identify lines from Fe XXI to XXIV. This identification work is better described in Landi & Phillips (2004).

An example of the improvement in synthetic spectra calculations is given in Figure 2 where the emission measure loci

$$EM_i = 4\pi d^2 \frac{F_i^{obs}}{G_i(T)} \tag{2}$$

is displayed, where F_i^{obs} is the measured flux of each observed Fe XIX line in the 10-15 Å, $G_i(T)$ is the contribution function of each line and d is the Sun-Earth distance. In case there are no atomic physics problems, blending or mis-identifications, all the EM loci of lines of the same ions should agree within errorbars. Figure 2 shows in both panels these EM loci obtained using the present version of CHIANTI, but in the left panel the $G_i(T)$ functions are calculated assuming the NIST energy level assignment, without any check on the consistency of the NIST energy level identification with CHIANTI models, and the right panel shows the EM loci obtained after the thorough re-assignment of levels done in the present work. All the rest of the atomic data and transition probabilities are the same. The improvement given by the re-assignment work is evident, as well as the presence of one blended line, while the incompatibility of the NIST energy level assignments to the CHIANTI models leads to the factor-of-4 spread in the EM loci in the left panel.

REFERENCES

1. Bautista, M.A., Mendoza, C., Kallman, T.R., & Palmeri, P., *A&A*, **418**, 1171 (2004)
2. Binello, A.M., et al. *A&A*, **370**, 1071 (2001)
3. Brown, et al., *ApJS*, **140**, 589 (2002)
4. Curdt, W., Landi, E., & Feldman U., *A&A*, **427**, 1045 (2004)
5. Dere, K.P., et al. *A&AS*, **125**, 149 (1997)
6. Gu, M.F. *ApJ*, **582**, 1241 (2003)
7. Kink & Engstrom, *Phys. Scr.*, **56**, 31 (1999)
8. Kink et al. *ApJ*, **487**, 956 (1997)
9. Landi, E., & Gu, M.F. *ApJ*, submitted
10. Landi, E., & Phillips, K. *ApJ*, submitted
11. Zhang, H.L., & Sampson D.H. *ADNDT*, **82**, 357 (2002)

Author Index

TABLE 4. Ionization Edges (in eV), from Verner et al. 1996, ApJ, 465, 487

Ion	**C**	**N**	**O**	**Ne**	**Na**	**Mg**	**Al**	**Si**	**S**	**Ar**	**Ca**	**Fe**
I	11.26	14.53	13.62	21.56	5.139	7.646	5.986	8.152	10.36	15.76	6.113	7.902
II	24.38	29.60	35.12	40.96	47.29	15.04	18.83	16.35	23.33	27.63	11.87	16.19
III	47.89	47.45	54.94	63.46	71.62	80.14	28.45	33.49	34.83	40.74	50.91	30.65
IV	64.49	77.47	77.41	97.12	98.92	109.3	120.0	45.14	47.31	59.81	67.27	54.80
V	392.1	97.89	113.9	12.62	138.4	141.3	153.8	166.8	72.68	75.02	84.51	75.01
VI	490.0	552.1	138.1	157.9	172.2	186.5	190.5	205.1	88.05	91.01	108.8	99.06
	V II	667.1	739.3	207.3	208.5	224.9	241.4	246.5	280.9	124.3	127.2	125.0
		VIII	871.4	239.1	264.2	266.0	284.6	303.2	328.2	143.5	147.2	151.1
			IX	1196.	299.9	328.2	330.1	351.1	379.1	422.5	188.3	233.6
			X	1362.	1465.	367.5	399.4	401.4	447.1	478.7	211.3	262.1
				XI	1649.	1762.	442.1	476.1	504.8	539.0	591.9	290.2
					XII	1963.	2086.	523.5	564.7	618.3	657.2	330.8
						XIII	2304.	2438.	651.7	686.1	726.7	361.0
							XIV	2673.	707.2	755.8	817.7	392.2
								XV	3224.	854.8	894.6	457.0
								XVI	3494.	918.0	974.5	489.3
									XVII	4121.	1087.	1262.
									XVIII	4426.	1157.	1358.
										XIX	5129.	1456.
										XX	5470.	1582.
											XXI	1689.
											XXII	1799.
											XXIII	1950.
											XXIV	2046.
											XXV	8829.
											XXVI	9278.